AQA Mathematics

for GCSE

Exclusively endorsed and approved by AQA

Series Editor
Paul Metcalf

Series Advisor
David Hodgson

Lead Author
Margaret Thornton

June Haighton
Anne Haworth
Janice Johns
Steven Lomax
Andrew Manning
Kathryn Scott
Chris Sherrington
Mark Willis

FOUNDATION

Module 5

Published in 2006 by:
Nelson Thornes Ltd
Delta Place
27 Bath Road
CHELTENHAM
GL53 7TH
United Kingdom

07 08 09 10 / 10 9 8 7 6 5 4 3 2 1

A catalogue record for this book is available from the British Library.

ISBN 978 0 7487 9758 5

Cover photograph: Gary and Terry Andrewartha/Survival Anglia/OSF/Photolibrary
Illustrations by Roger Penwill
Page make-up by MCS Publishing Services Ltd, Salisbury, Wiltshire

Printed in Great Britain by Scotprint

Acknowledgements

The authors and publishers wish to thank the following for their contribution:
David Bowles for providing the Assess questions
David Hodgson for reviewing draft manuscripts

Thank you to the following schools:
Little Heath School, Reading
The Kingswinford School, Dudley
Thorne Grammar School, Doncaster

The publishers thank the following for permission to reproduce copyright material:

Explore photos
Diver – Corel 55 (NT); Astronaut – Digital Vision 6 (NT);
Mountain climber – Digital Vision XA (NT); Desert explorer – Martin Harvey/Alamy.

Compass – Stockbyte 35 (NT); Alnwick Castle – Malcolm Fife/Zefa/Corbis; Rabbit G/V Hart/Photodisc 50 (NT); Discs – Nick Koudis/Photodisc 37 (NT); Coins – Corel 590 (NT); Electricity pylons – Photodisc 4 (NT); Rangoli – Dinodia/Alamy; Helicopter – Corel 415 (NT); Eurostar – Imagin London (NT); Marathon – Corel 76 (NT); Petrol pum – Photodisc 17 (NT); Lily Nosbaum – Matthew Nosbaum; Jogger – Photodisc 51 (NT); Motorway jam – Digital Vision 15 (NT); Vintage car – Corel 433 (NT); London Eye – Peter Adams/Digital Vision BP (NT); Eiffel Tower – Corel 397 (NT); Colosseum – Corel 600 (NT); Cheetah – Karl Ammann/Digital Vision AA (NT); Coach – Corel 422 (NT); Beach Corel 76 (NT); Arab dhow – Corel 58 (NT); Lightning – Jim Reed/Digital Vision WW (NT); Albert Einstein – Illustrated London News V2 (NT); Church – Corel 750 (NT); Ferry – Corel 707 (NT); Diver – Corel 184 (NT).

The publishers have made every effort to contact copyright holders but apologise if any have been overlooked.

Contents

Introduction

This book has been written by teachers and examiners who not only want you to get the best grade you can in your GCSE exam but also to enjoy maths.

Each chapter has the following stages:

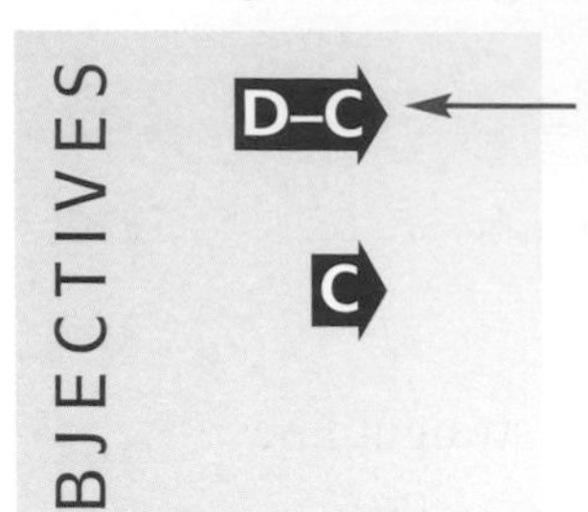

The objectives at the start of the chapter give you an idea of what you need to do to get each grade. Remember that the examiners expect you to perform well at the lower grade questions on the exam paper in order to get the higher grades. So, even if you are aiming for a C grade you will still need to do well on the G grade questions on the exam paper.

Key information and examples to show you how to do each topic. There are several Learn sections in each chapter.

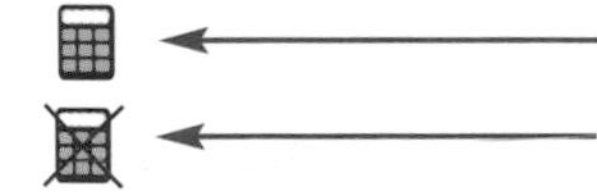

Questions that allow you to practise what you have just learned.

Means that these questions should be attempted with a calculator.

Means that these questions are practice for the non-calculator paper in the exam and should be attempted without a calculator.

These questions show how the maths in this topic can be used to solve real-life problems.

1 *Underlined questions are harder questions.*

Open-ended questions to extend what you have just learned.

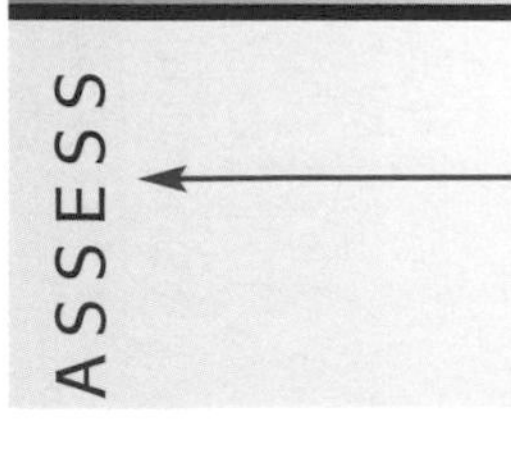

End of chapter questions written by an examiner.

Some chapters feature additional questions taken from real past papers to further your understanding.

1 Angles

OBJECTIVES

Examiners would normally expect students who get an F grade to be able to:

Express fractions of full turns in degrees and vice versa

Recognise acute, obtuse, reflex and right angles

Estimate angles and measure them accurately

Use properties of angles at a point and angles on a straight line

Understand the terms 'perpendicular lines' and 'parallel lines'

Examiners would normally expect students who get a D grade also to be able to:

Recognise corresponding angles and alternate angles

Understand and use three-figure bearings

What you should already know ...

- Addition and subtraction of whole numbers
- Simple fractions such as quarter, half, third, etc.

VOCABULARY

Revolution – one revolution is the same as a full turn or 360°

Right angle – an angle of 90°

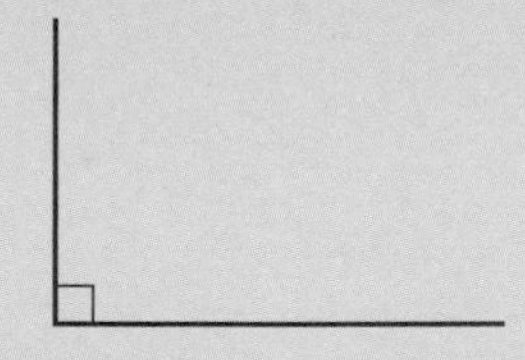

Acute angle – an angle between 0° and 90°

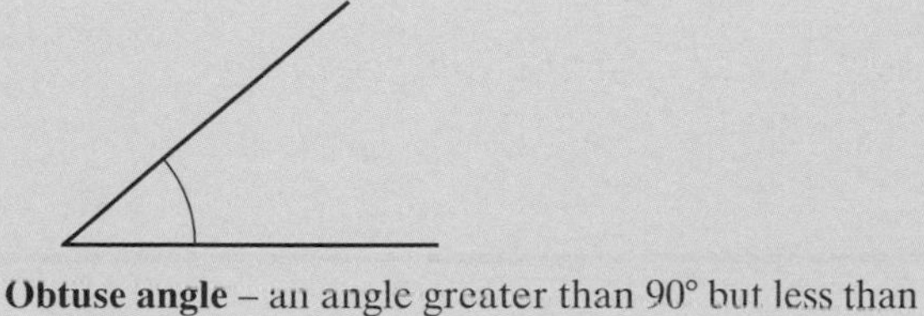

Obtuse angle – an angle greater than 90° but less than 180°

Reflex angle – an angle greater than 180° but less than 360°

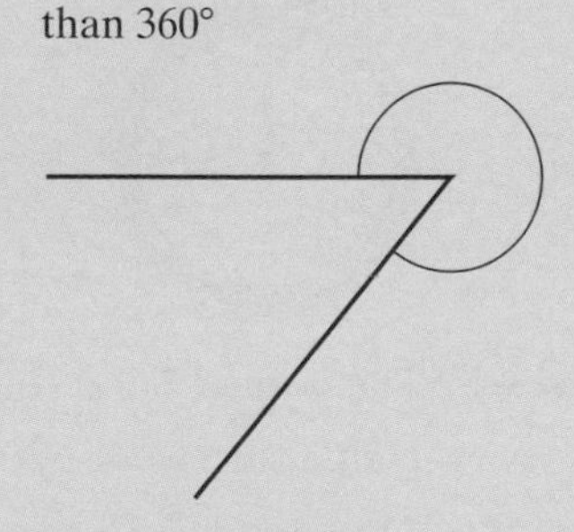

Perpendicular lines – two lines at right angles to each other

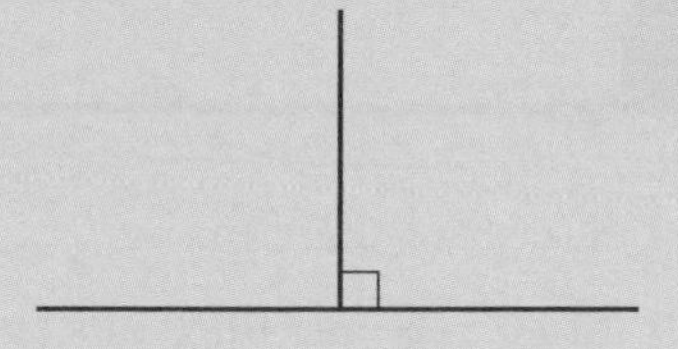

Parallel lines – two lines that never meet and are always the same distance apart

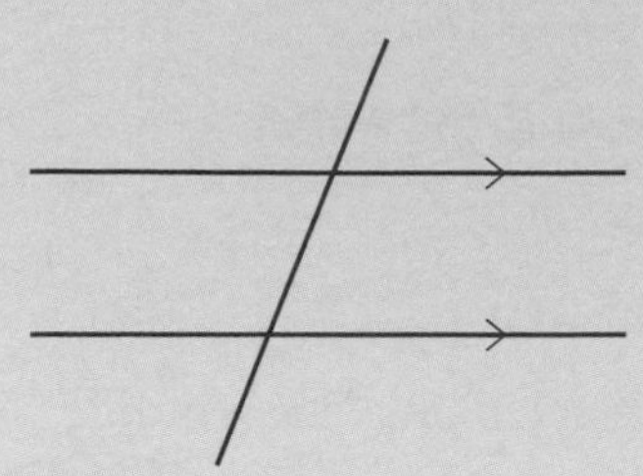

Transversal – a line drawn across parallel lines

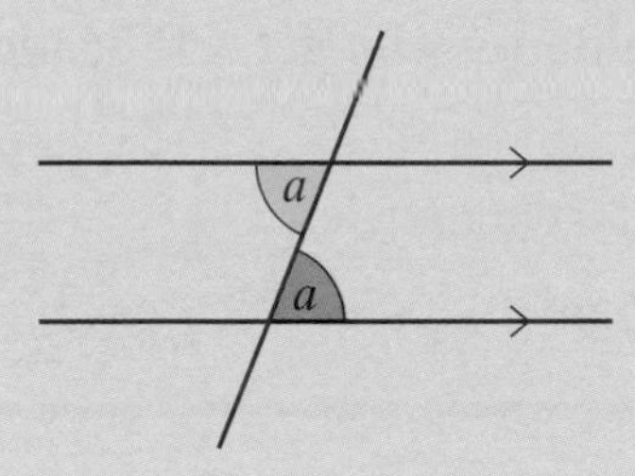

Alternate angles – the angles marked a, which appear on opposite sides of the transversal

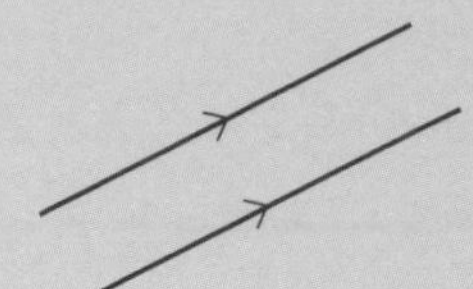

Corresponding angles – the angles marked c, which appear on the same side of the transversal

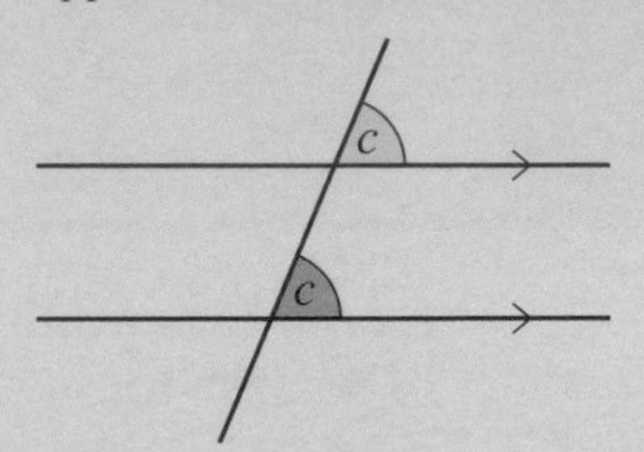

Bearing – an angle measured clockwise from North; all bearings should be written as three figure numbers, for example, 125° or 045°

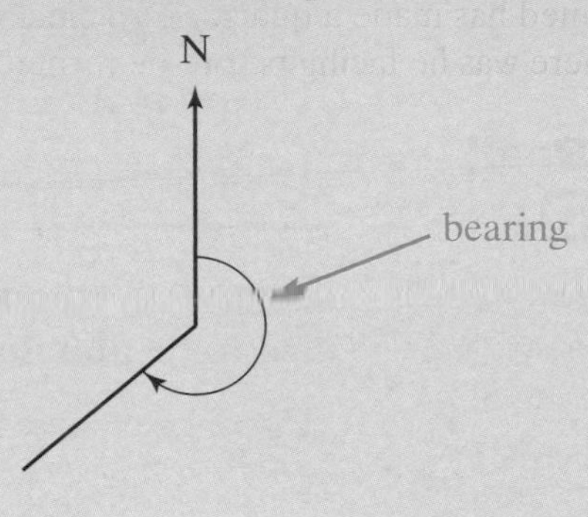

Learn 1 Full turns and part turns

Example: What angle does the minute hand of a clock move through between 12:00 and 12:30?

The minute hand travels through half of a turn.

Half a turn = $\frac{1}{2} \times 360° = 180°$

There are 360° in a full turn

Apply 1

1 How many degrees are there in:

a a quarter turn **b** one third of a turn **c** one twelfth of a turn **d** five twelfths of a turn?

2 What fraction of a turn is:

a 60° **b** 45° **c** 270° **d** 300°?

3 **a** Jack faces north and makes a quarter turn clockwise.
Which way is he facing now?

b Jill faces east and makes a half turn.
Which way is she facing now?

c Huw faces west and makes a quarter turn anticlockwise.
Which way is he facing now?

d Emma has made a half turn and is now facing east.
Where was she facing before she turned?

e Ahmed has made a quarter turn clockwise and is now facing west.
Where was he facing before he turned?

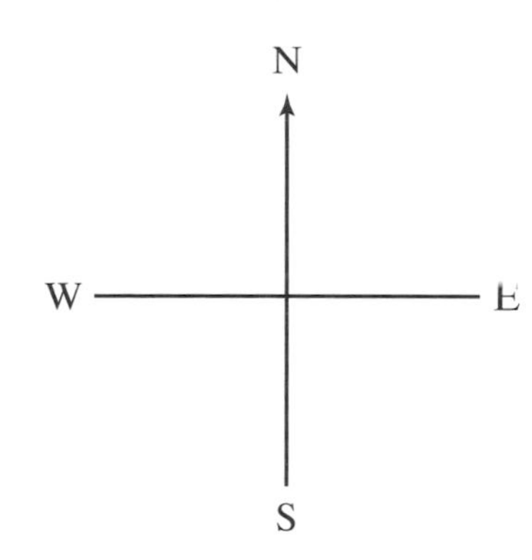

4 Get Real!

a What angle does the **minute** hand of a clock move through between 1200 and 1220?

b What angle does the **hour** hand of a clock move through between 0700 and 1400?

c What is the angle between the hour hand and the minute hand at half past one?

Learn 2 Types of angles

Examples: Describe these angles:

a

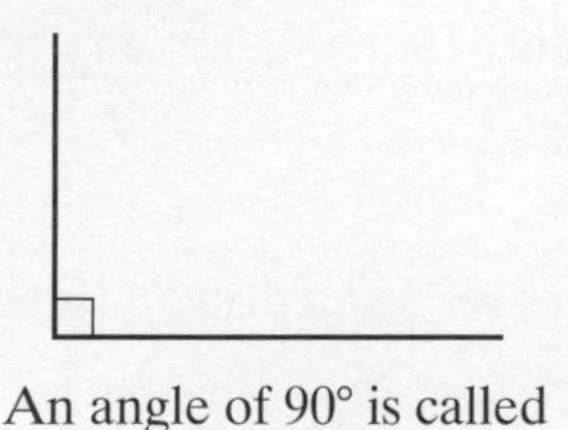

An angle of 90° is called a **right angle**.

It is marked with ∟ as shown above

c

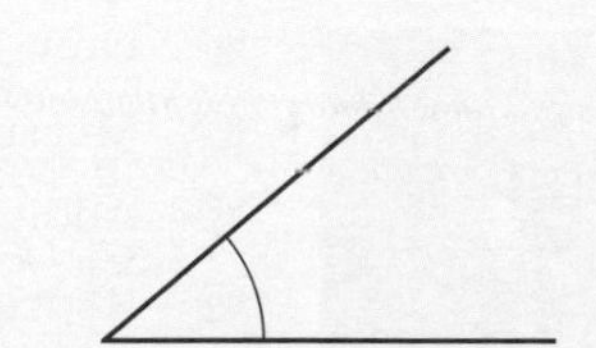

An angle less than 90° is called an **acute angle**.

b

An angle between 90° and 180° is called an **obtuse angle**.

d

An angle greater than 180° but less than 360° is called a **reflex angle**.

Apply 2

1 For each marked angle, write down whether it is an acute angle, an obtuse angle or a reflex angle.

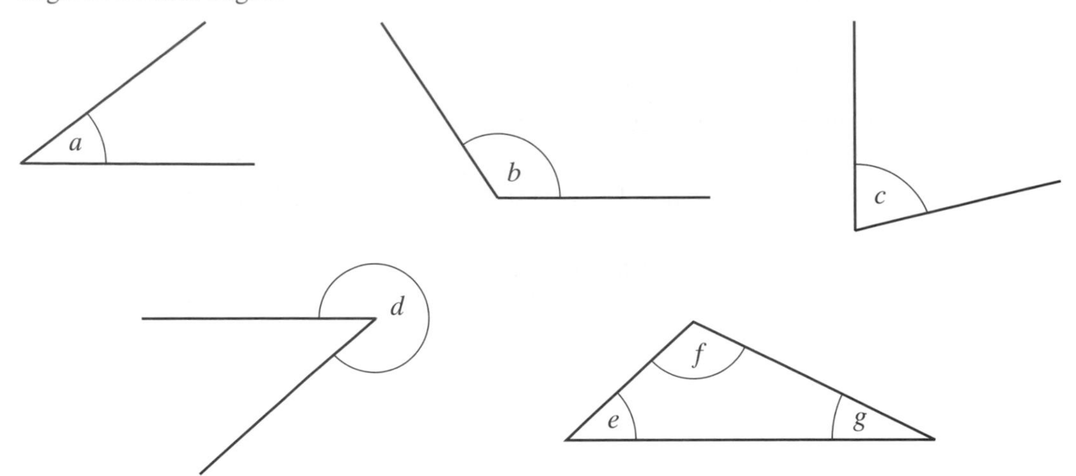

2 **Estimate** the size of each of the angles in question **1**.

3 **Measure** the size of each of the angles in question **1** with a protractor.
Check that your answers agree with your answers to question **1**.

4 What is $e + f + g$?

Explore

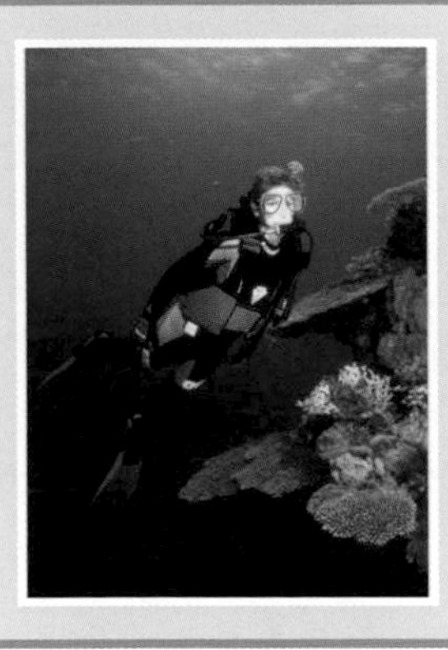

- Draw a 4-sided figure (not a square or a rectangle)
- Measure the angles
- Add them up
- Repeat this process for 5-sided and 6-sided figures. What do you notice?

Investigate further

Learn 3 Angles and lines

Examples: Find the missing angles.

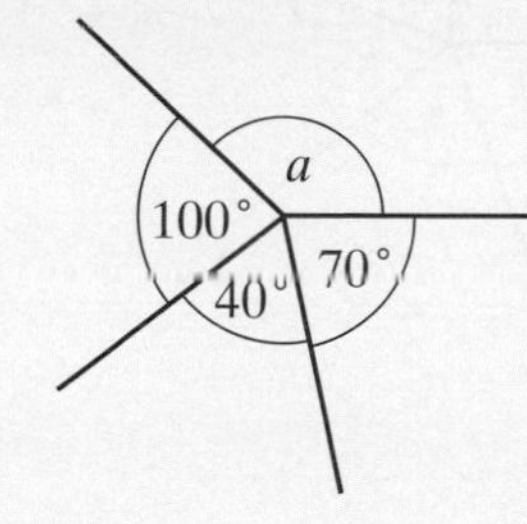

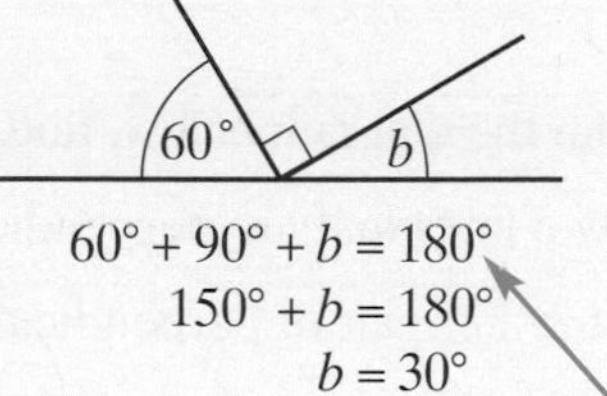

$100° + 40° + 70° + a = 360°$
$210° + a = 360°$
$a = 150°$

Angles at a point add up to 360°

$60° + 90° + b = 180°$
$150° + b = 180°$
$b = 30°$

Angles on a straight line add up to 180°

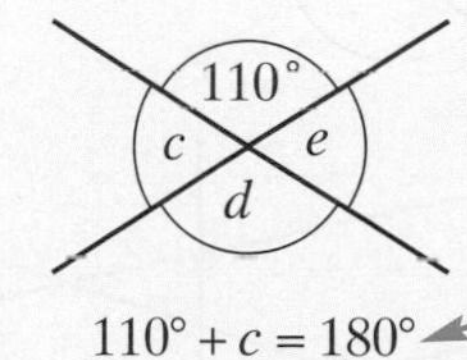

$110° + c = 180°$
$c = 70°$

Similarly
$c + d = 180°$
$70° + d = 180°$
$d = 110°$

Also
$d + e = 180°$
$110° + e = 180°$
$e = 70°$

Opposite angles are always the same

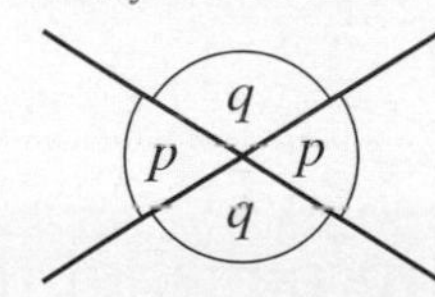

Apply 3

1 Calculate the size of each of the marked angles.
The diagrams are not drawn accurately.

a

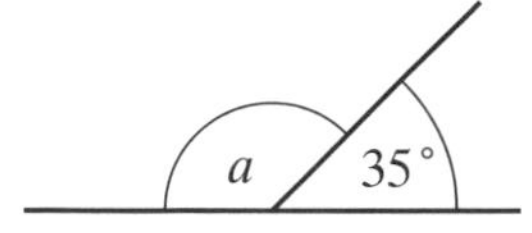

b

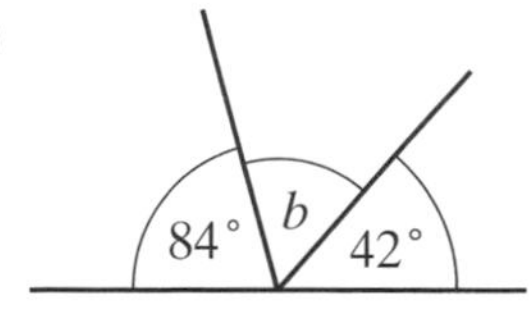

c

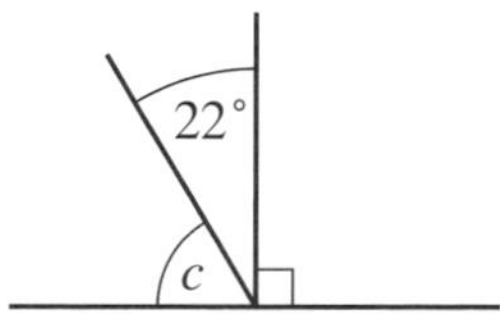

d

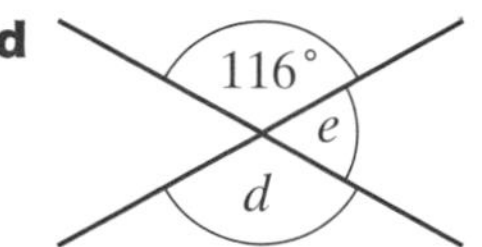

e

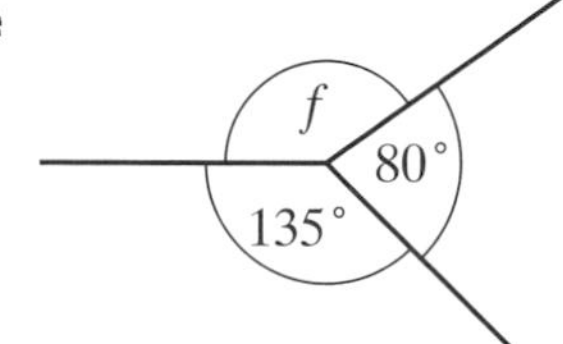

f

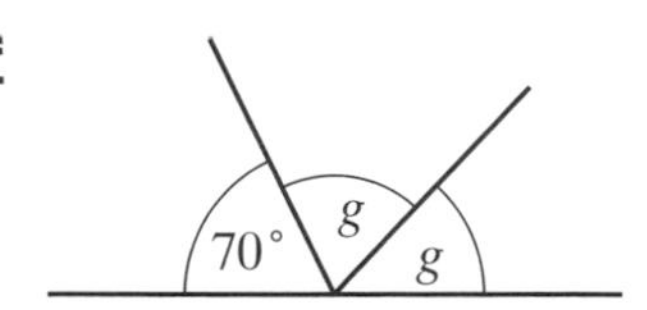

g

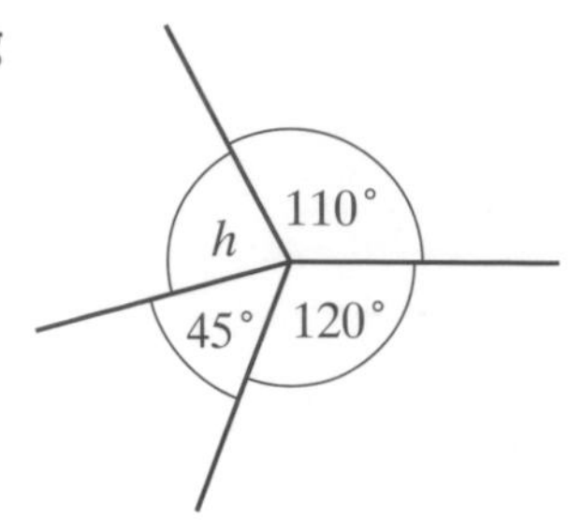

h

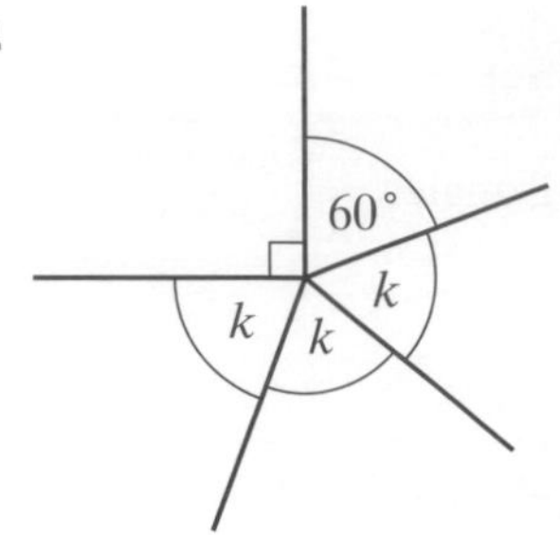

i

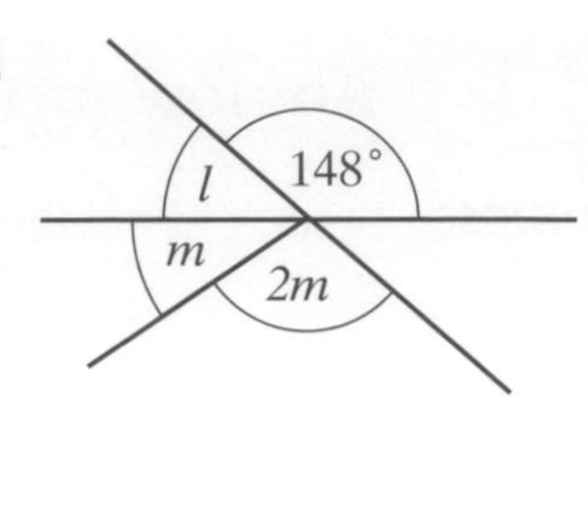

2 From the diagram below, find

a two lines that are perpendicular to LM,

b another pair of perpendicular lines,

c two pairs of parallel lines.

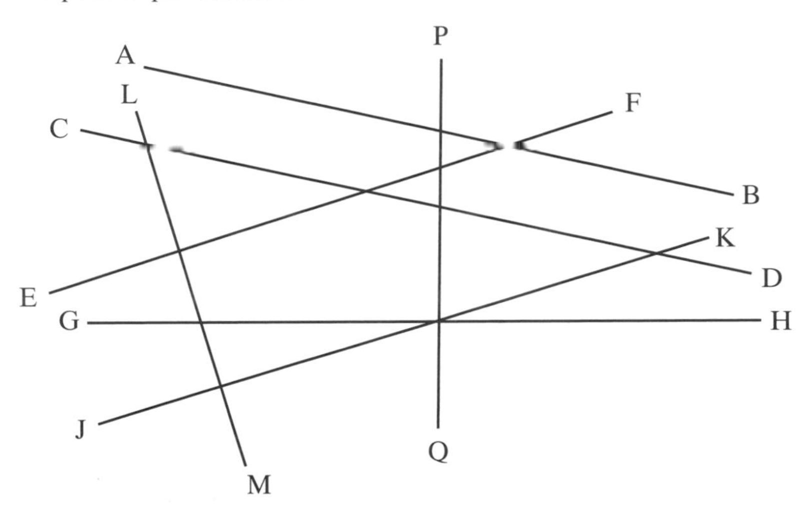

Learn 4 Angles and parallel lines

Examples: Find the missing angles.

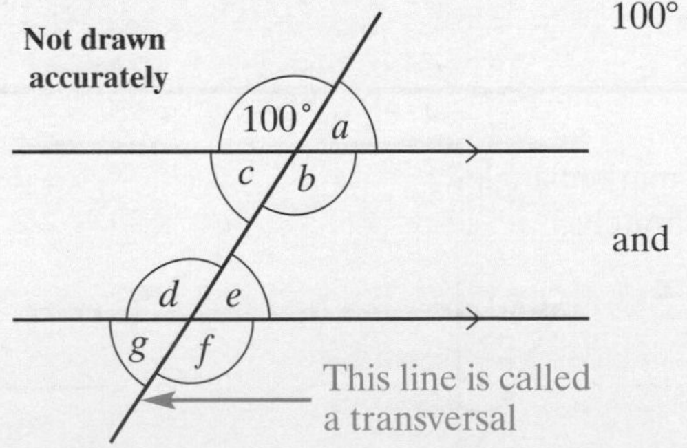

$100° + a = 180°$
$a = 80°$

Angles on a straight line add up to 180°

$b = 100°$

Opposite angles are equal

and

$c = a$
$c = 80°$

$d = 100°$ These angles are called corresponding angles

Corresponding angles are equal

Similarly

$e = a$ corresponding angles
$e = 80°$

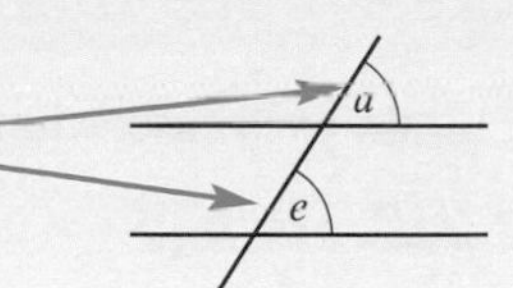

$f = b$ corresponding angles
$f = 100°$

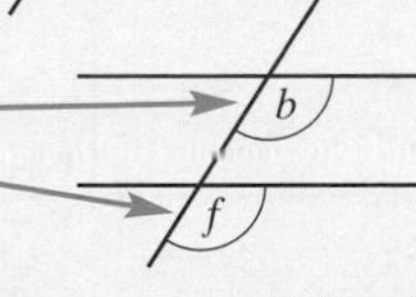

$g = c$ corresponding angles
$g = 80°$

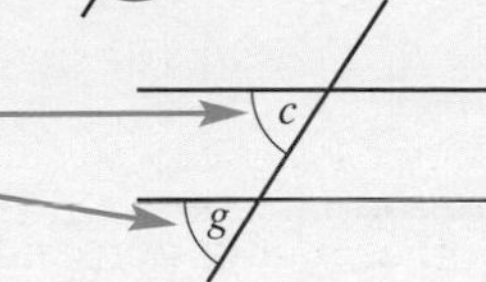

From the diagram $\left.\begin{array}{r} c = e \\ b = d \end{array}\right\}$ These are called alternate angles

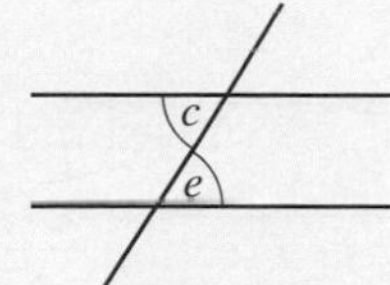

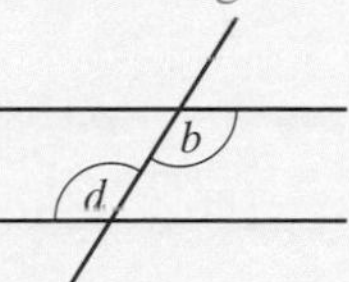

Apply 4

Work out the size of each of these angles.
Write down a reason beside each answer.

Your reason will be one of these: **alternate angles** **corresponding angles** **opposite angles** **angles on a straight line**

The diagrams are not drawn accurately.

1

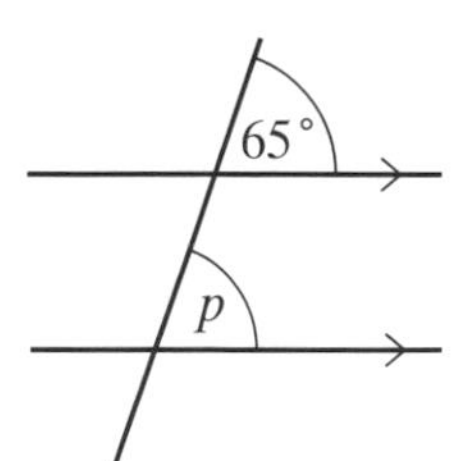

2

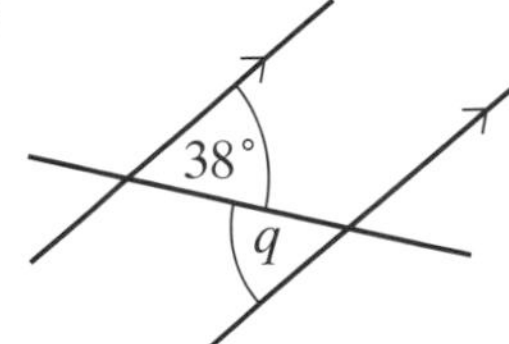

3

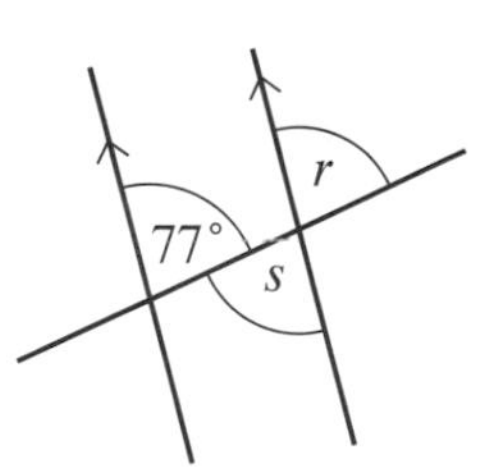

4

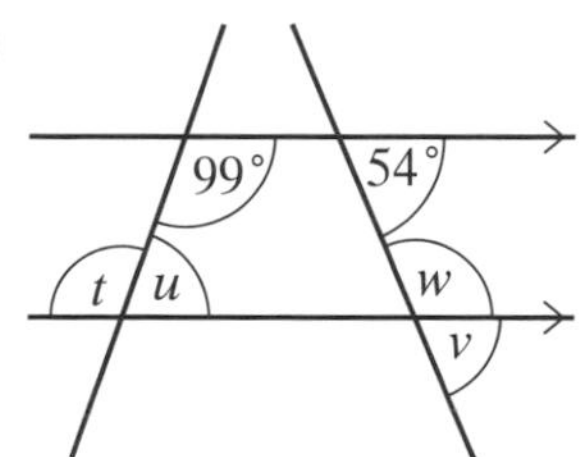

5

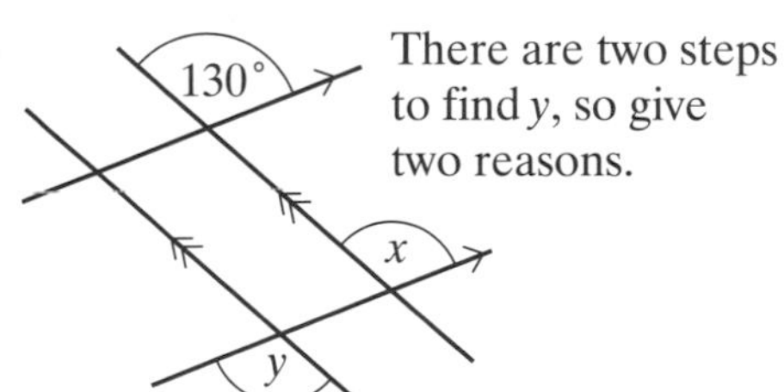

There are two steps to find y, so give two reasons.

6

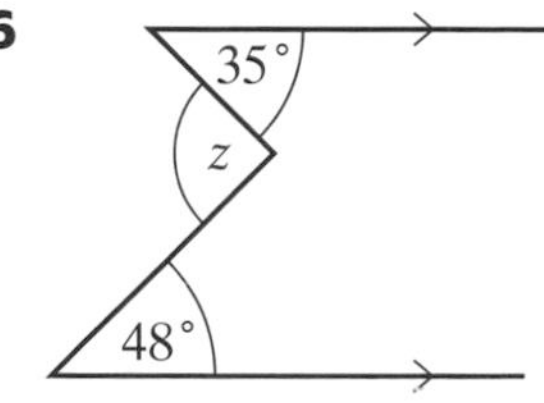

HINT To find z put in another parallel line.

Learn 5 Three-figure bearings

Examples: In each case, what is the bearing of A from B?

a

N

Not drawn accurately

144°

B

A

The bearing of A from B is 144°.

Bearings are measured from North in a clockwise direction

b

N

A

55°

B

The bearing of A from B is 055°.

Bearings are always written as 3 figures, so 55° is written 055°

c

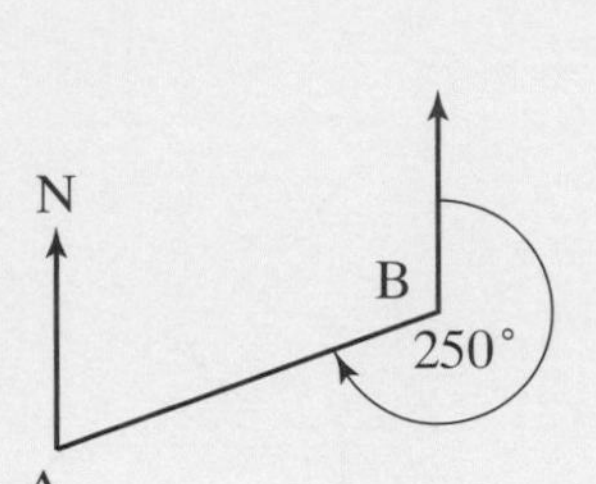

The bearing of A from B is 250°.

Apply 5

1 For each diagram, write down the 3-figure bearing of D from E.

a

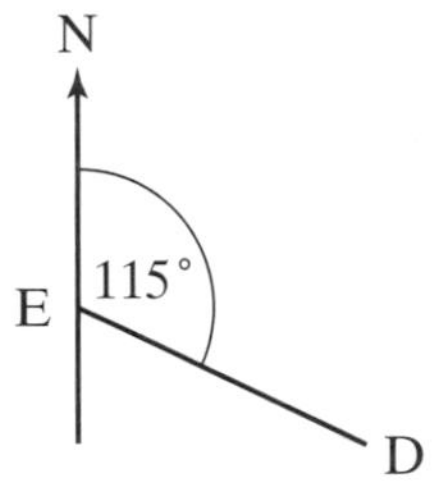

b

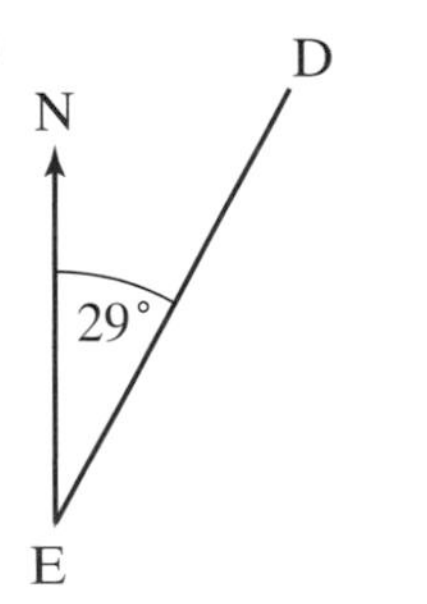

c

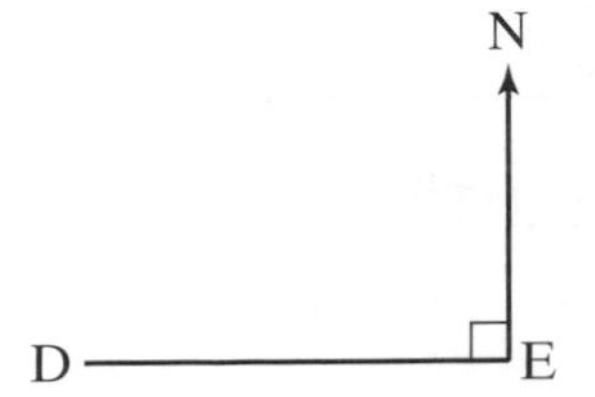

Not drawn accurately

d

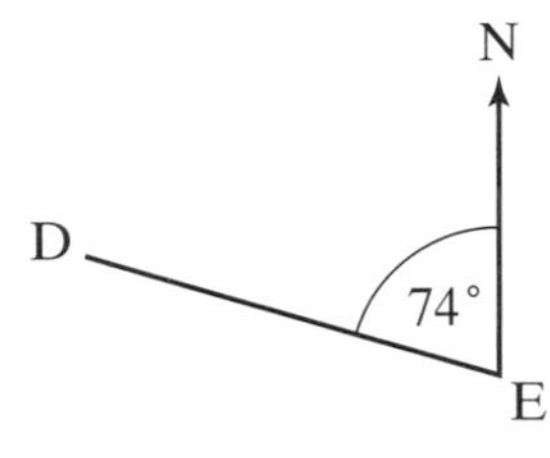

e

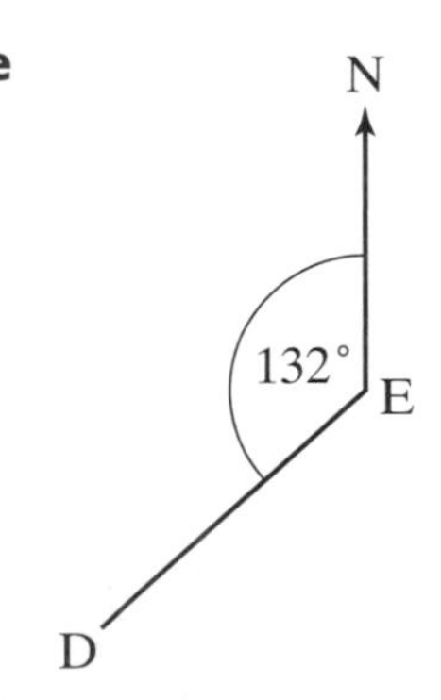

f

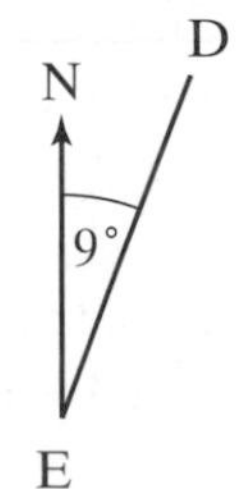

2 Get Real!

The diagram shows the three towns of Ashby, Derby and Loughborough.
Derby is due north of Ashby.
Loughborough is on a bearing of 080° from Ashby.
The bearing of Loughborough from Derby is 137°.

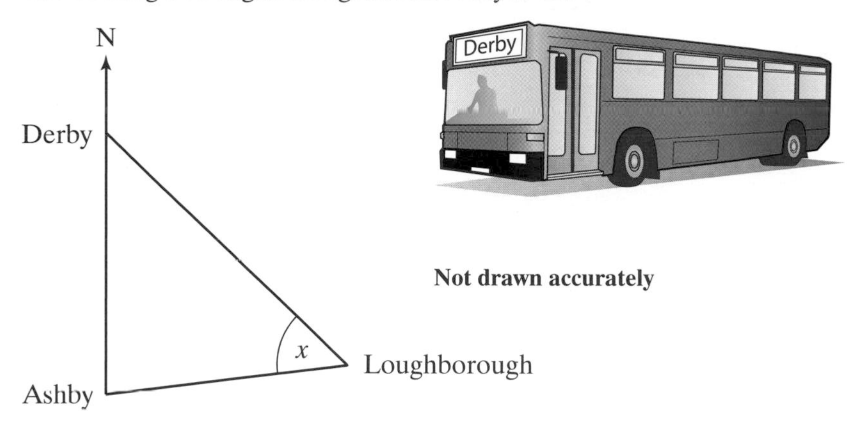

Copy the diagram, adding the bearings and working out the size of angle x.

3 Get Real!

The diagram shows the three towns of Nelson, Haworth and Todmorden.
Haworth is due east of Nelson.
The bearing of Todmorden from Nelson is 150°.
The bearing of Todmorden from Haworth is 220°.

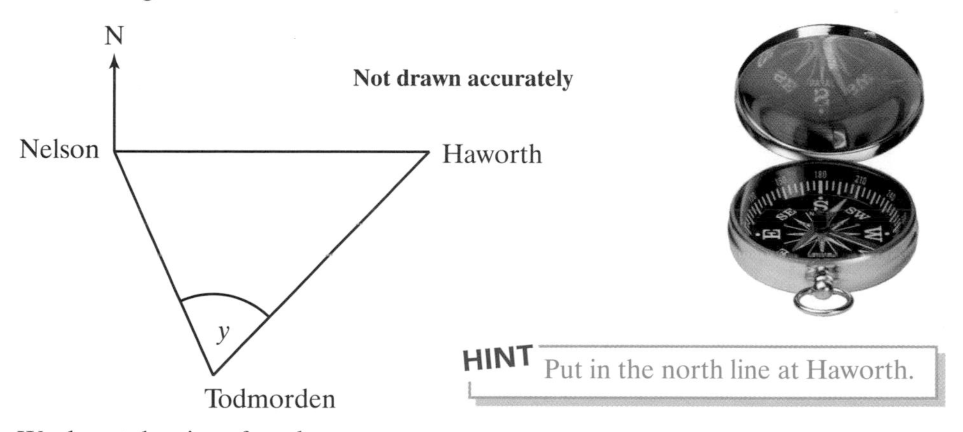

HINT Put in the north line at Haworth.

Work out the size of angle y.

Angles

ASSESS

The following exercise tests your understanding of this chapter, with the questions appearing in order of increasing difficulty.

1 Calculate the number of degrees in:

a $\frac{2}{3}$ of a revolution

b $\frac{1}{8}$ of a revolution

c 0.3 of a revolution

d $1\frac{1}{2}$ revolutions

2 The second hand on a clock revolves at 1 revolution per minute. How many degrees per second is this?

3 What is the angle between the hour hand and the minute hand of a clock

a at two o'clock

b at half past five?

4 Calculate the size of each of the marked angles.

a

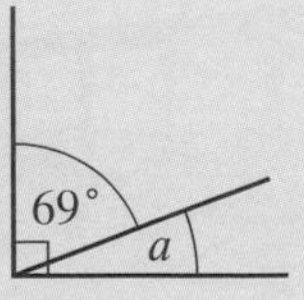

b

c

d

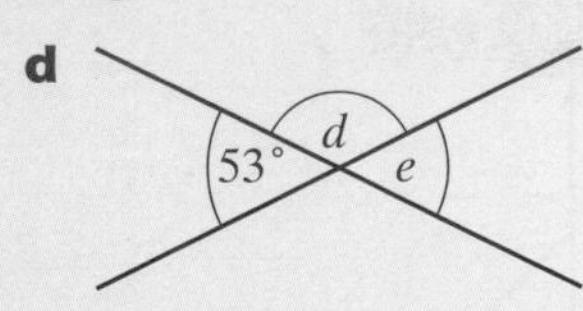

e

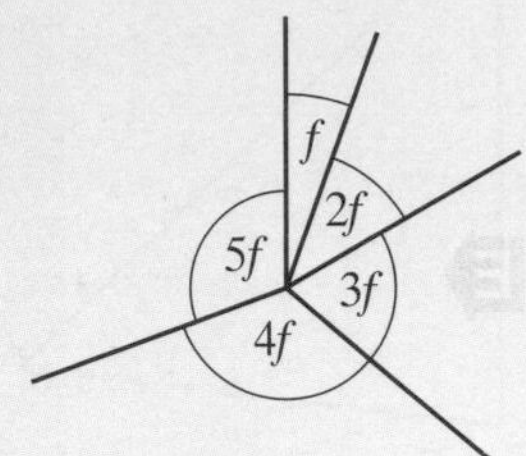

Not drawn accurately

5 Calculate the size of each of the marked angles.

a

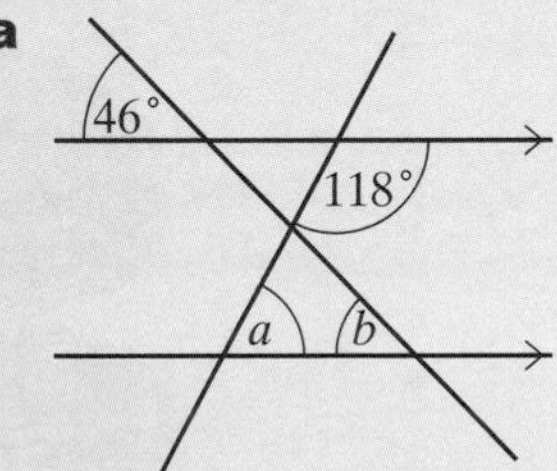

b

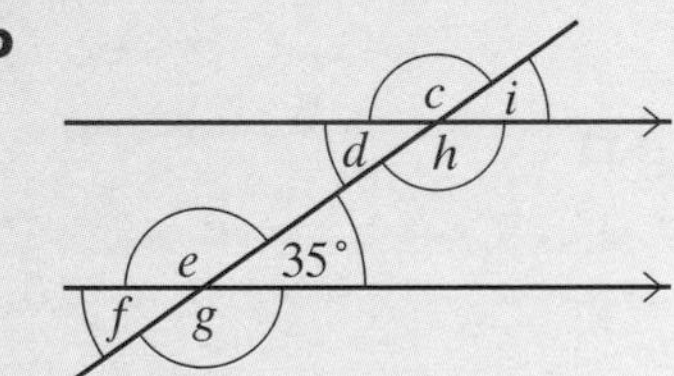

Not drawn accurately

c

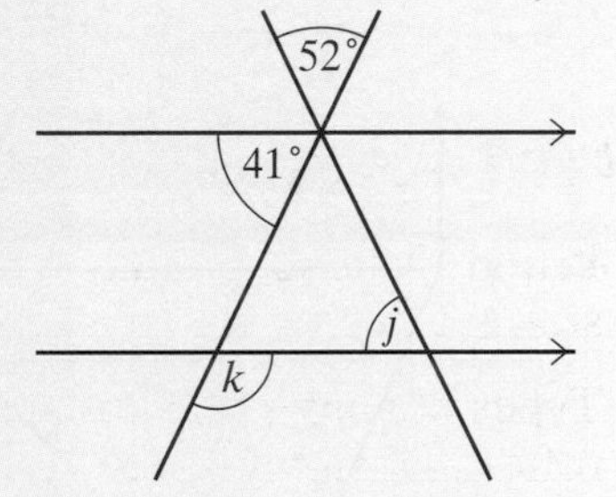

d

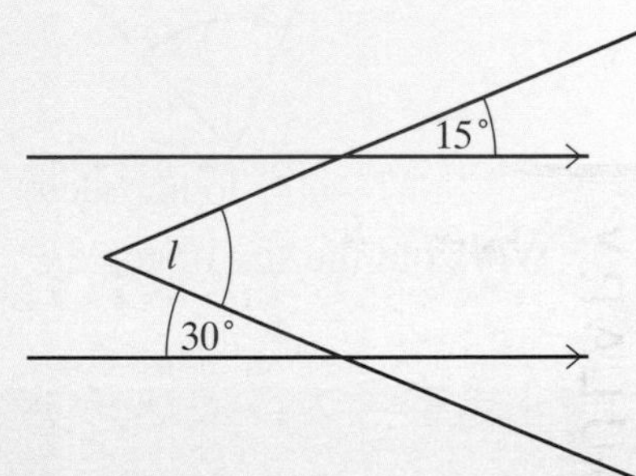

HINT Draw another parallel line.

6 The cruise liner 'Oriana' is sailing on a bearing of 060°.
To avoid a storm ahead it changes course to a bearing of 125°.
Through what angle has it turned?

7 Two aircraft take off from Birmingham International Airport, one on a bearing of 152° and the other on a bearing of 308°.
What is the angle between the two planes' flight paths?

2 Properties of triangles

OBJECTIVES

Examiners would normally expect students who get a G grade to be able to:

Identify isosceles, equilateral and right-angled triangles

Use the word ‘congruent’ when triangles are identical

E

Examiners would normally expect students who get an E grade also to be able to:

Show that the angles of a triangle add up to 180° and use this to find angles

Show that the exterior angle of a triangle is equal to the sum of the interior opposite angles

Use angle properties of isosceles, equilateral and right-angled triangles

What you should already know ...

- Properties of angles at a point and on a straight line
- Types of angles, including acute, obtuse, reflex and right angles
- Parallel lines, including opposite angles, corresponding angles and alternate angles

VOCABULARY

Exterior angle – if you make a side of a triangle longer, the angle between this and the next side is an exterior angle

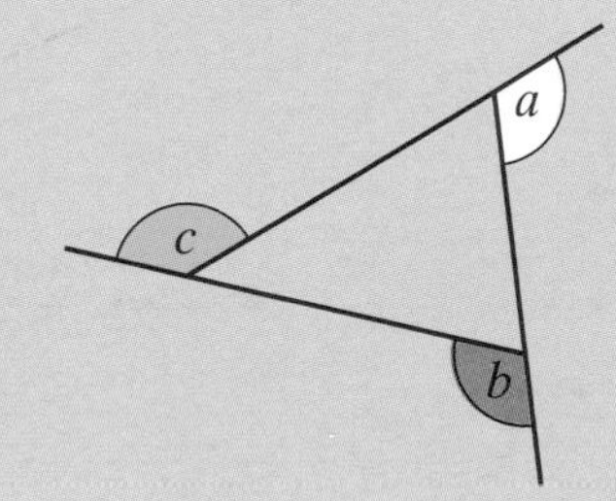

a, b and c are exterior angles

Isosceles triangle – a triangle with 2 equal sides and 2 equal angles; the equal angles are called **base angles**

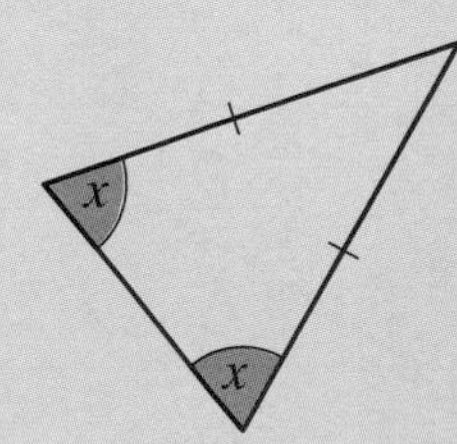

the x angles are base angles

Equilateral triangle – a triangle with 3 equal sides and 3 equal angles – each angle is 60°

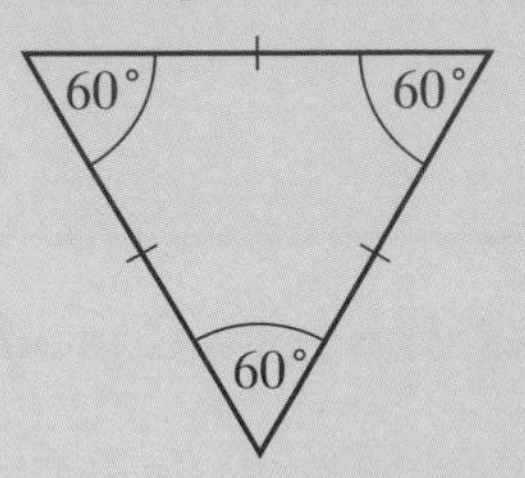

Right-angled triangle – a triangle with one angle of 90°

Congruent – exactly the same size and shape; one of the shapes might be rotated or flipped over

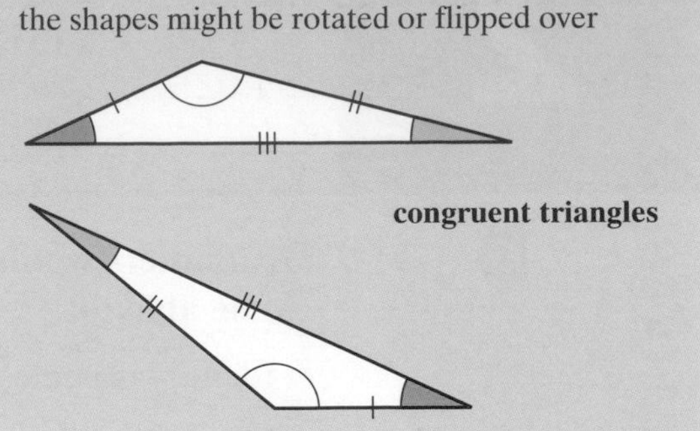

Learn 1 Angle sum of a triangle

Triangle ABC has angles a, b and c.

Suppose side AB is extended to D then BE is drawn from B parallel to AC.

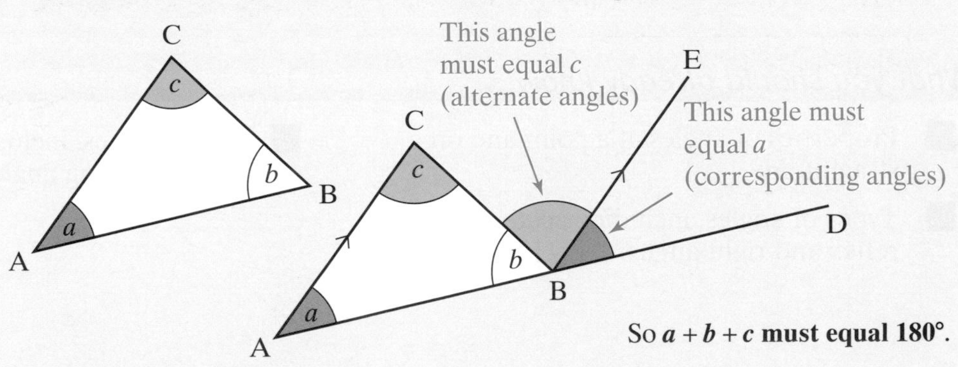

So $a + b + c$ **must equal 180°.**

Example:

In triangle PQR, angle P is 32° and angle Q is 54°.
Calculate the size of angle R and state what type of angle it is.

The sketch shows the triangle PQR.

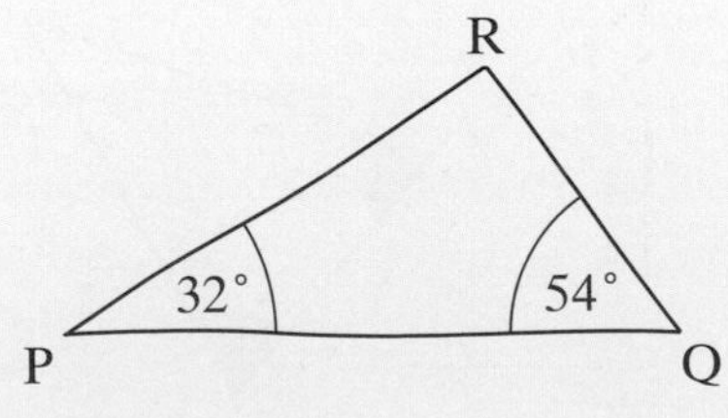

angle P + angle Q = 32° + 54° = 86°

angle R = 180° – 86° ← The sum of the angles of a triangle = 180°

angle R = 94°

The angle R is obtuse. ← The angle is greater than 90° but less than 180°

Apply 1

1 Work out the angle marked by each letter and state what type of angle it is.

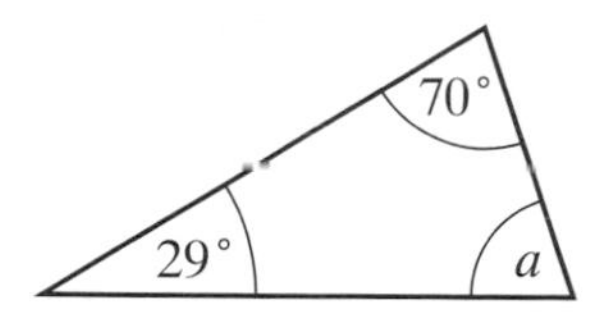

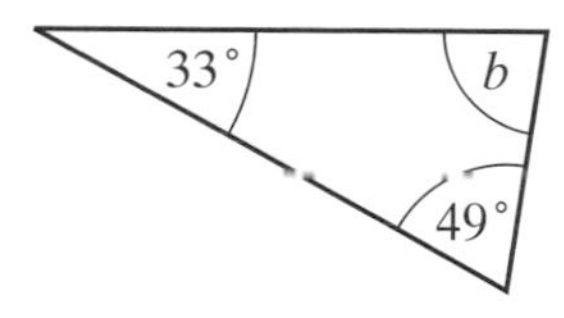

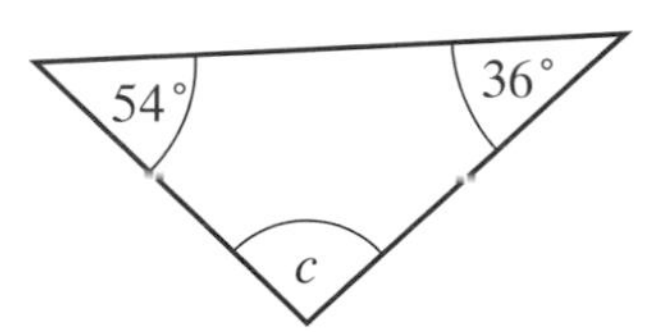

Not drawn accurately

2 Get Real!
The sketch shows the side of a ramp – it is not drawn accurately.
Calculate the angle marked x.

3 Get Real!
Both sides of a roof are inclined at 34° to the horizontal.
Calculate the angle, x, between the sides of the roof.

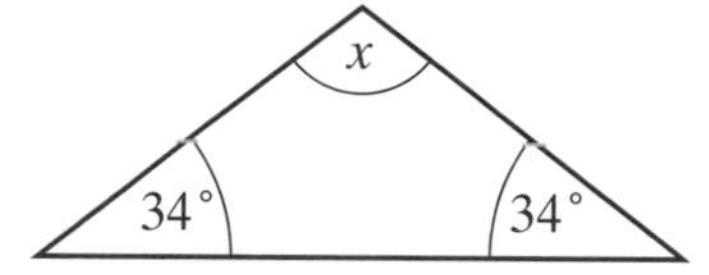

Not drawn accurately

4 In triangle LMN, angle L is 26° and angle M is 108°.
Sketch the triangle and calculate angle N.

5 Calculate the angles marked by letters. Give a reason for each answer.
Choose from: **angles on a straight line, angles of a triangle, opposite angles.**

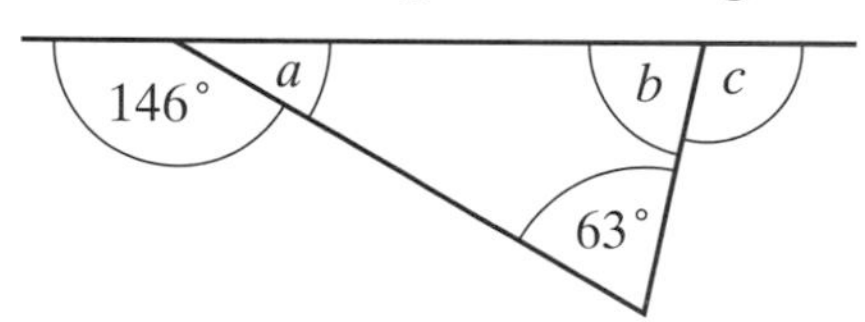

79°
q
57°
p
r

Not drawn accurately

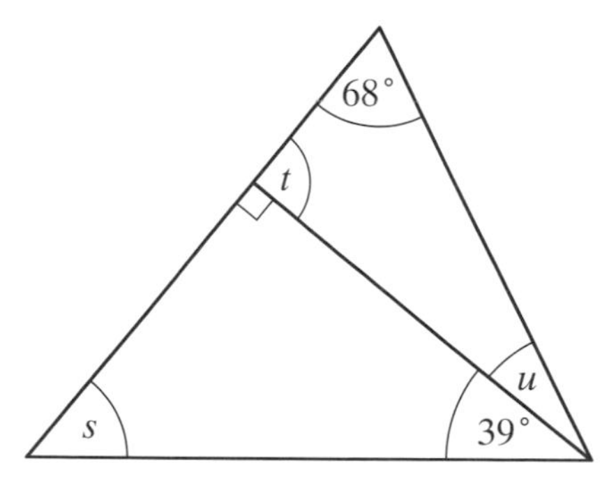

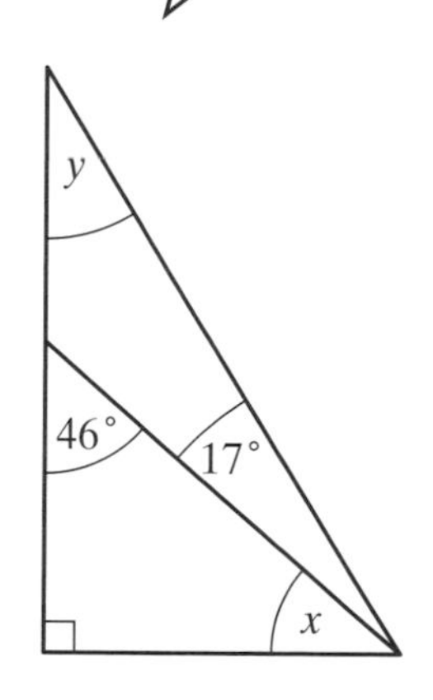

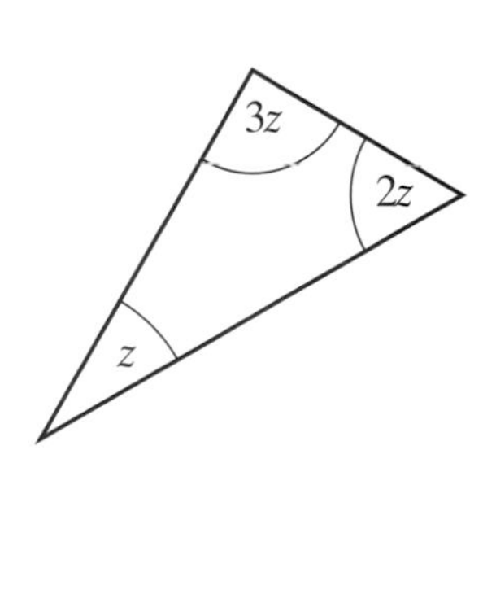

6 In triangle PQR, angle P is 80°. Give five possible pairs of values for angles Q and R.

7 Calculate the angles marked by letters. Give a reason for each answer.

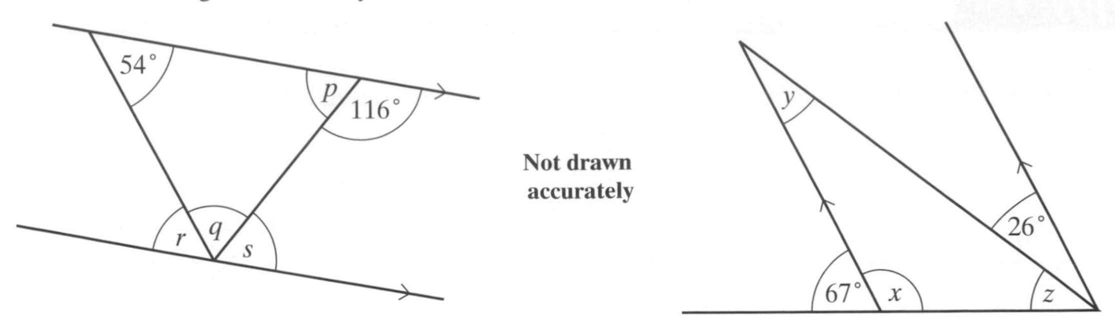

8 Triangle ABC has a right angle at A and angle B equals angle C.
Calculate angle B and angle C.

9 In triangle PQR, angle P is 10° larger than angle R and angle Q is 20° larger than angle R.
Calculate angles P, Q and R.

Learn 2 Exterior angle of a triangle

When one side of a triangle is extended you get an **exterior** angle.
The exterior angle **is equal to the sum of the interior opposite angles (at the other two vertices)**. Look at the diagram in Learn 1 to see the proof for this.

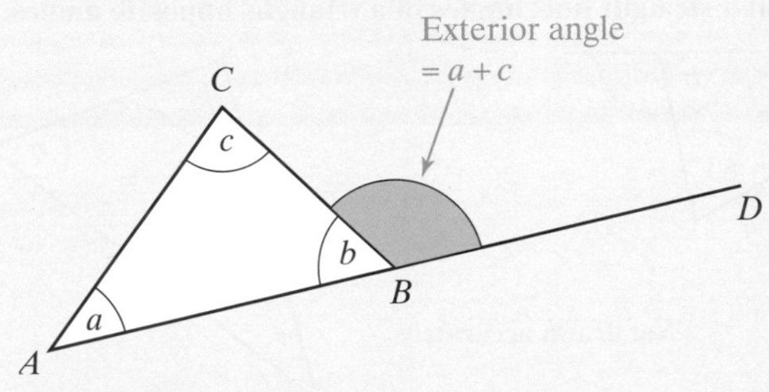

Example: Find the angles x and y.

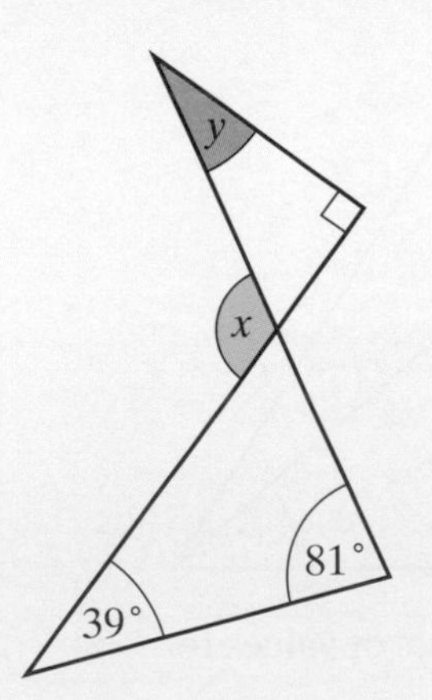

$x = 81° + 39°$ ← The angle x is an exterior angle for the lower triangle

$x = 120°$

$x = y + 90°$ ← The angle x is also an exterior angle for the upper triangle

$120° = y + 90°$

$y = 30°$

Apply 2

1 Work out the angles marked by letters.

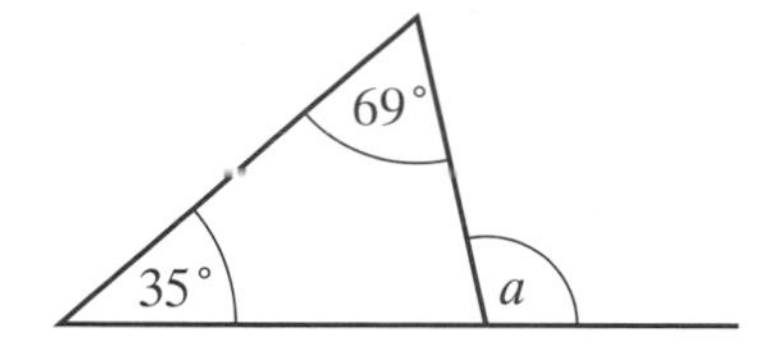

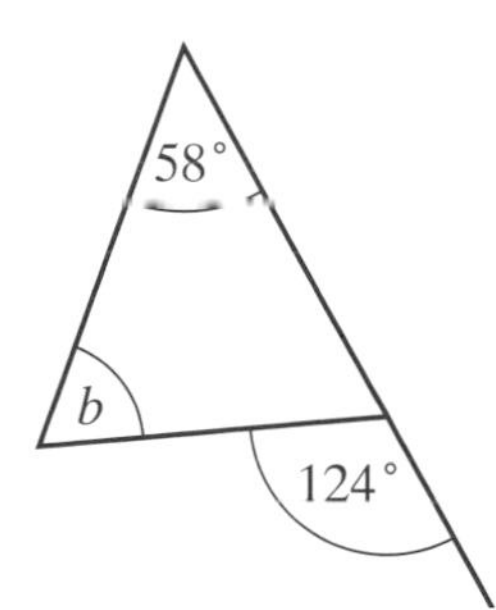

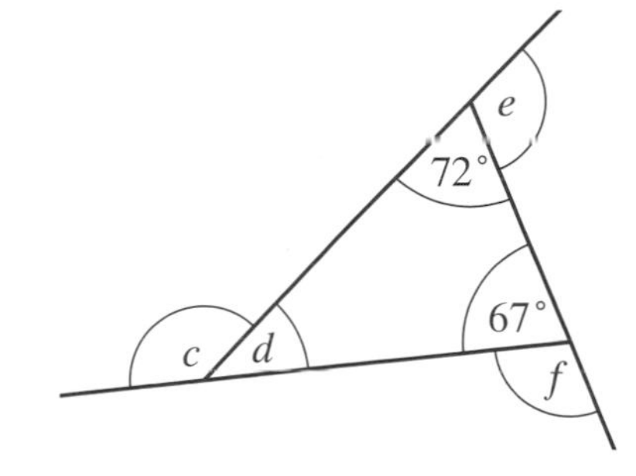

Not drawn accurately

2 Get Real!
The diagram shows the end view of a shelf.
Calculate the angle marked x.

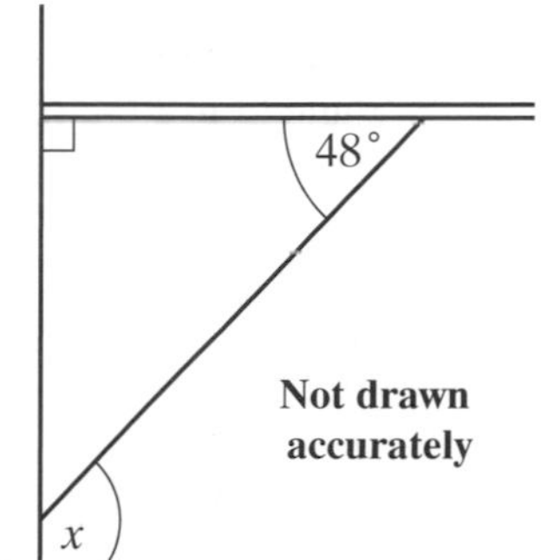

Not drawn accurately

3 Calculate the angles marked by letters. Give a reason for each answer.
Choose from: **exterior angle of a triangle, angles on a straight line.**

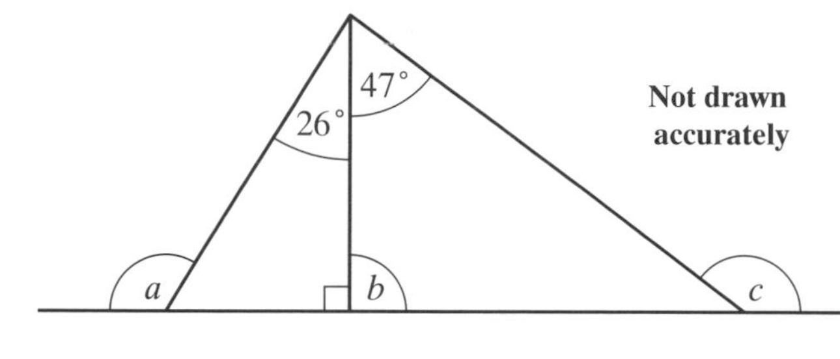

Not drawn accurately

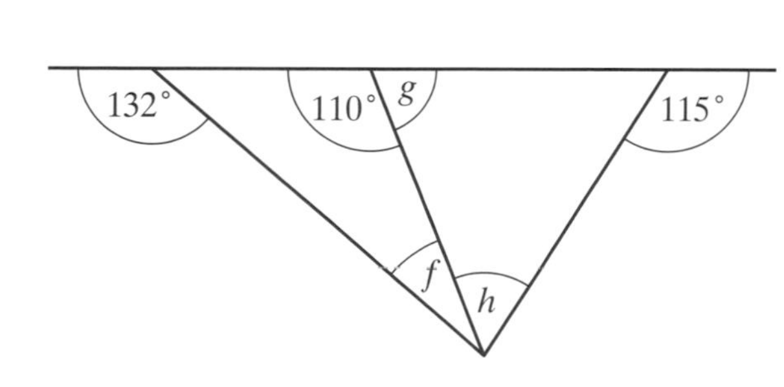

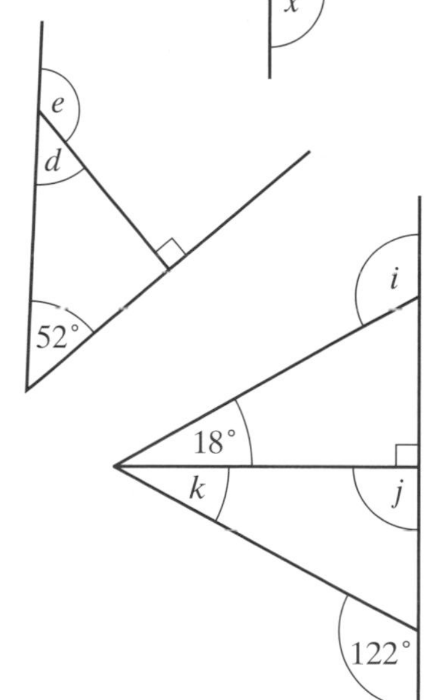

4 One of the exterior angles of a triangle is 115°.
Find two possible sets of values for the interior angles of the triangle.

5 Get Real!
The diagram shows a point P on one side of a river and two points, Q and R, on the other side.
The sides of the river are parallel.
Calculate the angles marked x and y.

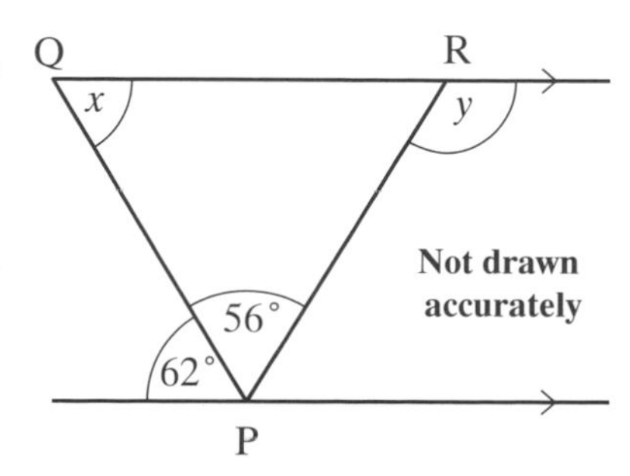

Not drawn accurately

6 Get Real!

The diagram shows the positions of three villages. Charfield is due north of Wickwar, the bearing of Hillesley from Wickwar is 076° and the bearing of Hillesley from Charfield is 120°.

a Calculate the angles marked x and y.

b What is the bearing of

i Wickwar from Hillesley

ii Charfield from Hillesley?

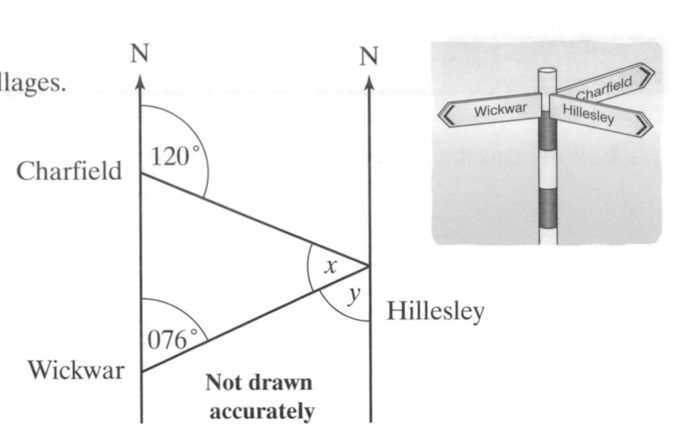

Explore

- Draw any triangle and extend the sides to make three exterior angles as shown in the diagram
- Measure the exterior angles and find their sum
- Draw any quadrilateral (4-sided shape) and extend the sides to make four exterior angles
- Measure the exterior angles and find their sum

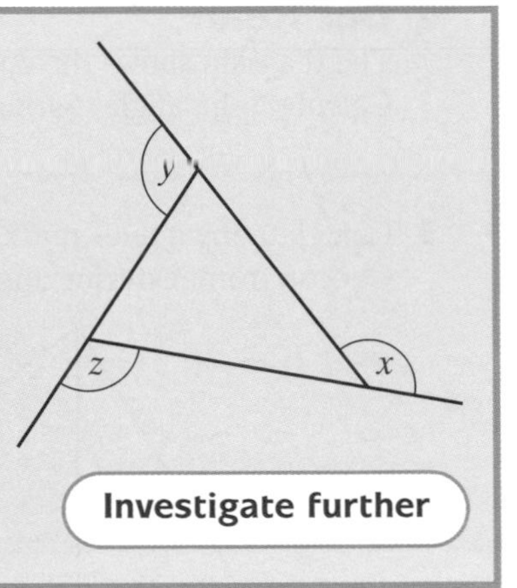

Investigate further

Learn 3 Describing triangles

Examples: Calculate the size of angles a, b, c and d.

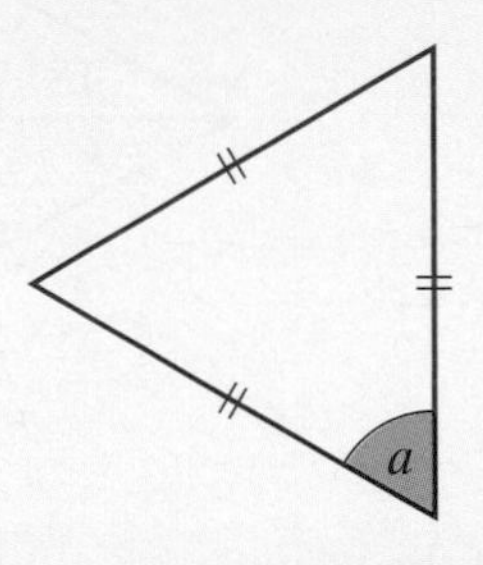

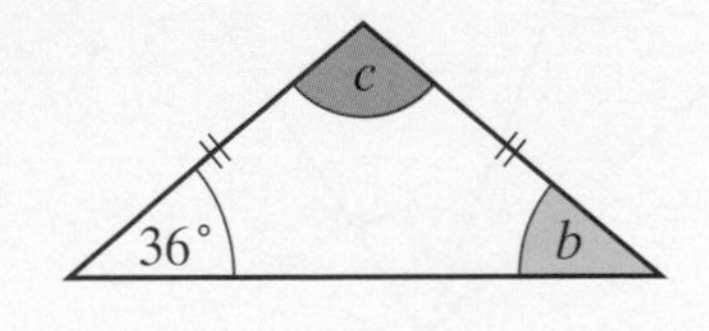

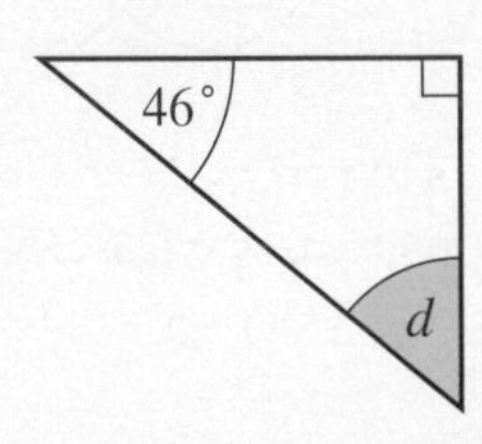

$a = 60°$	Angles of an equilateral triangle
$b = 36°$	Base angles of an isosceles triangle
$c = 108°$	Angles of a triangle add up to 180°
$d = 44°$	Angles of a right-angled triangle

Apply 3

1 Work out the angles marked by letters.

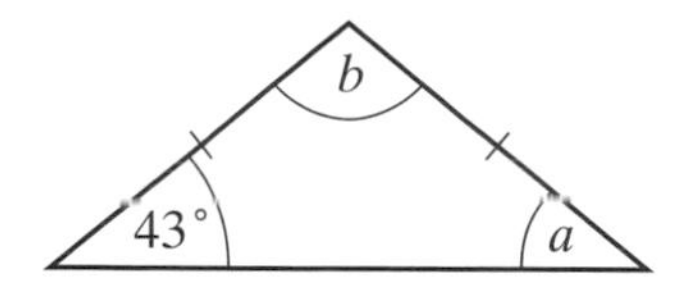

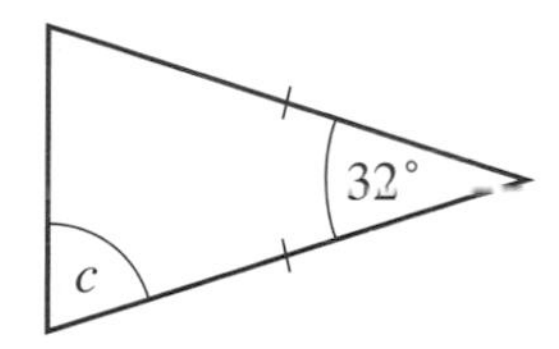

Not drawn accurately

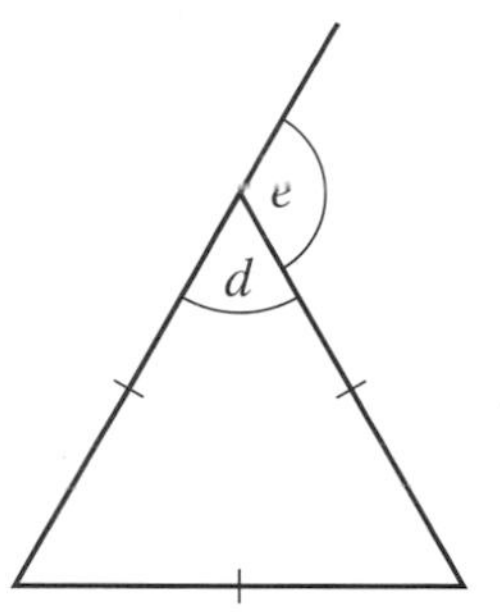

2 Calculate the angles marked by letters.

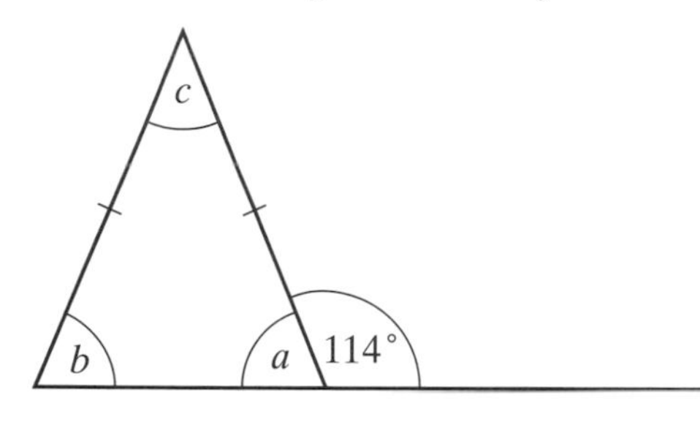

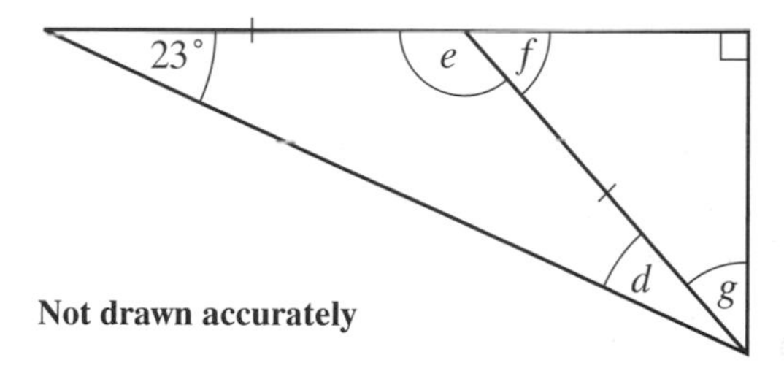

Not drawn accurately

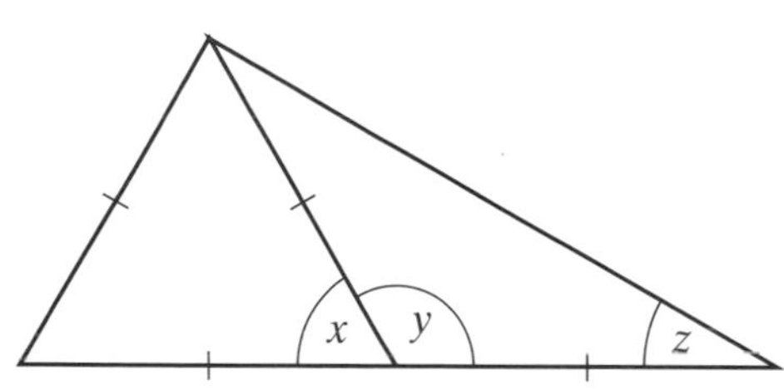

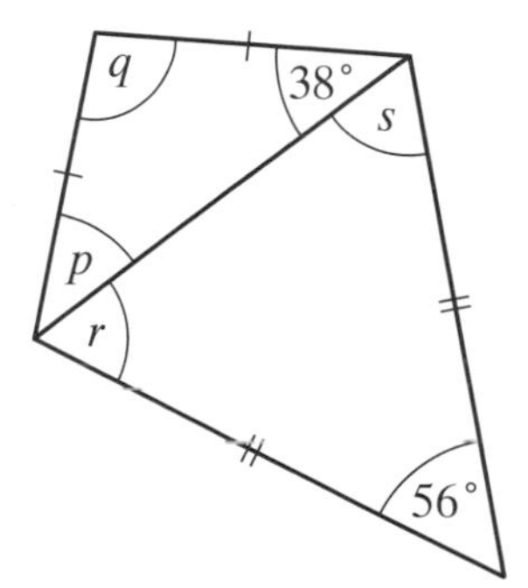

3 **Get Real!**

The diagram shows the beams in a roof.
Calculate the angles x and y.

Not drawn accurately

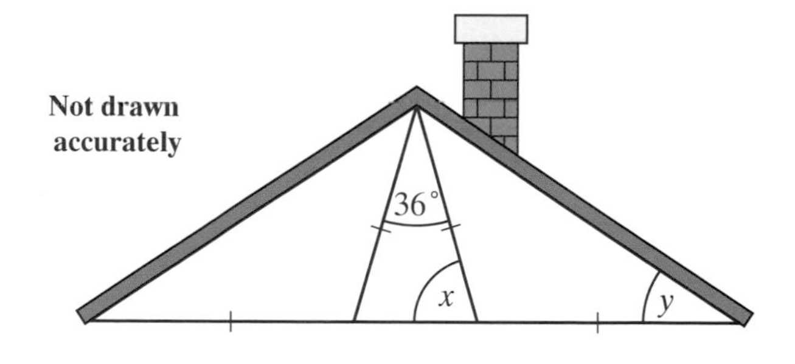

4 Calculate the angles marked by letters.
Give a reason for each answer.

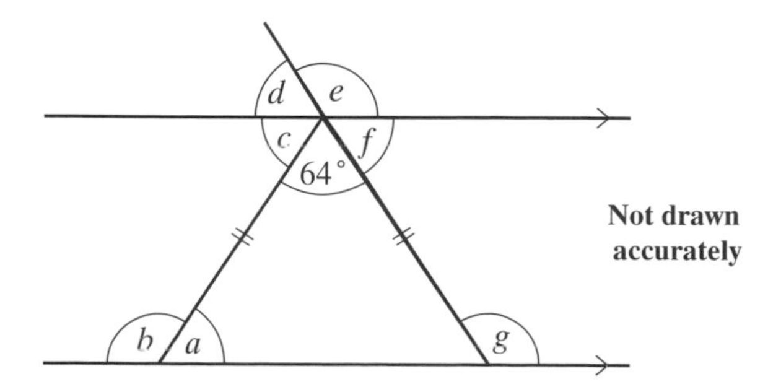

Not drawn accurately

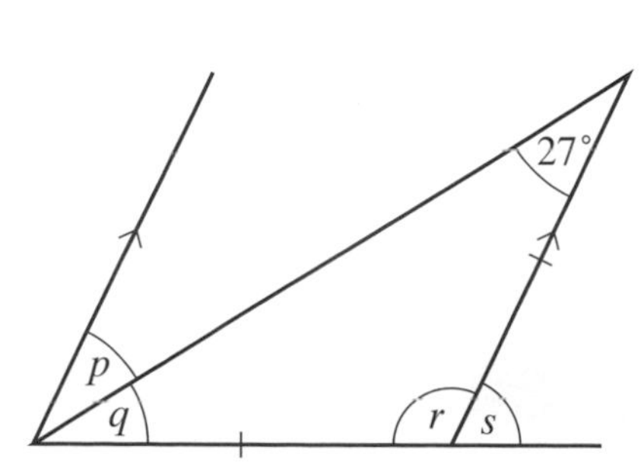

5 Measure the sides and angles of each triangle.
Identify two pairs of congruent triangles.

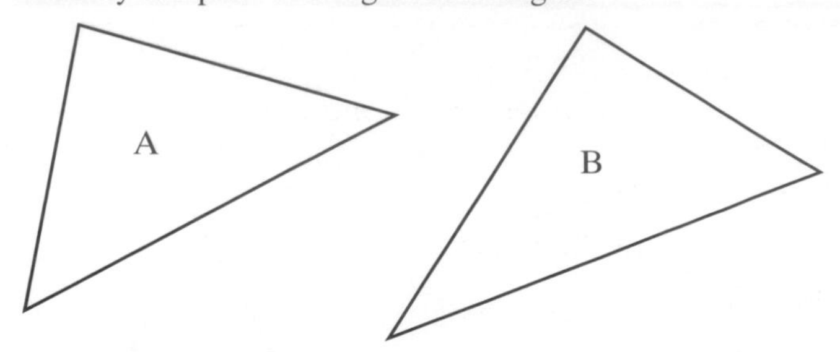

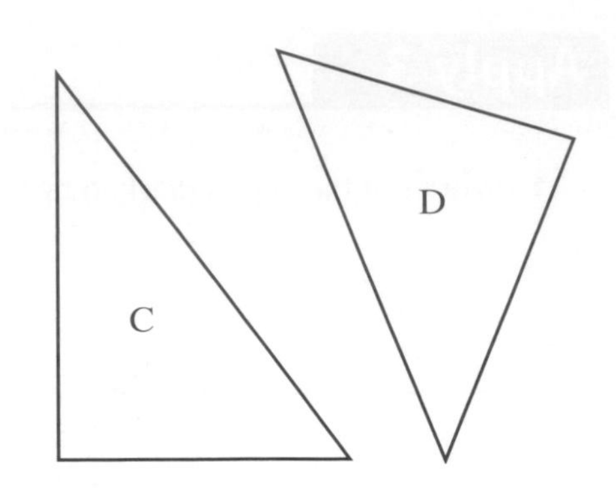

6 One angle of an isosceles triangle is 28°.
Draw two possible triangles and give their other angles.

7 **Get Real!**
The diagram shows a rectangular gate, ABCD.
The diagonals AC and BD meet at E.

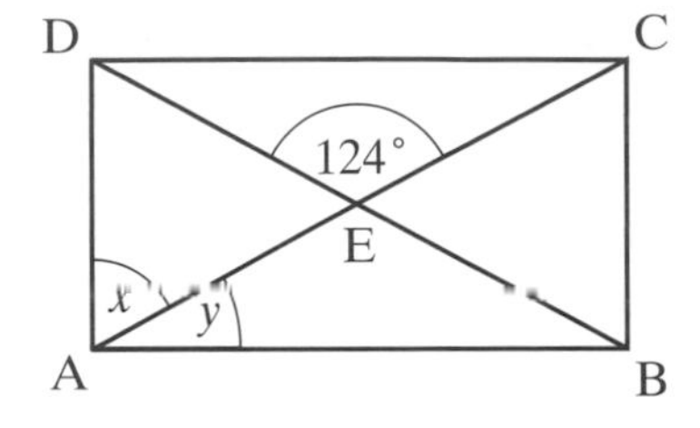

a Identify two pairs of congruent isosceles triangles.

b Identify four congruent right-angled triangles.

c The obtuse angle between the diagonals is 124°.
Calculate the angles x and y.

8 **a** Calculate all the angles in this star.

b How many right-angled triangles are there in the diagram?

c How many pairs of congruent triangles are there in the diagram?

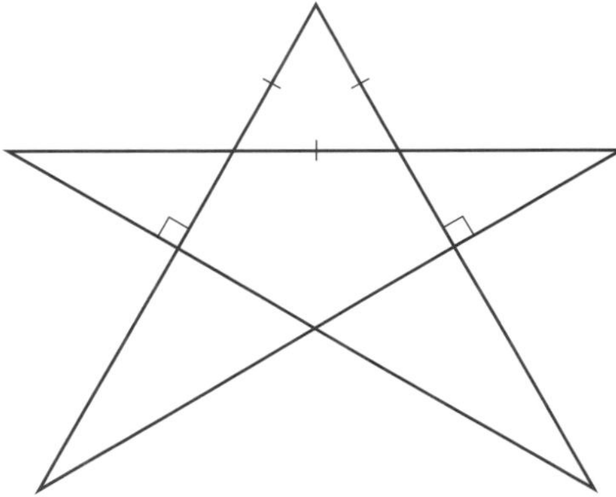

9 **a** The diagram shows how five congruent isosceles triangles fit together to make a pentagon.
Calculate the angles, x and y, of one of the isosceles triangles.

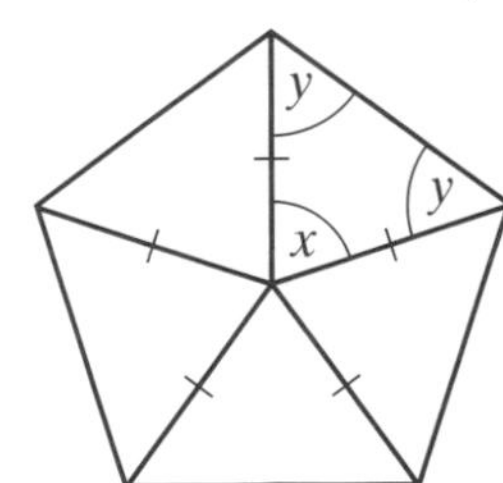

b Eight congruent isosceles triangles fit together in a similar way to make an octagon.
Calculate the angles of one of the isosceles triangles.

Explore

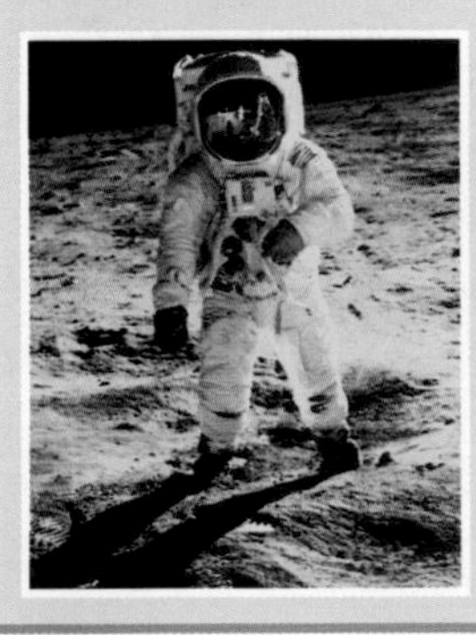

- How many different isosceles triangles can you draw that have an angle of 30°?
- How many different isosceles triangles can you draw that have an angle of 120°?

Investigate further

Properties of triangles

ASSESS

The following exercise tests your understanding of this chapter, with the questions appearing in order of increasing difficulty.

1 Copy and complete the following statements:

a A triangle with all sides equal is called ... .

b A triangle with two sides equal is called ... .

c A triangle with all angles equal is called ... .

d A triangle with two angles equal is called ... .

e A triangle with one angle of 90° is called ... .

f An equilateral triangle has got all equal ... and ... .

g An isosceles triangle has got two equal ... and ... .

h Two triangles that are identical in shape and size are called ... .

2 In each of the triangles below write down the letters of all triangles that are

a equilateral

b isosceles

c right-angled.

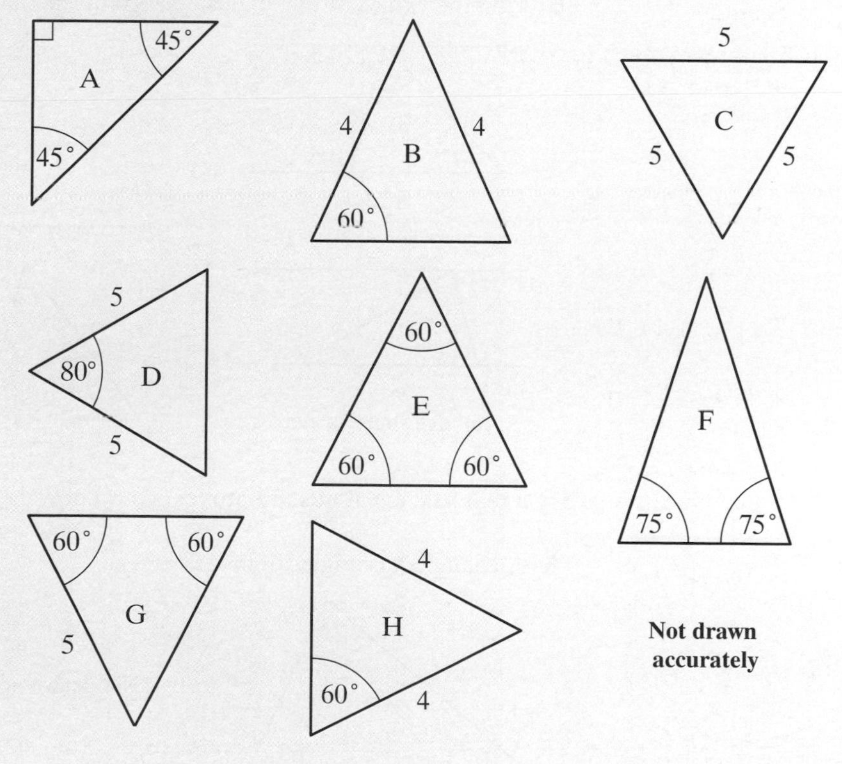

3 Find the values of the angles marked in the diagrams below.

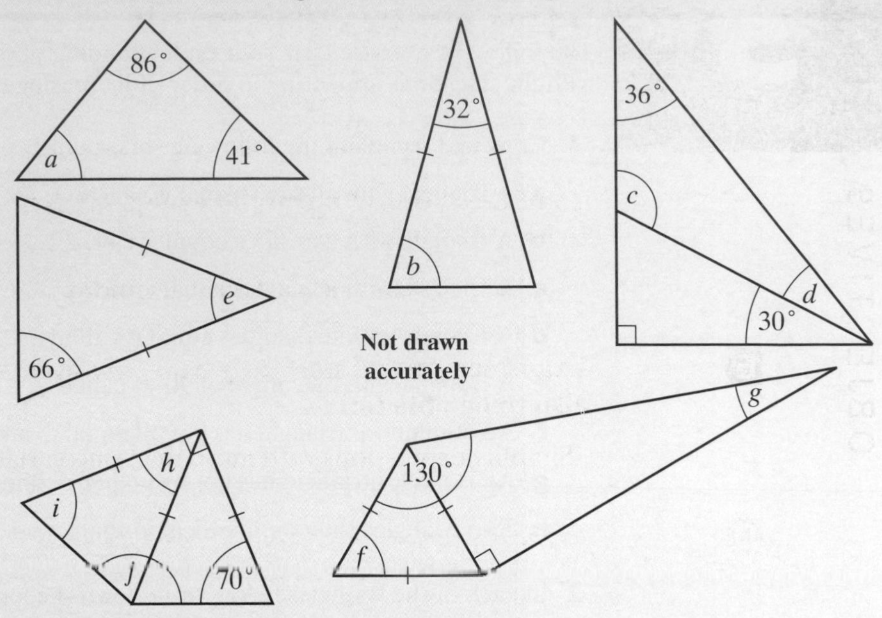

4 An isosceles triangle has a base angle of 65°.
What are the sizes of the other two angles?

5 Find the values of the angles marked in the diagrams below.

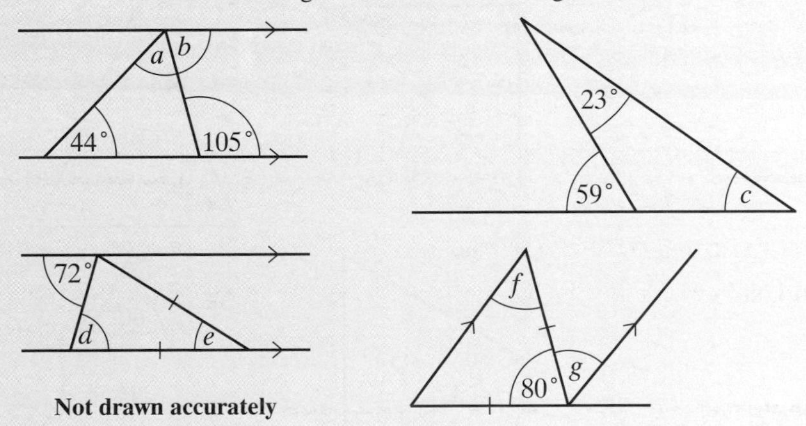

Not drawn accurately

Try a real past exam question to test your knowledge:

6 A triangle has angles of 63°, $2x$ and x.

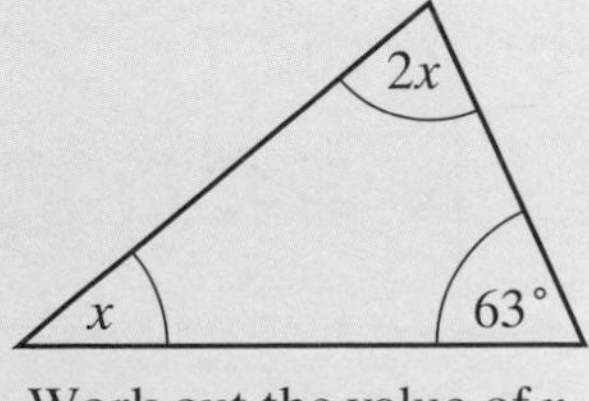

Not drawn accurately

Work out the value of x.

Spec A, Mod 5 Foundation Paper 1, June 03

3 Use of symbols

OBJECTIVES

F **Examiners would normally expect students who get an F grade to be able to:**

Simplify expressions with one variable such as $a + 2a + 3a$

E **Examiners would normally expect students who get an E grade also to be able to:**

Simplify expressions with more than one variable such as $2a + 5b + a - 2b$

D **Examiners would normally expect students who get a D grade also to be able to:**

Multiply out expressions with brackets such as $3(x + 2)$ or $5(x - 2)$

Factorise expressions such as $6a + 8$ and $x^2 - 3x$

C **Examiners would normally expect students who get a C grade also to be able to:**

Expand (and simplify) harder expressions such as $x(x^2 - 5)$ and $3(x + 2) - 5(2x - 1)$

What you should already know ...

- Add, subtract and multiply integers
- Multiply a two-digit number by a single-digit number

VOCABULARY

Variable – a symbol representing a quantity that can take different values such as x, y or z

Term – a number, variable or the product of a number and a variable(s) such as 3, x or $3x$

Algebraic expression – a collection of terms separated by + and – signs such as $x + 2y$ or $a^2 + 2ab + b^2$

Product – the result of multiplying together two (or more) numbers, variables, terms or expressions

Collect like terms – to group together terms of the same variable, for example, $2x + 4x + 3y = 6x + 3y$

Simplify – to make simpler by collecting like terms

Expand – to remove brackets to create an equivalent expression (expanding is the opposite of factorising)

Factorise – to include brackets by taking common factors (factorising is the opposite of expanding)

Linear expression – a combination of terms where the highest power of the variable is 1

Linear expressions	Non-linear expressions
x	x^2
$x + 2$	$\frac{1}{x}$
$3x + 2$	$3x^2 + 2$
$3x + 4y$	$(x + 1)(x + 2)$
$2a + 3b + 4c + \ldots$	x^3

Learn 1 Collecting like terms

Examples: Simplify the following expressions.

a $3f + 2g + 4f - 3g + 3 = 7f - g + 3$ ← $3f + 4f = 7f$

b $4ad + 3a^2 + 2da + a^2 = 6ad + 4a^2$ ← $+3a^2 + a^2 = 4a^2$

← $4ad + 2da = 4ad + 2ad = 6ad$
Note: $2da$ is the same as $2ad$

Apply 1

1 Simplify the following by collecting like terms.

a $a + a + a + a + a + a$
b $p + p + p + p$
c $k + k$
d $d + d + 2d + d$
e $3y + 2y$
f $2m + m + 4m$
g $2k + 4k + k + 2k$
h $2m + 3m + 2m + m$
i $5q + 2q + 2q + q$
j $t + t + t + t + t - t$
k $g + g + g + g + g - g - g - g$
l $x + x - x + x - x$
m $y - y - y + y + y$
n $2d + 5d - d - 3d + 2d$
o $4a - 6a$
p $a + a - a - a$
q $3p + 4p - 2p$
r $-2a + 4a$

2 Michelle writes the answer to $p + p + p + p \div p$ as p^5. Kwame writes the answer as $p5$. Is anyone correct?

3 Toby thinks the answer to $4a + 2a + 3d + d = 10ad$. Is he correct? Explain your answer.

4 Simplify the following by collecting like terms.

a $x + 2x + 3 + x$
b $y + y + 5 + y + 3$
c $2f + 6 + 3f - 2$
d $7y + 2y + 1$
e $3x - 2x + 6 + 4x + 2$
f $a + b + 2a$
g $p + 2q + 3p + 4q$
h $3t + 6g + 4t - 2g$
i $2f + 3g + 4 - 2g - f$
j $2p + 6 - p - 2 - p$
k $5y + 6 - 4y - 6$
l $4y - 6y - 2$
m $6p - 4 - 2p - 3 - 5p$
n $2w + 4t - 3t - w$

5 The answer is $5q$. Write down five questions that have this answer.

6 Get Real!

The diagram shows an L-shaped floor with dimensions as shown.
A carpenter is trying to work out the length of wood needed to make the skirting boards. Find an expression for the perimeter of the room in terms of x.

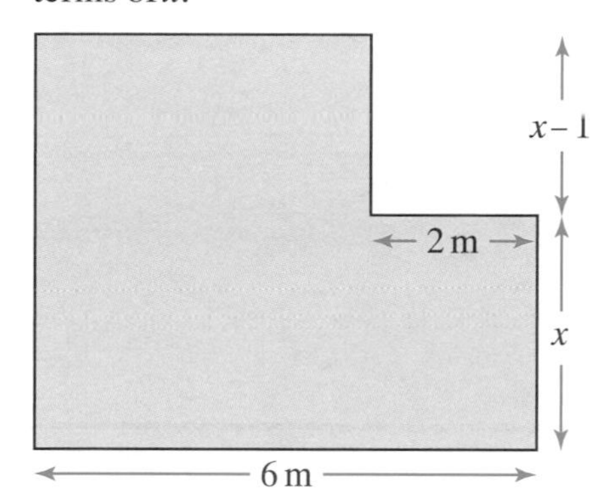

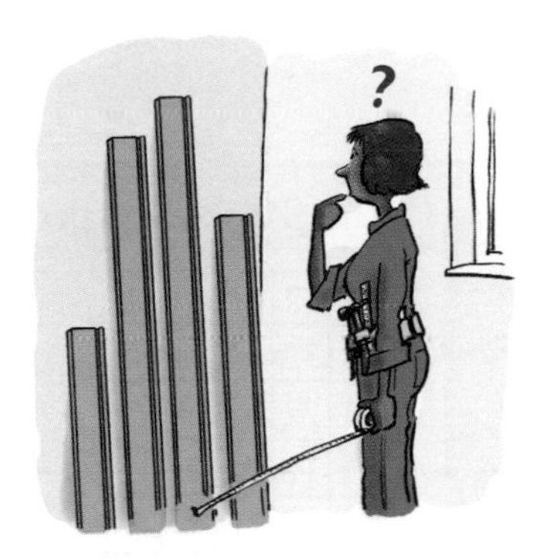

7 Louise is asked to simplify $4ab + 5ba$. She thinks it is impossible.
Is she correct?

8 Simplify the following by collecting like terms.

a $4ab + 2ab$
b $3pq + 2pq$
c $2ef + ef + 5ef$
d $3cd + 2cd + 4dc$
e $6gh + 2hg + gh$
f $6mn + mn - 3nm$
g $2baa + 3baa$
h $6tom + 2mot + 4omt$
i $2xy + 3xy - xy$
j $8mn + 2mn - 3mn$
k $7pq + 3qp - 2pq$
l $8bc - 2bc + 3cb$
m $3pqr + 2prq + 3qpr - rpq$
n $2ab + a + b$
o $7gh + 3g - 2g + hg$

9 The answer is $7ab + 2cd$. Write down five questions that have this answer.

10 $x^2 + x^2 + x^2 - 6x$
Do you agree with this statement? Give reasons for your answer.

11 Simplify the following by collecting like terms.

a $p^2 + p^2$
b $y^3 + 2y^3 + 3y^3$
c $4t^5 + 2t^5 + t^5$
d $5t^6 + 2t^6 - 3t^6$
e $3p^2 + p + 2p^2$
f $4a^2 + a^3 + 2a^2 + 3a^3$
g $4y^3 + 2y^2 + 3y^3 - y^2$
h $3d^2 + 2h^3 + d^3 - 4h^2$
i $5f^4 + g^6 - 2f^4 + 2g^6$
j $4x^3 + 2y^2 + x^2 + 3x^3$
k $2xy^2 + 3xy^2$
l $6pq^3 + 2pq^3 + 3q^3p$
m $4ab^2 + 2ab^2 - 3ab^2$
n $4gh^2 - 2hg^2 + gh^2$

12

A $2x + 3xy$	B $4x + 6yx$	C $2x^2 + 3xy^2$	D $2x + 3x^2y$

Simplify the following by collecting like terms.

a A + B
b B + A
c A − B
d A + B + C
e A + B + C + D
f C + D
g B + D
h Can you spot a link between **A** and **B**?

13 **a** The answer is $12p^2$. Find five questions with this answer.

b The answer is $12p^2 + 6q^3$. Find five questions with this answer.

c The answer is $12p^2q + 6pq^2$. Find five questions with this answer.

d The answer is $12p^2q + 6p + 6pq^2$. Find five questions with this answer.

Explore

The grid below is part of a 10×10 number square

41	42	43	44	45
31	32	33	34	35
21	22	23	24	25
11	12	13	14	15
1	2	3	4	5

The shaded cells form an L-shape

- Add together the four shaded cells and write down the total
- Multiply the bottom number (3) by 4 and add 31
- Is the total the same?
- Try it with a different L

Investigate further

Learn 2 Expanding brackets

Examples:

a Expand $5(2y - 1)$.

	2y	**−1**
5	10y	−5

$5 \times -1 = -5$

$5(2y - 1) = 10y - 5$ ← Write $10y - 5$ not $10y + -5$

$10y - 5$ is a linear expression because the highest power of the variable (y) is 1.

b Expand $p(p^2 - 5)$.

	p²	**−5**
p	p³	−5p

$p \times -5 = -5p$

$p(p^2 - 5) = p^3 - 5p$

$p \times p^2 = p \times p \times p = p^3$

Apply 2

1 Complete the following:

a $3(t + 4) = 3t + \ldots$

b $5(p + 2) = 5p + \ldots$

c $6(h + 3) = \ldots + 18$

d $4(2a + 3) = 8a + \ldots$

e $3(3y + 2) = 9y + \ldots$

f $5(y - 2) = 5y - \ldots$

g $4(b - 3) = 4b - \ldots$

h $7(d - 2) = \ldots - 14$

i $8(2d - 3) = 16d - \ldots$

j $5(4p - 4) = \ldots - 20$

k $-2(a + 6) = -2a \ldots$

l $-3(b - 4) = -3b \ldots$

2 Multiply these out:

a $4(x+2)$
b $6(y+3)$
c $3(m+2)$
d $5(b+3)$
e $6(t+4)$
f $3(2y+3)$
g $5(3p+6)$
h $\frac{1}{3}(6x-15)$
i $\frac{1}{4}(16f-4)$
j $7(g-4)$
k $\dfrac{35h+10}{5}$
l $-4(q+2)$
m $-2(a+3)$
n $-5(2m-3)$
o $-7(-4a-1)$
p $4(h-1)$
q $\frac{1}{2}(4b+6)$

3 Sam thinks the answer to $5(3x-2)$ is $15x-2$. Hannah says he is wrong.
Who is correct and why?

4 Complete the following:

a $x(x+2) = x^2 + \ldots$
b $p(p+4) = \ldots + 4p$
c $x(x-2) = x^2 \ldots$
d $f(f-3) = \ldots - 3f$
e $y(y^2-3) = \ldots - 3y$
f $2f(f-2) = \ldots$

5 Expand:

a $p(p+3)$
b $b(b-4)$
c $a(5+a)$
d $x(x^2+3)$
e $w(ig+am)$
f $x(x^3+4x)$
g $t(t^2-1)$
h $mu(fc-mu)$
i $h^2(h^3+4)$
j $p(p^2-7)$

6 The answer is $12y-36$.
Write down five questions of the form $a(by+c)$ with this answer.
(a, b, and c are integers – positive or negative numbers.)

Explore

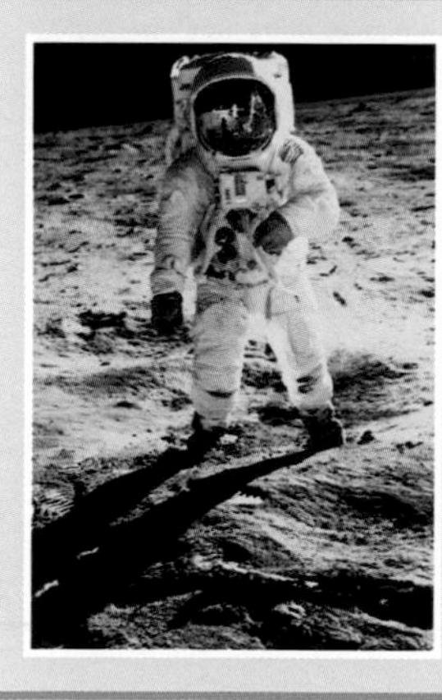

- Think of a number
- Add 2
- Multiply the new total by 4
- Halve your answer
- Subtract twice the original number
- The answer is 4

Investigate further

Learn 3 Expanding brackets and collecting terms

Example: Expand and simplify $3(x-2)-5(2x-1)$.

Treat this as two separate algebraic expressions, $3(x-2)$ and $-5(2x-1)$, and merge the answers together at the end

Step 1 Expand $3(x-2)$.

	x	-2
3	$3x$	-6

$3(x-2)=3x-6$

Step 2 Expand $-5(2x-1)$.

	$2x$	-1
-5	$-10x$	$+5$

$-5 \times -1 = +5$

$-5(2x-1)=-10x+5$

Step 3 Merge the two answers by collecting like terms.

$$3(x-2)-5(2x-1)=3x-6-10x+5$$
$$=-7x-1$$

Underlining or circling like terms (including their sign) helps when collecting them: $3x-10x=-7x$ and $-6+5=-1$

Apply 3

1 Complete the following.

a $2(y+3)+3(y+2)$
$= 2y+6+\ldots+\ldots$
$= \ldots y+\ldots$

b $3(d+2)+4(d-3)$
$= \ldots+\ldots+4d-\ldots$
$= \ldots d-\ldots$

c $5(2x+2)-3(2x-3)$
$= 10x+\ldots-\ldots x+\ldots$
$= \ldots x+\ldots$

2 Expand and simplify:

a $4(x+2)+2(x+3)$

b $2(p+3)+3(2p-4)$

c $6x-(2-x)$

d $3(m-1)-4(m-2)$

e $\frac{1}{2}(6y-3)+\frac{1}{4}(12-4y)$

f $4x-(x+2)$

g $4(t-2)-2(t+1)$

3 Simplify:

a $4(2m-3)+3(m-6)$

b $3(2a-1)-3(4-a)$

c $5(6x-3)+2(3-2x)$

d $4(2y-1)-4(3y-5)$

e $5(2t-4)-7(2-3t)$

f $2x(3y+1)+3y(2x-1)$

4 Josie thinks the answer to $3(2m-1)-4(m-2)$ is $2m-11$.
Explain what she has done wrong.

5 Find the integers a and b if $4(x-a)-b(x-1)=2x-14$.

6

A $3y+2$	B $2y-3$	C $5y-1$	D $y+2$

Expand and simplify:

a 2A + 3C

b C − 2D

c 2A − B

d Work out a combination of two cards that gives the answer 13.

7 Get Real!

The diagram shows an L-shaped floor with dimensions as shown.
The floor is to be covered with tiles, each measuring 1 m by 1 m.

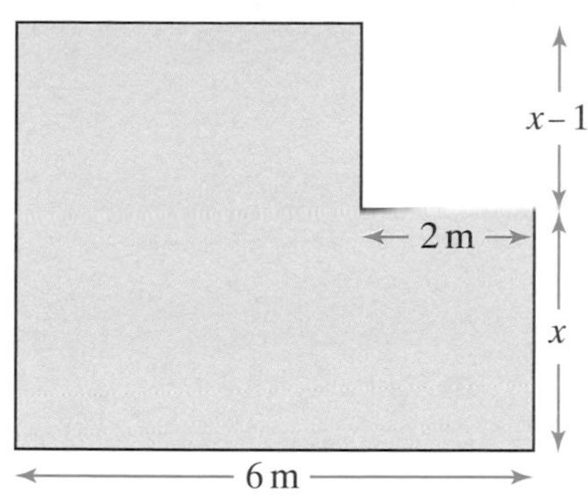

a By splitting the floor into two rectangles, calculate the area of the floor.

b By splitting the floor into two different rectangles, calculate the area of the floor.

c Are your answers the same each time?
Give a reason for your answer.

Explore

- Pick a blue card
- Double the number
- Add 1 to the new number
- Multiply the new number by 5
- Pick a white card
- Add this number to the previous result
- Subtract 5
- What do you notice?

5	6	7	8
1	2	3	4

Investigate further

Learn 4 Factorising expressions

Examples:

a Factorise $5a - 10$

$5a - 10 = 5(a - 2)$

$5a = (5) \times a$ and $10 = 2 \times (5)$
Both terms have 5 as a common factor

b Factorise $x^2 - 3x$

$x^2 - 3x = x(x - 3)$

$x^2 = x \times (x)$ and $3x = 3 \times (x)$
Both terms have x as a common factor

Treat factorising as the 'reverse' of expanding.

HINT You can always check if you have factorised correctly by multiplying the bracket out to make sure you get the question.

Apply 4

1 Complete the following.

a $2y + 10 = 2(y + ...)$

b $4g + 20 = ... (... + 5)$

c $12d + 42 = 6(... + ...)$

d $18xy - 30ab = 6(3xy - ...)$

e $20ab + 35\,cd - 10ef = 5(4ab + ...\,cd - ...)$

2 Factorise the following expressions.

a $2x + 6$

b $3y + 12$

c $7y - 63$

d $8y - 40$

e $14y + 20$

f $6t - 18$

g $21 - 42t$

h $44t - 55c$

i $13a + 23b$

j $20b + 35a + 15c$

k $-12a - 16$

l $-42ab + 70cd + 14ef$

m $18f - 36g + 12h$

n $20xy + 60tu - 80pq$

o $2f - 11g + 33h$

HINT Two of these expressions cannot be factorised.

3 Sandra thinks that $12a + 18$ factorised completely is $3(4a + 6)$.
Is she correct? Explain your answer.

4 The answer is $4(...\,p + ...)$ where $...\,p + ...$ is an expression.
Find five expressions that can be factorised completely with answers in this form.

5 Rashid thinks he has factorised $2x + 3y$ correctly as $2(x + 1\frac{1}{2}y)$.
Rana thinks the expression cannot be factorised? Who is correct and why?

6 Complete the following.

a $4cd + 7c = c(4d + ...)$

b $4ab + 9bc = b(... + 9c)$

c $x^2 + 5x = x(x + ...)$

d $5abc - 8bcd = bc(5a - ...)$

7 Factorise the following expressions.

a $2ab + 5b$

b $3cd + 7c$

c $2pq + p$

d $3xy + 5xz$

e $p^2 + 2p$

f $x^3 - 3x$

g $a^2 - a$

h $6xy + 7xyz$

i $3de + 5fg$

j $x^2 - 3x$

k $wig - wam$

l $bugs + bunny$

m $hocus - pocus$

n $silly + billy$

o $wonkey - donkey$

8 Copy these two tables.
Match the expression with the correct factorisation buddy.
Fill in the missing buddies.

Expression
$2x^2 + 8x$
$6x^2y - 3xy$

Factorisation
$2x(x + 4)$
$3xy(2y - 1)$
$x(x + 8)$

Explore

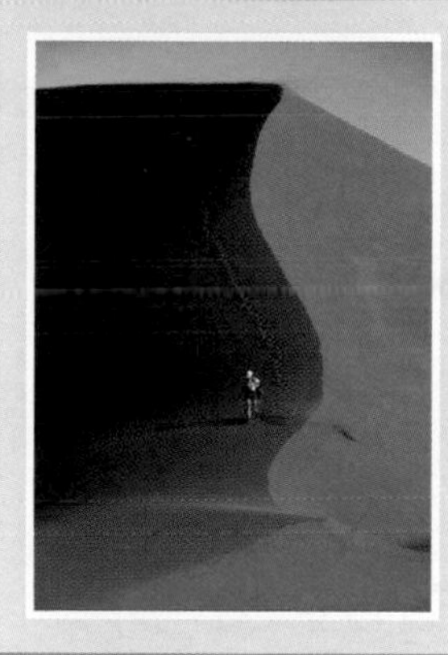

- Pick five consecutive numbers
- Add them together
- Is the answer a multiple of 5?
- Investigate further by picking five different consecutive numbers

Investigate further

Use of symbols

ASSESS

The following exercise tests your understanding of this chapter, with the questions appearing in order of increasing difficulty.

1 Simplify the following expressions.

a $4a + 7a + 9a$
b $5b - 9b + 6b$
c $4c - 7c$
d $3f - 4f - f$
e $4g + 6g - 10g$
f $3k - 4k + 5k - 6k + 8k + k - 4k - 2k$
g $2a + 3a + 4$

2 Simplify the following expressions.

a $2m + 3n + 4m + n$
b $9v - 3w + 6v - 5w$
c $5e - 3u - 2u - e$
d $4x + 4y - 4x - 3y$
e $w + z - w$
f $a - b - a + b + c$
g $6u + 3m - 2y - 5a + 3y - 2m$
h $q + u + e + u + e$
i $m + i + s + s + i + s + s + i + p + p + i$

3 **a** Simplify the following expressions where possible.

i $4w + 3w^2 - w^2 - 5w$
ii $2 - 3x + 4x^2 + 9x^3 - 5 + 4x - 4x^2 + 2x^3$
iii $3de + 5ef$
iv $3bil + 2ben - 2lbi - ben$
v $7t + 9u - 6u - t^2$

b Remove the brackets from the following expressions.

i $2(a + 3)$
ii $5(3a - 1)$
iii $-3(4a + 5)$
iv $-7(3a - 6)$
v $7(a + 2b)$
vi $5(6a - 3b)$
vii $-3(3a + 2b)$
viii $-7(-3a - 2b)$
ix $-6(-4a - 3b)$
x $-2(a - 2b + 3)$

4 Factorise the following expressions.

a $2a + 10$
b $10b - 12$
c $16 - 4c$
d $5d + 20e + 35f$
e $6g - 9h + 12i$
f $10j + 15k - 20l$
g $30p - 45q - 75r$
h $5x + 15y$
i $2ab - 3a$
j $x^2 + 7x$
k $4a^2 - 6a$
l $bil + ben$
m $abra + cadabra$
n $12cat + 3sat - 6mat$

5 Remove the brackets from the following expressions and simplify.

a $2c + 3(c + 5)$
b $2c - 3(c + 5)$
c $2c + 3(c - 5)$
d $2c - 3(c - 5)$
e $4(c - 3) + 2(c + 7)$
f $4(c - 3) - 2(c + 7)$
g $5(2c - 3) - 3(2c + 1)$
h $5(2c - 3) - 3(-2c - 1)$
i $3(2c - d + 3e) + (5c + 3d - 2e)$
j $3(2c - d + 3e) - (5c + 3d - 2e)$
k $y(y^2 - 7)$
l $2z(5z^2 + 4z - 8)$
m $x(x - 3) + 5(x - 3)$
n $p(p^2 + 3p - 4) - 6(p^2 + 3p - 4)$

6 **a** Simplify $2x + 3y + 5x - 2y - 4x$
b Factorise $4c + 12$
c Factorise $x^2 + 5x$

7 **a** Simplify $5p + 2q - q + 2p$
b Multiply out $4(r - 3)$
c Multiply out $s(s^2 + 6)$
d Simplify fully $(2t^3u) \times (3tu^2)$

4 Perimeter and area

OBJECTIVES

G **Examiners would normally expect students who get a G grade to be able to:**

Find the perimeter of a shape by counting sides of squares

Find the area of a shape by counting squares

Estimate the area of an irregular shape by counting squares and part squares

Name the parts of a circle

F **Examiners would normally expect students who get an F grade also to be able to:**

Work out the area and perimeter of a simple rectangle, such as 3 m by 8 m

E **Examiners would normally expect students who get an E grade also to be able to:**

Work out the area and perimeter of a harder rectangle, such as 3.6 m by 7.2 m

D **Examiners would normally expect students who get a D grade also to be able to:**

Find the area of a triangle, parallelogram, kite and trapezium

Find the area and perimeter of compound shapes

Calculate the circumference of a circle to an appropriate degree of accuracy

Calculate the area of a circle to an appropriate degree of accuracy

C **Examiners would normally expect students who get a C grade also to be able to:**

Find the perimeter of a semicircle

Find the area of a semicircle

What you should already know ...

- Multiply and divide one- and two-digit numbers
- Find one half of a number

VOCABULARY

Shape – an enclosed space

Polygon – a closed two-dimensional shape made from straight lines

Triangle – a polygon with three sides

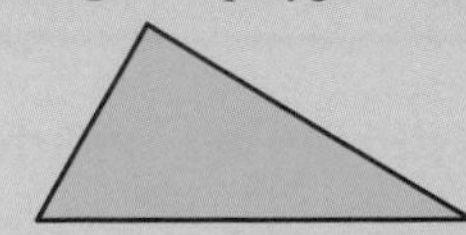

Quadrilateral – a polygon with four sides

Square – a quadrilateral with four equal sides and four right angles

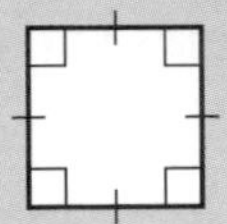

Rectangle – a quadrilateral with four right angles, and the opposite sides equal in length

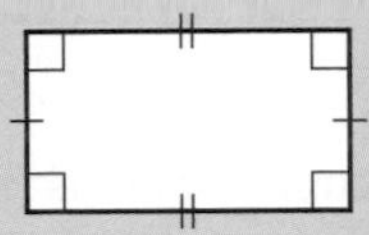

Area – the amount of enclosed space inside a shape

Perimeter – the distance around an enclosed shape

Dimension – the measurement between two points on the edge of a shape

Rhombus – a quadrilateral with four equal sides and opposite sides parallel

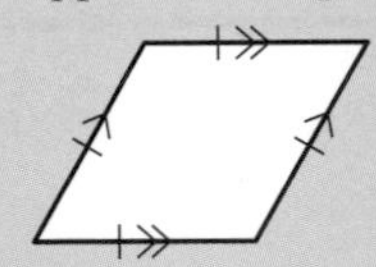

Parallelogram – a quadrilateral with opposite sides equal and parallel

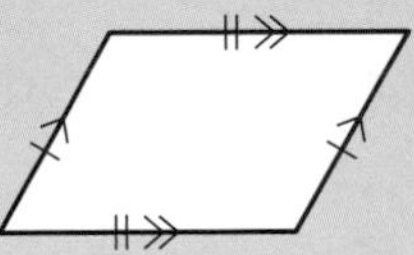

Trapezium (pl. **trapezia**) – a quadrilateral with one pair of parallel sides

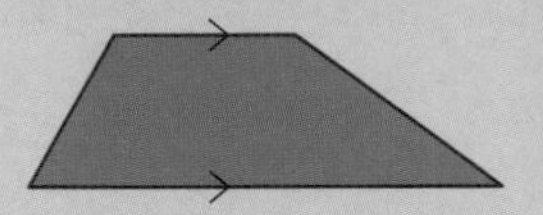

Kite – a quadrilateral with two pairs of equal adjacent sides

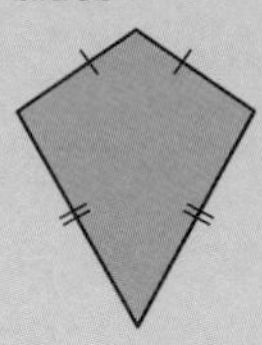

Pentagon – a polygon with five sides

Hexagon – a polygon with six sides

Octagon – a polygon with eight sides

Circle – a shape formed by a set of points that are always the same distance from a fixed point (the centre of the circle)

Diameter – a chord passing through the centre of a circle; the diameter is twice the length of the radius

Radius – the distance from the centre of a circle to any point on the circumference

Circumference – the perimeter of a circle

Chord – a straight line joining two points on the circumference of a circle

Semicircle – one half of a circle

Quadrant (of a circle) – one quarter of a circle

Learn 1 Perimeters and areas of rectangles

Examples: Find the perimeter and area of each of the following shapes.

a

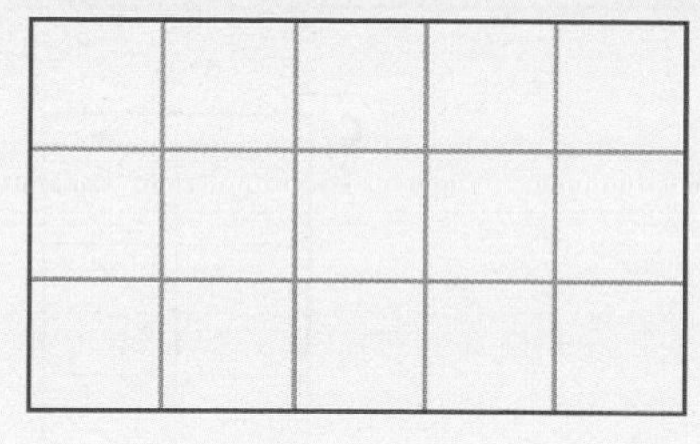

b

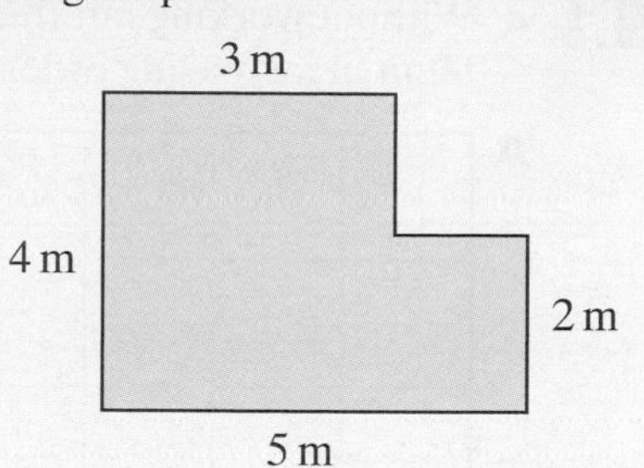

a

1	*2*	*3*	*4*	*5*
6	*7*	*8*	*9*	*10*
11	*12*	*13*	*14*	*15*

Perimeter $= 5 + 3 + 5 + 3 = 16$ cm ← Perimeter is the distance around the outside of a shape

Area (by counting the squares) $= 15\text{ cm}^2$ ← The units of area are squared

A quicker method than counting all the squares is to say that each row has 5 squares and there are 3 rows, so $5 \times 3 = 15$.

Area of a rectangle = length × width

b

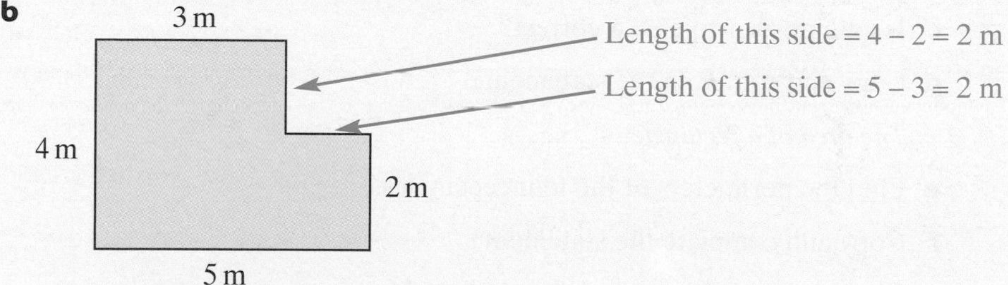

Perimeter $= 5 + 2 + 2 + 2 + 3 + 4 = 18$ m

To find the area, divide the shape into rectangles:

A: 4 m × 3 m   B: 2 m × 2 m ($5 - 3 = 2$ m)

Area of A $= 4 \times 3 = 12\text{ m}^2$

Area of B $= 2 \times 2 = 4\text{ m}^2$

Total area $= 12 + 4 = 16\text{ m}^2$

← The units of area are squared

Apply 1

1 **a** Without working out the areas look at these four rectangles and put them in increasing order of size.

A

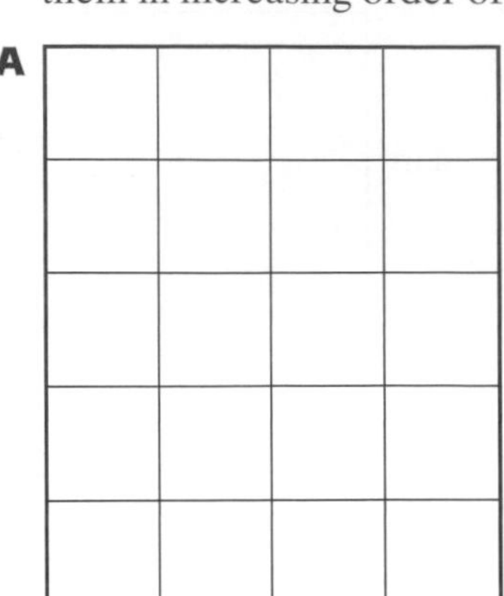

C

B

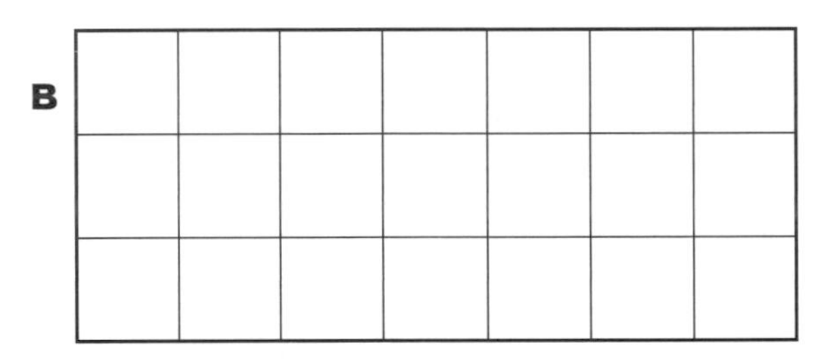

D

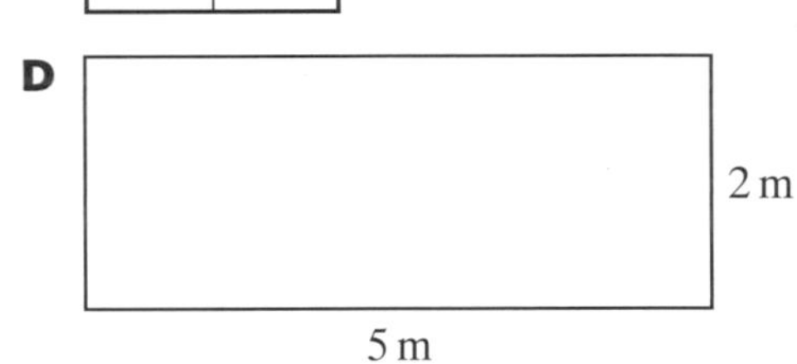

b Now work out the area of each rectangle.

c Is your answer to part **a** correct?

d Copy and complete the statement:

The area of a rectangle = ... × ...

e Find the perimeters of the four rectangles.

f Copy and complete the statement:

The perimeter of a rectangle can be found by ...

2 Apalia thinks that the area of a rectangle with length 4 cm and width 7 cm is 28 cm. Is she correct? Give reasons for your answer.

3 How many different rectangles can you make using 100 one-centimetre squares? Write down the dimensions of your rectangles. (All dimensions must be whole numbers.)

4 **Get Real!**

Each student has a locker to put books in.

a What is the area of the base of the locker?

Christian turns each locker onto its side.

b What is the new area of the base of the locker?

c Which layout gives students more floor space to put things in their lockers?

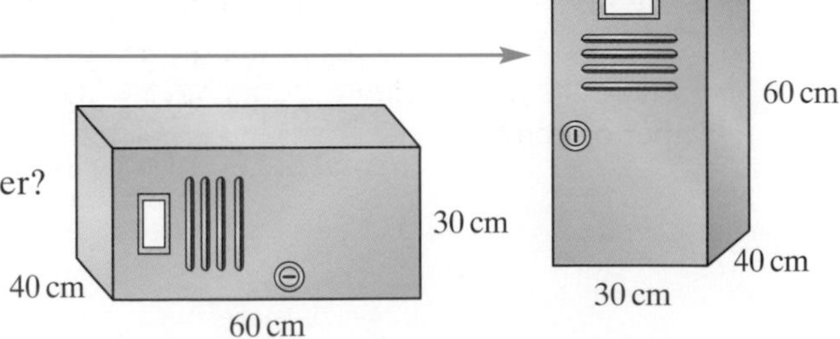

5 Find the perimeter of each of the following shapes:

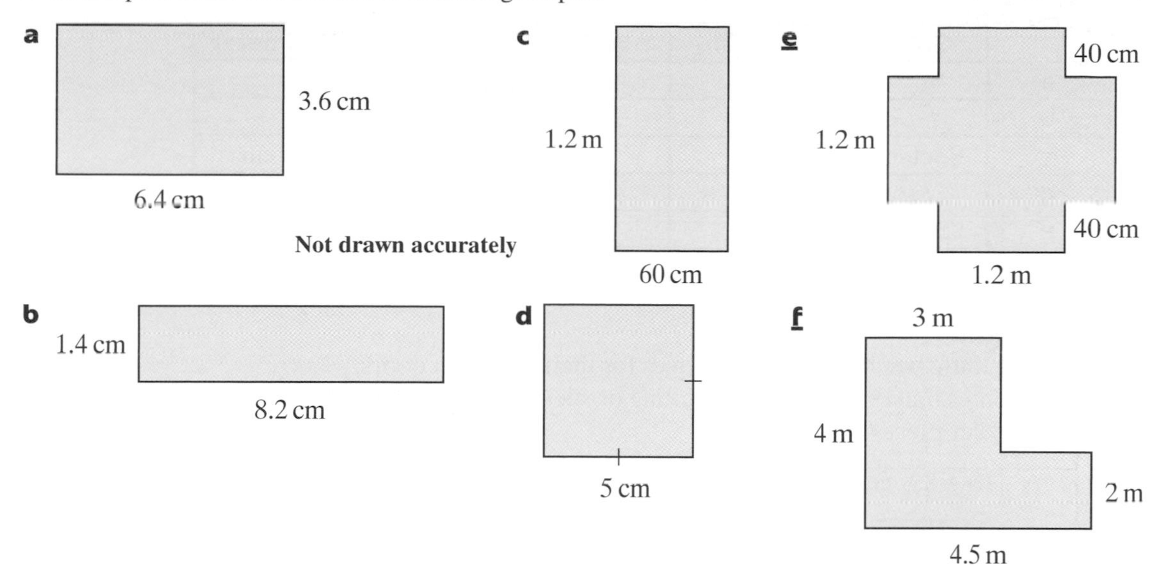

6 Find the area of each of the following shapes:

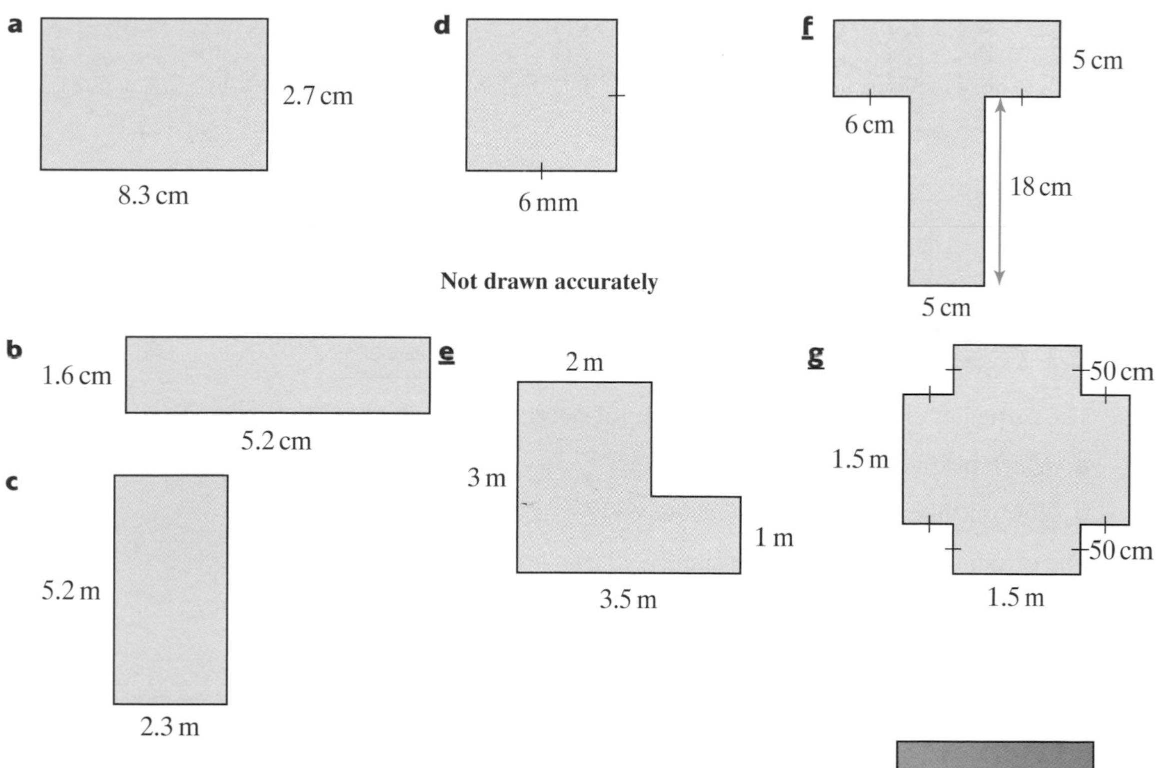

7 Get Real!

Anna is varnishing her door.
A small tin of varnish covers 1.15 m^2 and costs £1.20
A medium tin of varnish covers 3.4 m^2 and costs £3.20
Which tins of varnish should Anna buy for the cheaper option?

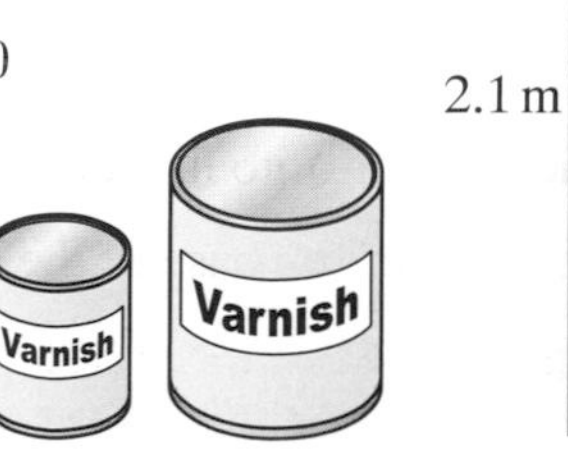

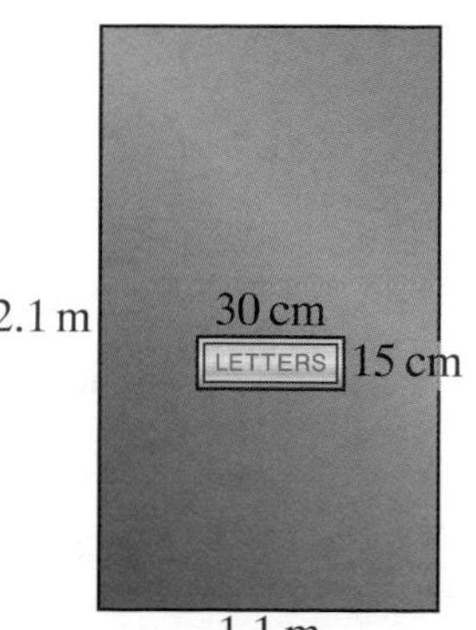

8 Copy the following table and fill in the gaps.

	Shape	Length	Width	Area	Perimeter
a		4 cm	2 cm	8 cm^2	
b		5 cm		15 cm^2	
c	Rectangle	10 cm			26 cm
d		4 cm		16 cm^2	
e	Square				20 cm
f		0.5 m	20 cm		

9 Get Real!

Andy and Katie would like name plaques for their bedroom doors.
The carpenter makes the letters by cutting or sticking together two types of rectangular pieces of wood.

Piece A – length 8 cm, width 4 cm   Piece B – length 6 cm, width 4 cm

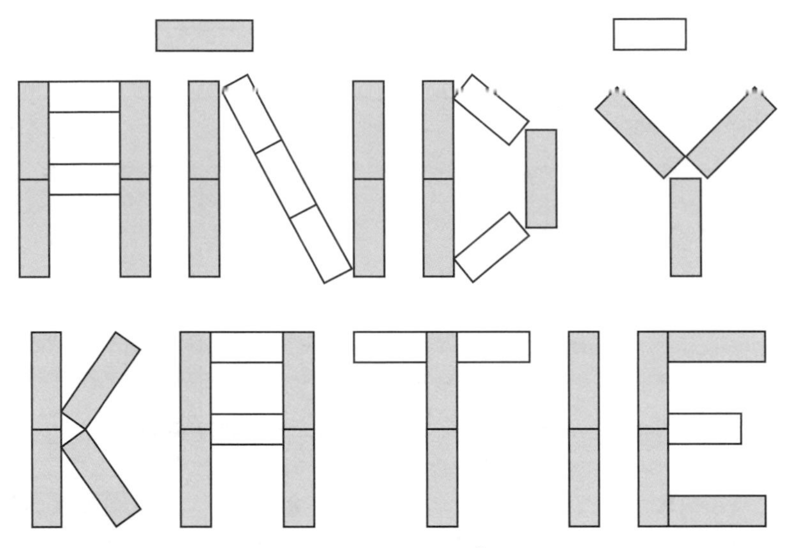

The carpenter charges 1p for every 2 cm^2 of wood.

a Which plaque is the cheapest to make?

b How much would your name plaque cost?

The carpenter can also put a finishing gold trim around the letters at a charge of 1p per 2 cm.

c How much would your name plaque cost to trim?

Explore

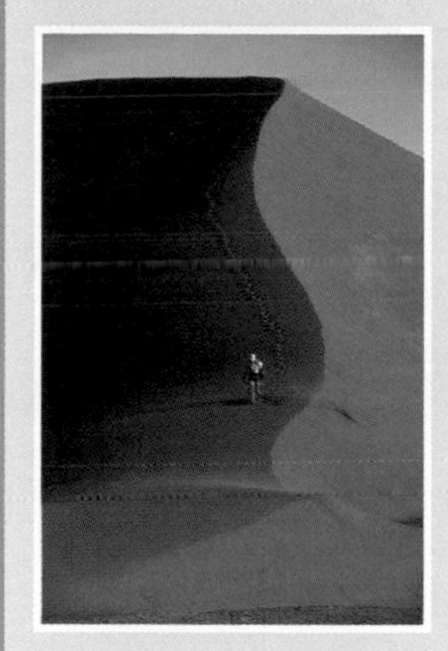

Sally's hallway ceiling is made of 1 m square tiles. They have fallen down but some tiles have stayed together. The pieces lying on the floor are:

A 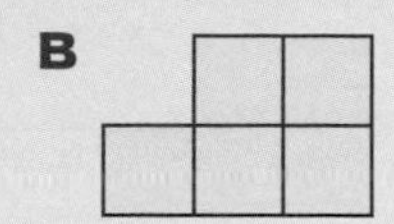 **B** 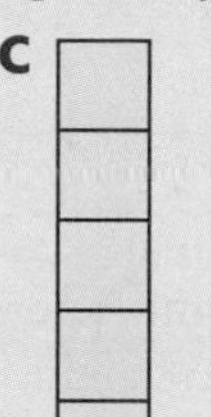 **C** 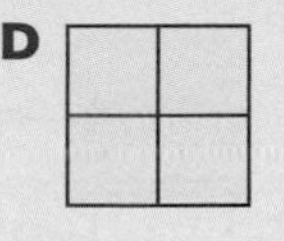 **D** 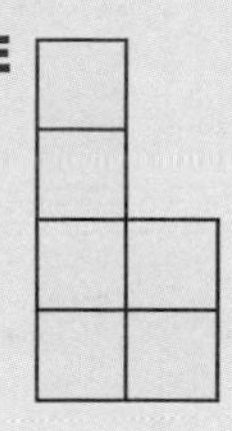 **E**

- Write down the area of each piece
- What is the total area of all five pieces?
- Make the pieces out of squared paper
- Try and fit them together to make the ceiling
- What are the dimensions of the rectangle?
- What is the area of this rectangle?
- Is it the same as you got for all five pieces?

HINT The ceiling is rectangular – use your last answer to help you!

Investigate further

Explore

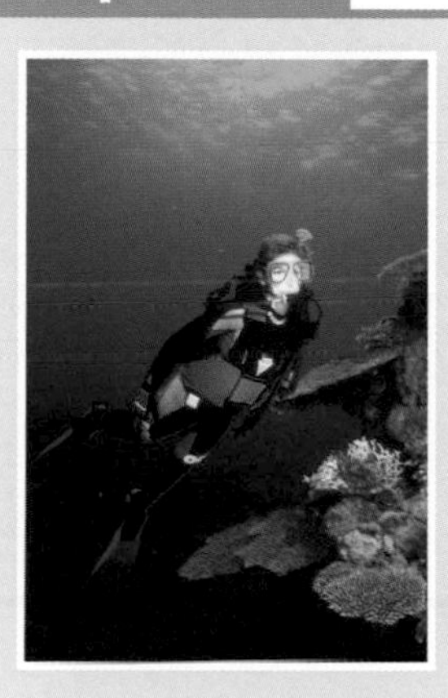

- Betty's rabbit needs 3 m^2 of space
- Betty wants the space to be rectangular
- Betty also needs to buy wire fencing to keep foxes away from her rabbit
- Betty is running out of pocket money and wants to buy the smallest amount of wire fencing possible
- Betty thinks that the space will have to be 1 m × 3 m
  What is the perimeter?
- Betty's dad thinks a rectangle measuring 6 m × 0.5 m is better
  Does this have the same perimeter?

Investigate further

Learn 2 Perimeters and areas of triangles and parallelograms

Examples: Find the perimeter and area of the following shapes.

a Triangle

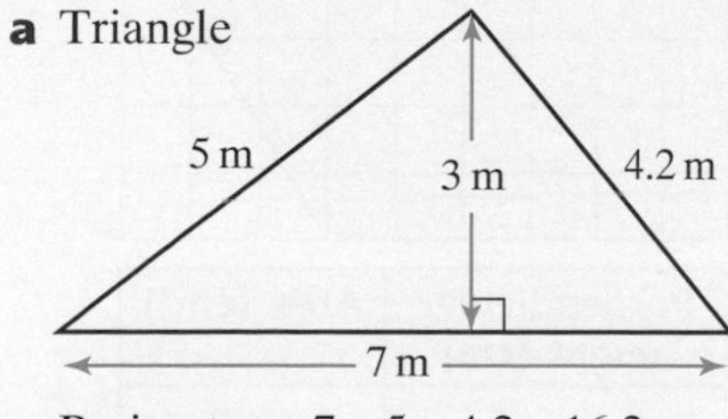

Not drawn accurately

Perimeter = 7 + 5 + 4.2 = 16.2 m
Area = $\frac{1}{2} \times 7 \times 3 = 10.5\ m^2$

Area of a triangle = $\frac{1}{2}$ × base × perpendicular height Area $= \frac{1}{2} \times b \times h$

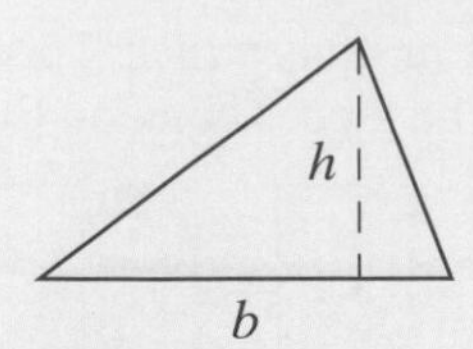

+ h b =

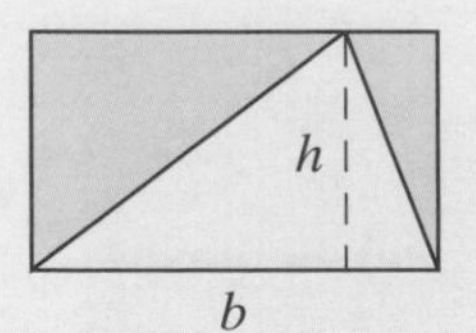

Two triangles can be joined together to make a rectangle with the same base and height. The area of a triangle is half the area of a rectangle with the same base and height.

b Parallelogram

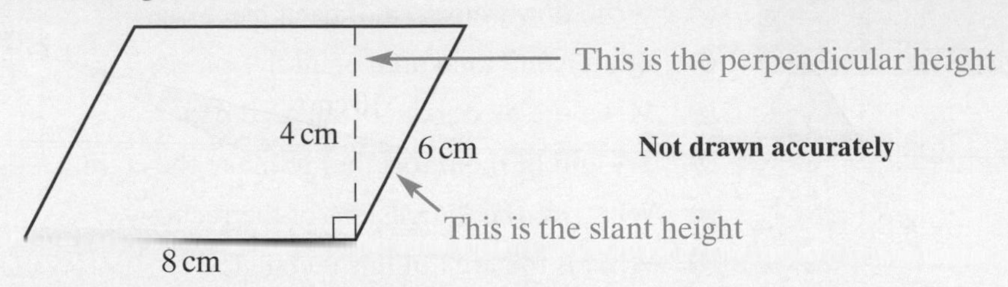

Perimeter = 8 + 6 + 8 + 6 = 28 cm
Area = 8 × 4 = 32 cm^2

Area of a parallelogram = base × perpendicular height

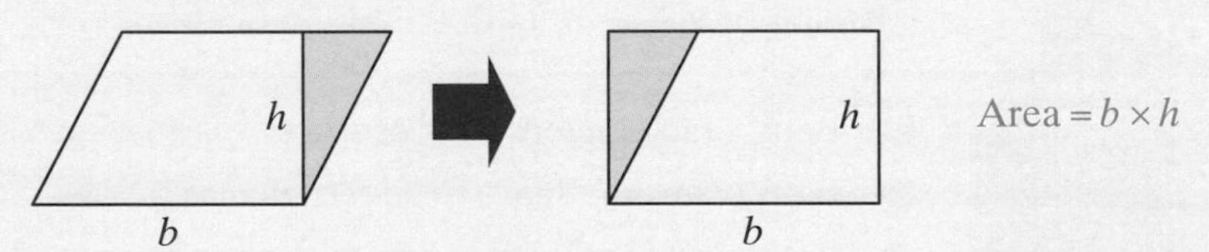

A parallelogram can be transformed into a rectangle as shown above.
They both have the same area.

Apply 2

1 **a** Copy and complete the table for the following shapes made from 1 cm squares.

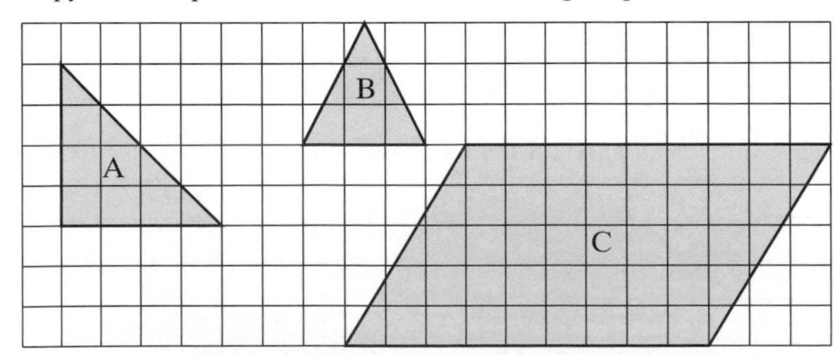

Shape	Name	Base (cm)	Perpendicular height (cm)	Area (cm^2)
A				
B				
C				

b Copy and complete the statements:

The area of a triangle = ... × ... × ...

The area of a parallelogram = ... × ...

2 Find the area of each of the following shapes:

a

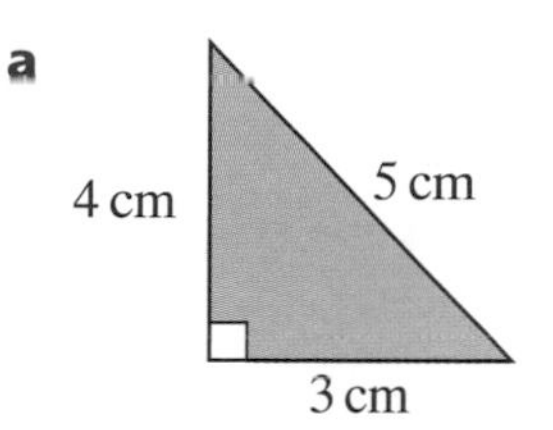

c

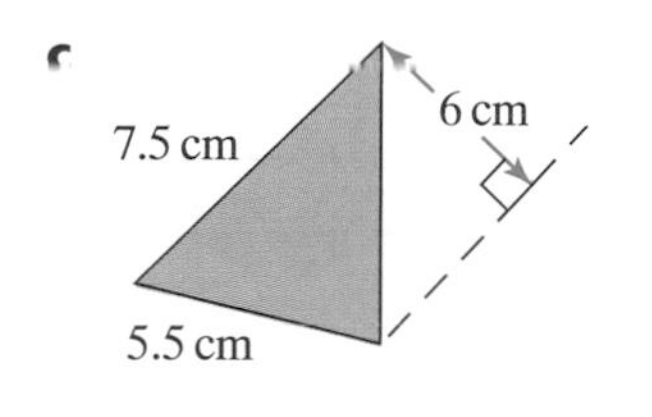

e

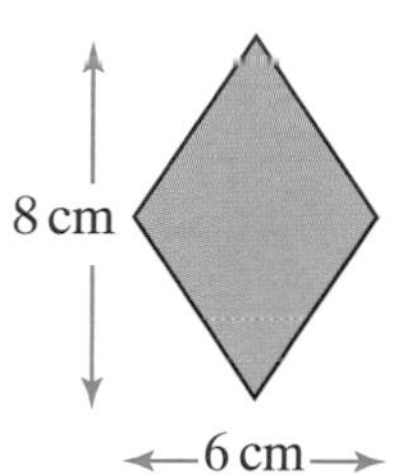

b

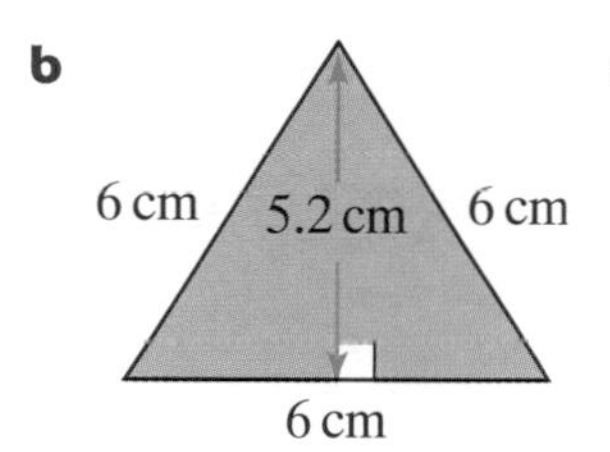

d

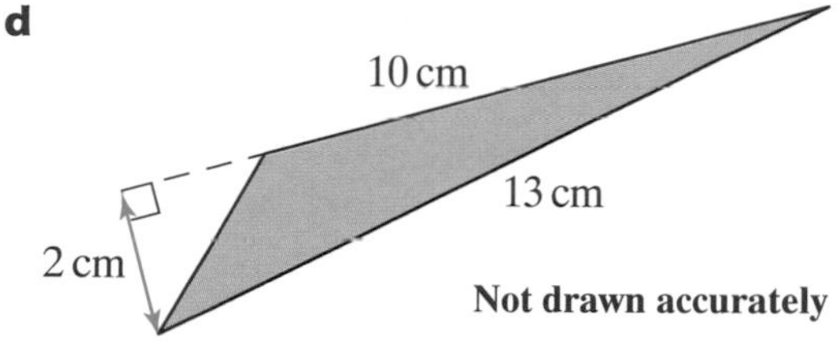

f

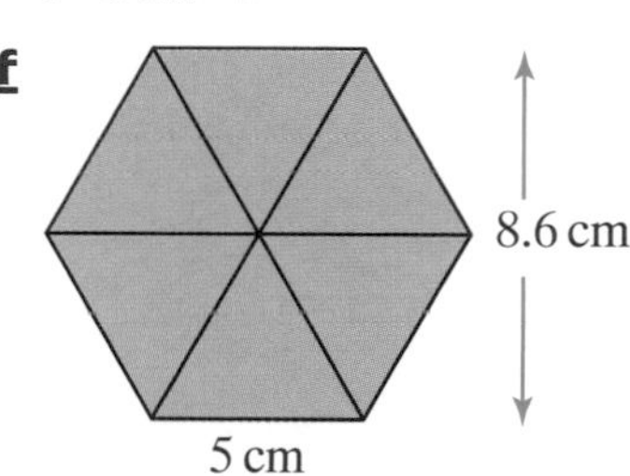

3 Five students are trying to find the area of the following triangle:

- Sameera thinks the answer is 48 cm^2 because $6 \times 8 = 48$
- Bruce thinks the answer is 30 cm^2 because $\frac{1}{2} \times 6 \times 10 = 30$
- Cassie thinks the answer is 24 cm^2 because $6 + 8 + 10 = 24$
- Des thinks the answer is because 40 cm^2 because $\frac{1}{2} \times 8 \times 10 = 40$
- Elliot thinks the answer is 24 cm^2 because $\frac{1}{2} \times 6 \times 8 = 24$

Who is correct? What mistakes have the other students made?

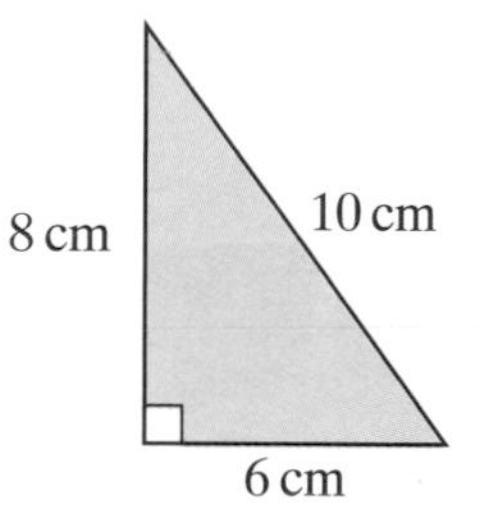

4 **a** The area of a parallelogram is 60 cm^2. Sketch five parallelograms with that area, showing the dimensions in each case.

b The area of a triangle is 30 cm^2. Sketch five triangles with that area, showing the dimensions in each case.

5 Find the area of each of the following parallelograms:

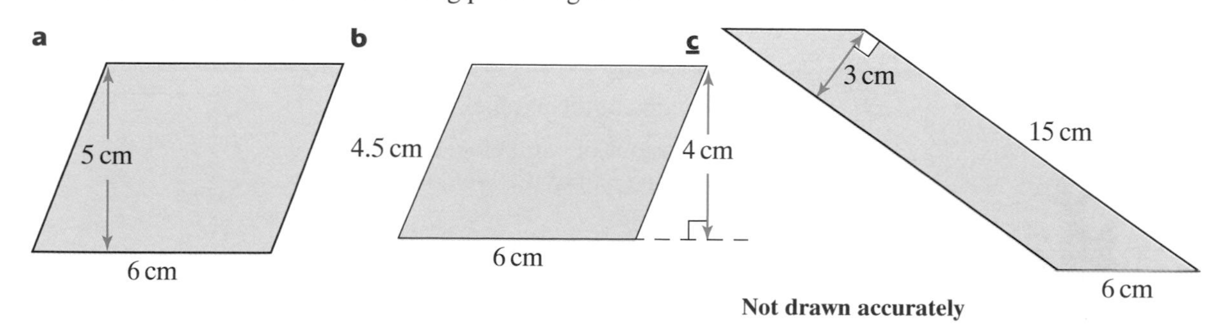

6 Copy and fill in the gaps in the table.

	Shape (Parallelogram/Triangle)	Base	Perpendicular height	Area
a	Parallelogram	5 cm	4 cm	
b	Triangle	5 cm	4 cm	
c		10 cm	2 cm	10 cm^2
d	Parallelogram	4 cm		8 cm^2
e	Triangle	4 cm		8 cm^2
f	Parallelogram	0.5 m	20 cm	

7 Get Real!

Reece wants to make a kite. The yellow silk costs £5 per square metre and the green silk costs £7 per square metre. Will he be able to buy enough silk with £3?

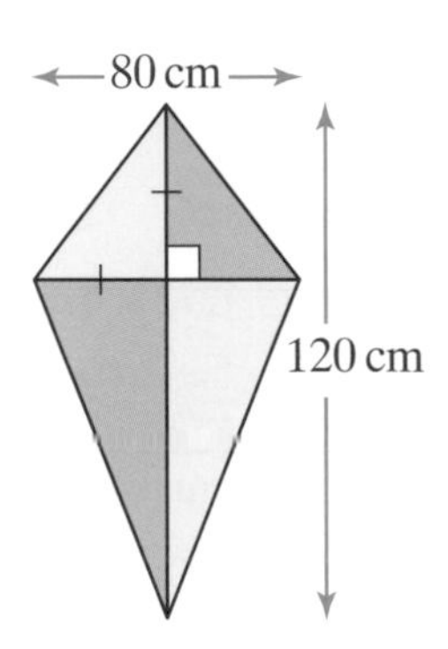

Explore

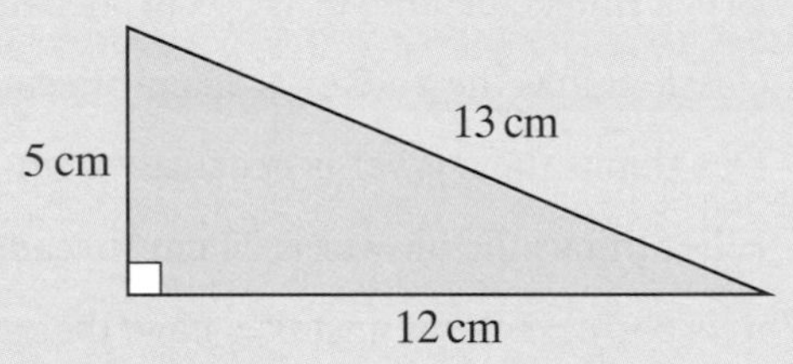

- Calculate the perimeters of the two triangles
- Calculate the areas of the two triangles
- What do you notice?

Investigate further

Explore

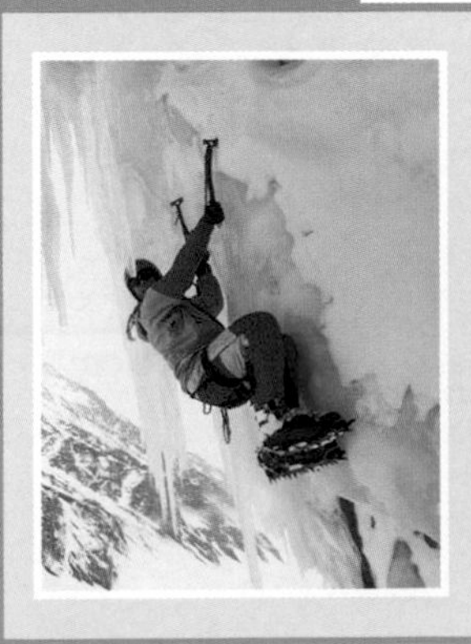

- Calculate the area of the parallelogram
- Find the dimensions of parallelograms with an area that is one half of the area of the parallelogram shown
- Find the dimensions of parallelograms with an area that is one quarter of the area of the parallelogram shown

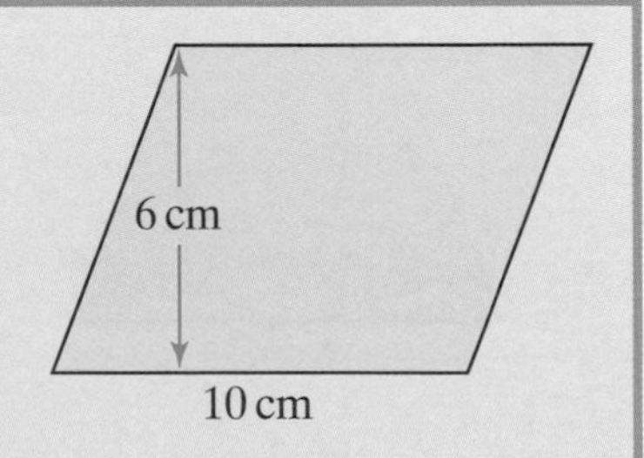

Investigate further

Learn 3 Areas of compound shapes

Example: Find the area of the following shape:

a Parallelogram

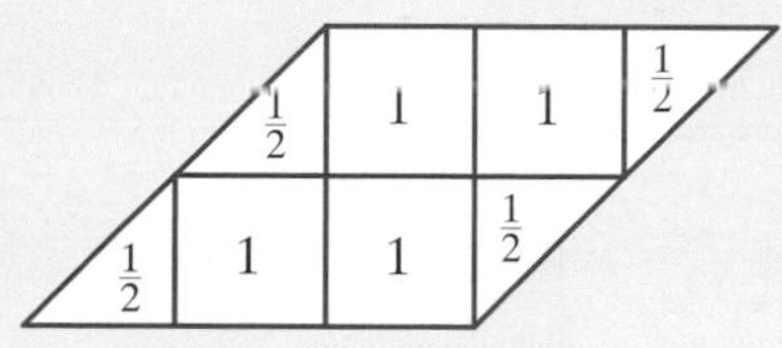

Area $= \frac{1}{2} + 1 + 1 + \frac{1}{2} + \frac{1}{2} + 1 + 1 + \frac{1}{2}$
$= 6 \text{ cm}^2$

Remember that the area of a parallelogram = base × perpendicular height = 3 × 2 = 6 cm^2

b Kite

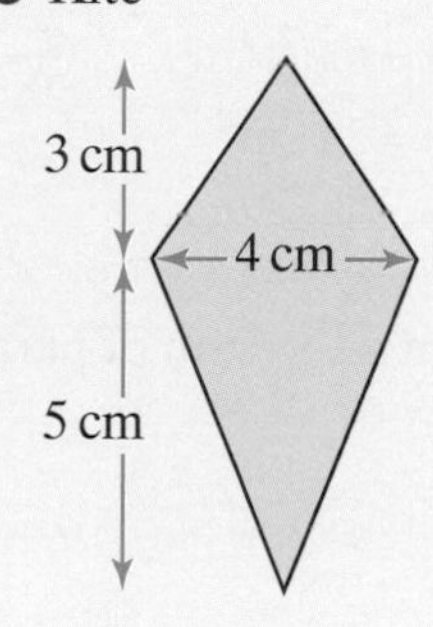

Area =

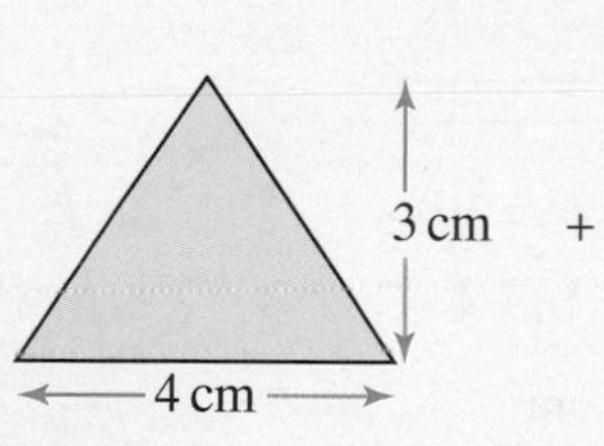

+

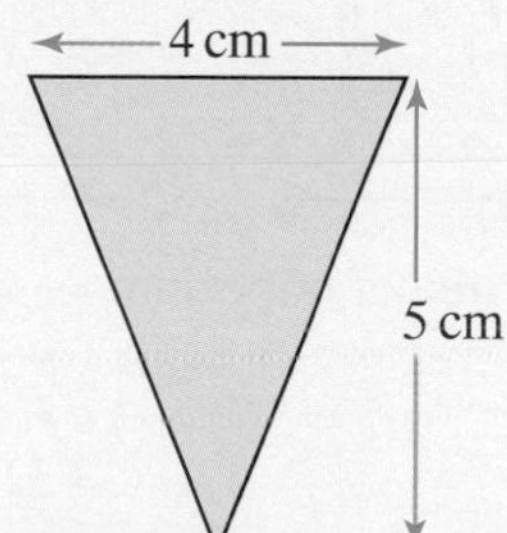

Area $= \frac{1}{2} \times 4 \times 3$
$= 6 \text{ cm}^2$

Area $= \frac{1}{2} \times 4 \times 5$
$= 10 \text{ cm}^2$

Total area = 6 + 10 = 16 cm^2

In general the area of a kite $= \frac{1}{2}$ height × width
$= \frac{1}{2} \times (3 + 5) \times 4$
$= \frac{1}{2} \times 8 \times 4$
$= 16 \text{ cm}^2$

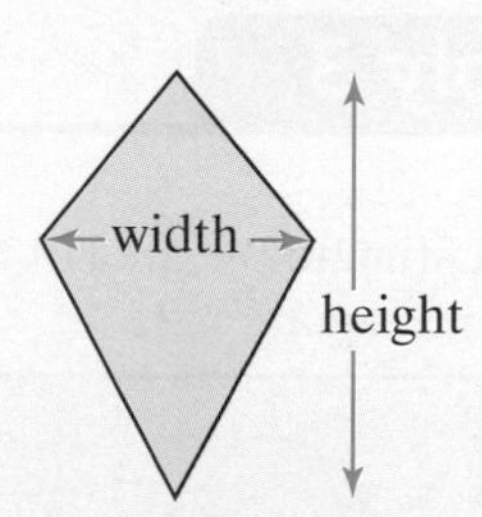

c Trapezium

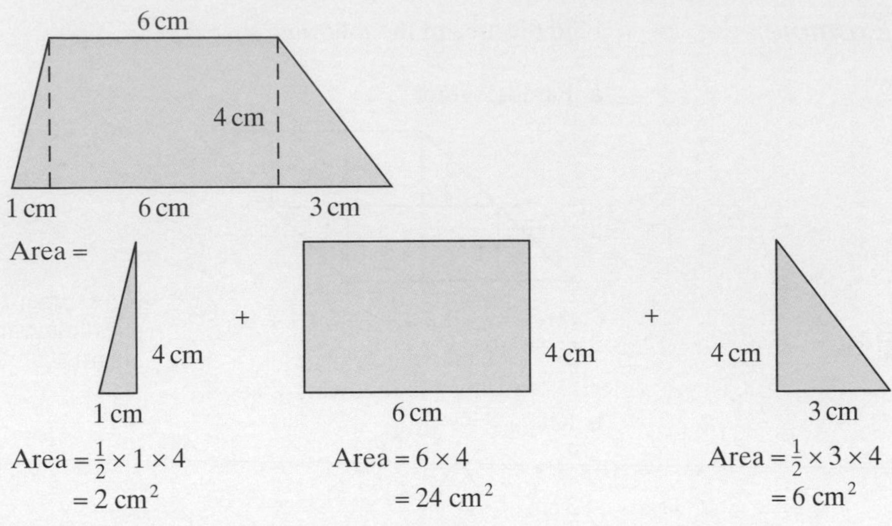

Area $= \frac{1}{2} \times 1 \times 4$
$= 2\ \text{cm}^2$

Area $= 6 \times 4$
$= 24\ \text{cm}^2$

Area $= \frac{1}{2} \times 3 \times 4$
$= 6\ \text{cm}^2$

Total area $= 2 + 24 + 6 = 32\ \text{cm}^2$

In general the area of a trapezium $= \frac{1}{2} \times$ (sum of parallel sides) $\times h$

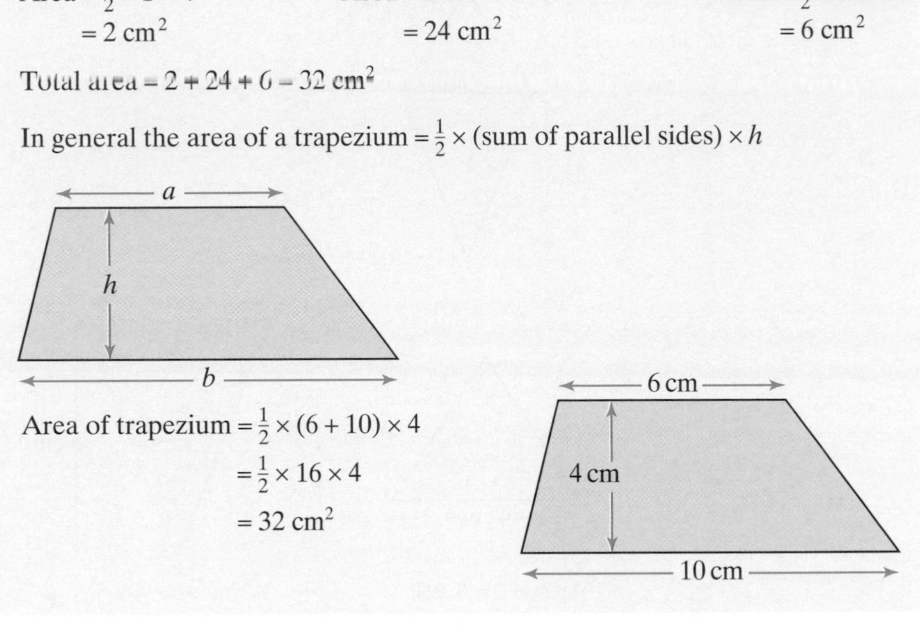

Area of trapezium $= \frac{1}{2} \times (6 + 10) \times 4$
$= \frac{1}{2} \times 16 \times 4$
$= 32\ \text{cm}^2$

Apply 3

1 Estimate the area of the island where each square represents one square mile.

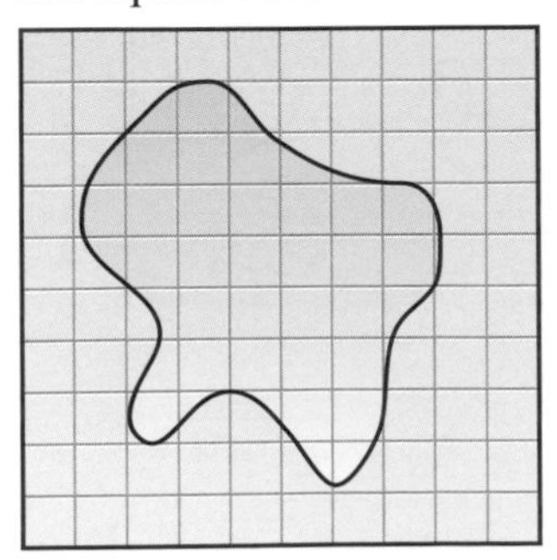

2

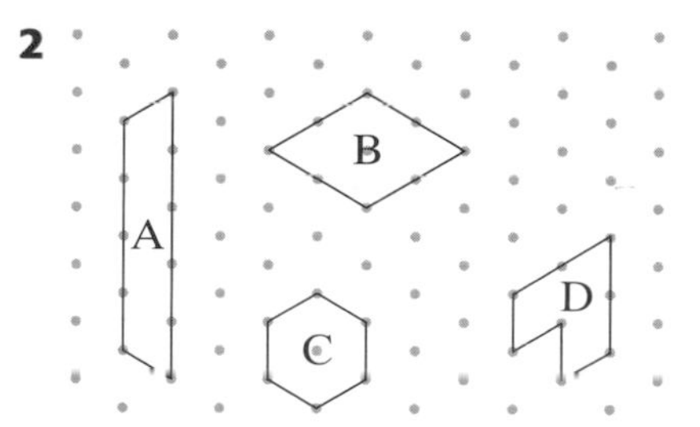

a Which of the shapes have the same area?

b Which of the shapes has the largest area?

3 **a** Find the area of each of the following shapes.

b Find the perimeters of each of the following shapes.

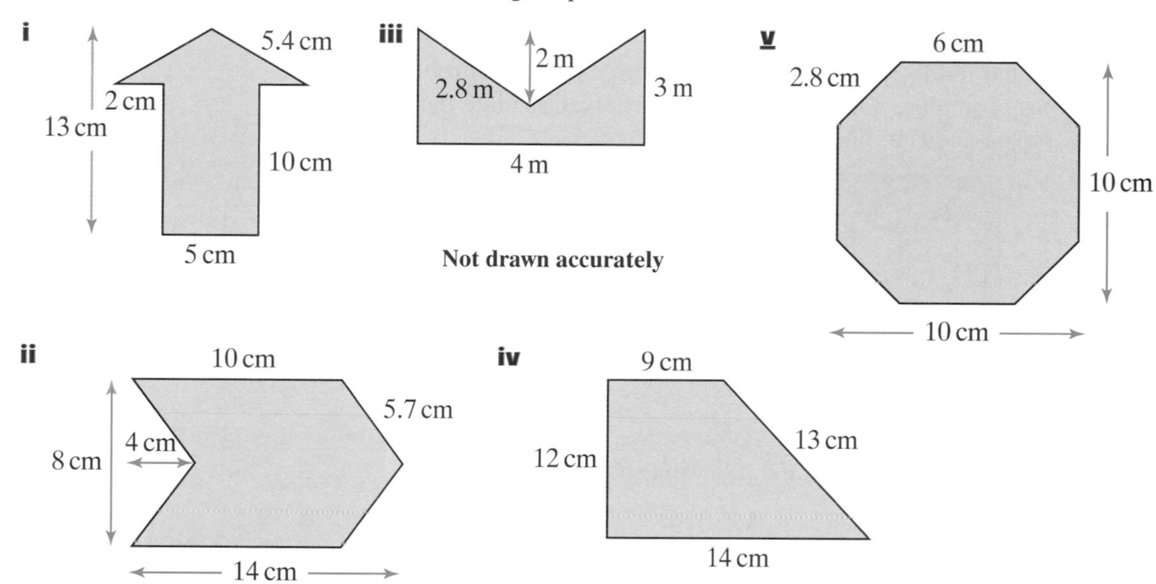

4 Roberta is finding the area of a hexagon. She spots that she can split it into a rectangle and two triangles:

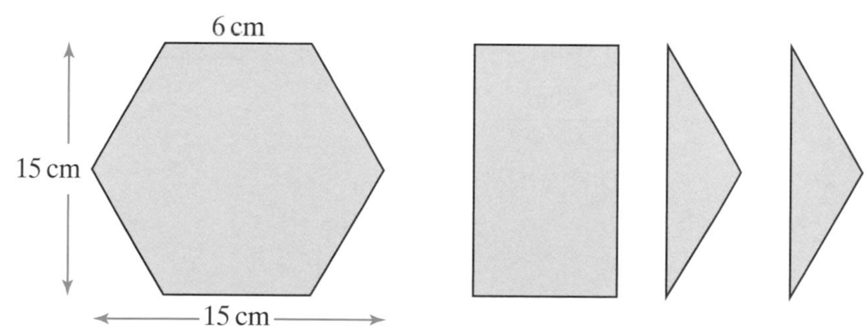

Area of rectangle $= 6 \times 15 = 90\ \text{cm}^2$
Area of triangle $= 15 \times 4.5 = 67.5\ \text{cm}^2$

Area of hexagon $= 90 + 2 \times 67.5 = 225\ \text{cm}^2$

Do you agree with Roberta? Give reasons for your answer.

5 Get Real!

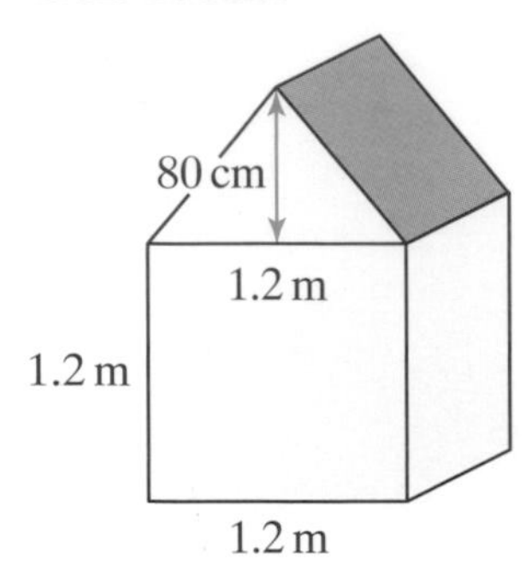

George is varnishing the front of Fred's doghouse. How many tins of varnish does he need? (The label on the tin has instructions that 1 litre of paint covers $0.5\ m^2$.)

6 Get Real!

Jane has made an 'EXIT' sign using a large piece of cardboard and two types of small rectangles to make the letters. Calculate the area of grey card she needs to paint.

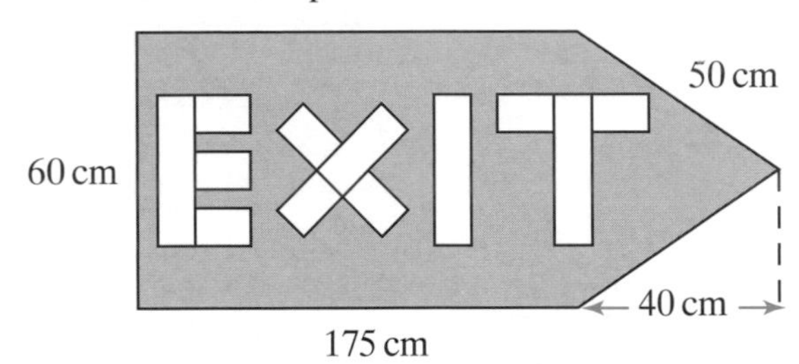

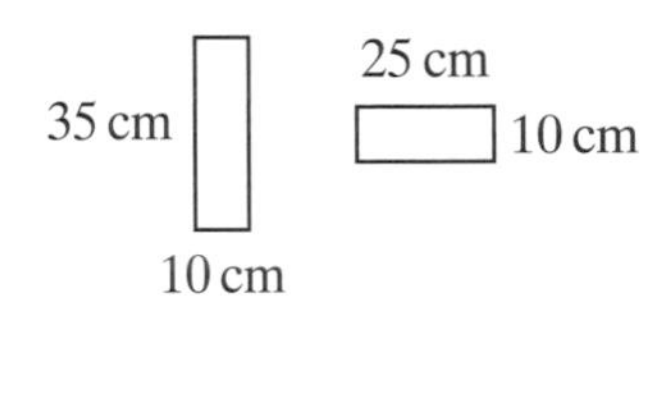

7 Find the area of each of the following shapes:

a

6 cm
4 cm
5 cm
12 cm

b

10 cm
15 cm
20 cm

c

16 cm
12 cm
6 cm
13 cm

d

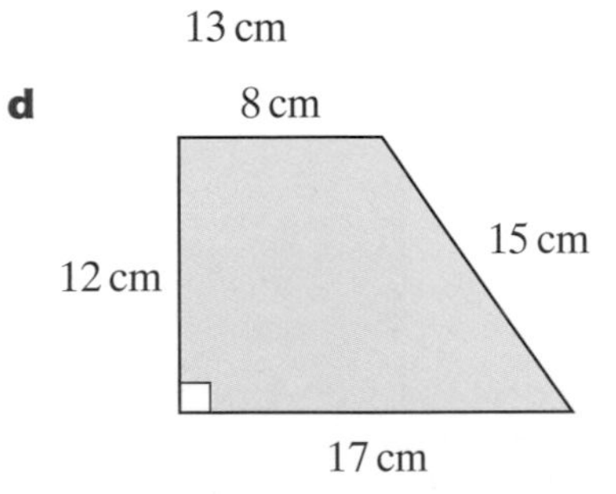

e

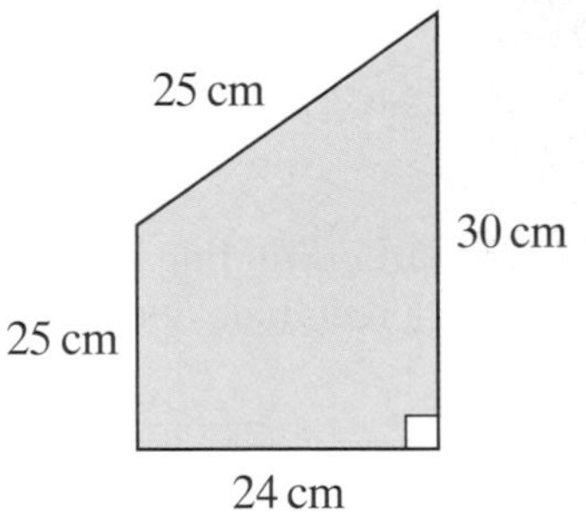

8 Sketch five trapezia with the area $25\ cm^2$, stating clearly the dimensions in each case.

Explore

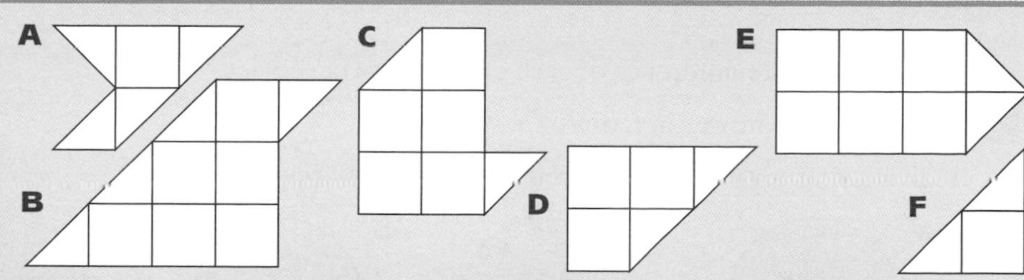

- Write down the area of each piece
- What is the total area of all six pieces?
- Make the pieces out of squared paper
- Try and fit them together to make a parallelogram
- What are the dimensions of the parallelogram?
- What is the area of this parallelogram?
- Is it the same as you got for all six pieces?

Investigate further

Explore

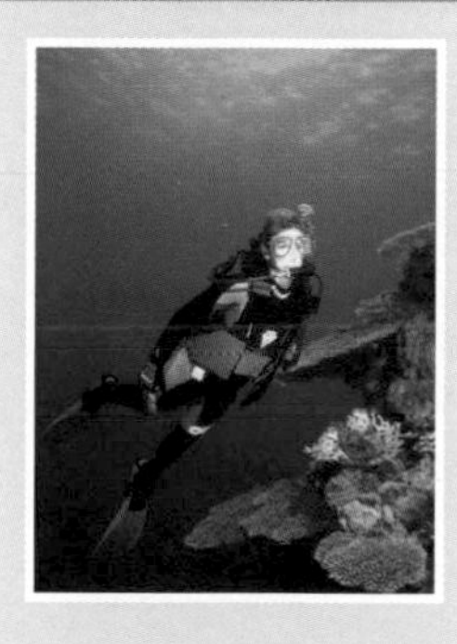

- Make two copies of the trapezium shown
- Try and fit the two pieces together to make a rectangle or parallelogram
- What is the area of the shape you have made?
- Deduce the area of the trapezium

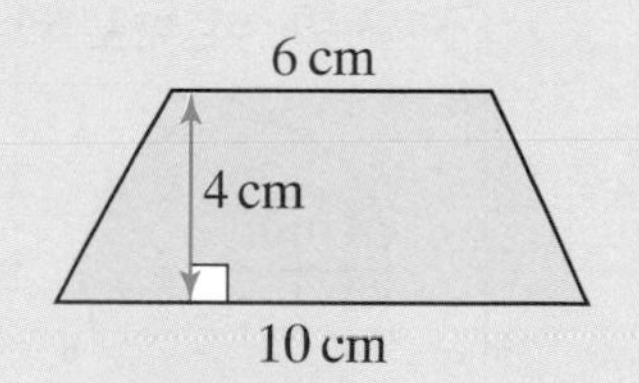

Investigate further

Learn 4 Circumferences of circles

Examples:

Calculate the circumference of this circle:

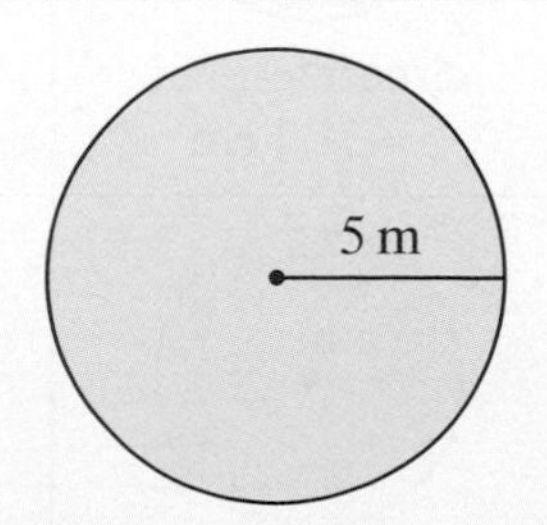

a leaving your answer in terms of π

b giving your answer to 3 significant figures.

Use the fact that $c = \pi \times d$ where c is the circumference and d is the diameter

Diameter = 2 × radius

a Circumference $= \pi d = \pi \times 10 = 10\pi$ m

You may be asked to leave your answers in terms of π on the non-calculator paper

b Circumference $= \pi d = \pi \times 10 = 31.4$ m (3 s.f.)

The calculator gives more decimal places but you need to round to an appropriate degree of accuracy (usually 3 s.f. or 2 d.p.)

Apply 4

1 Calculate the circumference of each of the following circles:

i leaving your answer in terms of π

ii giving your answer to an appropriate degree of accuracy.

a 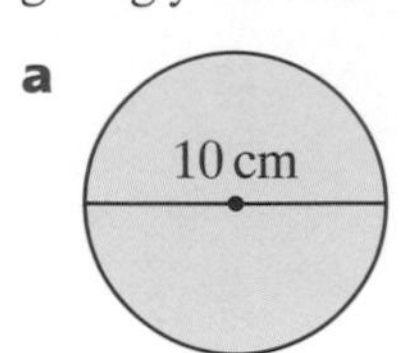

b 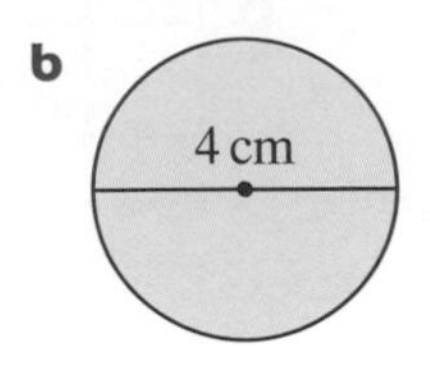

c 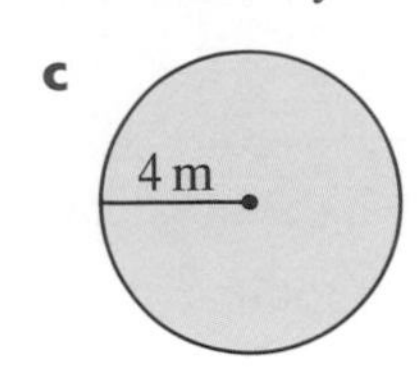

d 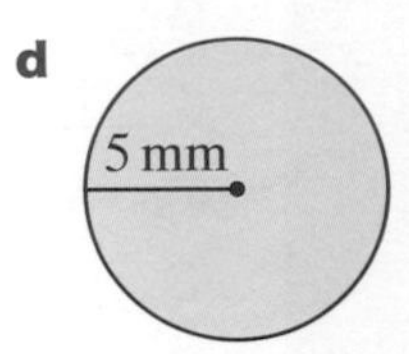

2 Get Real!

The London Eye has a diameter of 135 m and takes approximately half an hour to make one complete revolution. How far has the base of a capsule travelled in:

a 30 minutes

b 15 minutes

c 1 hour?

3 The circumference of a circle is π times d.
Can you find a correct line of three?
(Answers are in terms of π or to 1 decimal place.)

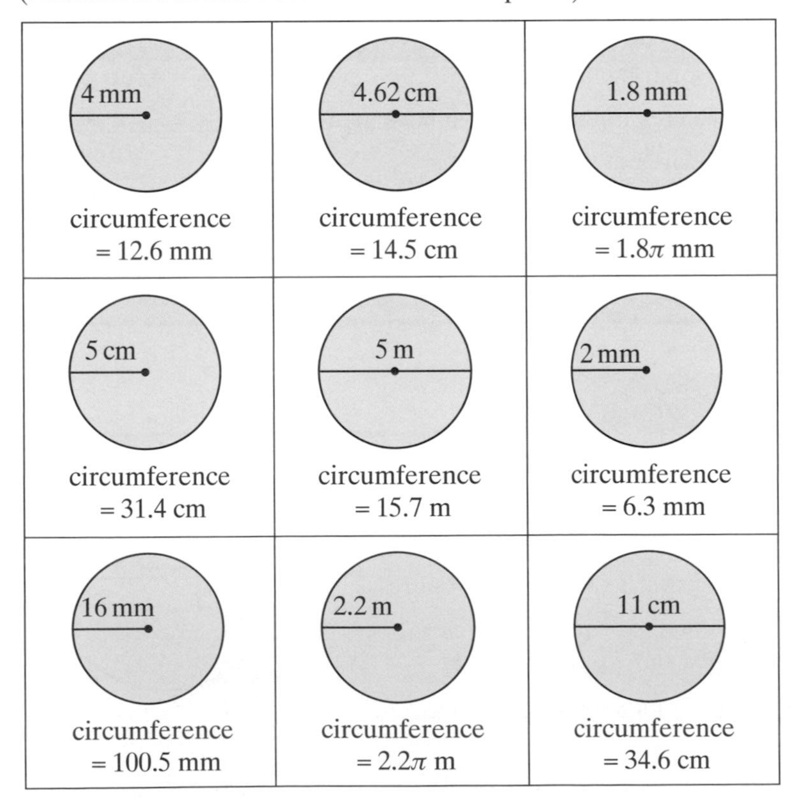

4 mm (radius) circumference = 12.6 mm	4.62 cm (diameter) circumference = 14.5 cm	1.8 mm (diameter) circumference = 1.8π mm
5 cm (radius) circumference = 31.4 cm	5 m (diameter) circumference = 15.7 m	2 mm (radius) circumference = 6.3 mm
16 mm (radius) circumference = 100.5 mm	2.2 m (radius) circumference = 2.2π m	11 cm (diameter) circumference = 34.6 cm

4 Calculate the total perimeter of the following shapes:

i leaving your answers in terms of π

ii giving your answers to an appropriate degree of accuracy.

a

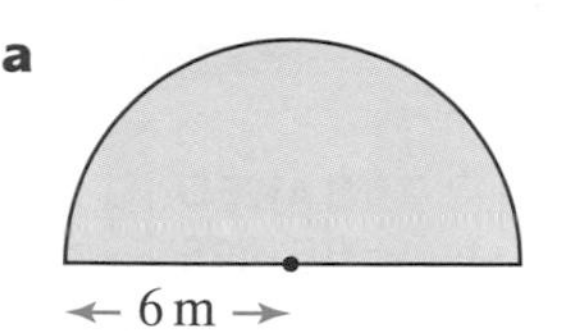

b

66 cm

1 m

c

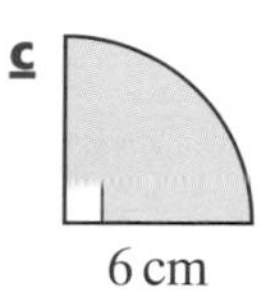

d

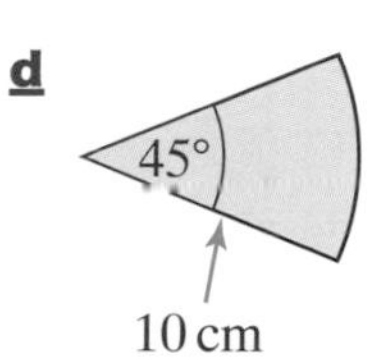

5 Get Real!

Ahmed lives 1500 m from school. The diameter of his bike's wheel is 80 cm.

a Find the circumference of the wheel.

b How many complete revolutions does the wheel turn during Ahmed's journey to school?

6 Get Real!

Jack and Susan have a race around the track shown below.
Jack is in lane 1 and Susan in lane 8.

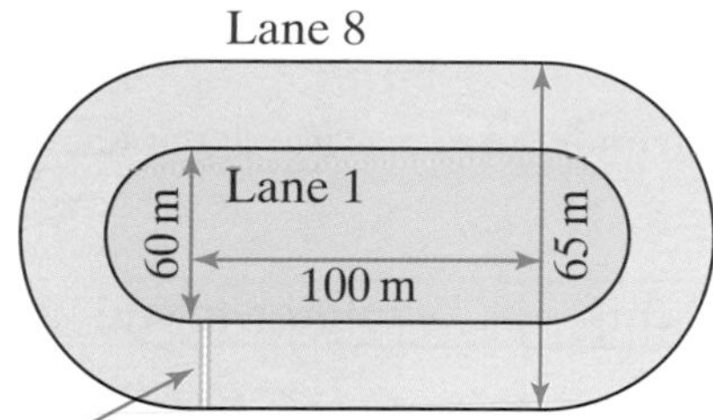

a How far does Jack run?

b How far does Susan run?

c How can you make the race fair?

7 Copy and complete the following table, giving your answers to an appropriate degree of accuracy.

Radius	Diameter	Circumference
	5 cm	
4 m		
		10 mm
		15π cm

8 Get Real!

Ahmed's CDs have a circumference of 40 cm.
He wants to make square covers for them.
Find, correct to 1 d.p., the dimensions of the smallest square into which the CDs will fit.

9 Get Real!

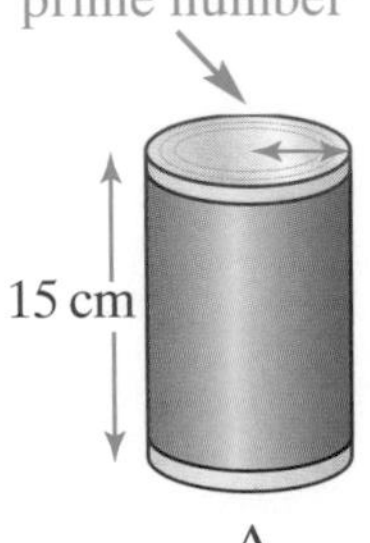

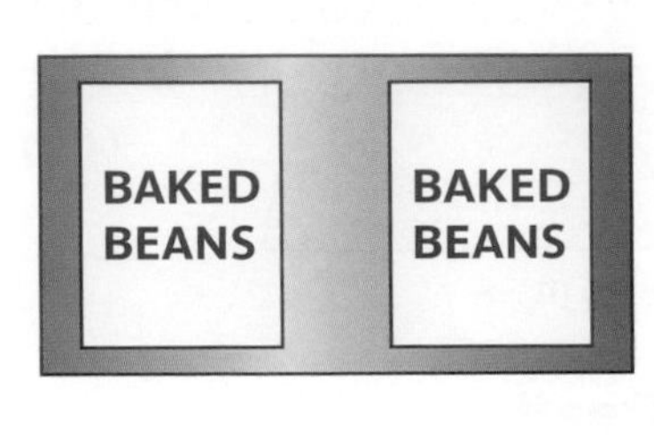

The labels have fallen off three tins.
The area of the 'Baked Beans' label is 942.5 cm^2 (1 d.p.).

a Which tin does it fit – A, B or C?

b Is the area of the label approximately 1 square metre?
Give a reason for your answer.

Explore

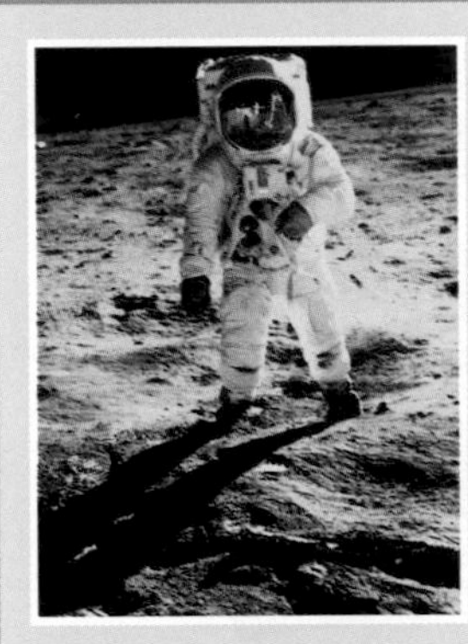

- Find five circular objects
- For each object, measure the circumference and the diameter
- Copy and complete the table

Object	Circumference (c cm)	Diameter (d cm)	$c \div d$

Investigate further

Learn 5 Areas of circles

Examples:

Calculate the area of this circle:

a leaving your answer in terms of π

b giving your answer to 3 significant figures.

Use the fact that $A = \pi \times r^2$ where A is the area and r is the radius

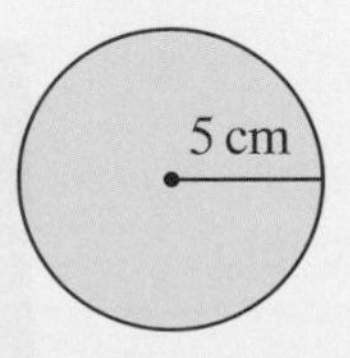

You may be asked to leave your answers in terms of π on the non-calculator paper

a Area $= \pi r^2 = \pi \times 5^2 = 25\pi$ cm^2

b Area $= \pi r^2 = \pi \times 5^2 = \pi \times 25 = 78.5$ cm^2 (3 s.f.)

The units are squared for area

The calculator gives more decimal places but you need to round to an appropriate degree of accuracy (usually 3 s.f. or 2 d.p.)

Apply 5

1 Calculate the area of each of the following circles:

 **i** leaving your answer in terms of π

ii giving your answer to 2 d.p.

a 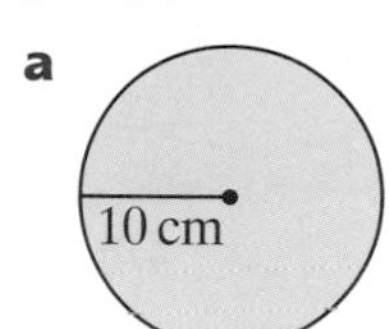

b 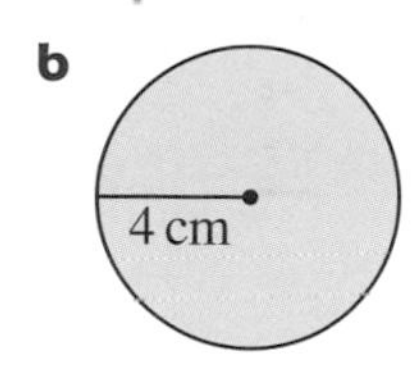

c 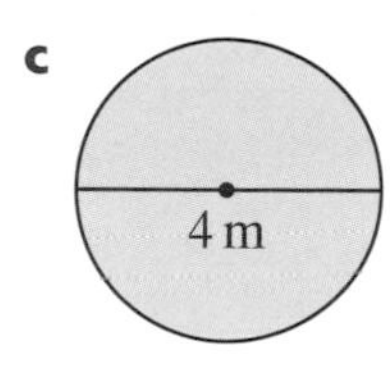

d 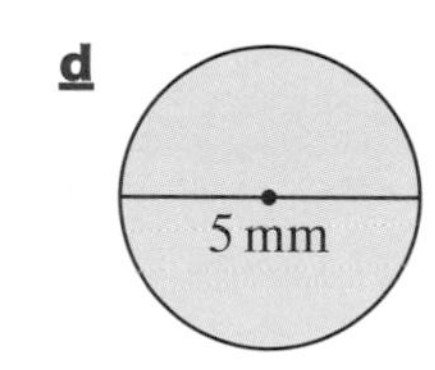

2 Joy and Jan are finding the area of a CD.

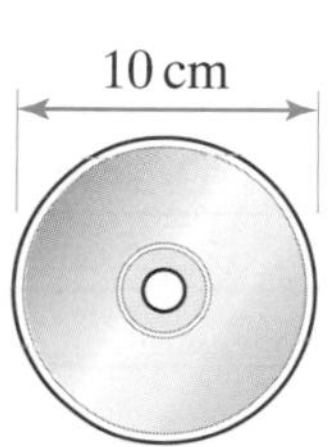

a This is Joy's method:

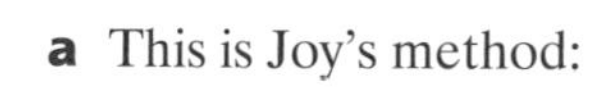

Area $= \pi \times d = \pi \times 10 = 31.42\ \text{cm}^2$ (to 2 d.p.)

Do you agree with Joy's answer? Justify your answer.

b This is Jan's method:

Area $= \pi r^2 = \pi \times 5^2 = 246.74\ \text{cm}^2$ (to 2 d.p.)

Calculator buttons used: $\boxed{\pi}\ \boxed{\times}\ \boxed{5}\ \boxed{=}\ \boxed{x^2}\ \boxed{=}$

Do you agree with Jan's answer? Justify your answer.

3 Copy this and match each circle with the correct area.
Fill in the missing areas and the missing radius.

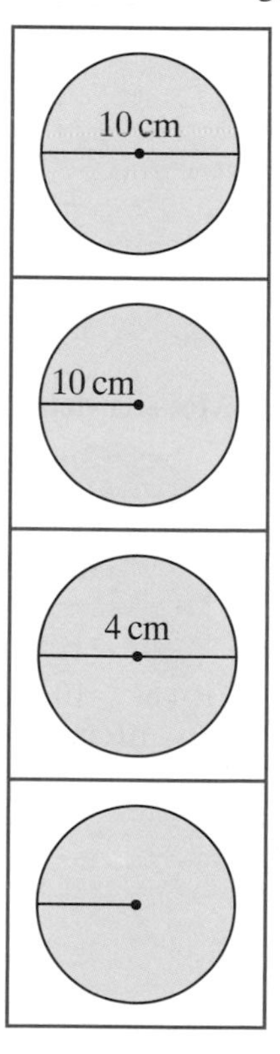

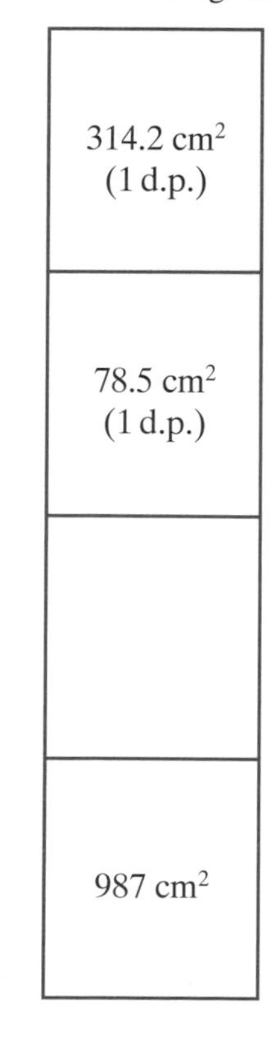

4 Calculate the total area of each of the following shapes:

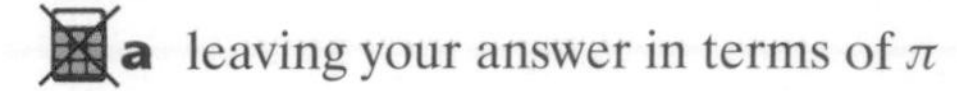
a leaving your answer in terms of π

b giving your answer to an appropriate degree of accuracy.

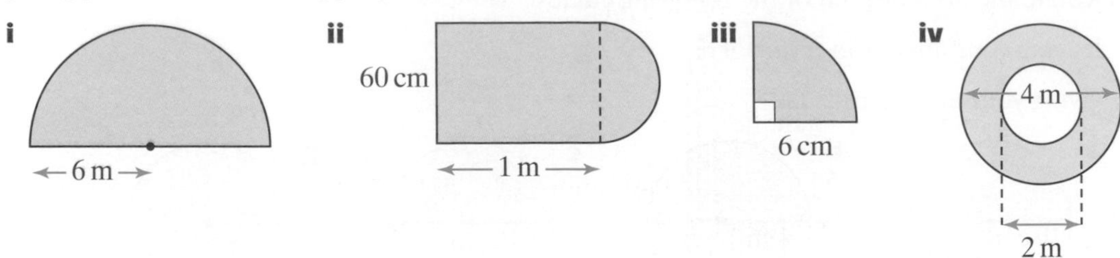

5 Complete the following table, giving your answers to an appropriate degree of accuracy.

Radius	Diameter	Area
	5 cm	
4 m		
		$10\ mm^2$
		$16\pi\ cm^2$

6 **Get Real!**

Steve wants to paint his favourite computer game on his bedroom wall. His parents aren't so keen! They will only allow him to do it if the black paint covers no more than two thirds of the wall.

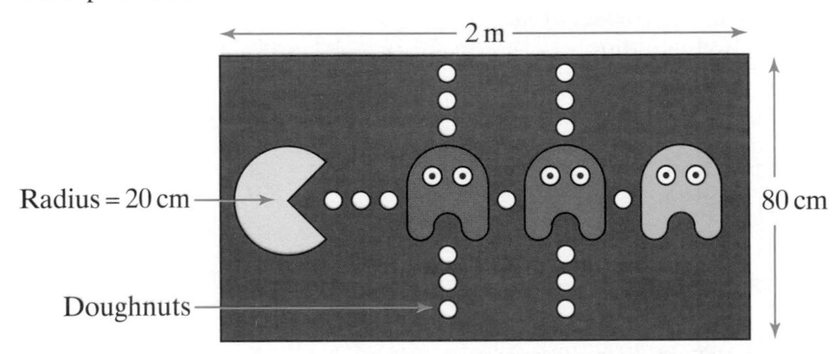

a Calculate the area of one monster (including the eyes).

b Calculate the total area of the doughnuts if each one has a radius of 5 cm.

c Calculate the area covered by the yellow doughnut eater.

HINT Note the right angle.

d Calculate the total area covered by the monsters, doughnuts and doughnut eater.

e Calculate the area which will be black?

f Will Steve's parents allow him to paint his wall?

Monster details

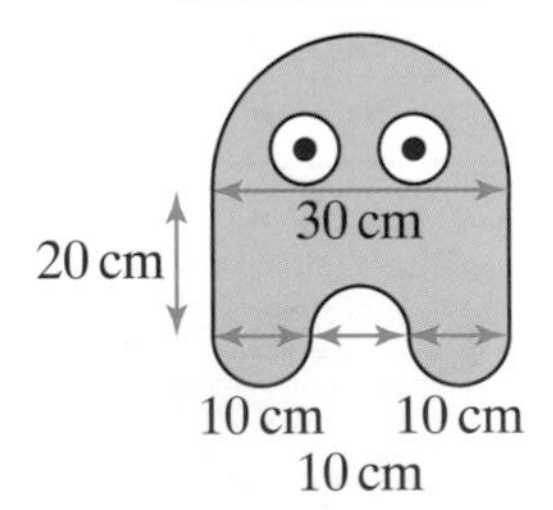

Explore

- Divide a circle into six equal pieces and cut out the pieces
- Try to make the following shape:

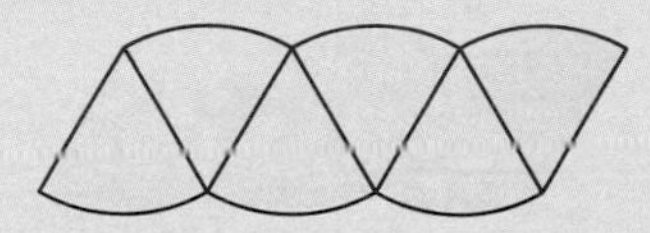

- Measure the length and height of your shape
- Work out the approximate area of the shape
- Repeat with the same size circle but this time divided into 10 pieces

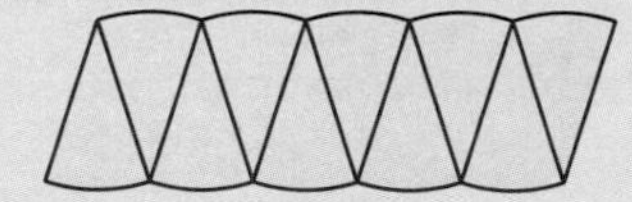

Investigate further

Perimeter and area

ASSESS

The following exercise tests your understanding of this chapter, with the questions appearing in order of increasing difficulty.

1 **a** This coffee stain is divided up into centimetre squares.
Estimate its area.

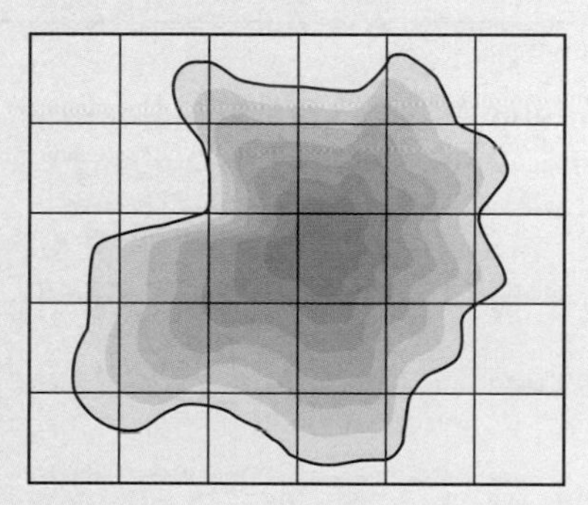

b The diagram shows a map of Jersey.
Each full square represents 1 square mile.
Estimate the area of Jersey to the nearest 5 square miles.

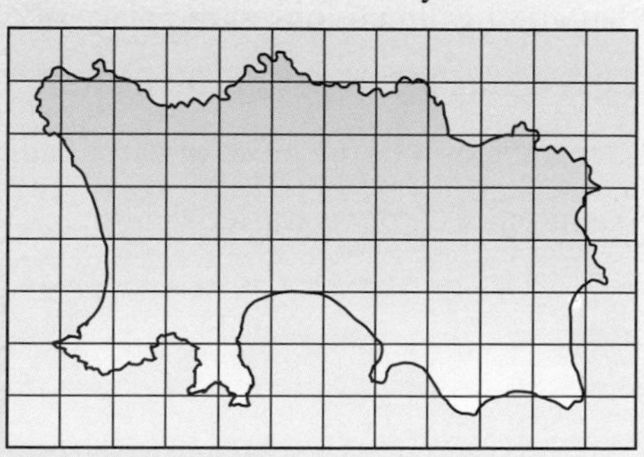

2 Accurately draw the diagram below on a piece of card or paper.

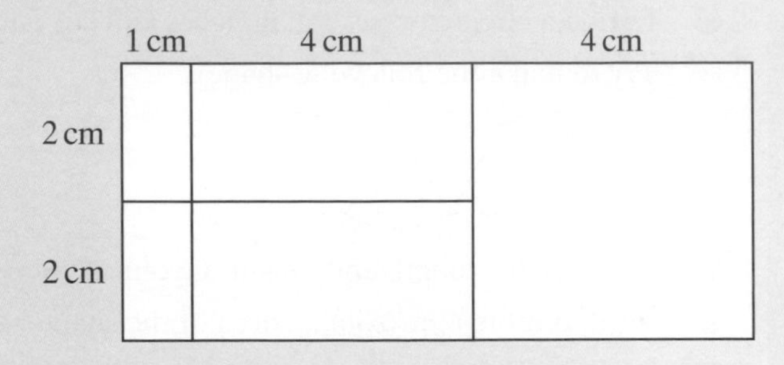

a Write down the area and perimeter of the large rectangle.

Cut the shape into the five pieces shown.
Rearrange the five pieces to make a square.

b Write down the area and perimeter of the square.

c What do you notice?

3

Flower bed
3 m Barbecue
Path
5 m Patio Lawn
6 m 8 m

The diagram shows Tom and Ali's garden. The barbecue area is a square and the path is 1 m wide. Find the area and perimeter of:

a the patio

b the lawn

c the barbecue

d the path

e the flower bed

f the whole garden.

4 A rectangular room measuring 3 m 81 cm by 4 m 65 cm is to be carpeted. The carpet is cut from a roll 4 m wide and the length can only be cut to an exact number of metres.

a What are the dimensions of the piece of carpet cut from the roll?

The carpet costs £17.95 per square metre.

b What is the cost of the piece of carpet cut from the roll?

c What area of carpet is wasted?

5 Find the areas of the following shapes.

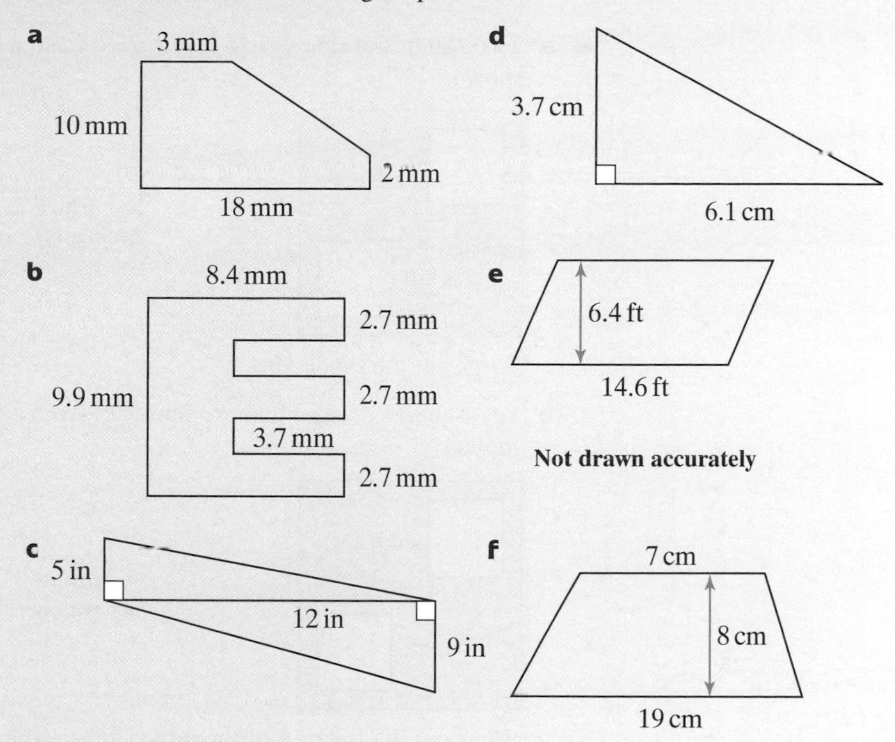

6 a Find the circumference of a circle of diameter 12.6 cm. (Take $\pi = 3.14$)

b A garden reel contains 30 m of hosepipe. The reel has a diameter of 20 cm. Calculate the number of times the reel rotates when the complete length of the hosepipe is unwound.
(You may ignore the thickness of the hosepipe.)

7 a A washer has an outer radius of 3.6 cm and a hole of radius 0.4 cm.

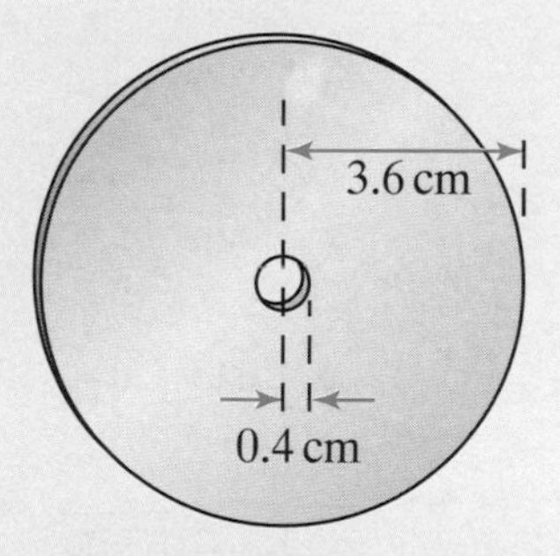

Calculate the area of the face of the washer, giving your answer:

i in terms of π

ii to 1 decimal place.

b Three thin silver discs, of radii 4, 7 and 10 cm, are melted down and recast into another disc of the same thickness. Find the radius of this disc.

Try a real past exam question to test your knowledge:

8 a Two squares of side 4 cm are removed from a square of side 12 cm as shown.

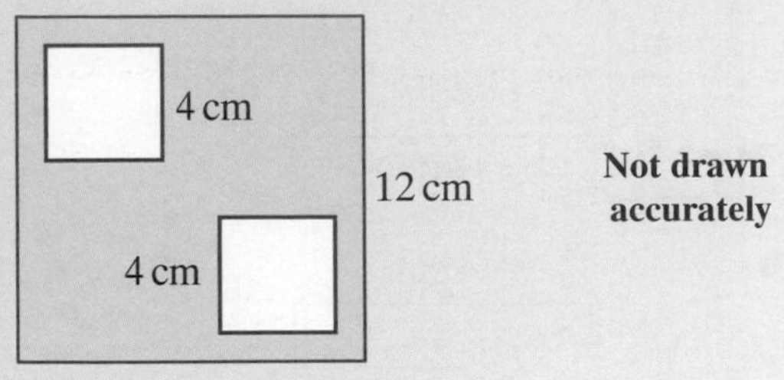

Work out the shaded area.

b Two squares of side x cm are removed from a square of side $3x$ cm as shown.

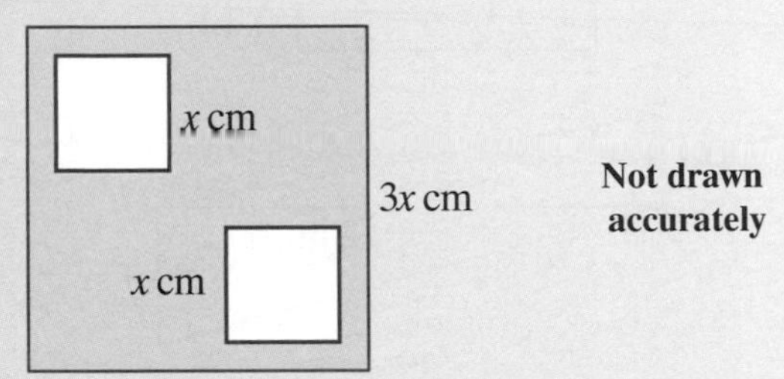

Work out the fraction of the large square which remains.
Give your answer in its simplest form.

Spec B Modular, Module 5, Nov 04

5 Properties of polygons

OBJECTIVES

G **Examiners would normally expect students who get a G grade to be able to:**

Recognise and name shapes such as isosceles triangle, parallelogram, rhombus, trapezium and hexagon

Examiners would normally expect students who get an E grade also to be able to:

Calculate interior and exterior angles of a quadrilateral

Investigate tessellations

Examiners would normally expect students who get a C grade also to be able to:

Classify a quadrilateral by geometric properties

Calculate exterior and interior angles of a regular polygon

What you should already know ...

- Recall and use properties of angles at a point, angles on a straight line, perpendicular lines, and opposite angles at a vertex
- Distinguish between acute, obtuse, reflex and right angles
- Use parallel lines, alternate angles and corresponding angles
- Prove that the angle sum of a triangle is 180°
- Prove that the exterior angle of a triangle is equal to the sum of the interior opposite angles
- Use angle properties of equilateral, isosceles and right-angled triangles
- Understand simple congruence

VOCABULARY

Equilateral triangle – a triangle with 3 equal sides and 3 equal angles – each angle is 60°

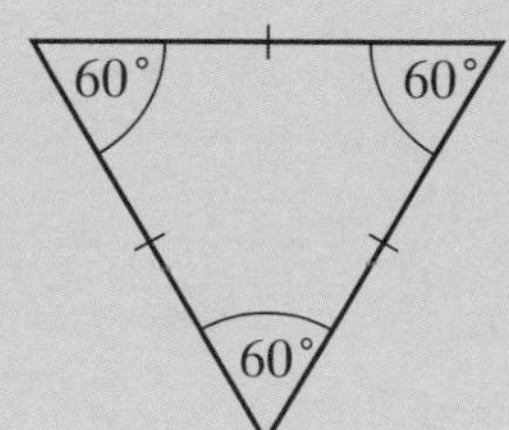

Isosceles triangle – a triangle with 2 equal sides and 2 equal angles; the equal angles are called **base angles**

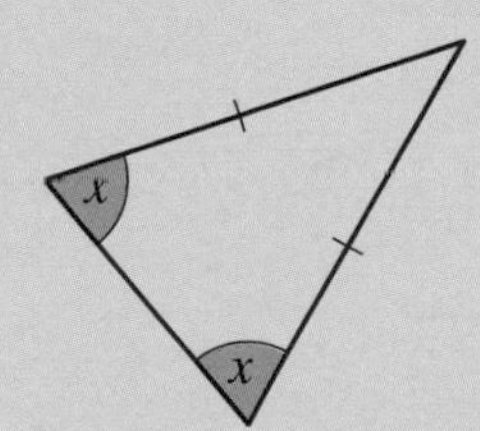

the x angles are base angles

Square – a quadrilateral with four equal sides and four right angles

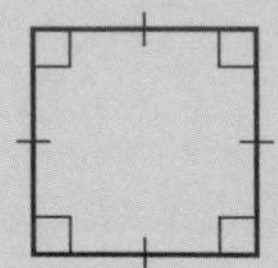

Rectangle – a quadrilateral with four right angles, and opposite sides equal in length

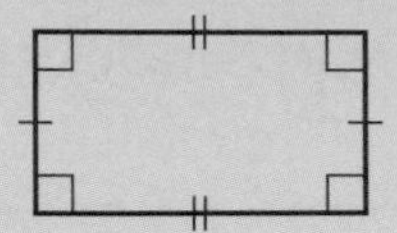

Kite – a quadrilateral with two pairs of equal adjacent sides

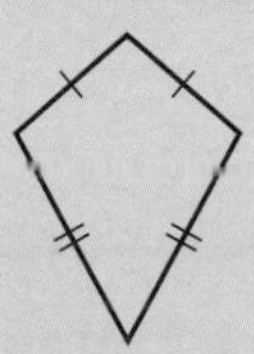

Trapezium – a quadrilateral with one pair of parallel sides

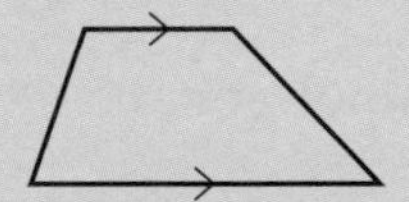

Isosceles trapezium – a quadrilateral with one pair of parallel sides. Non-parallel sides are equal

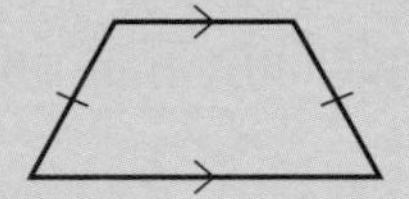

Parallelogram – a quadrilateral with opposite sides equal and parallel

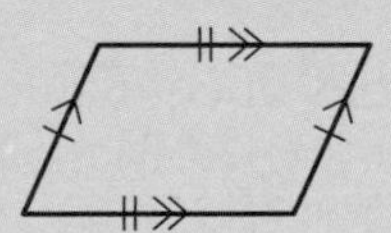

Rhombus – a quadrilateral with four equal sides and opposite sides parallel

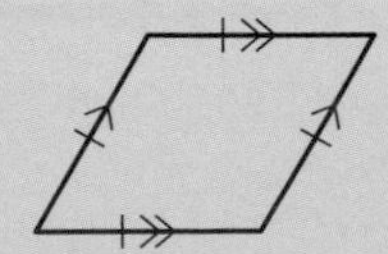

Polygon – a closed two-dimensional shape made from straight lines

Pentagon – a polygon with five sides

Hexagon – a polygon with six sides

Heptagon – a polygon with seven sides

Octagon – a polygon with eight sides

Nonagon – a polygon with nine sides

Decagon – a polygon with ten sides

Regular polygon – a polygon with all sides and all angles equal

Regular pentagon – a pentagon with all sides and all angles equal

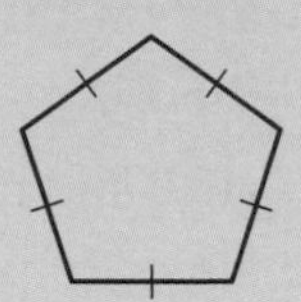

Irregular polygon – a polygon whose sides and angles are not all equal (they do not all have to be different)

Interior angle – an angle inside a polygon

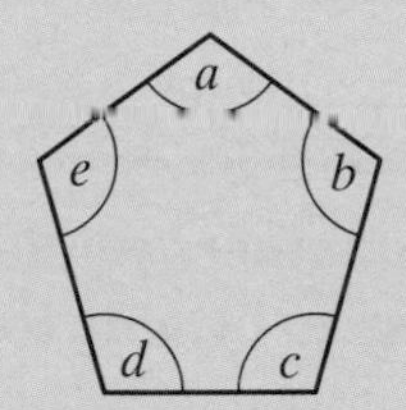

a, b, c, d and e are interior angles

Exterior angle – if you make a side of a triangle longer, the angle between this and the next side is an exterior angle

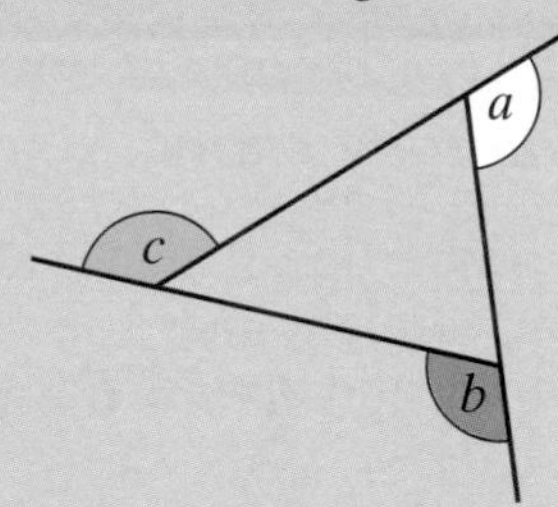

a, b and c are exterior angles

Convex polygon – a polygon with no interior reflex angles

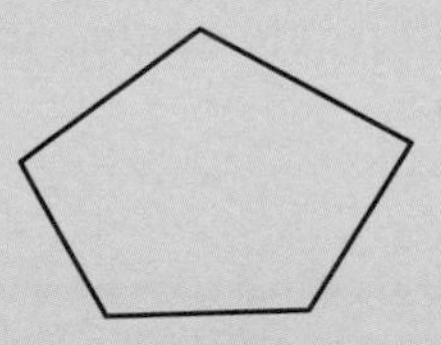

Concave polygon – a polygon with at least one interior reflex angle

Tessellation – a pattern where one or more shapes are fitted together repeatedly leaving no gaps

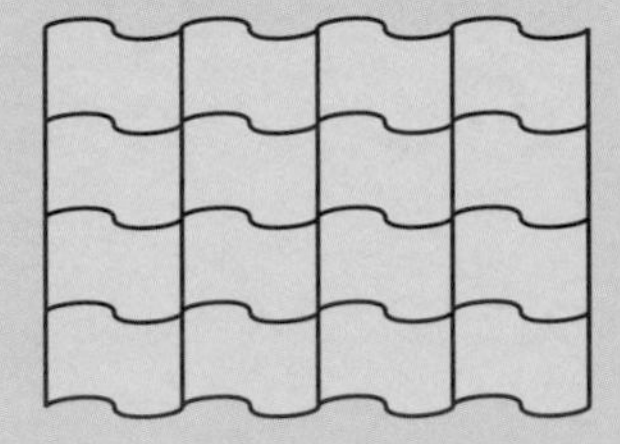

Learn 1 Angle properties of quadrilaterals

Example: Calculate the angles marked with letters in the shape below.

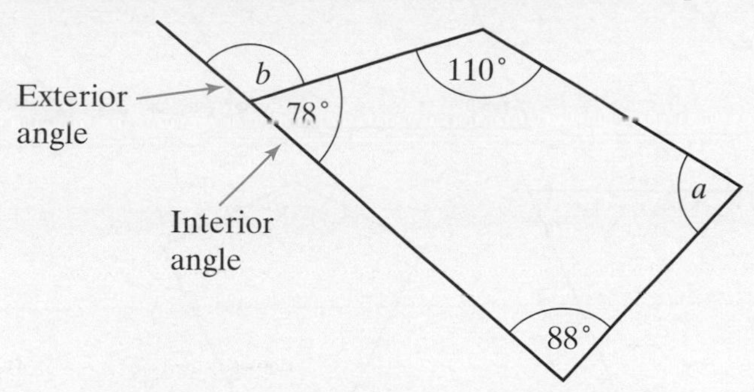

The angles in the quadrilateral add up to 360°, so

$$a = 360° - (78° + 88° + 110°)$$
$$= 360° - 276°$$
$$= 84°$$

The exterior and interior angles add up to 180°, so

$$b = 180° - 78°$$
$$= 102°$$

Apply 1

1 Name these quadrilaterals.

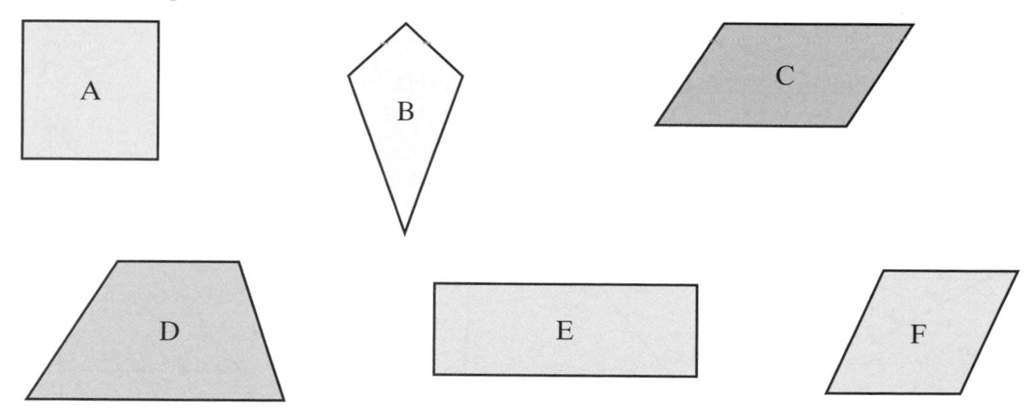

2 Calculate the size of the angles marked with letters.

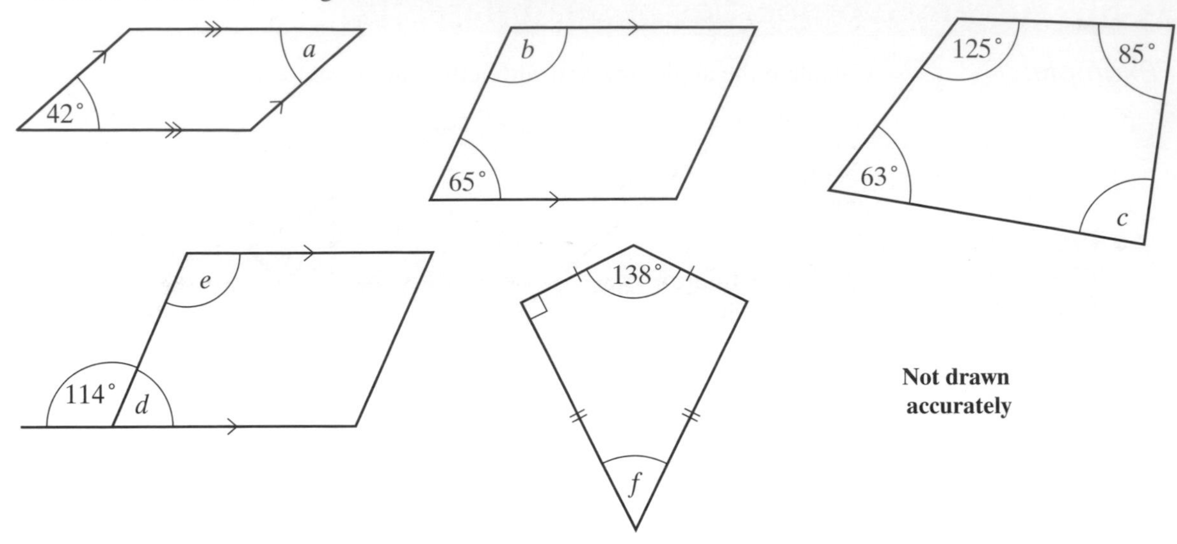

3 **a** Three angles of a quadrilateral are 60°, 65° and 113°.
Find the size of the fourth angle.

b Two angles of a quadrilateral are 74° and 116° and the other two angles are equal. What size are the other two angles?

c If all four angles of a quadrilateral are equal, what size are they?
What sort of quadrilateral could it be?

4 A quadrilateral has three angles of 84°.
What is the size of the fourth angle?

5 An isosceles trapezium has an angle of 83°.
What are the sizes of the other three angles?

6 Calculate the sizes of the marked angles in these quadrilaterals.

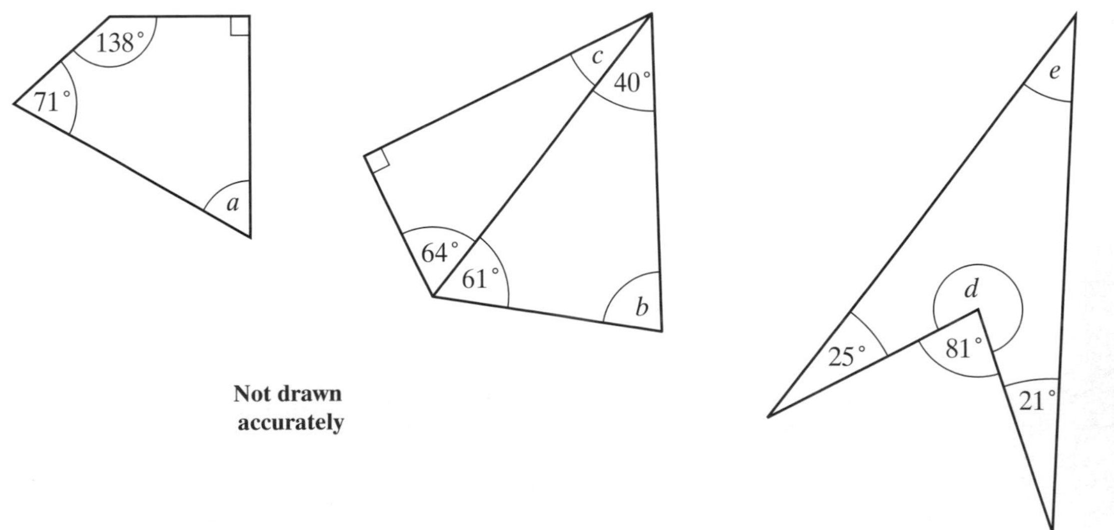

7 Barry measures the angles of a quadrilateral. He says that three of the angles are 82° and the other one is 124°. Could he be right?
Explain your answer.

8 Harry measures the angles of a quadrilateral. He says that the angles are 72°, 66°, 114° and 108°. He says the shape is a trapezium. Could he be right? Explain your answer.

9 Larry measures the angles of a quadrilateral. He says that two of them are 78°, and two of them are 102°. What quadrilateral might he have been measuring? (There are three possible answers; can you find them all?)

10 Get Real!

Khadija has two pieces of card in the shape of equilateral triangles. Each side is 4 cm long.
She makes a puzzle for her brother like this.

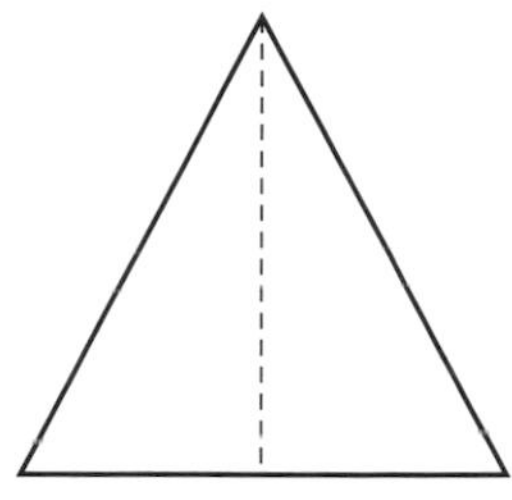

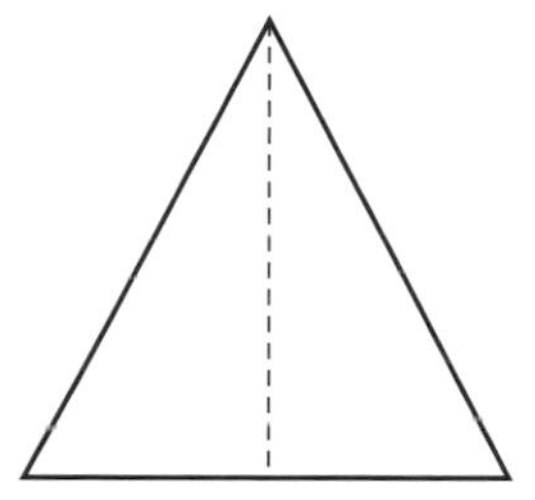

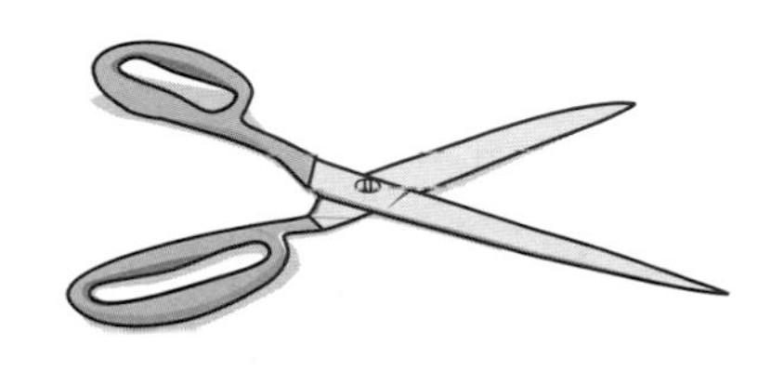

She cuts each piece of card in half along the dotted lines as shown.
She now has four congruent right-angled triangles.

a She asks her brother to rearrange the four triangles to make:

i a rectangle

ii a trapezium

iii a parallelogram

iv a rhombus.

Draw a picture to show her brother's answers.

b Work out the size of the angles in one of the right-angled triangles.

11 Sally's teacher wants the class to draw a trapezium with two sides of 5 cm, one side of 7 cm and one side of 10 cm. Sally says there is more than one answer. Is she right? Give a reason for your answer.

Explore

- A kite always has an obtuse angle
  True or false?
- Can a kite have two obtuse angles?

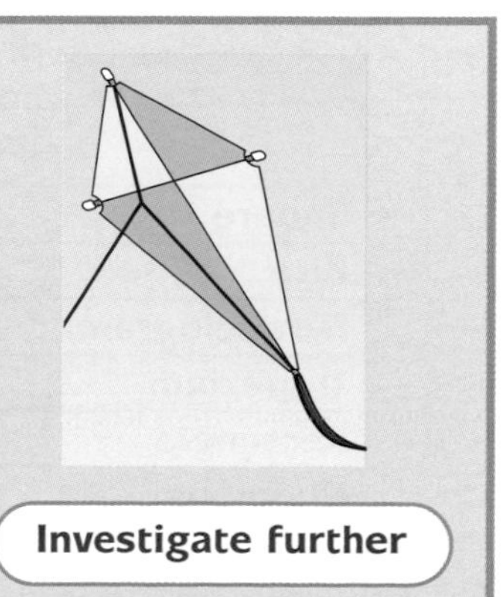

Investigate further

Learn 2 Diagonal properties of quadrilaterals

Example: The diagonals of a quadrilateral are different lengths, but one is the perpendicular bisector of the other.
What is the mathematical name of the quadrilateral?

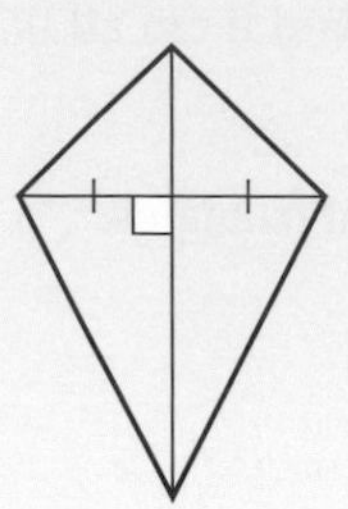

The quadrilateral is a kite.

Apply 2

1 **a** Copy these shapes and draw in their diagonals.

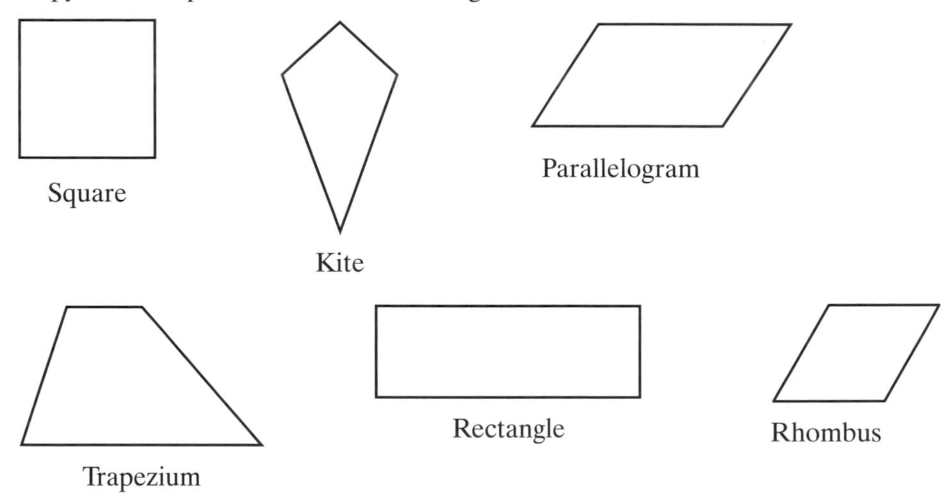

b Use your diagrams to complete a copy of this table.

Shape	Are the diagonals equal? (Yes/No)	Do the diagonals bisect each other? (Both/One only/No)	Do the diagonals cross at right angles? (Yes/No)	Do the diagonals bisect the angles of the quadrilateral? (Yes/Two only/No)
Square				
Kite				
Parallelogram				
Trapezium				
Rectangle				
Rhombus				

2 Rajesh says that he has drawn a quadrilateral. Its diagonals are equal.
What quadrilaterals might he have drawn?
(Use the table from question **1** to help you.)

3 Michelle says that the diagonals of a rectangle bisect the corner angles.
So angles a and c are both 45°, and angle b must be 90°.
Is she right? Explain your answer.

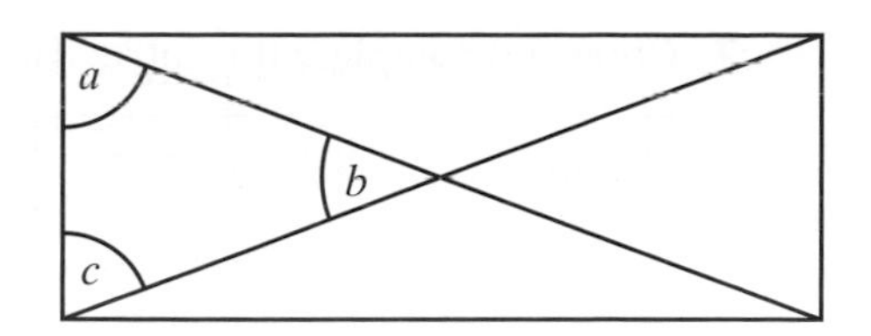

4 The diagram below shows a rhombus ABCD. AC and BD are the diagonals.
Angle ADB = 32°.

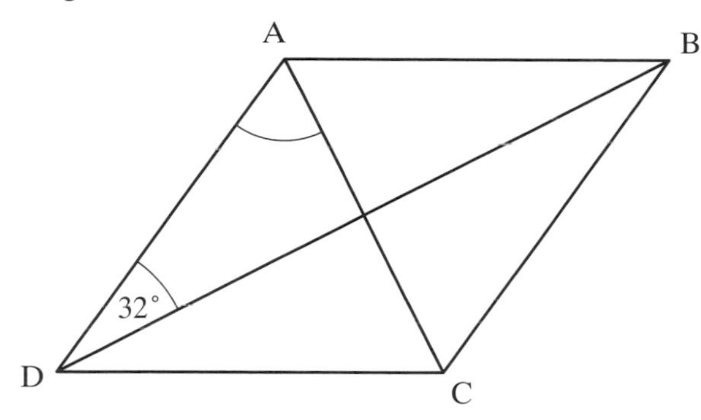

Not drawn accurately

Calculate angle DAC.

5 Calculate the angles marked with letters in the diagrams below.
You will need to use the properties of diagonals.
Give a reason for each of the angles.

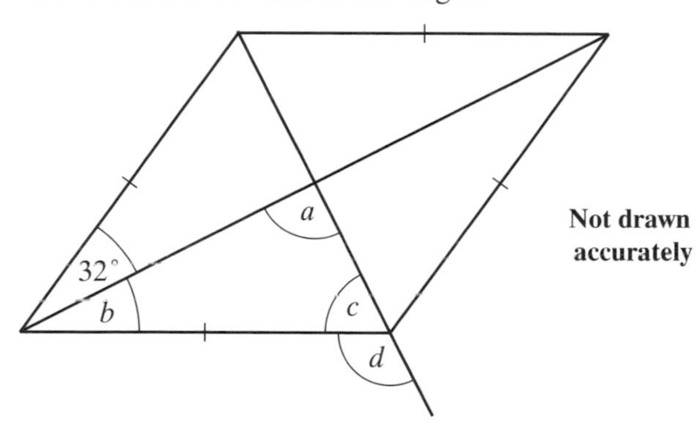

Not drawn accurately

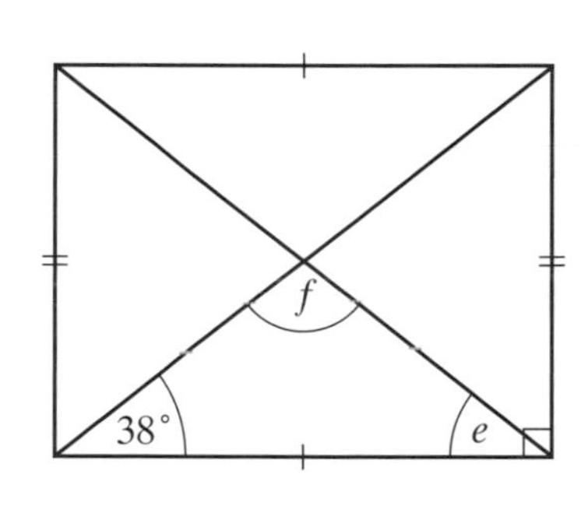

6 **Get Real!**
A builder has two types of wall tile. One is an isosceles trapezium, the other is a rhombus. He creates this tessellation.

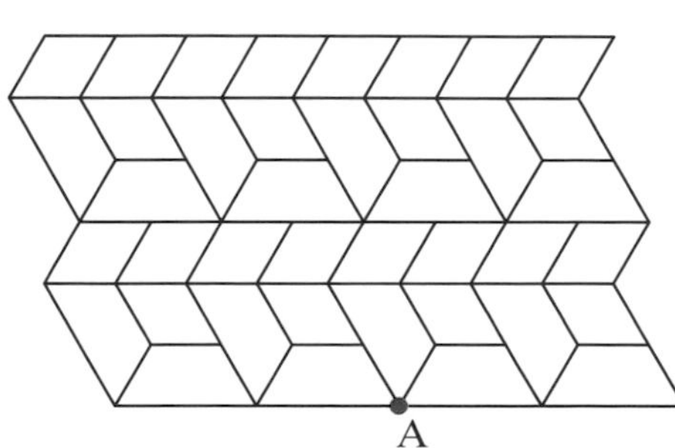

a By looking at the angles around point A, calculate the angles of the trapezium.

b Now calculate the angles of the rhombus.

7 Copy and complete this table. The top line has been done for you.

Shape	Number of different length sides (at most)	Number of right angles (at least)	Pairs of opposite sides parallel	Diagonals must be equal	Diagonals bisect each other	Diagonals cross at right angles
Square	1	4	Both	Yes	Yes	Yes
Rectangle						
Trapezium						
Rhombus						
Parallelogram						
Kite						
Isosceles trapezium						

Explore

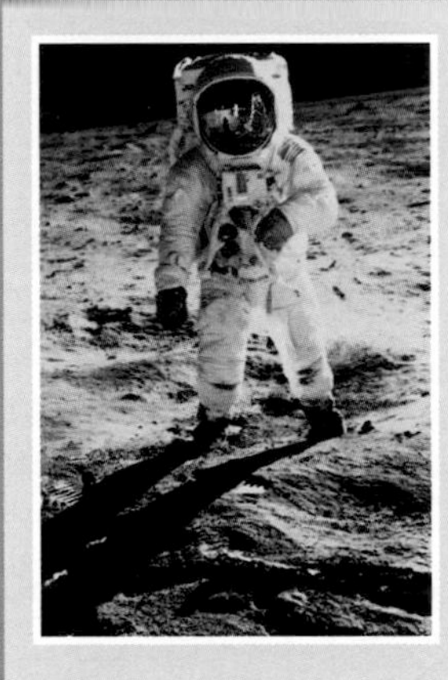

- Roger draws a quadrilateral
  He uses it to make this tessellation pattern

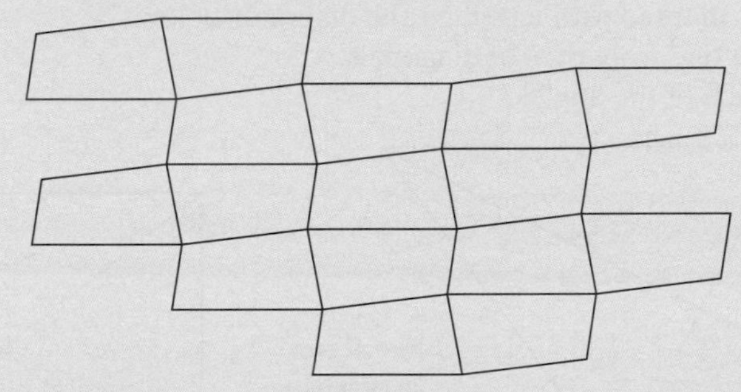

- He says you can make a tessellation with any quadrilateral

Investigate further

Learn 3 Angle properties of polygons

Examples:

a Calculate the sum of the interior angles of a:

i pentagon **ii** hexagon.

i The pentagon can be divided into three triangles by drawing diagonals from a start point.

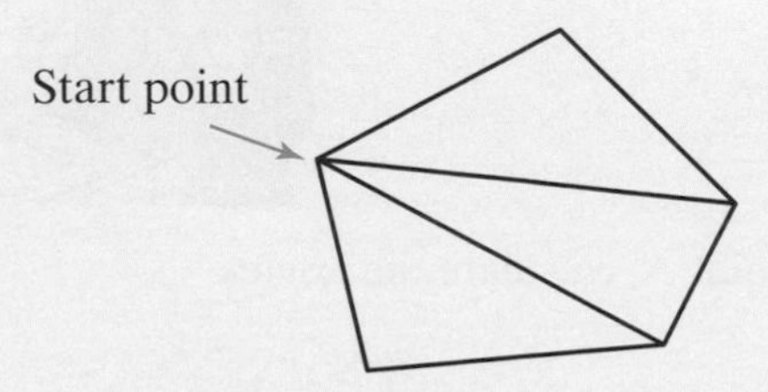

The sum of the angles is
$(5 - 2) \times 180° = 540°$

ii The hexagon can be divided into four triangles in the same way.

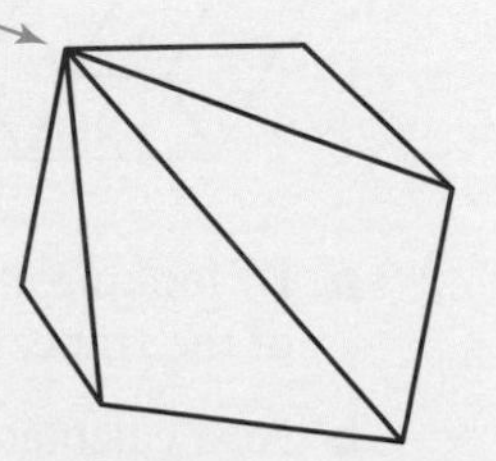

The interior angles of a polygon add up to (number of sides − 2) × 180°

The sum of the angles is
$(6 - 2) \times 180° = 720°$

b Calculate the interior angle of a regular octagon.

To calculate the interior angle of a regular octagon you can use one of the following methods.

Either:
An octagon has eight sides.
The sum of the angles can be found by dividing the octagon into six triangles as shown.
So the sum of the angles is $(8 - 2) \times 180° = 1080°$.
A regular octagon has all angles equal, so each angle is $1080° \div 8 = 135°$.

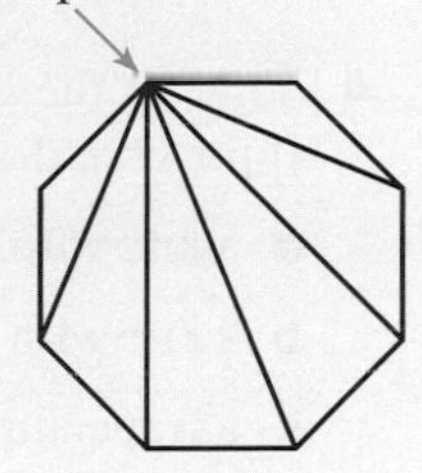

Or:
A regular octagon has eight equal exterior angles.
So each exterior angle is $360° \div 8 = 45°$.
So each interior angle is $180° - 45° = 135°$.

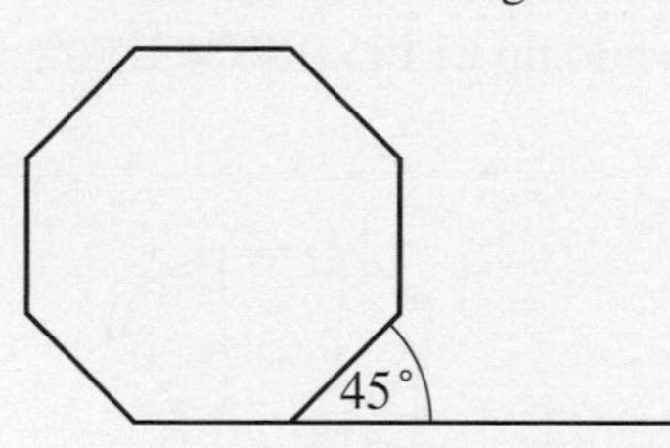

c A regular polygon has interior angles of 144°.
How many sides does it have?

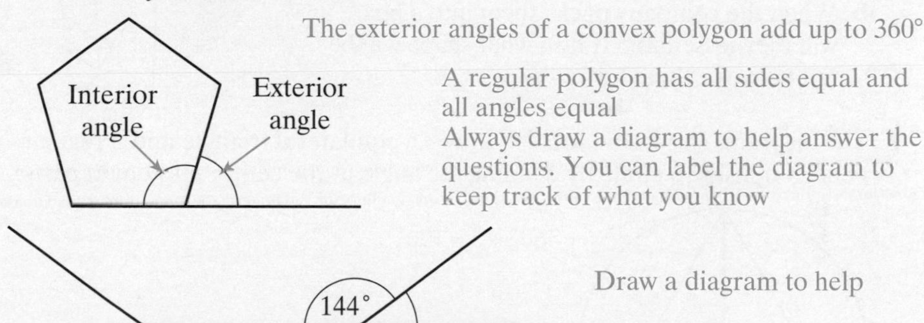

The exterior angles of a convex polygon add up to 360°

A regular polygon has all sides equal and all angles equal
Always draw a diagram to help answer the questions. You can label the diagram to keep track of what you know

Draw a diagram to help

Each exterior angle must be $180° - 144° = 36°$.
Exterior angles add up to 360°, so there must be $360 \div 36 = 10$ exterior angles, and 10 sides.

Apply 3

1 Calculate the angles marked with letters in the diagram.
Explain how you worked them out.

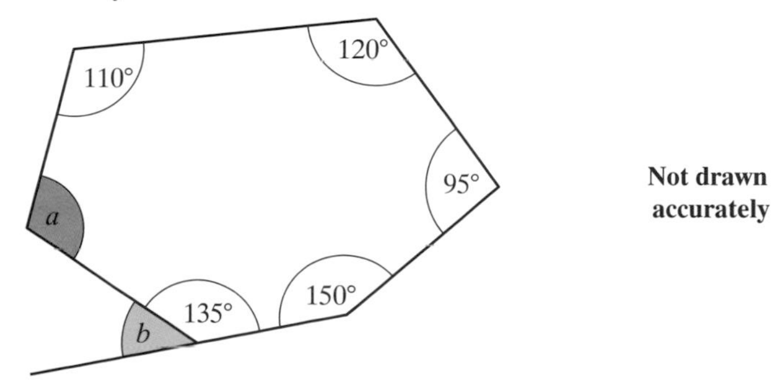

Not drawn accurately

2 Four of the angles of a pentagon are 110°, 130°, 102° and 97°.
Calculate the fifth angle.

3 **a** A regular polygon has an exterior angle of 40°. How many sides has it?

b Will this polygon tessellate?

4 Some of these quadrilaterals are possible, and others are not.
If possible draw:

a a kite with a right angle

b a kite with two right angles

c a trapezium with two right angles

d a trapezium with only one right angle

e a triangle with a right angle

f a triangle with two right angles

g a pentagon with one right angle

h a pentagon with two right angles

i a pentagon with three right angles

j a pentagon with four right angles.

5 Elena says that the interior angles of a decagon add up to $10 \times 180° = 1800°$.
Is she right? Give a reason for your answer.

6 **Get Real!**
A company makes containers as shown.
The top is in the shape of a regular octagon.

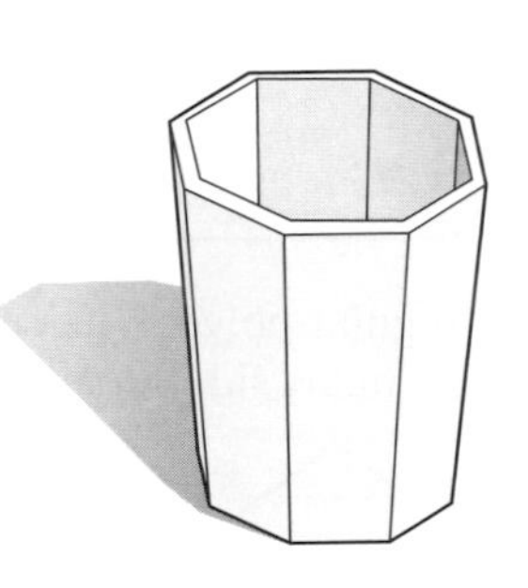

a What is the size of each interior angle?

b When the company packs them into a box, will they tessellate? If not, what shape will be left between them?

7 The diagrams show how you can draw an equilateral triangle and a regular pentagon inside a circle by dividing the angle at the centre into equal parts.

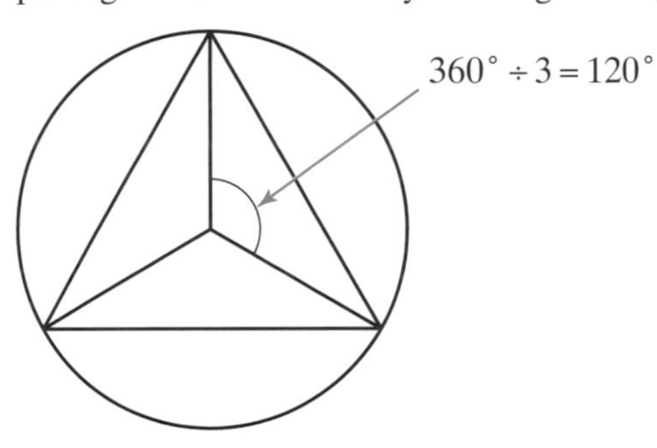

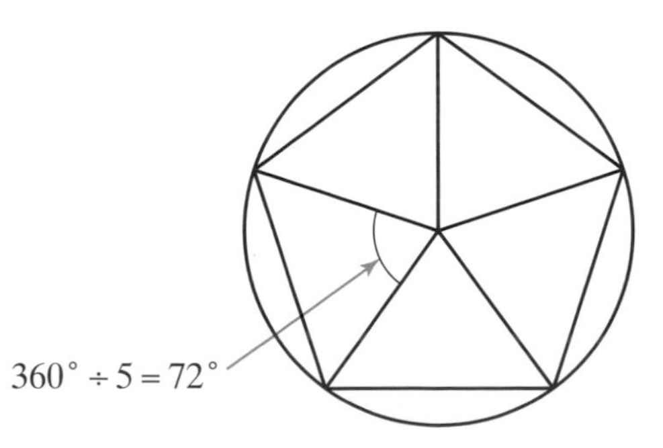

Use the same method to draw a regular hexagon and a regular nonagon inside a circle.

8 The diagram shows a regular pentagon ABCDE and a regular hexagon DEFGHI.

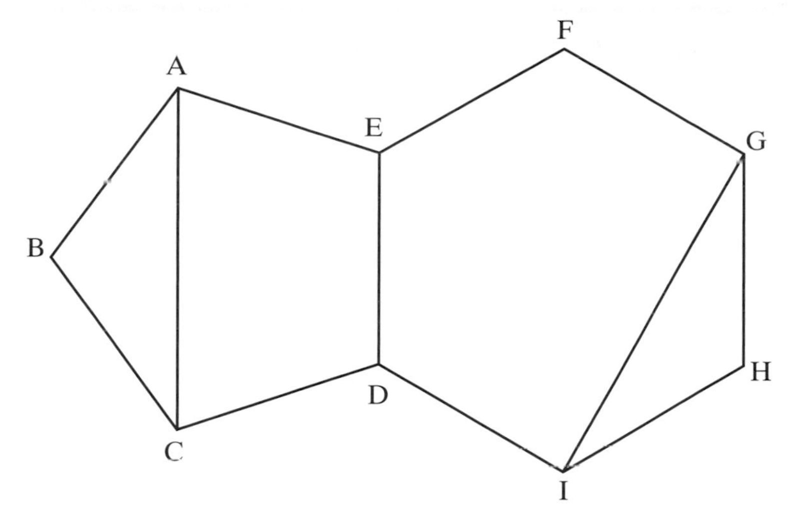

Calculate:

a angle EDC

b angle EDI

c obtuse angle CDI

d angle BAC

e angle CAE

f angle HIG

g angle DIG

Explore

- Which regular polygons tessellate?
- What are the interior angles of the polygons that tessellate?
- Which tessellations can be made using two different regular polygons?

Investigate further

Properties of polygons

ASSESS

The following exercise tests your understanding of this chapter, with the questions appearing in order of increasing difficulty.

1 a Name these shapes.

A
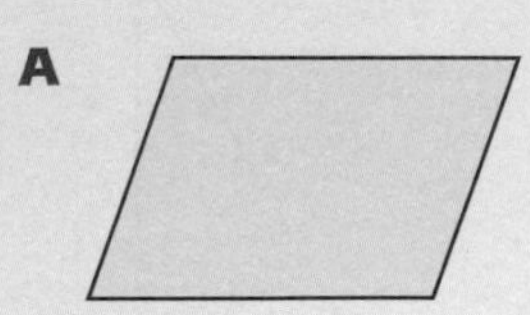

D
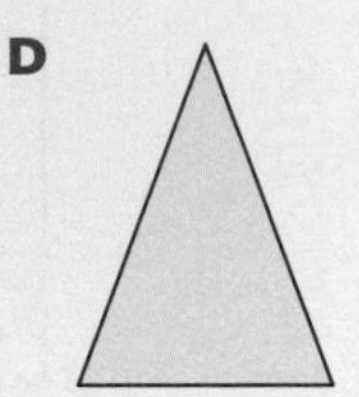

B

E
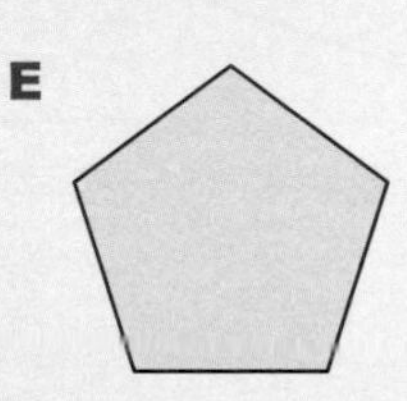

C
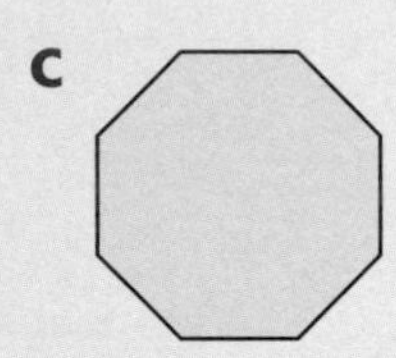

F
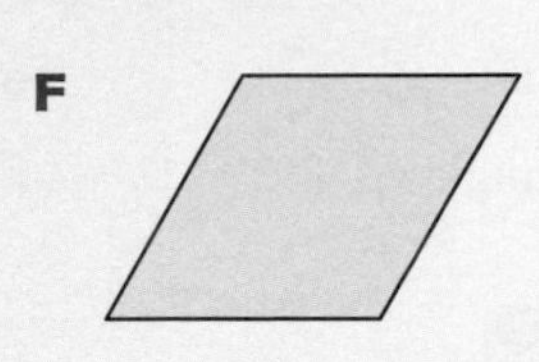

b Write down the letters of the diagrams that have:

i some sides equal

ii all sides equal

iii any acute angles

iv some equal angles

v any adjacent sides equal

vi all diagonals equal

vii diagonals perpendicular to each other

viii any adjacent angles equal.

2 a Will **any** parallelogram tessellate?
Explain your answer.

b Will **any** rhombus tessellate?
Explain your answer.

3 a Draw a quadrilateral that can be cut into two pieces by drawing a straight line across it.

b Draw a quadrilateral that can be cut into three pieces by drawing a straight line across it.

4 Find the values of the marked angles in these diagrams.

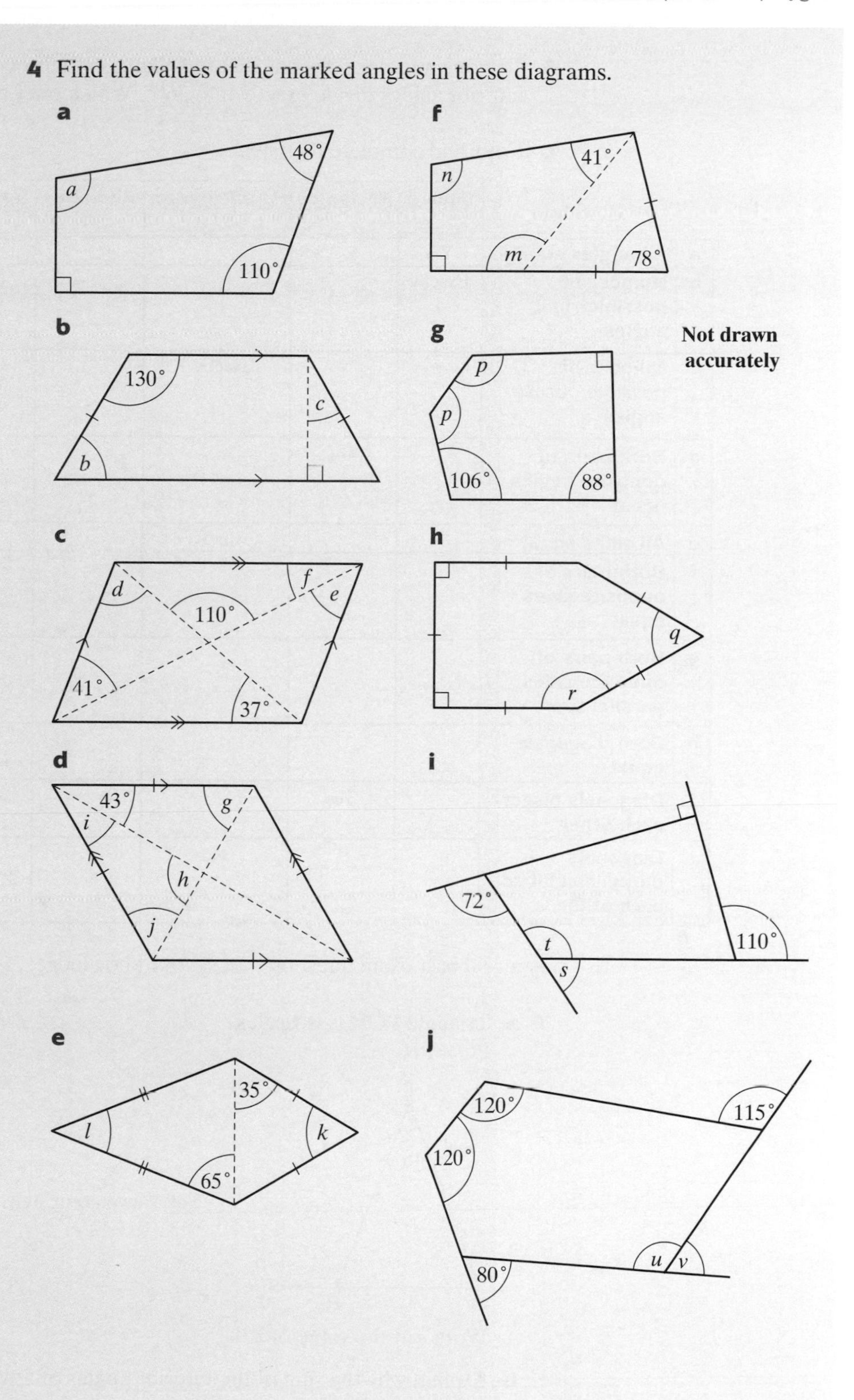

5 The only regular polygons that tessellate on their own are those whose interior angles divide exactly into 360°. Which ones are they?

6 Copy and complete the table.

		Square	Rectangle	Parallelogram	Rhombus	Trapezium	Isosceles trapezium	Kite
a	**All angles equal**	Yes						
b	**Number of possible right angles**	Exactly 4				0 or 2		1 or 2
c	**Number of possible obtuse angles**	0		Exactly 2				
d	**Both pairs of opposite angles equal**		Yes					
e	**All sides equal**			No				
f	**Both pairs of opposite sides equal**					No		
g	**Both pairs of opposite sides parallel**						No	
h	**Both diagonals equal**							No
i	**Diagonals bisect each other**		Yes					
j	**Diagonals perpendicular to each other**			No				

Try a real past exam question to test your knowledge:

7 a Triangle PQR is isosceles.
PQ = PR.

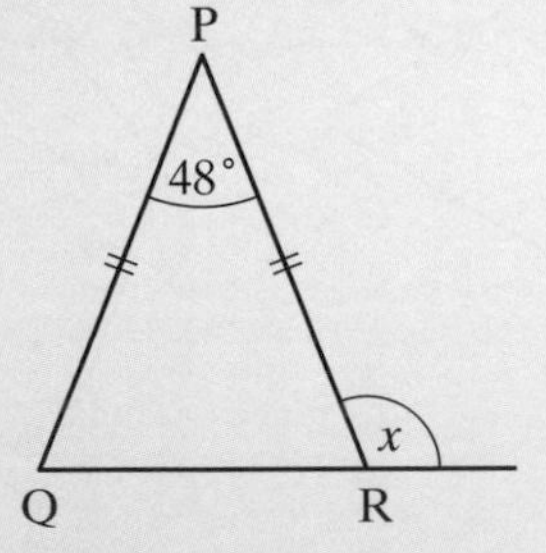

Not drawn accurately

Work out the value of x.

b Explain why the sum of the interior angles of any quadrilateral is 360°.

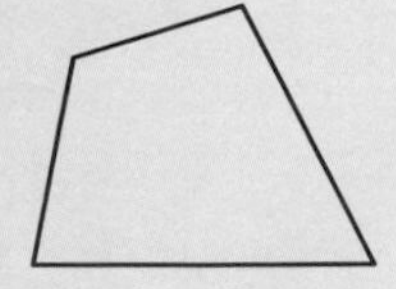

Spec A, Int Paper 2, Nov 04

6 Sequences

OBJECTIVES

G **Examiners would normally expect students who get a G grade to be able to:**

Continue a sequence of numbers or diagrams

Write the terms of a simple sequence

F **Examiners would normally expect students who get an F grade also to be able to:**

Find a particular term in a sequence involving positive numbers

Write the term-to-term rule in a sequence involving positive numbers

E **Examiners would normally expect students who get an E grade also to be able to:**

Find a particular term in a sequence involving negative or fractional numbers

Write the term-to-term rule in a sequence involving negative or fractional numbers

D **Examiners would normally expect students who get a D grade also to be able to:**

Write the terms of a sequence or a series of diagrams given the nth term

C **Examiners would normally expect students who get a C grade also to be able to:**

Write the nth term of a sequence or a series of diagrams

What you should already know ...

- Identify odd and even numbers

VOCABULARY

Sequence – a list of numbers or diagrams that are connected in some way

In this sequence of diagrams, the number of squares is increased by one each time:

The dots are included to show that the sequence continues

...

Term – a number, variable or the product of a number and variable(s) such as 3, x or $3x$

nth term – this phrase is often used to describe a 'general' term in a sequence; if you are given the nth term, you can use this to find the terms of a sequence

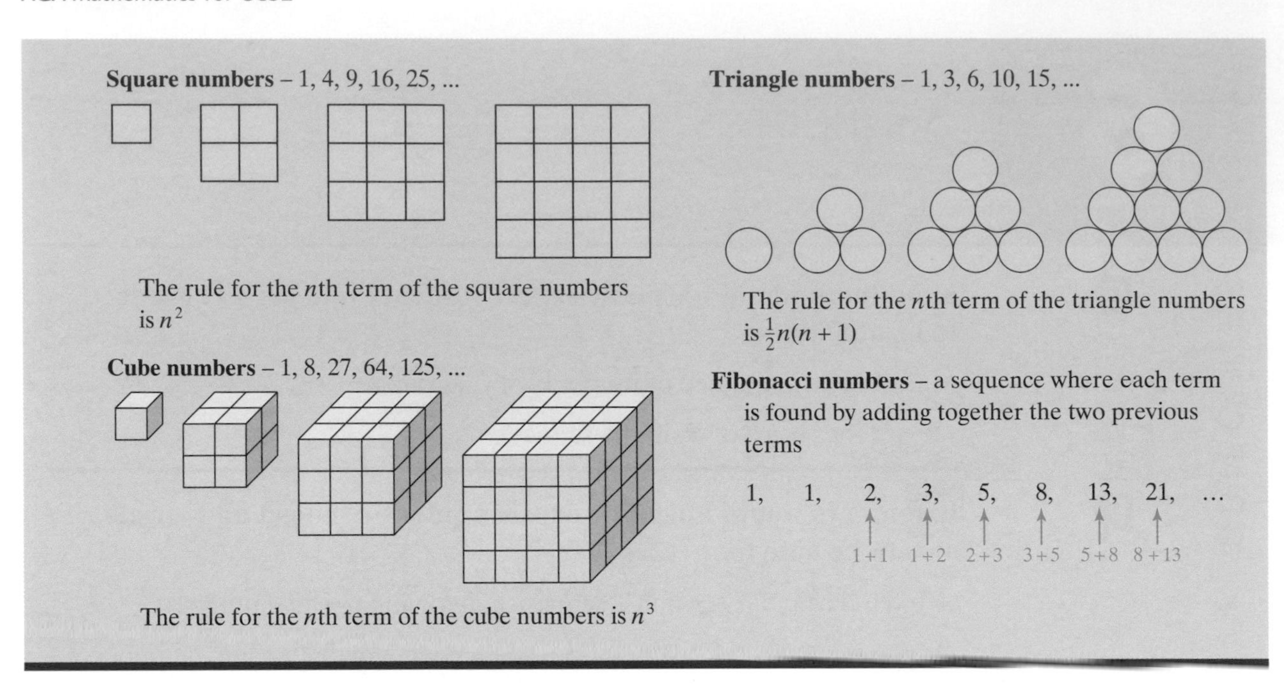

Learn 1 The rules of a sequence

Example: Write the next two terms in the sequence:

5, 10, 20, 40, ...

In this sequence of numbers, the number is multiplied by two to give the next number.

5, 10, 20, 40, ...
×2 ×2 ×2 ×2

The numbers of the sequence are usually called the terms of the sequence

In the sequence above, the number 5 is the first term, the number 10 is the second term, the number 20 is the third term, the number 40 is the fourth term, etc

The rule for the sequence (called the term-to-term rule) can be used to find the fifth and subsequent terms.

In this example, the term-to-term rule is 'multiply by 2' so the fifth term is 80 (40 × 2) and the sixth term is 160 (80 × 2) and so on.

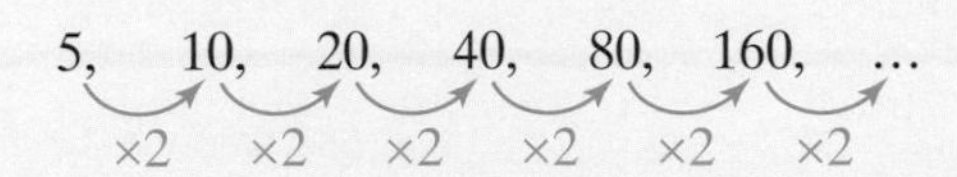

Apply 1

1 Draw the next two diagrams in the following sequences.

a

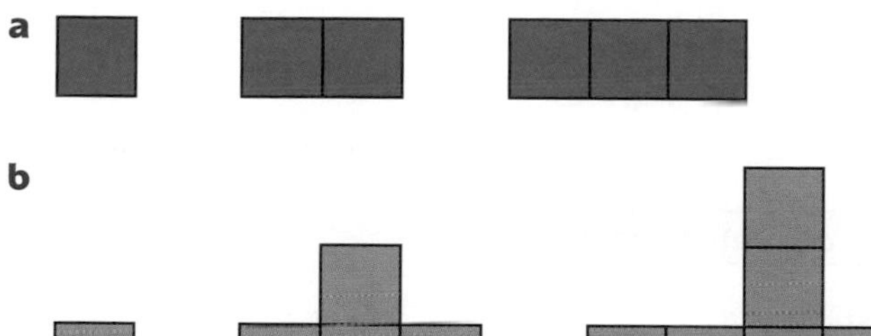

b

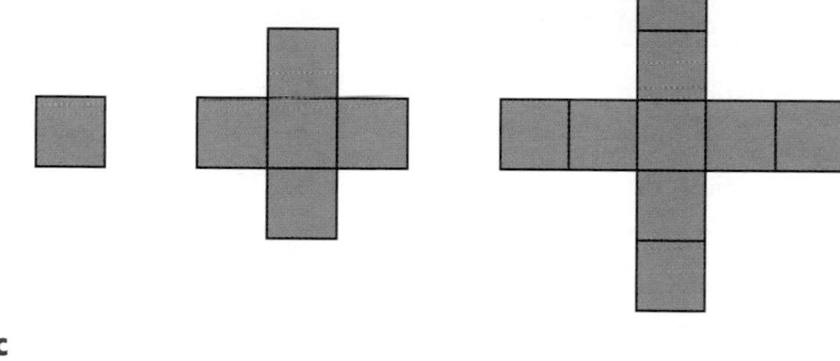

c

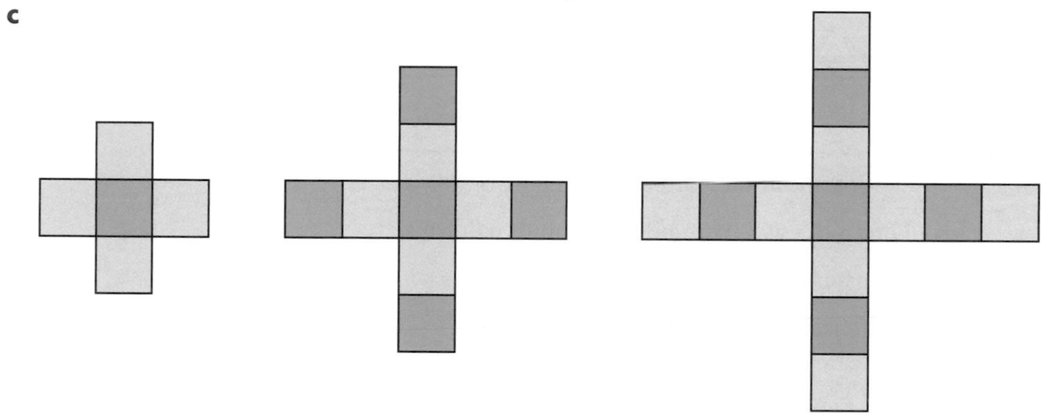

d

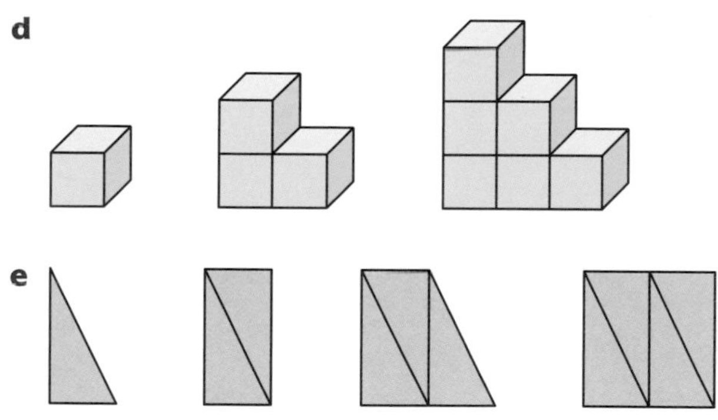

e

2 Write the next two terms in the following sequences.

a 3, 7, 11, 15, ...

b 6, 9, 12, 15, ...

c 2, 10, 18, 26, ...

d 0, 5, 10, 15, ...

e 1, 2, 4, 8, 16, ...

f 10, 100, 1000, 10 000, ...

3 Write the ninth and tenth terms in the following sequences.

a 2, 4, 6, 8, ...

b 3, 5, 7, 9, ...

c 10, 20, 30, 40, ...

d 101, 102, 103, 104, ...

4 Copy and complete this table.

Pattern (n)	Diagram	Number of matchsticks (m)
1		3 matchsticks
2		5 matchsticks
3		7 matchsticks
4		
5		

a What do you notice about the pattern of matchsticks?

b How many matchsticks will there be in the ninth pattern?

c How many matchsticks will there be in the tenth pattern?

d There are 41 matchsticks in the 20th pattern.
How many matchsticks are there in the 21st pattern?
Give a reason for your answer.

5 Fill in the missing numbers in the following sequences.

a 4, 6, 8, ..., 12, 14

b 3, 6, 12, ..., 48, 96

c 25, 16, 7, ..., −11

d 2, ..., 20, 29, 38

6 Write the term-to-term rule for the following sequences.

a 3, 7, 11, 15, ...

b 0, 5, 10, 15, ...

c 1, 2, 4, 8, 16, ...

d 3, 4.5, 6, 7.5, ...

e 2, 3, 4.5, 6.75, ...

f 100, 1000, 10 000, ...

g 20, 16, 12, 8, ...

h 54, 18, 6, 2, ...

7 The term-to-term rule is +6.
Write five different sequences that fit this rule.

8 Here is a sequence of numbers:

2 3 5 9 17

The rule for continuing this sequence is: **multiply by 2 and subtract 1**.

a What are the next **two** numbers in this sequence?

The same rule is used for a sequence that starts with the number -3.

b What are the next **two** numbers in this sequence?

9 Farukh is exploring number patterns.
He writes down the following products in a table.

1×1	1
11×11	121
111×111	12 321
1111×1111	1 234 321
11111×11111	
111111×111111	

Copy and complete the last two rows in the table.

Farukh says he can use the table to work out $1\,111\,111\,111 \times 1\,111\,111\,111$
Is he correct? Give a reason for your answer.

Explore

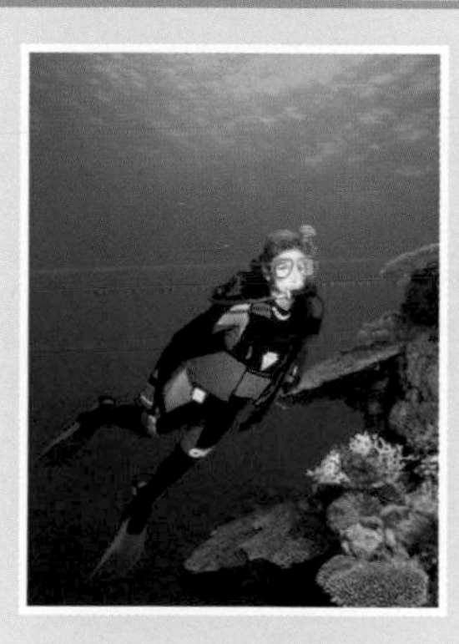

- Investigate the following non-linear sequences:
  1, 4, 9, 16, ...
  1, 8, 27, 64, ...
  1, 3, 6, 10, 15, ...
  1, 1, 2, 3, 5, 8, ...
  2, 3, 5, 7, 11, ...
- Invent your own non-linear sequences

Investigate further

Explore

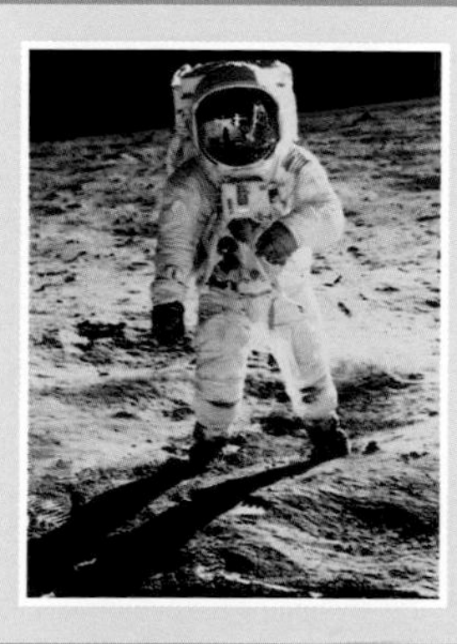

- Workers are paid 1p on their first day, 2p on the second day, 4p on the third day, 8p on the fourth day and so on
- How much will a worker get paid on the fifth day?
- How much money will a worker get for the first five days altogether?
- How much money will the worker get for the first ten days altogether?

Investigate further

Learn 2 The *n*th term of a sequence

Example:

Given the nth term of a sequence is $2n + 3$, use this to find the first four terms of the sequence.

If the nth term is $2n + 3$, you can use this to find the first term by replacing n with 1 in the formula.
Similarly, you can find the second term by replacing n with 2 in the formula and the 100th term by replacing n with 100 in the formula.
So the sequence whose nth term is $2n + 3$ is:

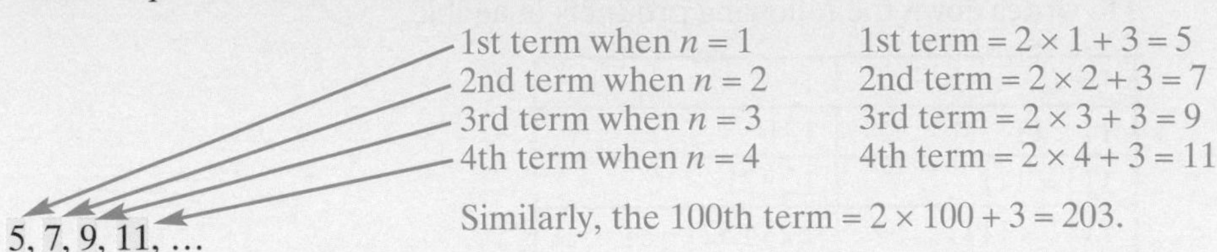

5, 7, 9, 11, …

Similarly, the 100th term $= 2 \times 100 + 3 = 203$.

The above sequence is called a linear sequence because the differences between terms are all the same. In this example, the differences are all +2. We say that the term-to-term rule is +2.

5, 7, 9, 11, …
+2 +2 +2

In general, to find the nth term of a linear sequence, you can use the formula:

nth term = difference × n + (first term − difference)
$= dn + (a - d)$

a is the first term = 5
d is the difference between terms = 2

So for 5, 7, 9, 11, …
+2 +2 +2

nth term = difference × n + (first term − difference)
$= 2 \times n + (5 - 2)$
$= 2n + 3$

Apply 2

Apart from question 9, this is a non-calculator exercise.

1 Write the first five terms of the sequence whose nth term is:

a $n + 3$
b $5n$
c $5n - 3$
d $2n - 5$
e $3n^2$
f $n^2 - 5$
g $n^2 + 3$
h $\frac{n}{n + 2}$

2 Write the 100th and the 101st terms of the sequence whose nth term is:

a $n + 5$
b $3n - 7$
c $100 - 2n$
d $n^2 + 1$

3 Jenny writes the sequence 3, 7, 11, 15, ...
She says that the nth term is $n + 4$.
Is she correct?
Give a reason for your answer.

4 Write the nth term in these linear sequences.

a 3, 7, 11, 15, ...
b 0, 5, 10, 15, ...
c 8, 14, 20, 26, ...
d −5, −1, 3, 7, ...
e 100, 95, 90, ...
f 23, 21, 19, 17, ...
g 4, 6.5, 9, 11.5, ...
h −5, 3, 11, 19, ...

5 Write the formula for the number of squares in the nth pattern.

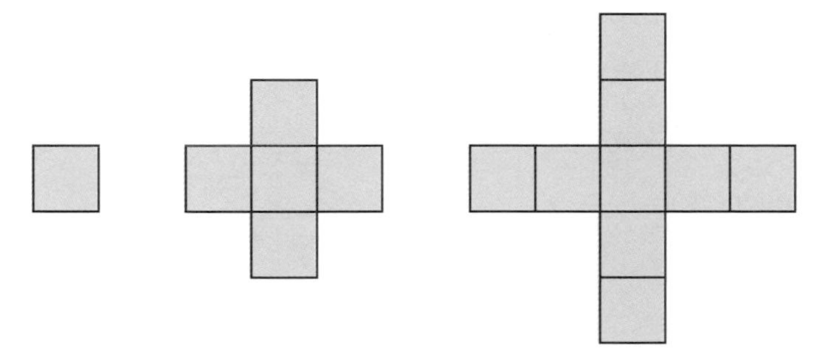

6

Pattern (n)	Diagram	Number of matchsticks (m)
1		3 matchsticks
2		5 matchsticks
3		7 matchsticks

Write the formula for the number of matchsticks (m) in the nth pattern.

7 **Get Real!**
Jackie builds fencing from pieces of wood as shown below.

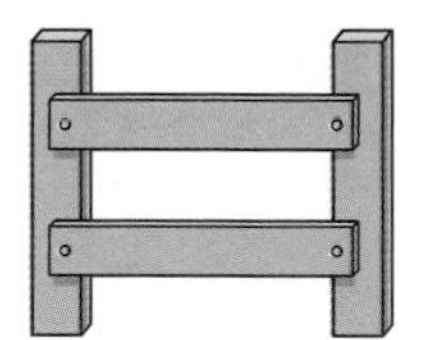

Diagram 1
4 pieces of wood

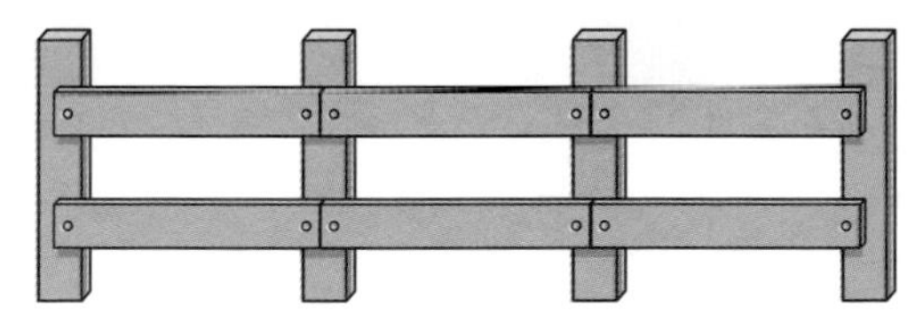

Diagram 2
7 pieces of wood

Diagram 3
10 pieces of wood

How many pieces of wood will there be in Diagram n?

8

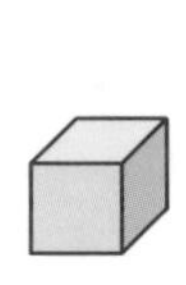
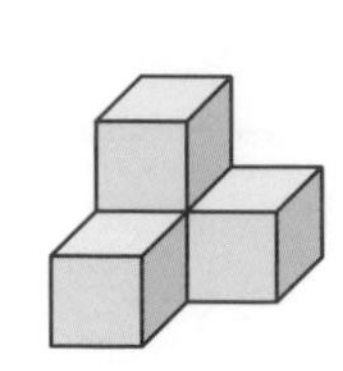
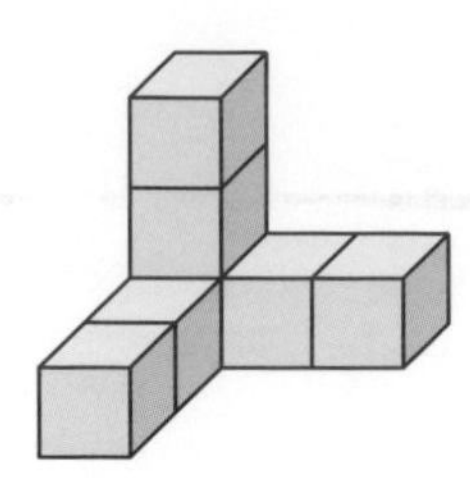

Stuart says that the number of cubes in the 100th pattern is 300.
How can you tell that Stuart is wrong?
Give a reason for your answer.

9 Write the nth term in these non-linear sequences.

a 1, 4, 9, 16, ...

b 2, 5, 10, 17, ...

c 2, 8, 18, 32, ...

d 1, 8, 27, 64, 125, ...

e 0, 7, 26, 63, 124, ..

f 10, 100, 1000, ...

10 Write the nth term in the following sequences.

a $1 \times 2, 2 \times 3, 3 \times 4, \ldots$

b $\frac{2}{3}, \frac{3}{4}, \frac{4}{5}, \frac{5}{6}, \ldots$

c $1 \times 2 \times 5, 2 \times 3 \times 6, 3 \times 4 \times 7, 4 \times 5 \times 8, \ldots$

d 0.1, 0.2, 0.3, 0.4, ...

Explore

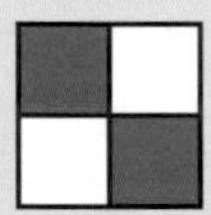
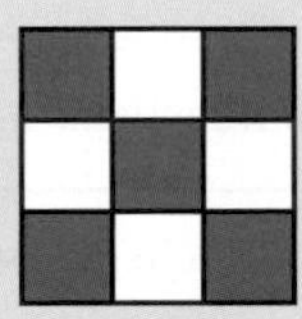
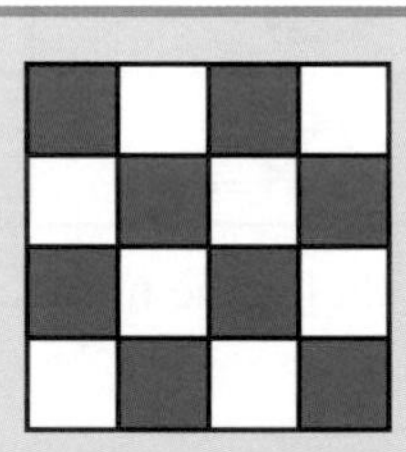

- Write the formula for the number of white tiles in the nth pattern
- Write the formula for the number of red tiles in the nth pattern

Investigate further

Sequences

ASSESS

The following exercise tests your understanding of this chapter, with the questions appearing in order of increasing difficulty.

1 a Draw the next two diagrams in the following sequences.

i

ii

b Write the next two terms in the following sequences.

i Sunday, Tuesday, Thursday, ...

ii A, E, I, M, ...

iii 2, 7, 12, 17, ...

iv 1.2, 1.4, 1.6, 1.8, ...

v 2, 6, 18, 54, ...

vi 3, 30, 300, 3000, ...

2 a Write the 6th and 10th terms in the following sequences.

i 2, 5, 8, 11, ...

ii 1, 6, 11, 16, ...

iii 3, 6, 12, 24, ...

b Write the term-to-term rule for the following sequences.

i 1, 5, 9, 13, ...

ii 2, 10, 50, 250, ...

iii 3, 8, 13, 18, 23, ...

3 a Write the 6th and 10th terms in the following sequences.

i 20, 17, 14, 11, ...

ii 64, 32, 16, ...

iii 1, −2, −5, −8, ...

b Write the term-to-term rule for the following sequences.

i 8, 5, 2, −1, ...

ii −1, −4, −7, −10, ...

iii 6, 3, 1.5, 0.75, 0.375, ...

4 Write the first three terms and the 5th, 20th and 50th terms of the sequences with nth term.

a $2n + 1$ **b** $5n - 2$ **c** $n^2 + 1$.

5 Find the nth term of the following sequences.

a 6, 8, 10, 12, ...
b 3, 13, 23, 33, ...
c 8, 6, 4, 2, ...
d −2, 5, 12, 19, ...
e 4, 7, 12, 19, ...

Try a real past exam question to test your knowledge:

6 The nth term of a sequence is $3n - 1$.

a Write down the first and second terms of the sequence.

b Which term of the sequence is equal to 32?

c Explain why 85 is not a term in this sequence.

Spec A, Int Paper 2, Nov 04

7 Coordinates

OBJECTIVES

G **Examiners would normally expect students who get a G grade to be able to:**

Use coordinates in the first quadrant, such as plot the point (3, 2)

F **Examiners would normally expect students who get an F grade also to be able to:**

Use coordinates in all four quadrants, such as plot the points (3, −2), (−2, 1) and (−4, −3)

E **Examiners would normally expect students who get an E grade also to be able to:**

Draw lines such as $x = 3$, and $y = x + 2$

D **Examiners would normally expect students who get a D grade also to be able to:**

Solve problems involving straight lines

Draw lines, such as $y = 2x + 3$

C **Examiners would normally expect students who get a C grade also to be able to:**

Find the midpoint of a line segment

Use and understand coordinates in three dimensions

What you should already know ...

- Negative numbers and the number line

VOCABULARY

Coordinates – a system used to identify a point; an x-coordinate and a y-coordinate give the horizontal and vertical positions

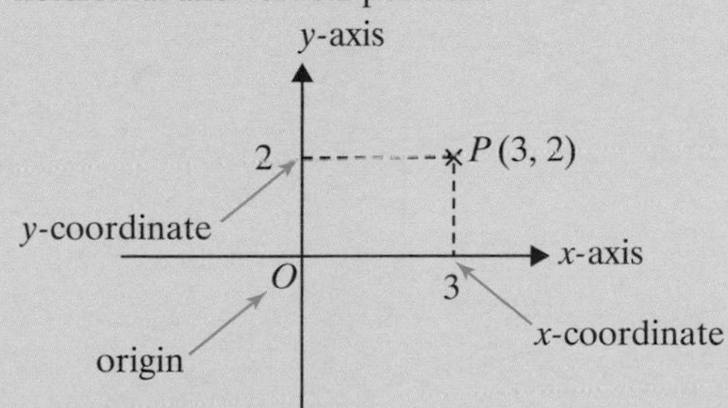

Origin – the point (0, 0) on a coordinate grid

Axis (pl. **axes**) – the lines used to locate a point in the coordinates system; in two dimensions, the x-axis is horizontal, and the y-axis is vertical. This system of Cartesian coordinates was devised by the French mathematician and philosopher René Descartes

In three dimensions, the x- and y-axes are horizontal and at right angles to each other and the z-axis is vertical

Horizontal – from left to right; parallel to the horizon

Horizontal

Vertical – directly up and down; perpendicular to the horizontal

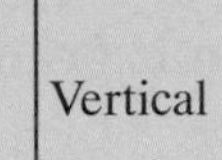

Gradient – a measure of how steep a line is

$$\text{Gradient} = \frac{\text{change in vertical distance}}{\text{change in horizontal distance}} = \frac{y}{x}$$

Intercept – the y-coordinate of the point at which the line crosses the y-axis

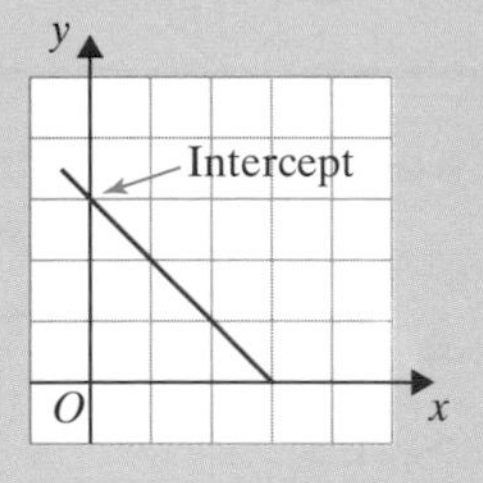

Equidistant – the same distance; if A is equidistant from B and C, then AB and AC are the same length

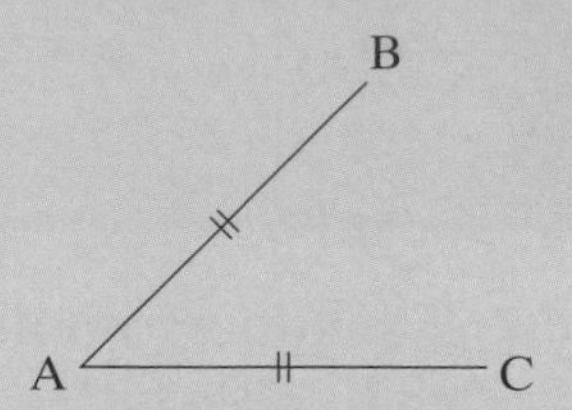

Line segment – the part of a line joining two points

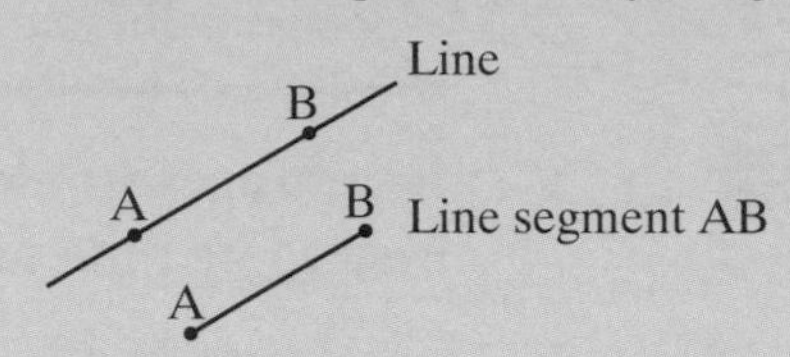

Midpoint – the middle point of a line

Quadrant – one of the four regions formed by the x- and y-axes in the Cartesian coordinate system

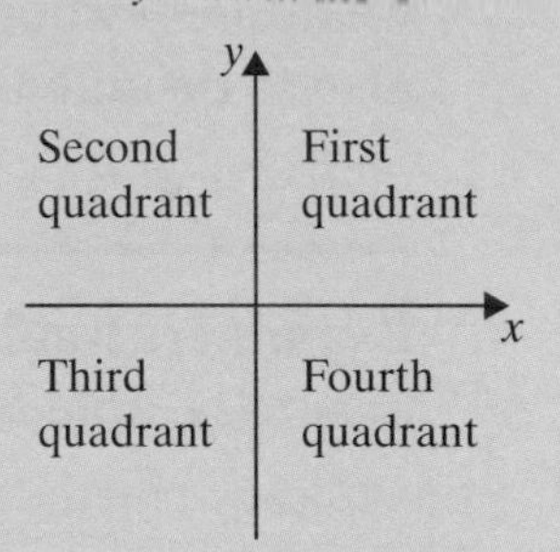

Learn 1 Coordinates in four quadrants

Example: Name the point drawn on the diagram.

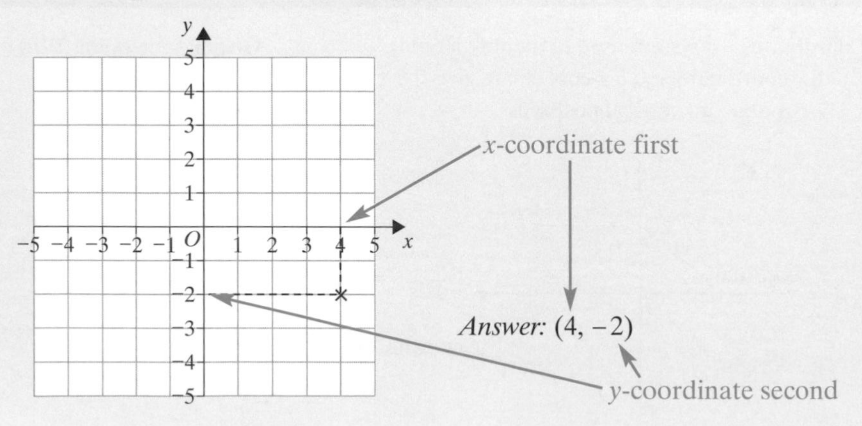

Apply 1

1 Write down the coordinates of points A, B, C, D and E, which are marked on the grid on the right.

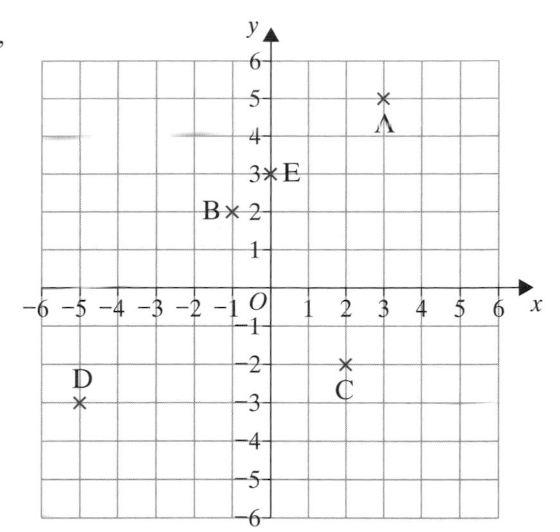

2 Draw a grid like the one in question **1**, with the x-axis and y-axis labelled from −6 to 6.

a On your grid, mark the points A(3, 2), B(4, −1), C(1, −2) and D(0, 1).

b Join A to B, B to C, C to D and D back to A.

c What is the name of the shape you have drawn?

3 A plot of land 10 metres square contains 25 anthills that have to be cleared.

The anthills are located at

(1.5, 0.5)	(4.5, 0)	(3.5, 1.5)	(6, 1.5)	(7.5, 1.5)
(9, 2)	(4.5, 2.5)	(1, 3)	(3.5, 3.5)	(9.5, 4)
(1, 4.5)	(7, 4.5)	(2.5, 6)	(4, 6)	(5, 6)
(6.5, 6)	(1, 6.5)	(10, 7)	(0, 8)	(3, 8)
(5, 8)	(8.5, 8.5)	(2, 9)	(5, 9)	(3.5, 10)

Draw axes from 0 to 10 and mark the positions of the anthills with a 'O'.

Starting at (0, 0), plot a path through the anthills from bottom to top that would keep you at least 1 metre from any anthill.

Write down the coordinates of each point you move to in a straight line from the previous point.

4 **a** Pip says that the two points, A and B, marked on the grid on the right are (1, 1) and (−3, 1). Is she right?

b A and B are two corners of a square. Write down the coordinates of two other points that could be the other two corners of the square. (There are three possible answers to this question. Can you find them all?)

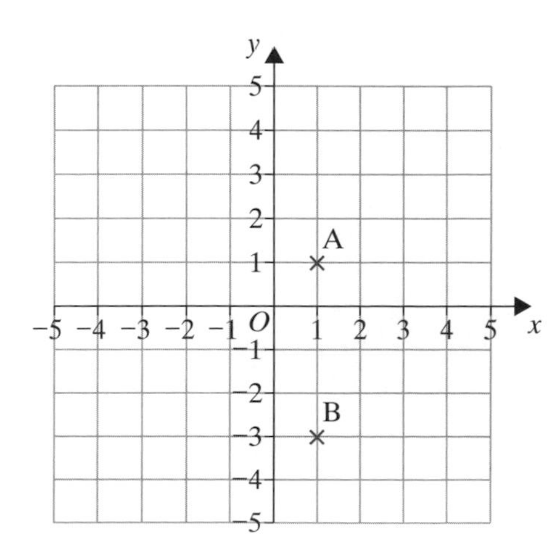

5 Holly says it is further from $(4, -2)$ to $(3, 5)$ than it is from $(4, 2)$ to $(-3, 5)$.
Lisa says it is further from $(4, 2)$ to $(-3, 5)$ than it is from $(4, -2)$ to $(3, 5)$.
Farid says the distances are the same.
Plot the points on a grid and measure the distances to find out who is correct.

6 Write down the coordinates of

a the bottom left-hand corner of the house

b the top right-hand corner of the chimney

c all four corners of the door.

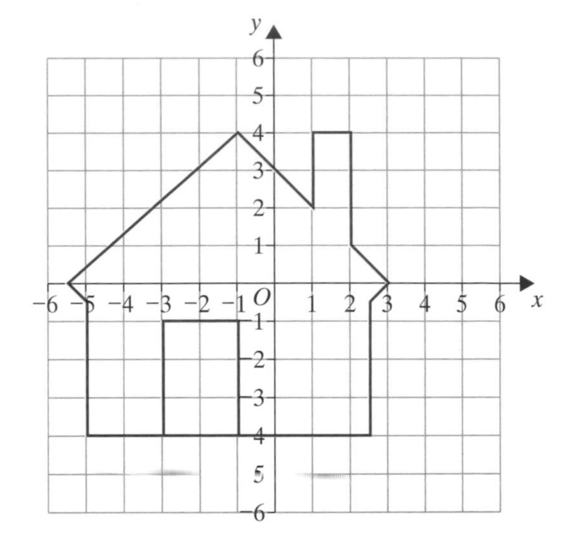

7 **Get Real!**
A boat receives a call for help from a yacht.

a Draw a grid with the x-axis and y-axis labelled from -6 to 6.
Use a scale of 1 cm to represent 1 km.

b The boat is at the point $(0, 0)$. Mark the position of the boat.

c The yacht is at the position $(3, -4)$. Mark the position of the yacht.

d The boat sails to the yacht. What is the bearing and the distance from the boat to the yacht?

Explore

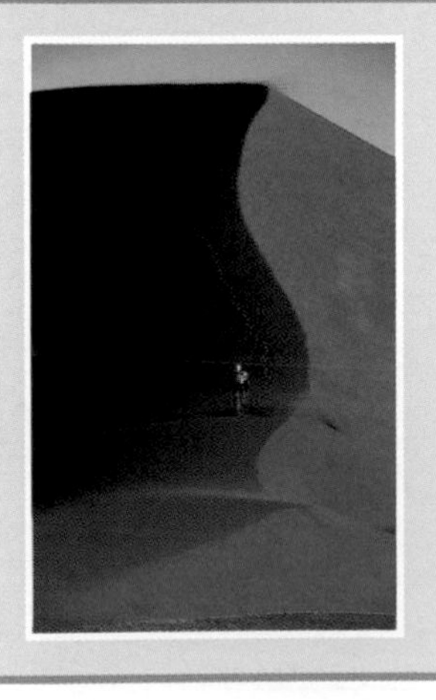

- Copy the grid and mark the points $(1, 3)$, $(1, 1)$, and $(3, 1)$
- You are not allowed to mark a point that would complete the four corners of a square
  For example, you must not mark the point $(3, 3)$
- Mark any other point and write down its coordinates
- Carry on marking points and writing down their coordinates, making sure you don't mark the four corners of a square of any size
- What is the greatest number of points you can mark?

Learn 2 Equations and straight lines

Example: Draw the line that has an equation of $y = 3x - 4$

A straight line is named by an equation which fits all pairs of coordinates on the line. For example, the line $y = 3x - 4$ joins all the points whose y-coordinate is four less than three times the x-coordinate ($3 \times x - 4$).

Choose three x-coordinates, such as (0,), (1,) and (3,).

Calculate the y-coordinates: $x = 0, y = 3 \times 0 - 4 = -4$ The point is (0, −4).
$x = 1, y = 3 \times 1 - 4 = -1$ The point is (1, −1).
$x = 3, y = 3 \times 3 - 4 = 5$ The point is (3, 5).

Plot the points and join them up with a straight line using a ruler.

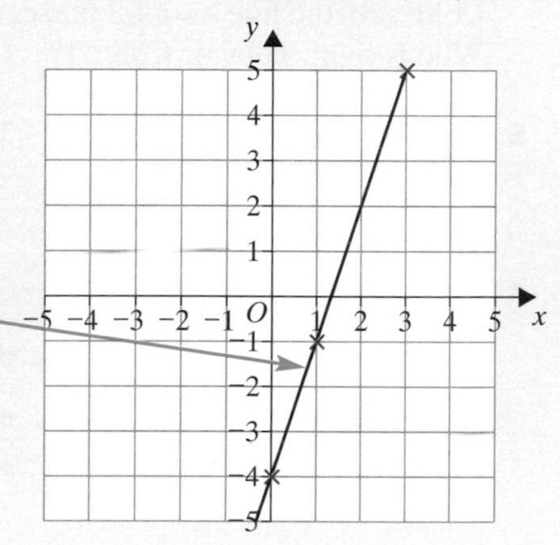

If your three points are not in a straight line, check your working

Apply 2

1 **a** Draw a grid with the x-axis and y-axis labelled from −6 to 6.

b Mark four points with an x-coordinate of 3, for example (3, −2). Draw a straight line through all four points.

c What is the equation of the line you have drawn?

d Now draw four points with a y-coordinate of −2. Draw a straight line through all four points.

e What is the equation of the line you have drawn?

f Where do the two straight lines cross?

2 **a** Draw a grid with the x-axis and y-axis labelled from −6 to 6.
Draw the lines $x = 2$ and $y = 4$.
Write down the coordinates of the points where the lines cross each other.

Now write down where these lines cross each other. You may be able to work them out without drawing the lines.

b $x = 5$ and $y = -4$

c $x = -3$ and $y = -2$

d $x = 0$ and $y = 1$

e $x = 2.5$ and $y = -3.5$

f $x = 2$ and $y = x$

g $x = 4$ and $y = x - 1$

3 The equation $y = x + 2$ means that y is 2 more than x. For example, the point (1, 3) has a y-coordinate which is 2 more than the x-coordinate.

a Write down five more points that fit the equation, that is, write down five more pairs of coordinates where the y-coordinate is two more than the x-coordinate.

b Plot these points on a grid.

c Draw a straight line through the points.

d Repeat steps **a**, **b** and **c** for the equation $y = x - 1$

4 Toby says the line $y = x + 3$ passes through the point (4, 1).
Colin says the line $y = x + 3$ passes through the point (1, 4).
Who is right, Toby or Colin?

5 Javindra says that the line $y = x + 3$ is the same line as $x = y - 3$.
Is she right?
Give a reason for your answer.

6 Where do these pairs of lines cross?

a $x = 4$ and $y = x + 2$

b $x = -2$ and $y = x - 1$

c $x = 3$ and $y = 2x$

d $x = -2$ and $y = 3x - 1$

e $y = 4$ and $y = x + 3$

f $y = -2$ and $y = 3x + 1$

7 For each grid below

i write down the coordinates of three points on the line

ii use your answers to part **i** to help you write down the equation of the line.

a

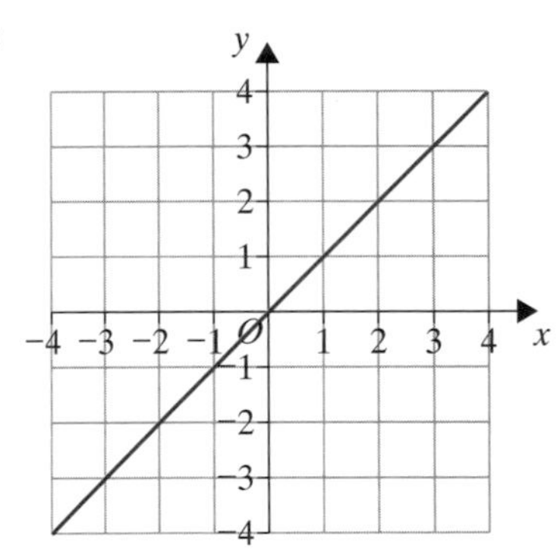

c

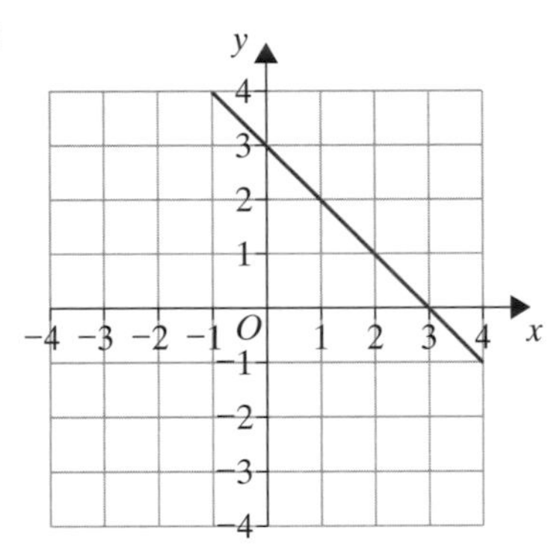

b

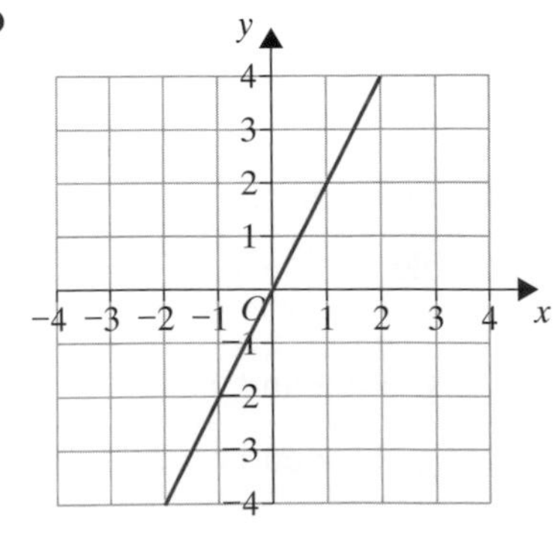

8 For each grid below

i write down the coordinates of three points on the line

ii use your answers to part **i** to help you write down the equation of the line.

a

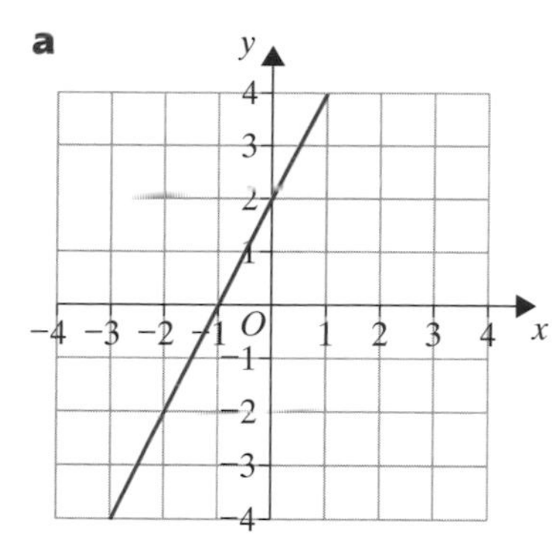

b

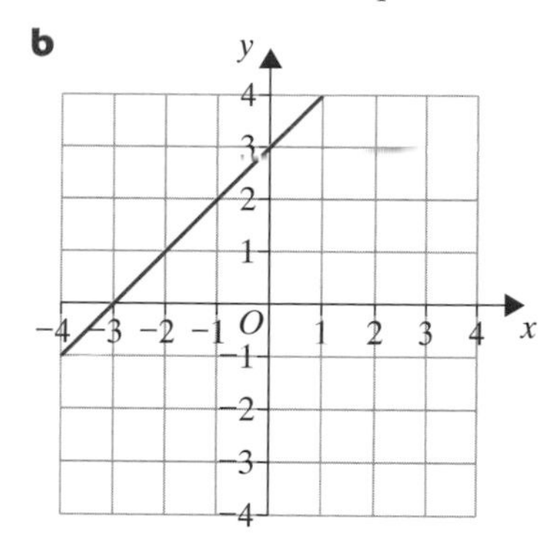

c

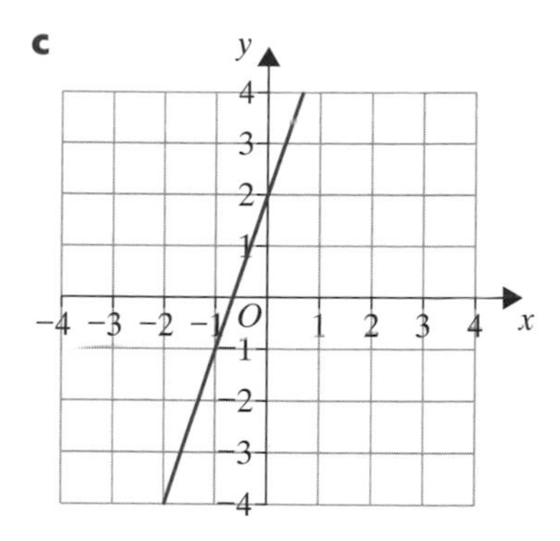

9 **Get Real!**

A farmer owns a field, and wants to put a fence around it.
He describes the shape as follows:

'It's a four-sided field, but it isn't a rectangle.
You can draw a plan of it like this:
Draw a coordinate grid with the x-axis and y-axis labelled from -5 to $+5$.
Draw in the four lines $x = 4$, $y = 5$, $y = 2x + 3$, and $x + y = -2$.'

Draw a plan of the farmer's field.

Explore

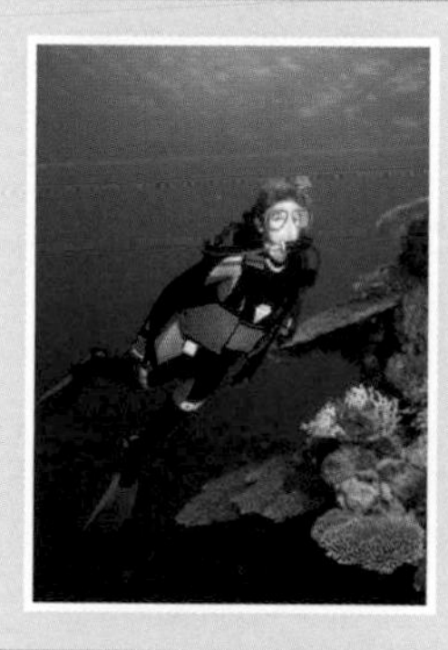

- Draw a coordinate grid with the x-axis and y-axis labelled from -6 to 6
- Draw a flag by joining (1, 3) to (1, 5) to (2, 4) to (1, 4)
- Double the x-coordinates and draw it again; (2, 3) to (2, 5) to (4, 4) to (2, 4)
- What happens to the flag?
- Try other manipulations, for example swap the coordinates over so (1, 3) becomes (3, 1), or subtract the y-coordinates from 8, or change the signs of the x-coordinate.

Investigate further

Explore

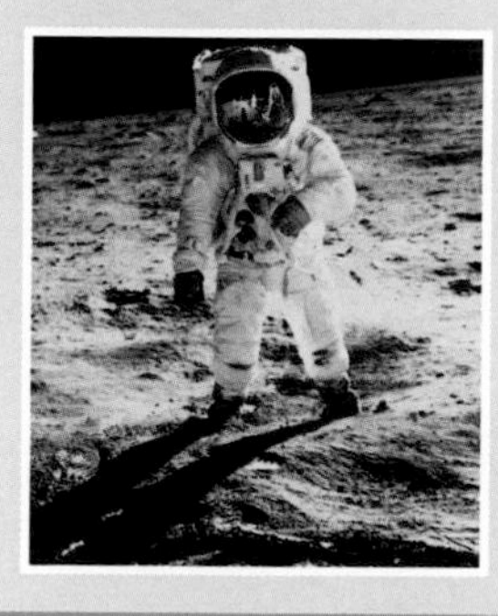

- You *may* want to use graph-plotting software for this
- Draw lines of equations such as $y = 2x + 1$
- What happens if you change the +1?
- What happens if you change the 2?

Investigate further

Learn 3 The midpoint of a line segment

The midpoint of the line segment from (a, b) to (c, d) is $\left(\frac{a+c}{2}, \frac{b+d}{2}\right)$.

The mean of the x-coordinates — The mean of the y-coordinates

Example:

Write down the coordinates of the point halfway between A and B.

A is the point $(-4, 3)$; B is $(1, 2)$.

The midpoint is at

$$\left(\frac{-4+1}{2}, \frac{3+2}{2}\right)$$

$$= \left(\frac{-3}{2}, \frac{5}{2}\right)$$

$$= (-1.5, 2.5)$$

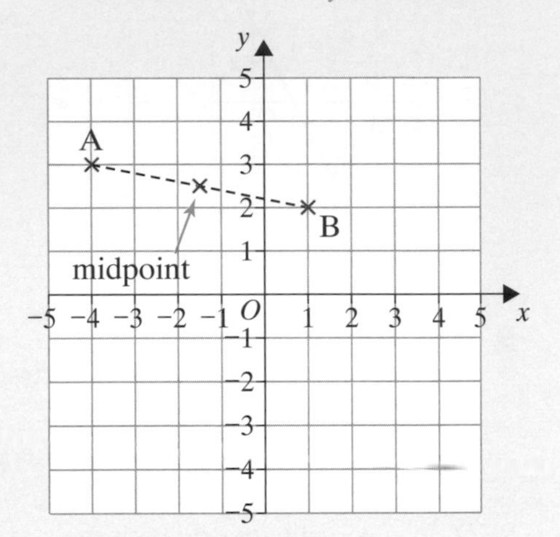

Apply 3

1 **a** Write down the coordinates of the point halfway between $(1, 6)$ and $(7, 2)$.

b Draw a grid with the x-axis and y-axis labelled from 0 to 8. Plot the points $(1, 6)$ and $(7, 2)$ and use your diagram to check your answer to part **a**.

2 A triangle ABC has vertices at A$(8, -4)$, B$(2, 6)$ and C$(-4, 2)$.

a Find the coordinates of X, the midpoint of AB.

b Find the coordinates of Y, the midpoint of AC.

c Draw a grid with the x-axis and y-axis labelled from -4 to 8. Plot the points A, B, C, X and Y.

d Draw the lines XY and BC. What do you notice about them?

3 Work out the coordinates of the point halfway between $(4, -5)$ and $(1, 3)$.

4 If A is the point $(5, -1)$ and B is the point $(-7, -4)$, what are the coordinates of the midpoint of the line AB?

5 Liam says that the point $(2, 1.5)$ is halfway between $(-4, 2)$ and $(8, -5)$.
Is he correct?
Give a reason for your answer.

6 C is the mid-point of the line AB.
The coordinates of C are (4, –1). B is the point (2, 5).
What are the coordinates of A?

7 A quadrilateral ABCD has coordinates as follows:
A(4, –2); B(7, –4); C(–2, 4); D(–5, 6).

a Find the midpoint of the diagonal AC.

b Find the midpoint of the diagonal BD.

c What can you say about the quadrilateral ABCD?

8 a (6, –8) is the midpoint of a line segment AB. A is the point (3, –6).
Find the coordinates of B.

b Write five other sets of coordinates that have a midpoint at (6, –8) on line segment AB.

9 Get Real!

A computer programmer uses coordinates to plot points in the screen when designing a computer game.

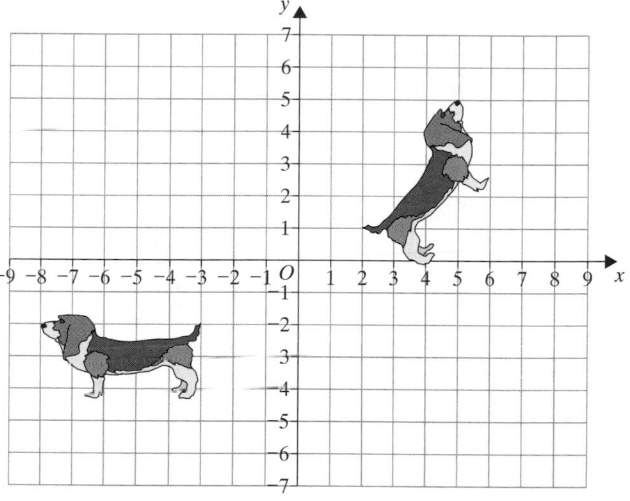

She draws a picture of a dog, with its nose at (–8, –2) and its tail at (–3, –2).
She draws a reflection of the dog with its nose at (5, 5) and its tail at (2, 1).
She wants to draw the mirror for the reflection. She knows the mirror must go halfway between the dog and its reflection.

a What are the coordinates of the point halfway between the dog's nose and its reflection?

b What are the coordinates of the point halfway between the dog's tail and its reflection?

10 Lori draws a circle with its centre at (–3, –2). She draws a diameter from the point (2, –8). Find the coordinates of the other end of the diameter.

11 Find your way through the maze by only occupying squares where M is the midpoint of AB. (No diagonal moves allowed)

1 **START**	2 A (3, 2) B (3, 7) M (6, 4.5)	3 A (−2, 11) B (4, −4) M (−1, 3.5)	4 A (6, −9) B (3, 9) M (4.5, 9)	5 A (−11, 29) B (−15, 22) M (−13, 25.5)	6 **END**
7 A (4, 5) B (2, 9) M (3, 7)	8 A (1, 2) B (2, 3) M (3, 5)	9 A (4, 2) B (−4, −2) M (0, 2)	10 A (−1.4, 0.2) B (1.3, 0.6) M (−0.05, 0.4)	11 A (2.3, −0.8) B (−1.3, 0.3) M (0.5, −0.25)	12 A (3, 1) B (4, 2) M (5, 3)
13 A (4, −4) B (−2, 6) M (1, 1)	14 A (8, −2) B (3, −4) M (5.5, −3)	15 A (−13, 24) B (−12, 9) M (−12.5, 17)	16 A (0, 7) B (−7, 0) M (−3.5, 3.5)	17 A (−1, 9) B (9, −1) M (4, −4)	18 A (−12, 19) B (−12, 2) M (0, 10.5)
19 A (0, 0) B (2, 3) M (2, 3)	20 A (−4, −4) B (−2, 7) M (−3, 1.5)	21 A (4, −11) B (−4, −11) M (0, 0)	22 A (−8.5, 3.4) B (4.1, 2.2) M (−2.2, 2.8)	23 A (23, −13) B (13, −12) M (18, −12.5)	24 A (4.2, −3.7) B (−2.4, 1.3) M (0.9, −1.2)
25 A (4, −12) B (−11, 8) M (−4.5, −2)	26 A (−4, 7) B (2, 5) M (−1, 6)	27 A (−3, 5) B (2, −6) M (−0.5, −0.5)	28 A (0, 9.2) B (−2.4, 3.2) M (−1.2, 6.4)	29 A (1, 5) B (3, 1) M (4, 3)	30 A (23, 17) B (−14, 11) M (4.5, 14)
31 A (4.2, 3) B (1.9, 8) M (3.1, 5.5)	32 A (−9, −8) B (−3, 9) M (6, 0.5)	33 A (−1, 4) B (−3, −4) M (−2, 0)	34 A (−3, 4) B (−2, −3) M (−2.5, 0.5)	35 A (4.2, −3.3) B (2.4, 1.3) M (3.3, −1)	36 A (−3, 2) B (2, −3) M (−0.5, −0.5)

Explore

- Draw a coordinate grid, labelled from −8 to 8 on both axes
- Plot the points A(6, 2), B(−4, 2), C(−4, −8) and D(6, −8)
- Join A to B, B to C, C to D and D to A
- Write down the name of the shape you have drawn
- Work out the midpoint of each side of the shape, and mark them on the grid
- Join these midpoints up.
  What shape have you drawn?

Investigate further

Learn 4 Coordinates in three dimensions

Remember to write coordinates alphabetically: (x, y, z)

Example: In the diagram of the cuboid
$OA = 2$ units, AB = 5 units and AD = 3 units.
O is the origin.
Write down the coordinates of A, B, C and D.

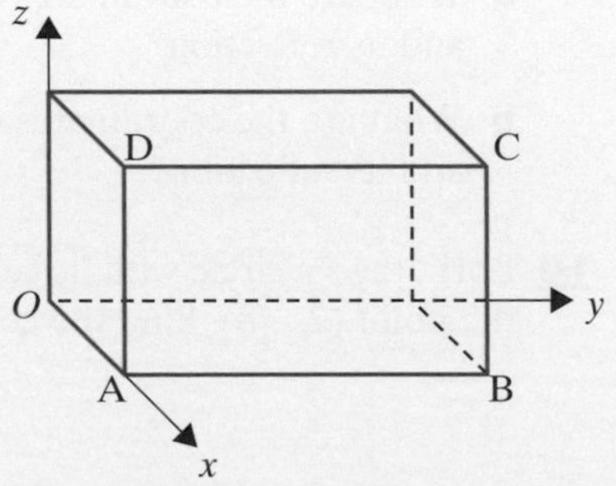

A = (2, 0, 0) C = (2, 5, 3)
B = (2, 5, 0) D = (2, 0, 3)

x-coordinate first, then y, then z

Draw and label diagrams to make the work easier

Apply 4

Questions 1 to 2 are about the diagram on the right.

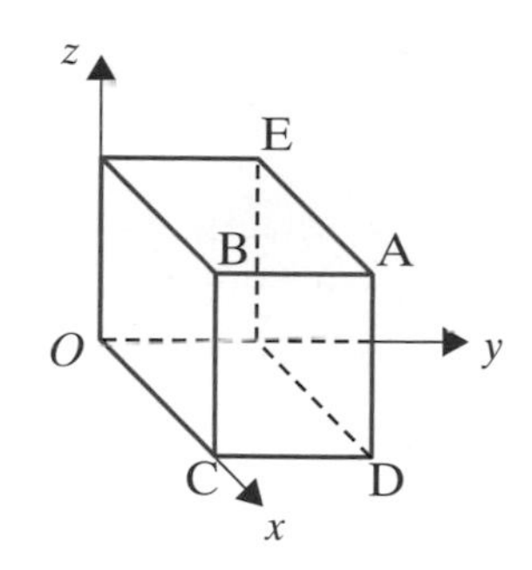

1 In the diagram, A is the point (6, 4, 5).
Write down the coordinates of points B, C, D and E.

2 An identical cuboid is placed on top of the one in the diagram.
Write down the coordinates of the top corner directly above A.

3 What is the midpoint of the line joining (6, 3, 7) to (2, 5, 9)?

4 What is the midpoint of the line joining (−2, 4, −5) to (6, 2, −6)?

5 Alex is designing a toy street. A plan of it looks like this:

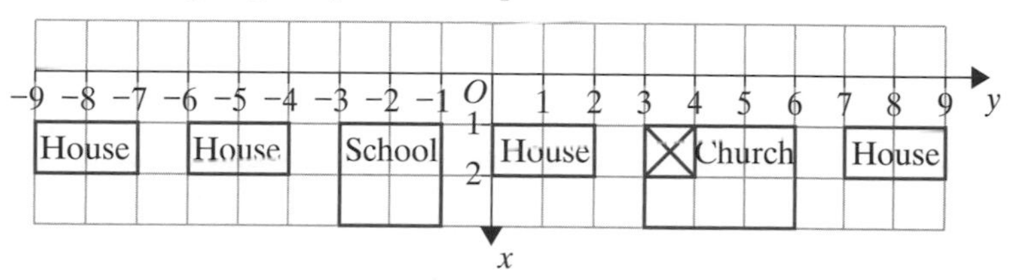

The front elevation looks like this:

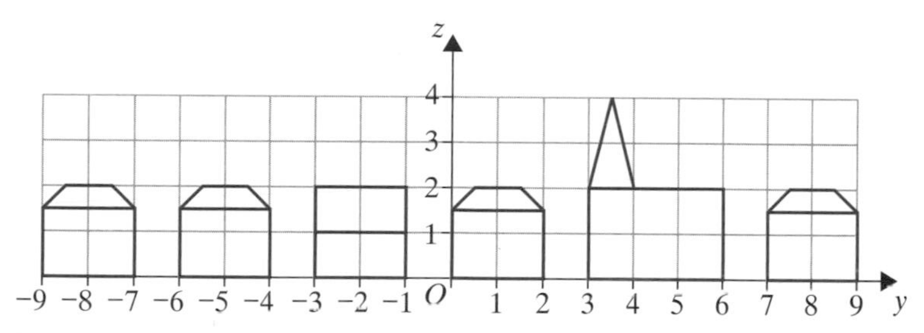

Write down the coordinates of the top of the church spire.

6 An electricity company plans where to put its electricity pylons on a map.
The results are shown below.

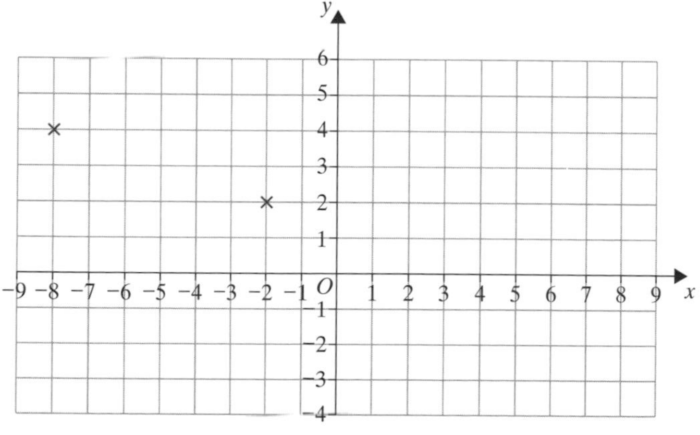

They plan to put the first one at (−8, 4), and the second at (−2, 2).

a If they space them equally and in a straight line, where will they place the third?

b The top of the first one is (−8, 4, 1). Assuming they are all the same height and on level ground, what are the three-dimensional coordinates of the top of the fourth pylon?

Explore

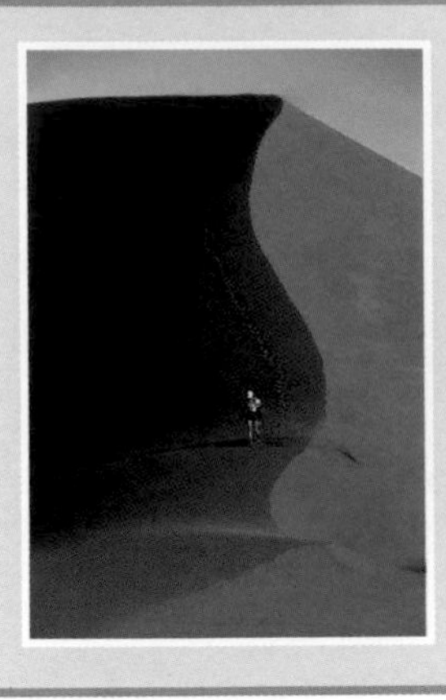

- A $2 \times 2 \times 2$ cube is placed with a corner at O, the origin, and other corners at A(2, 0, 0), B(0, 2, 0), C(2, 2, 0), D(0, 0, 2), E(0, 2, 2), F(2, 0, 2) and G(2, 2, 2)
- How many right angles can you find by joining up corners of the cube?
- How many angles of 45° can you find?
- How many angles of 60° can you find?

Investigate further

Coordinates

ASSESS

The following exercise tests your understanding of this chapter, with the questions appearing in order of increasing difficulty.

1 David has copied part of a map onto the square grid shown below.

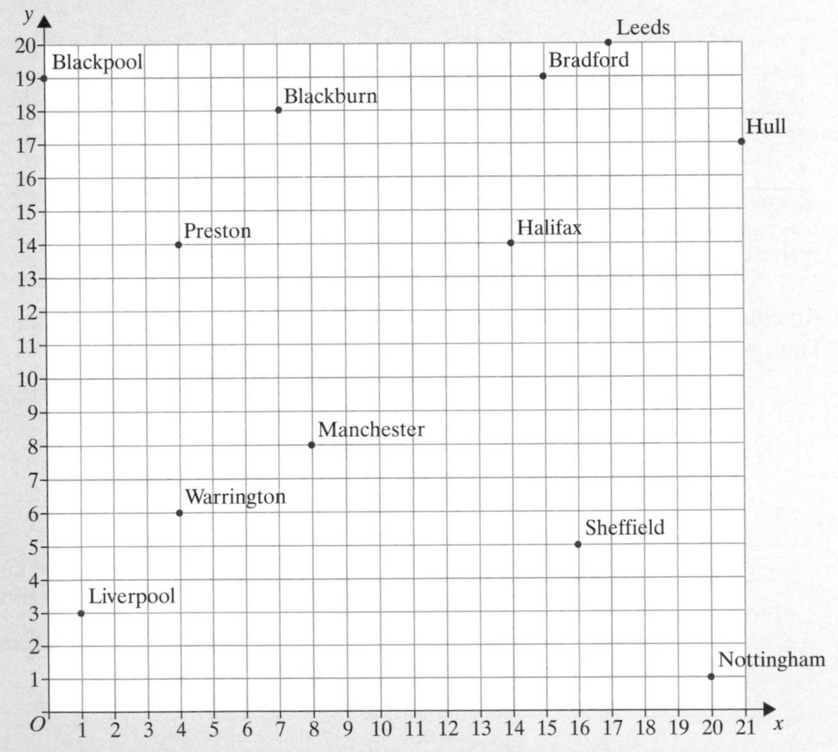

a Write down the coordinates of the 12 places shown.

b The following places also have coordinates on the grid. Copy the grid and mark in the places on it.

Chester (1, 0)
Southport (0, 12)
Clitheroe (8, 20)
Macclesfield (9, 1)
Burnley (11, 19)
Scunthorpe (20, 11)
Mansfield (17, 3)
Barnsley (16, 9)

2 Draw axes on a piece of graph paper. Take values of x from -9 to $+8$ and values of y from -8 to $+10$. Plot the following sets of points on the same grid and join the points together, resulting in a well known shape.

A: $(-9, -3)$ $(-6, -8)$ $(8, -8)$ $(8, -2)$ $(7, -3)$ $(-9, -3)$
B: $(-8, -2)$ $(7, -2)$ $(1, 10)$ $(1, -2)$
C: $(1, 7)$ $(-6, 10)$ $(-8, -2)$ $(1, -2)$

3 For each grid write the equation of the straight line.

a

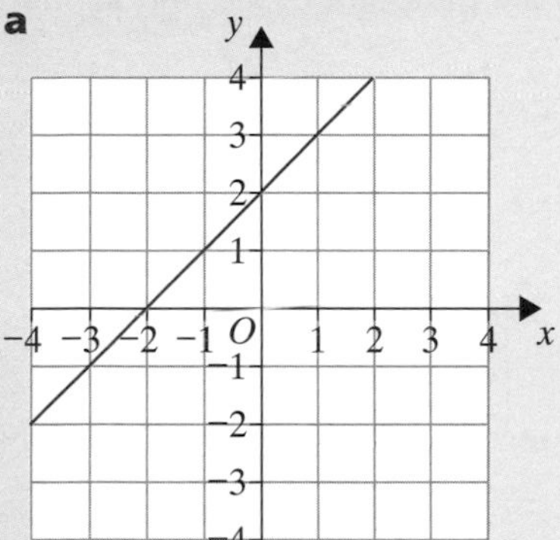

b

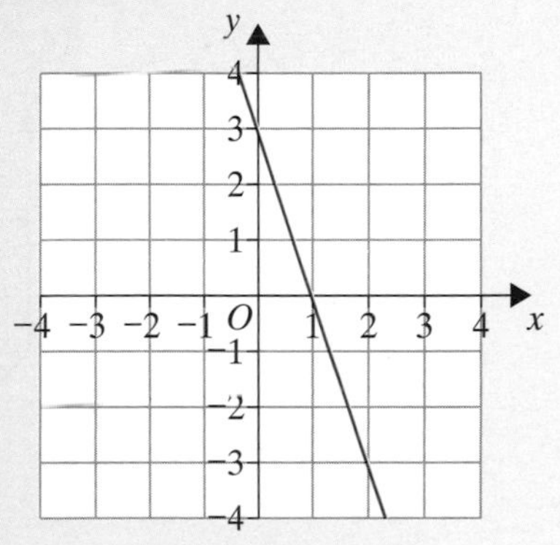

c

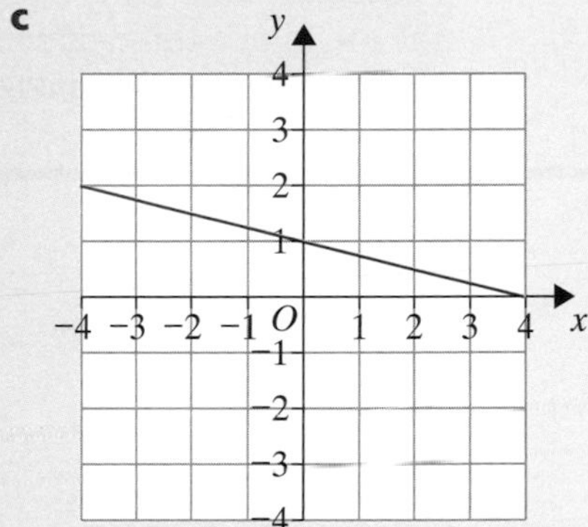

4 Draw a coordinate grid on x- and y-axes labelled from -6 to $+6$.
Draw and label these lines:

a $y = 3x + 2$

b $y = 5 - 3x$

c $y + 2x = 4$

5 Find the midpoints of the following line segments.
Draw sketch diagrams to show your answers.

a $(2, 3)$ and $(6, 7)$

b $(9, -9)$ and $(3, 4)$

c $(-4, 4)$ and $(6, 10)$

d $(2, 9)$ and $(-2, -9)$

e $(-4, 10)$ and $(6, -8)$

6 a On graph paper, plot the coordinates of the quadrilateral ABCD, given by the points $(-6, -7)$, $(-4, 8)$, $(2, 3)$ and $(5, -5)$.

b Calculate the coordinates of P, Q, R and S, the midpoints of AB, BC, CD and DA.

c Plot P, Q, R, S and draw the quadrilateral PQRS.

d What do you notice about the quadrilateral PQRS?

7 a Write down the midpoint of the line segment joining $(3, -1, 2)$ and $(4, 5, -5)$.

b

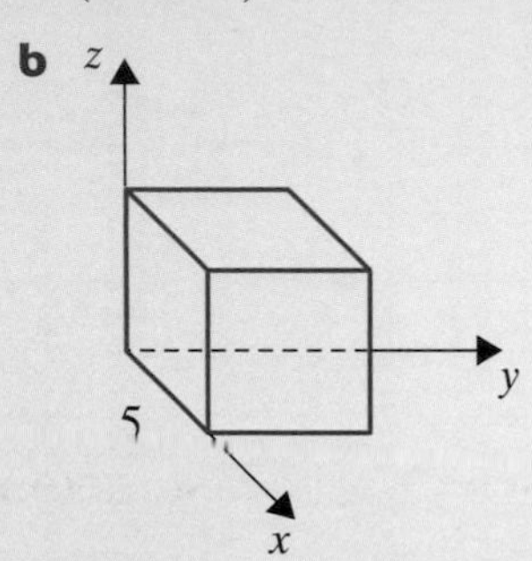

A cube of side 5 units is placed with one corner of its base at the origin of a three-dimensional grid, as shown.
Write down the coordinates of the other seven vertices.

8 Area and volume

OBJECTIVES

G **Examiners would normally expect students who get a G grade to be able to:**

Find the volume of a solid by counting cubes and stating units

E **Examiners would normally expect students who get an E grade also to be able to:**

Find the volume of a cube or cuboid

Find the height of a cuboid, given volume, length and breadth

D **Examiners would normally expect students who get a D grade also to be able to:**

Change between area measures, such as m^2 to cm^2

C **Examiners would normally expect students who get a C grade also to be able to:**

Calculate the surface areas of prisms

Calculate volumes of prisms

Change between volume measures, such as m^3 to cm^3 or cm^3 to litres

What you should already know ...

- Find the areas of rectangles, triangles and parallelograms
- Draw nets of solids
- Multiply and divide two-digit numbers
- Multiply and divide by powers of 10

VOCABULARY

Solid – a three-dimensional shape

Face – one of the flat surfaces of a solid

Cube – a solid with six identical square faces

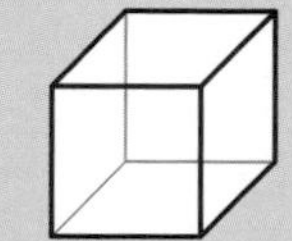

Cuboid – a solid with six rectangular faces (two or four of the faces can be squares)

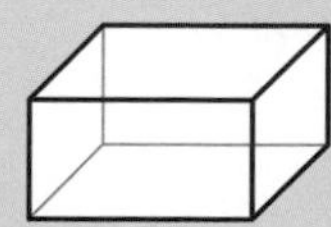

Vertex (pl. **vertices**) – the point where two or more edges meet

Edge – a line segment that joins two vertices of a solid

Volume – a measure of how much space fills a solid, commonly measured in cubic centimetres (cm^3) or cubic metres (m^3)

Cubic centimetre (cm^3) and **cubic metre (m^3)** – commonly used units of measurement for volume; $1\ cm^3 = 1000\ mm^3$, $1\ m^3 = 1\ 000\ 000\ cm^3$

Capacity – the amount of liquid a hollow container can hold, commonly measured in litres (1 litre $= 1000\ cm^3$)

Cross-section – a cut at right angles to a face and usually at right angles to the length of a prism

Prism – a three-dimensional solid with two cross-sectional faces that are identical polygons, parallel to each other; all other faces are either parallelograms or rectangles

Prisms are named according to the cross-sectional face; for example,

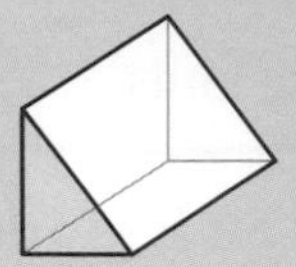

Triangular prism

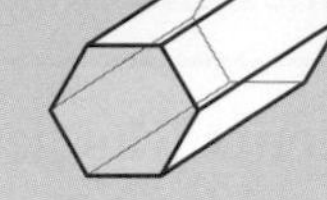

Hexagonal prism

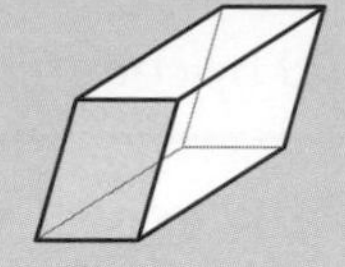

Parallelogram prism

The volume of a prism can be found by multiplying the area of the cross-section by the length

Cylinder – a prism with a circle as a cross-sectional face

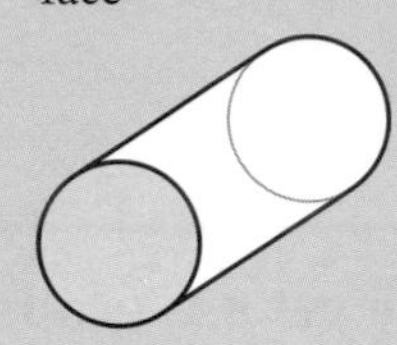

Net – a two-dimensional shape made of polygons that can be folded to make a three-dimensional solid, for example,

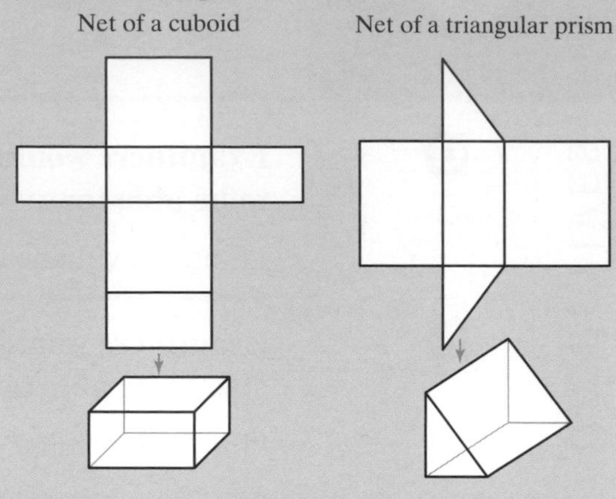

Surface area – the sum of all the areas of the faces of a solid

Square centimetre (cm^2) and square metre (m^2) – commonly used units of measurement for area; $10\,000\text{ cm}^2 = 1\text{ m}^2$ (1 square metre)

Learn 1 Volumes of cubes and cuboids

Examples:

a This cuboid is made from 1 cm cubes. Find its volume.

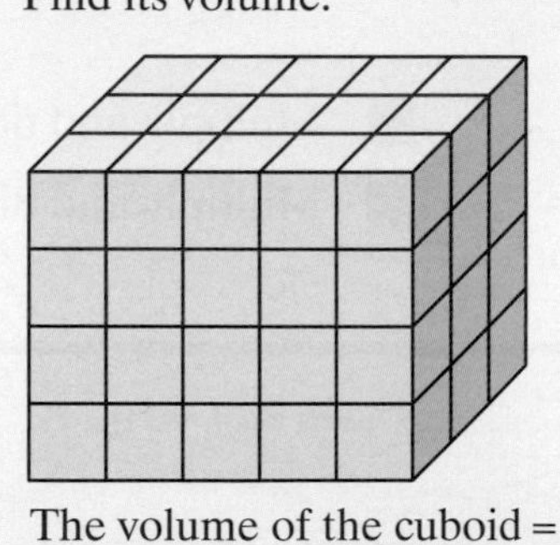

The volume of the cuboid = $5 \times 3 \times 4 = 60\text{ cm}^3$

This is the number of cubes on the bottom layer

The height tells you how many layers

The units of volume are cubed

b Calculate the volume of this cuboid.

Not drawn accurately

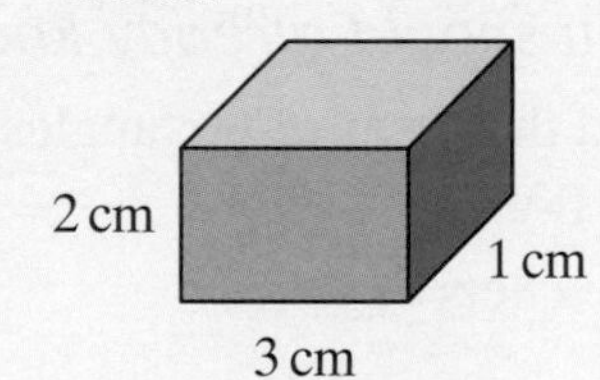

Volume = $3 \times 2 \times 1$
$= 6\text{ cm}^3$

Remember:
Volume of a cuboid = length × width × height

Apply 1

1 Ahmed makes his initials using multilink cubes.

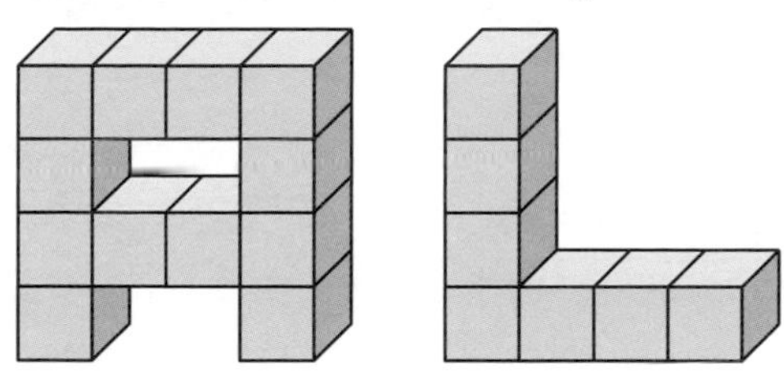

a What is the volume of Ahmed's initials?

b Make your initials from multilink cubes. What is the volume of your initials?

2 Each of the cuboids shown below is made from 1 cm cubes.

i 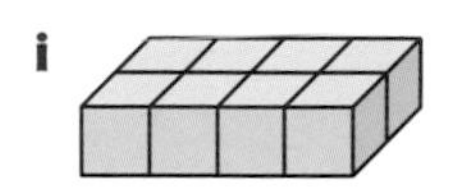 **ii** 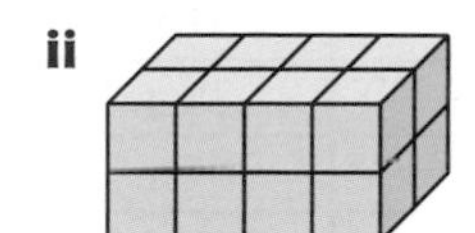 **iii** 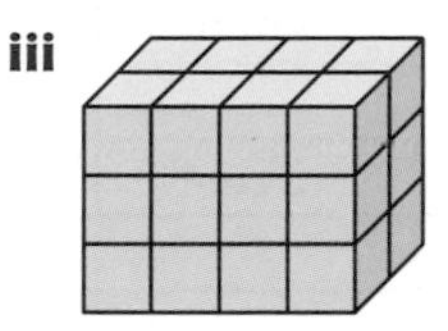 **iv** 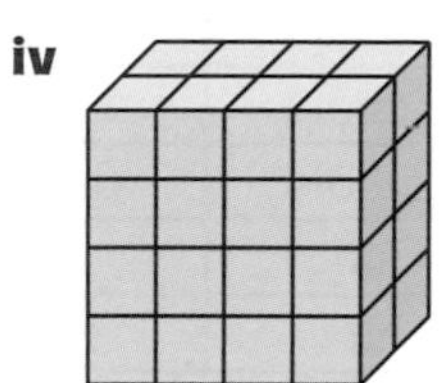

a Copy and complete the table by counting cubes:

	Length (cm)	Width (cm)	Height (cm)	Volume (cm^3)
i	4	2	1	
ii	4		2	
iii			3	
iv		2		

b Write down the rule for finding the volume of a cuboid using the length, width and height. Use the rule to check your answers to part **a**.

3 How many different cuboids can you make using twenty-four 1 cm cubes? Write down the dimensions of your cuboids.

4 Find the volume of each of the following solids. State the units in your answer.

a

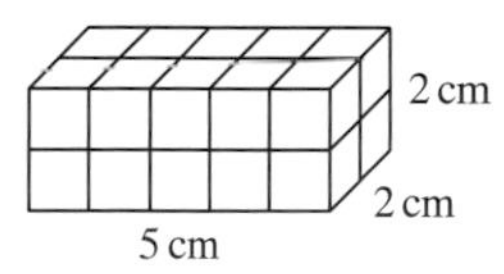

b

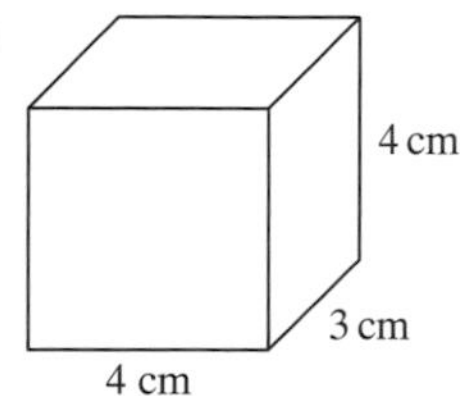

c

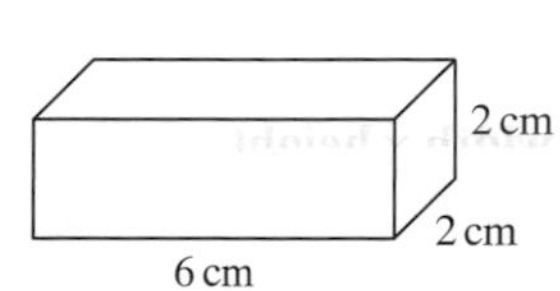

d

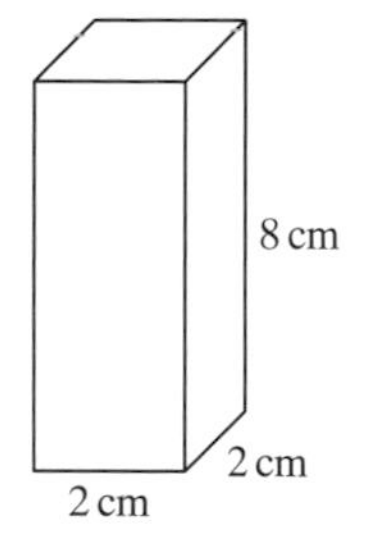

e

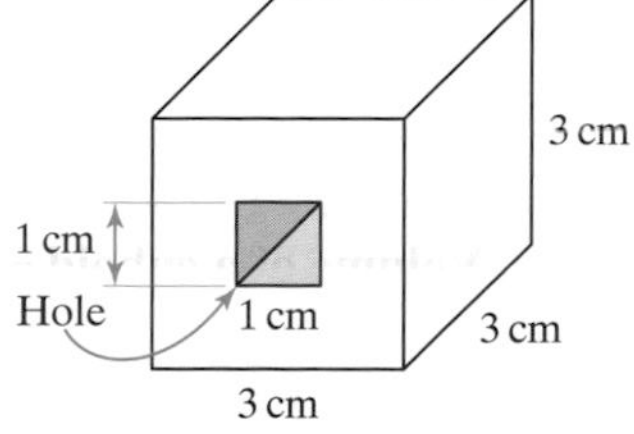

f

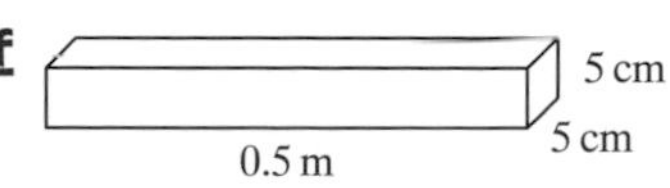

g

50 cm
2 m
2 m

h

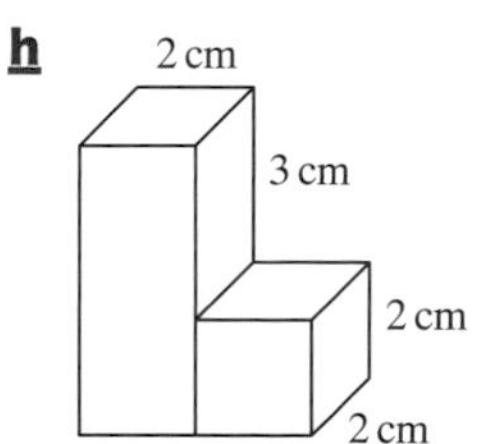

i

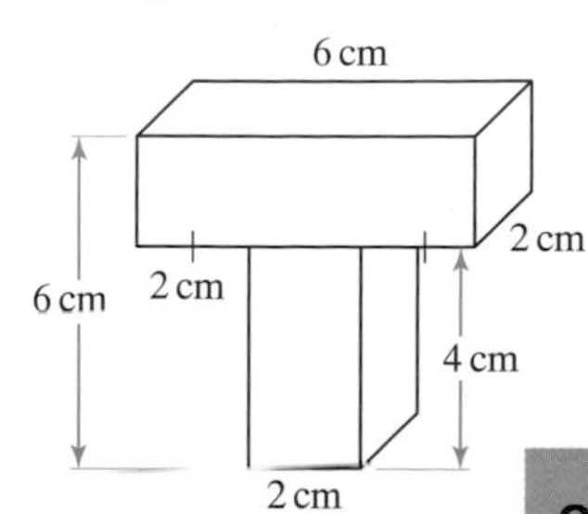

5 Bill thinks the volume of a cuboid measuring 3 cm × 3 cm × 4 cm is 36 cm.
Is he correct? Give a reason for your answer.

6 Get Real!
Anna is making ice cubes for the drinks for her party. If the dimensions of the holes in the ice cube tray are all 2 cm, how many ice cubes can Anna make using 1 litre of water? (1000 cm^3 = 1 litre)

7 The volume of a cuboid is 60 cm^3. Find the dimensions of five different cuboids with this volume.

8 Copy this table and fill in the gaps.

	Solid	Length	Width	Height	Volume
a	Cuboid	4 cm	2 cm	10 cm	
b	Cuboid	5 cm		3 cm	45 cm^3
c	Cuboid	10 mm	4 mm		120 mm^3
d	Cuboid	2 m		4 m	28 m^3
e	Cuboid	0.5 cm	2 cm		6 cm^3
f	Cube				27 cm^3
g	Cube	5 cm			

9 Get Real!
Steve has made a swing set for his children in the garden.
The swing set has 4 legs.
He must cement the legs of the frame into the ground.
Each hole must measure 30 cm × 30 cm × 30 cm.
How many litres of cement does Steve need?
(Remember: 1000 cm^3 = 1 litre)

10 a Calculate the volumes of these cubes.

i

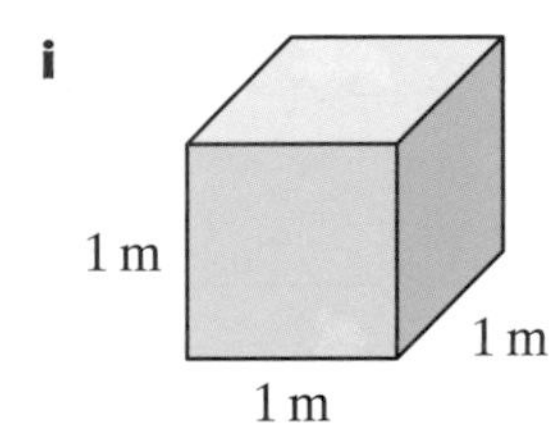

ii

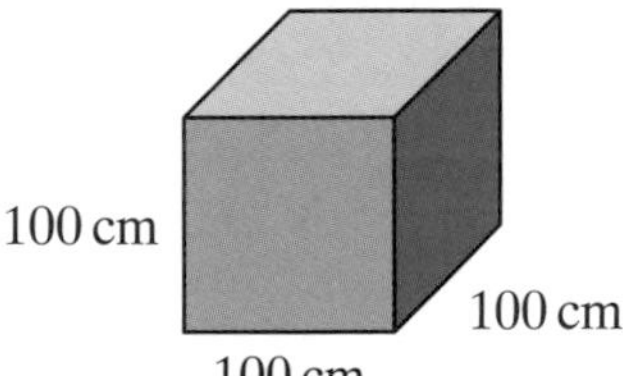

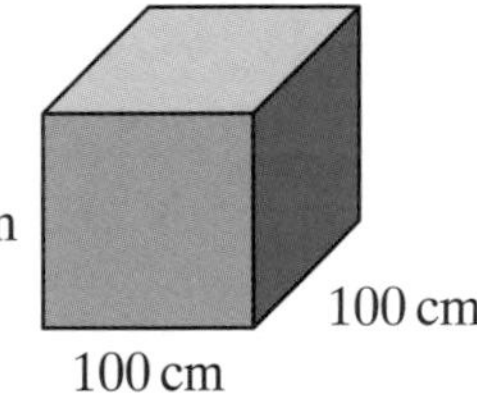

b What do you notice about the two cubes?

c Look at your answers to part **a** and complete the statement:

$1\ m^3 = \ldots\ cm^3$.

11 A fish tank at the local aquarium measures 2 m × 4.5 m × 4 m.

a What is the volume of the fish tank in m^3?

b Convert the answer to litres.

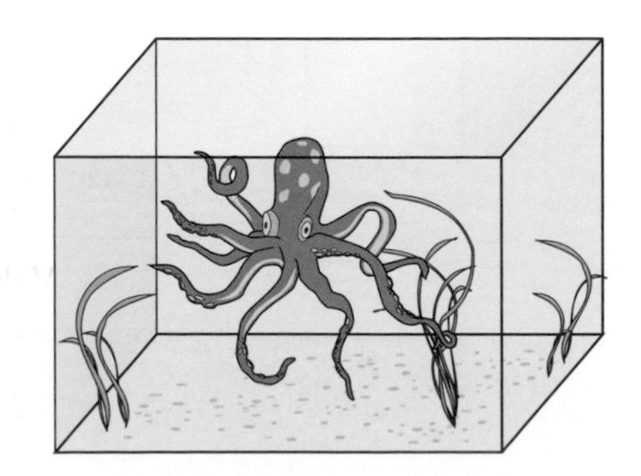

Explore

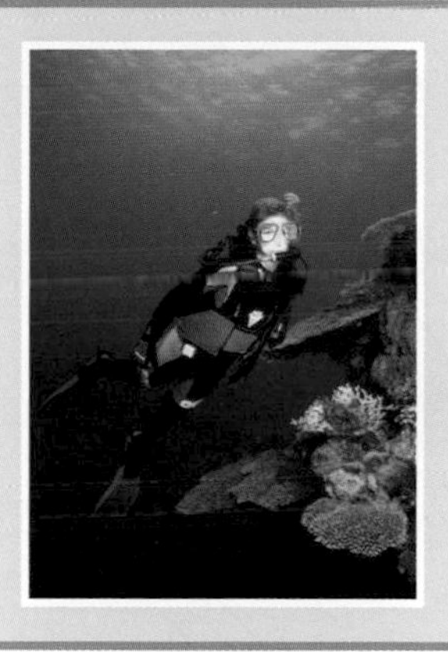

- The diagram shows part of one wall of the Great Pyramid in Egypt
- The Great Pyramid was constructed with approximately 2.3 million blocks of limestone
- Using the person as a guide and assuming the blocks are cubes, find the volume of one block
- Estimate the volume of the Great Pyramid
- Did the Egyptians have cranes to lift these blocks?

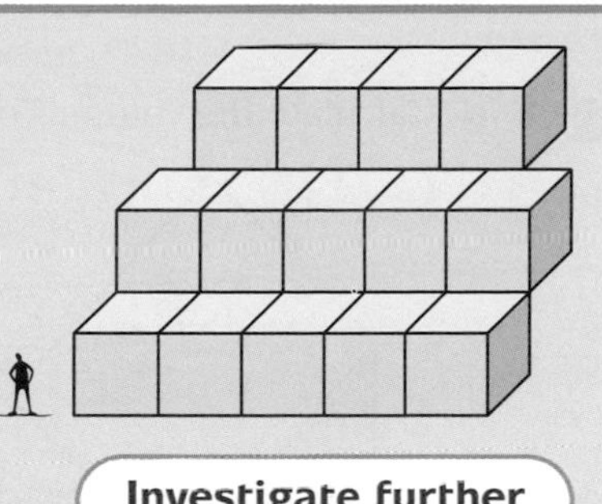

Investigate further

Explore

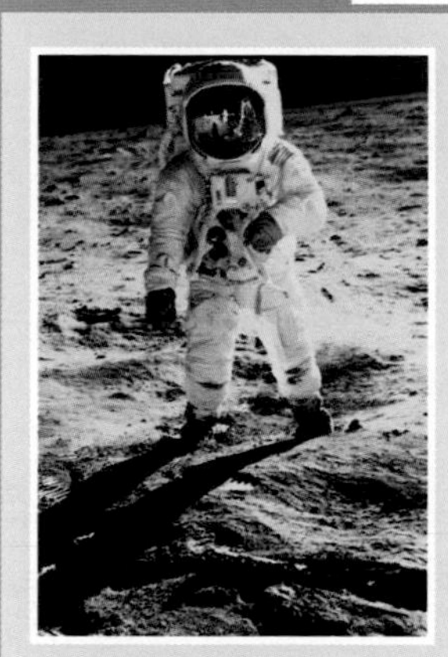

Alan has tried to make a cuboid but unfortunately all the pieces have fallen over and got muddled up.

A 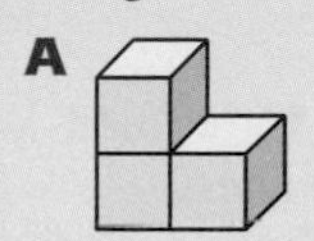 **B**  **C** 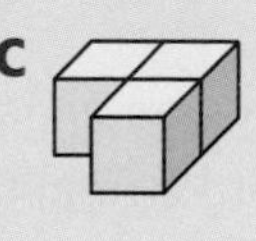 **D** 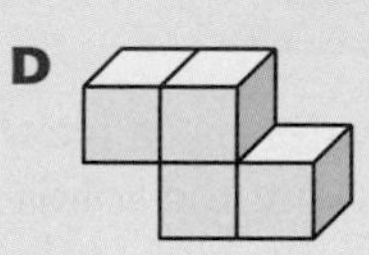 **E** 

Each of the five pieces is made from multilink cubes.

- Write down the volume of each piece
- What is the total volume of all five pieces?
- Use multilink cubes to make the pieces
- Try and fit them together to make Alan's cuboid
- What are the dimensions of his cuboid?
- What is the volume of his cuboid?
- Is it the same as you got for all five pieces?

HINT You are trying to make a cuboid – use your last answer to help you!

Investigate further

Learn 2 Volumes of prisms

Examples:

Calculate the volumes of these prisms.

a Trapezium prism **b** Triangular prism

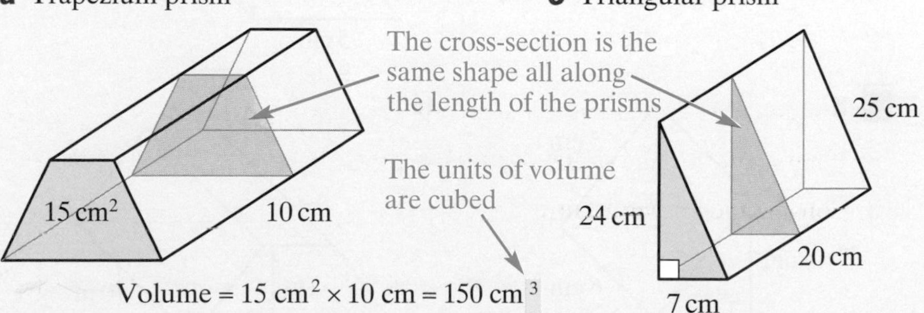

Volume = $15 \text{ cm}^2 \times 10 \text{ cm} = 150 \text{ cm}^3$

Volume of a prism = area of cross-section × length

Area of cross-section = $\frac{1}{2} \times 7 \times 24 = 84 \text{ cm}^2$

Volume = $84 \times 20 = 1680 \text{ cm}^3$

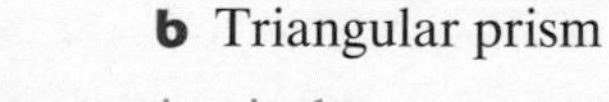

Apply 2

1 **a** Calculate the volumes of the following prisms:

i 15 cm² 10 cm

iii 25 cm² 5 cm

v 50 cm² 6 cm

ii

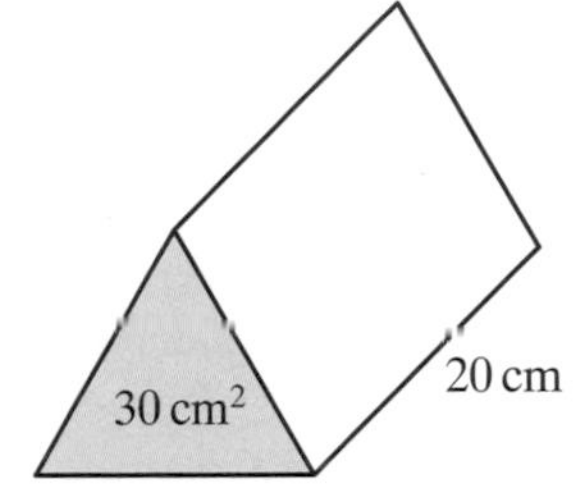

iv

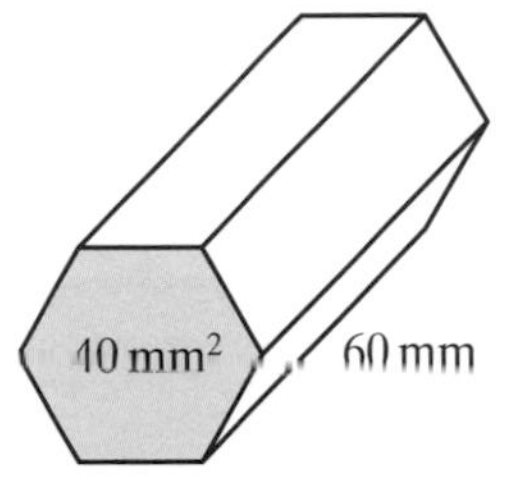

vi

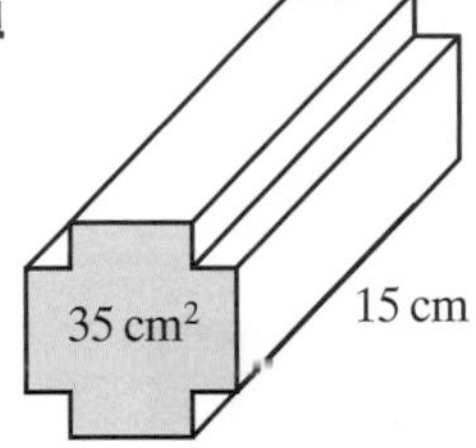

b 1 cm^3 of water has a mass of 1 g. If all the solids in question **1** are filled with water, put the solids in order from the lightest to the heaviest.

2 Jean calculates the volume of the parallelogram prism as follows: $12 \times 6 \times 5 \times 10 = 3600$ cm^2

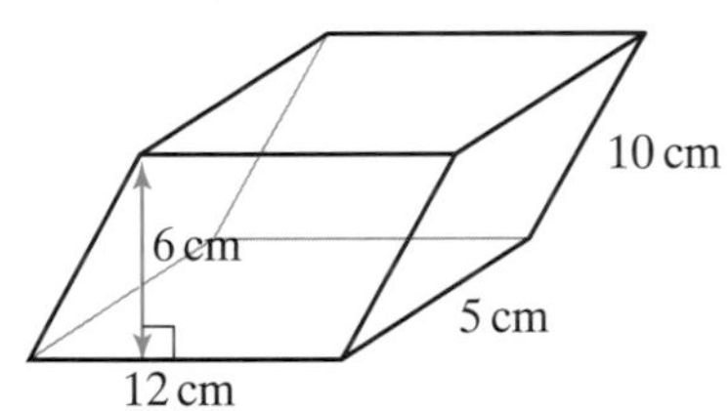

Is she correct? Give reasons for your answer.

3 **a** Calculate the volume of each of the following solids:

i

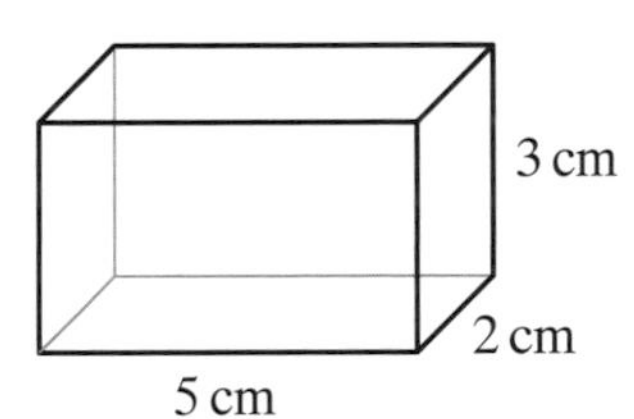

iii

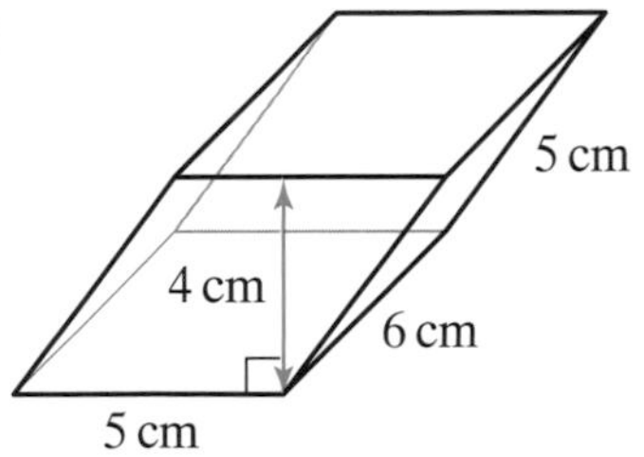

v

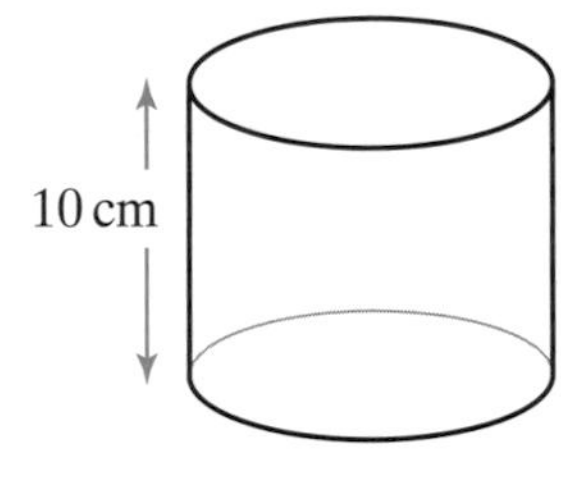

Area of circular base = 71.6 cm^2

You may be asked to leave your answer in terms of π on a non-calculator paper

ii

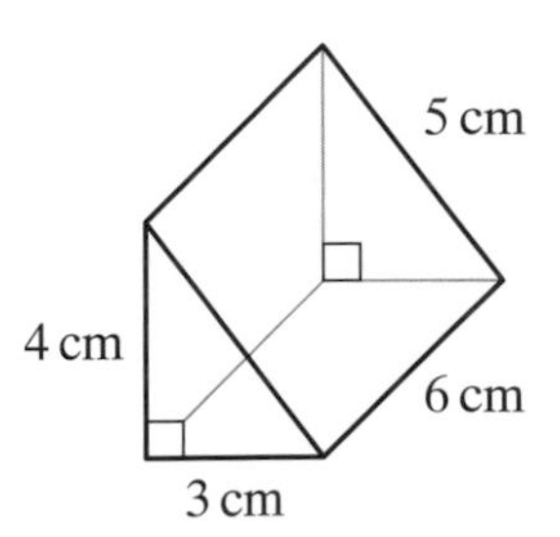

iv

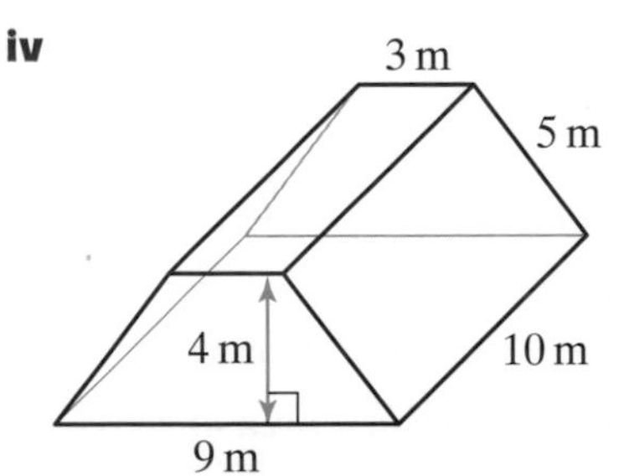

b Calculate the capacity of the shapes in part **a i**, **ii** and **iii** in litres.

4 Convert the answers to question **3** parts **iv** and **v** to:

a cm^3

b litres.

5 If 1 cm^3 of water has a mass of 1 g, what is the mass of 1 m^3 of water?

6 The volume of a prism is 60 cm^3. Find four different types of prism with this volume.

7 **Get Real!**
Wayne calculates the volume of chocolate in a packet as follows:
Volume of a prism = area of cross-section × length
Area of cross-section = $10 \times 13 = 130$ cm^2
Volume = $130 \times 20 = 2600$ cm^3 of chocolate
Is he correct? Give reasons for your answer.

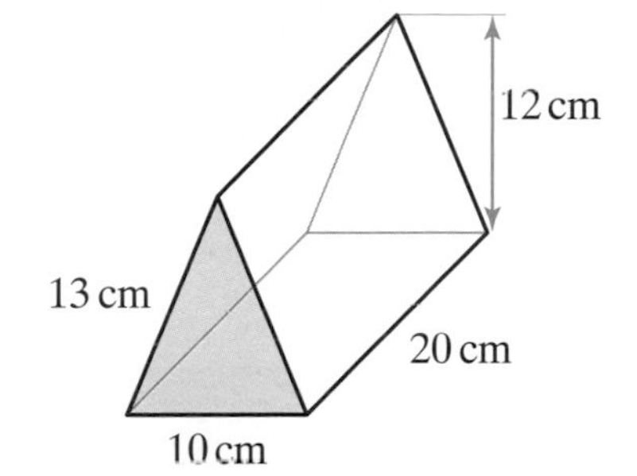

8 **Get Real!**
Maximus the Millionaire Mathematician (or M^3 to his friends!) will only play games with his personalised gold dice.

How many dice (side 3 cm) can he make by melting down one of his gold bars shown below?

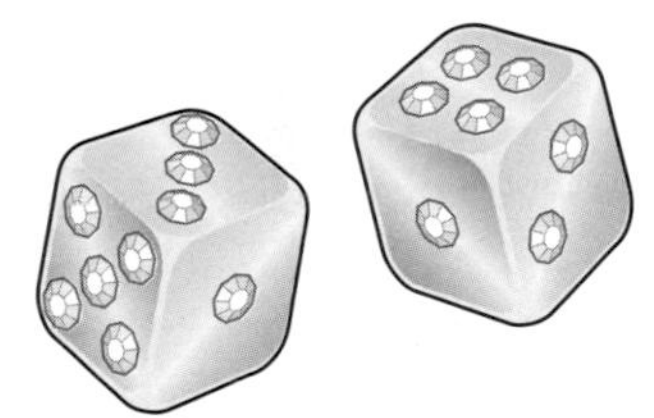

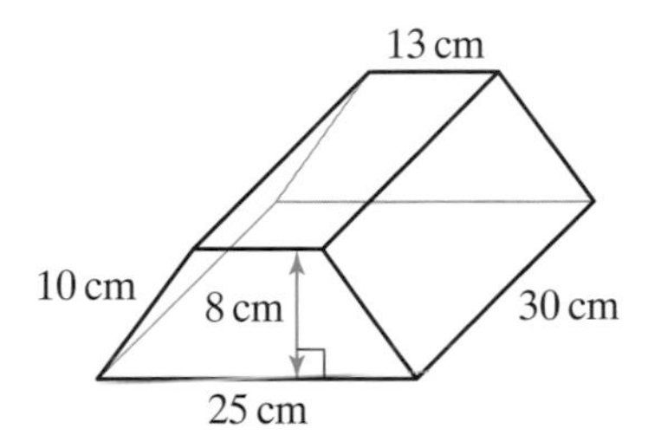

9 **Get Real!**
The cross-section of a school swimming pool is given below.
The pool is filled at a rate of 400 litres/min.
How long will it take to fill the pool? Give your answer in hours and minutes.

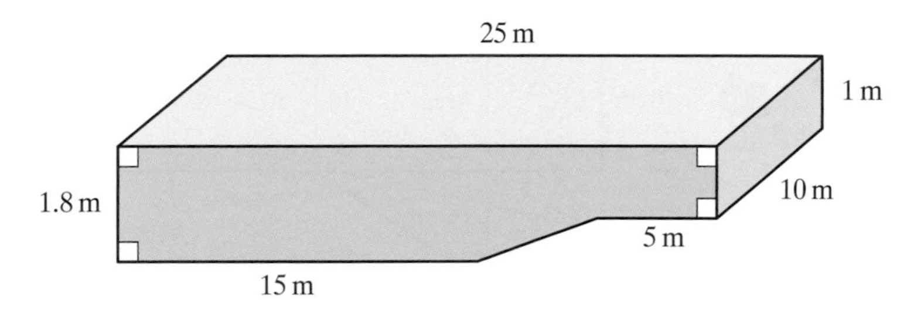

Explore

- Calculate the volume of this cuboid
- Find the dimensions of some cuboids with a volume that is one half of the volume of this cuboid
- Find the dimensions of some cuboids with a volume that is one quarter of the volume of this cuboid

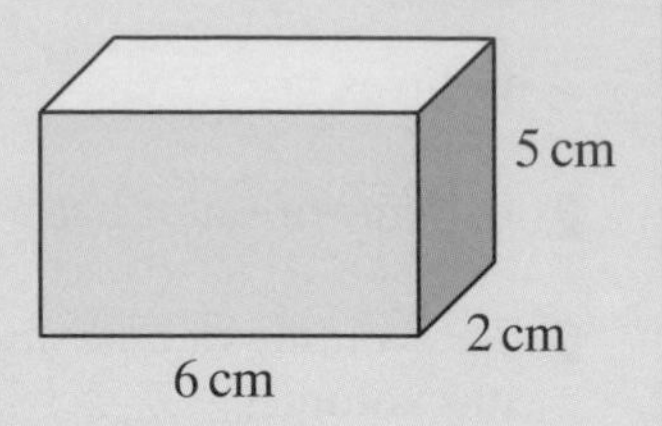

Investigate further

Learn 3 Surface areas of prisms

Examples:

Find the surface area of each of the following prisms.

a Cube

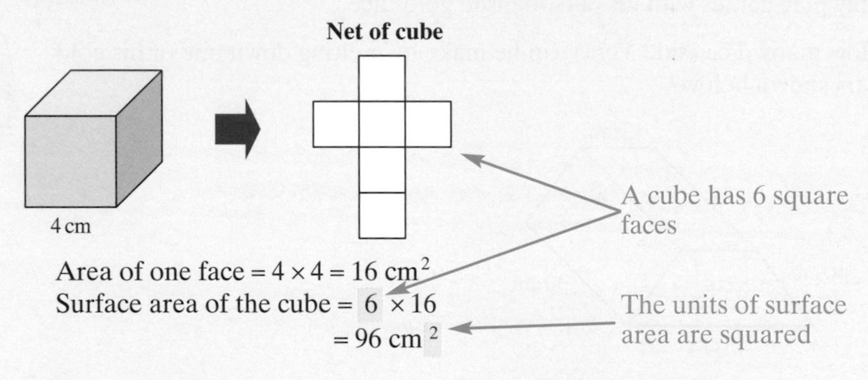

Area of one face = $4 \times 4 = 16\ \text{cm}^2$

Surface area of the cube = 6×16

$= 96\ \text{cm}^2$

b Triangular prism

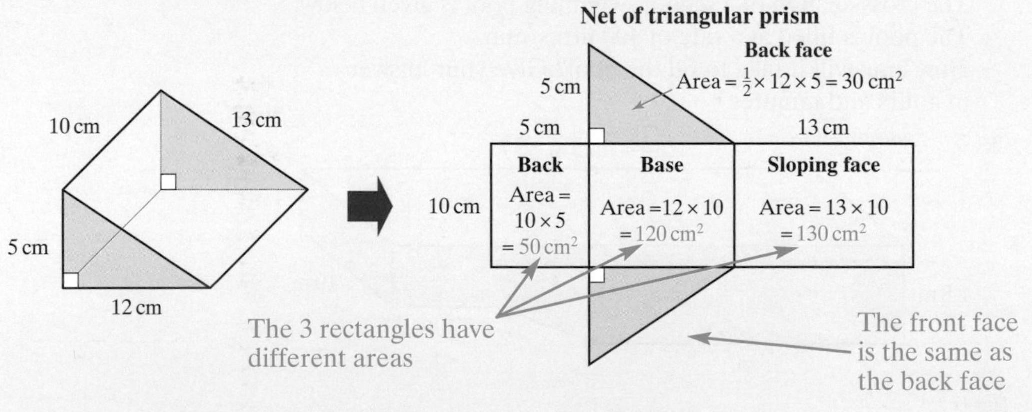

Surface area = $50 + 120 + 130 + 30 + 30 = 360\ \text{cm}^2$

Apply 3

1 Copy and complete the following diagrams by:

i adding the dimensions to the net
ii finding the surface area of the solid.

a

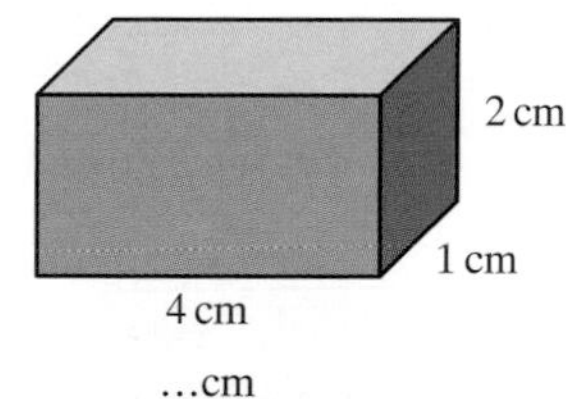

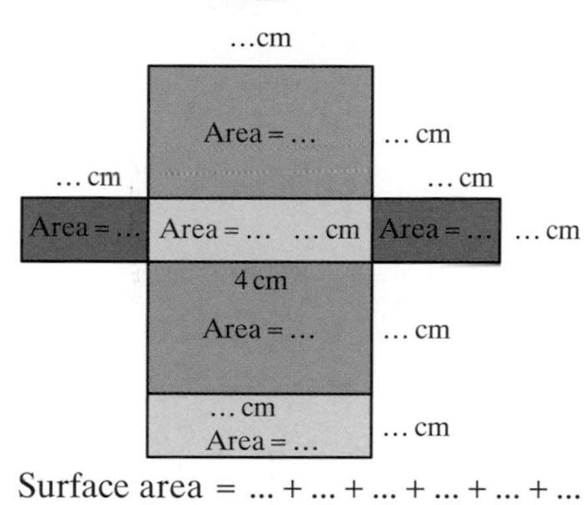

Surface area = ... + ... + ... + ... + ... + ...
= ... cm^2

b

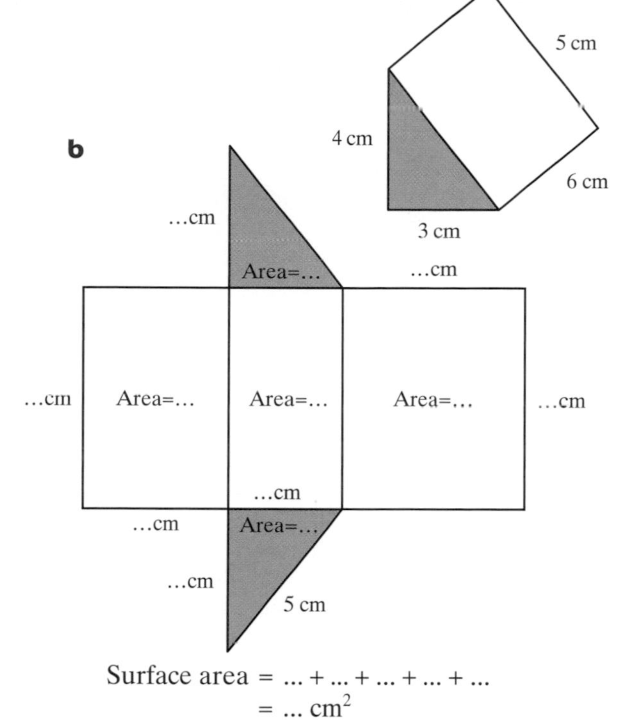

Surface area = ... + ... + ... + ... + ...
= ... cm^2

2 Albert finds the surface area of this triangular prism as follows:

Triangular end area $= \frac{1}{2} \times 4 \times 4 = 8$ cm^2
Rectangular base area $= 10 \times 4 = 40$ cm^2

Albert notices that the triangular prism is made of two triangles and three rectangles so:

Surface area = 2 ends + 3 sides $= (2 \times 8) + (3 \times 40) = 16 + 120 = 136$ cm^2

Is Albert correct? Give a reason for your answer.

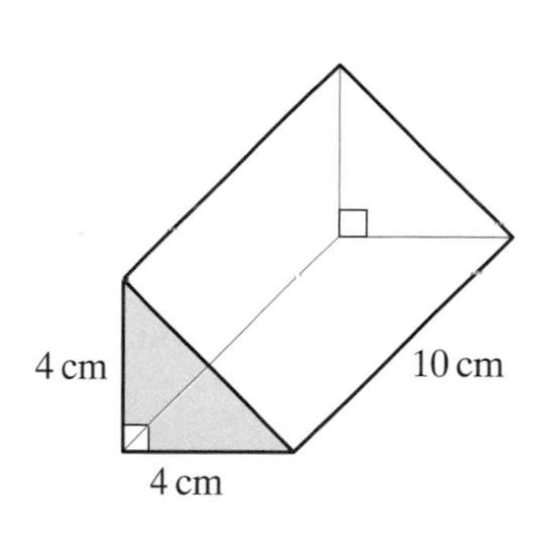

3 Calculate the surface areas of these prisms.

a

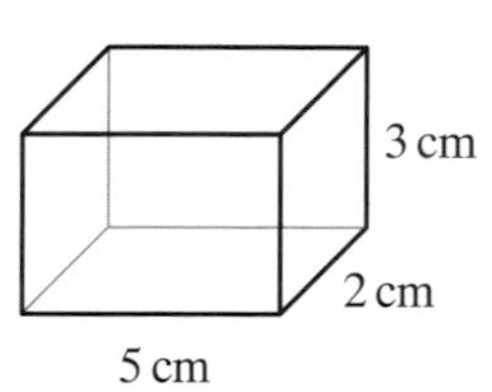

Not drawn accurately

b

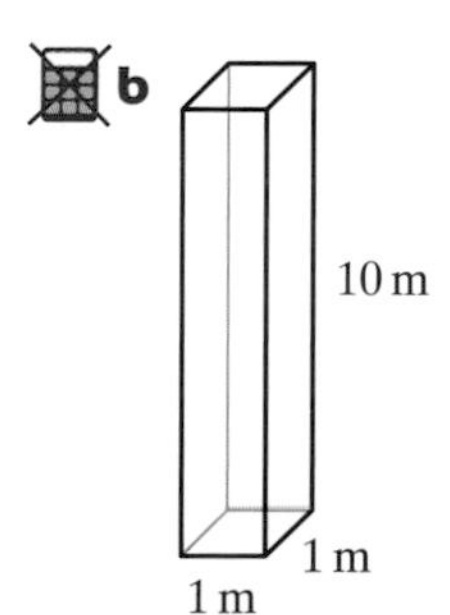

c

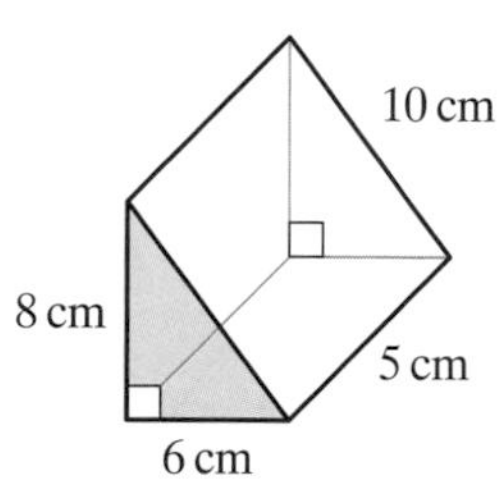

d

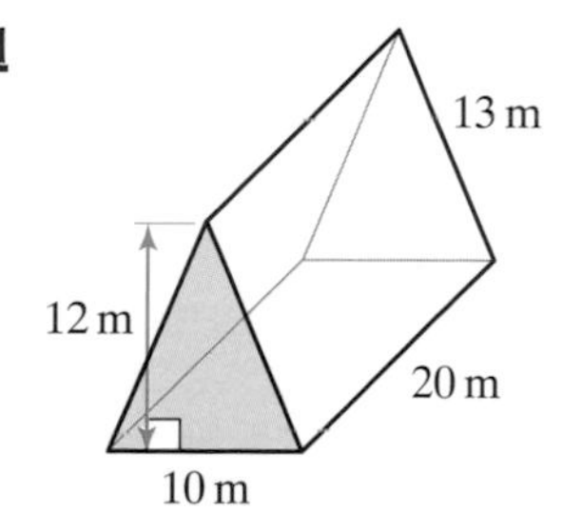

e

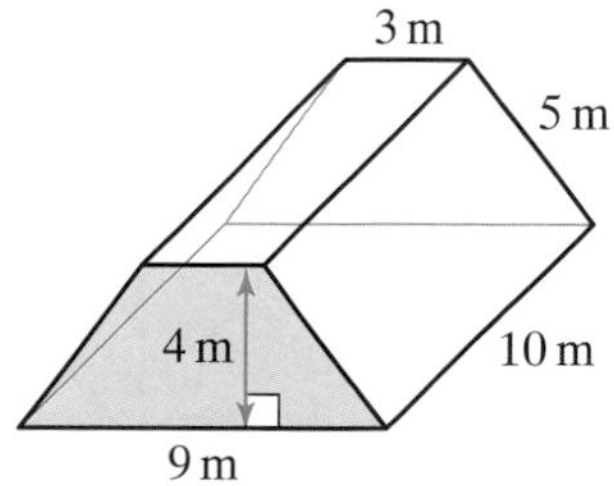

4 The surface area of a cuboid is between 60 cm^2 and 100 cm^2.
Find five possible cuboids.

5 Get Real!
Trevor wants to paint the outside of his play house.

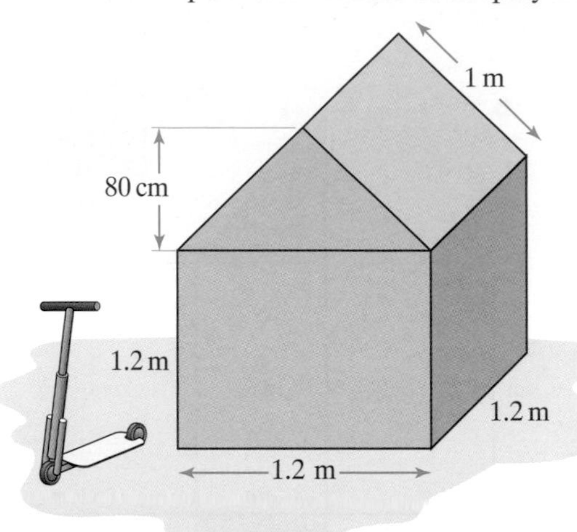

The label on the tin says that 1 litre of paint covers 4 m^2.
How many 1ℓ tins of paint does he need?

6 **a** Calculate the surface areas of these cubes:

i

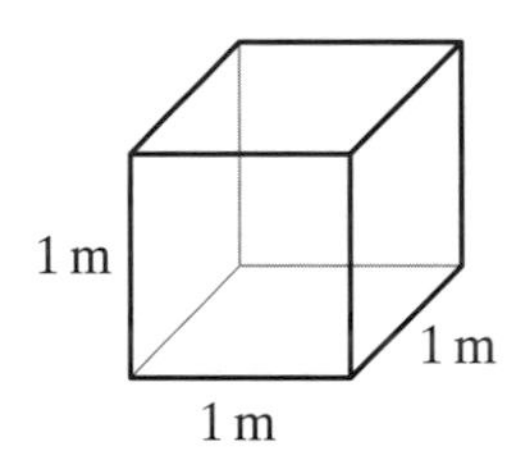

ii

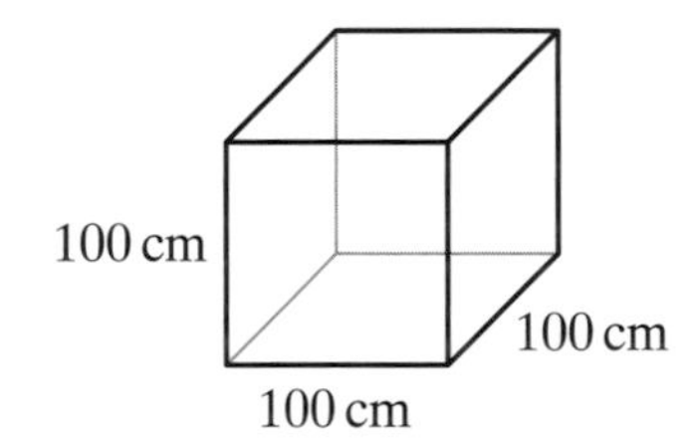

b What do you notice about the two cubes?

c Looking at your answers to part **a**, copy and complete the statement:
1 m^2 = ... cm^2.

d Using your findings from part **c**, convert all your answers to **3** parts **b**, **d** and **e** to cm^2.

7 Find expressions for the volume and surface area of each of the following solids:

a

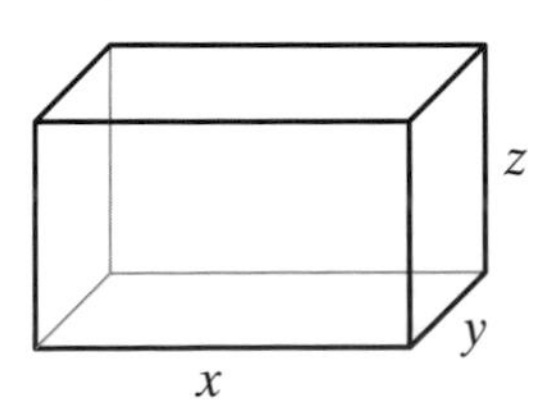

b

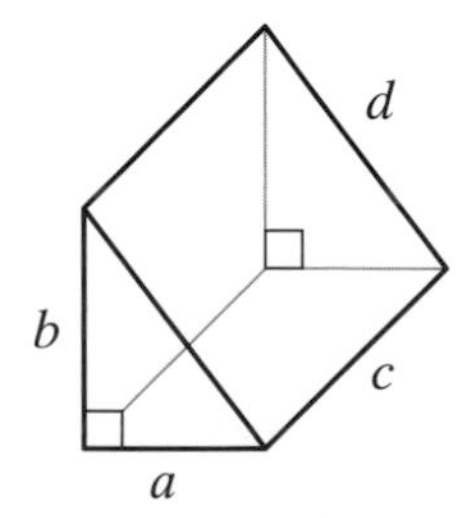

Explore

Most fast food restaurants now serve chunky chips as well as french fries

Does the shape of a chunky chip make it a healthier option?

Suppose the chips are cuboids with a volume of 24 cm^3, for example, 1 cm × 1 cm × 24 cm models a very long french fry whereas 2 cm × 3 cm × 4 cm makes a chunky chip

- Find five different size chips that have a volume of 24 cm^3
- Calculate the surface area of each chip
- Healthier chips have a smaller surface area to which the fat globules can attach themselves

  Which of your chips has the smallest surface area?

Investigate further

Explore

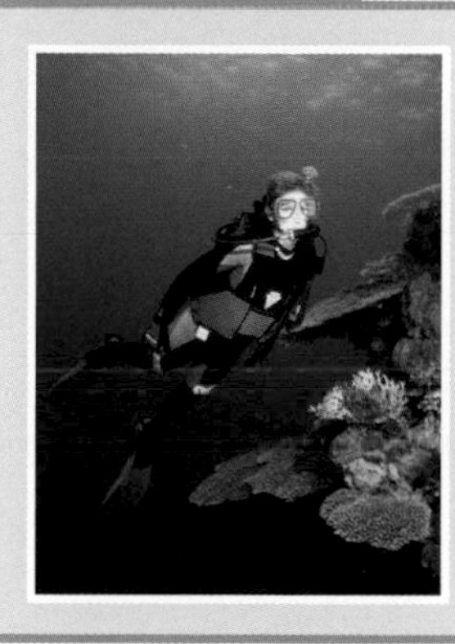

- A 6 cm × 6 cm × 6 cm cube is 'perfect' because the surface area has the same value as the volume (check it)
- Can you find another perfect cube?

Investigate further

Area and volume

ASSESS

The following exercise tests your understanding of this chapter, with the questions appearing in order of increasing difficulty.

1 Find the volume of the following shapes.
State the units of your answer.

a

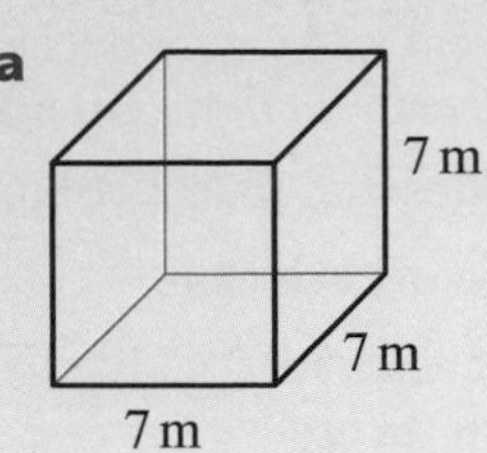

b

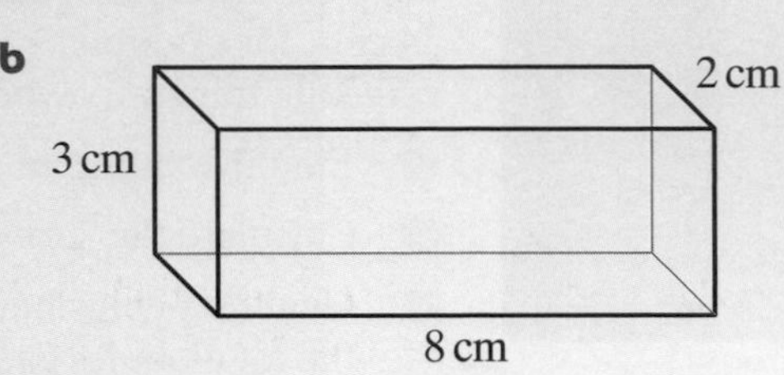

2 A cuboid contains 36 interlocking 1 cm cubes. Write down the dimensions of all the possible cuboids that can be made from these cubes.

3 A cuboid is 7 cm long, 2 cm wide and 6 mm high.
Calculate its volume

a in cm^3

b in mm^3

4 A door wedge is in the shape of a triangular prism 3 cm wide.
The triangular cross-section has a base of 6.4 cm and a height of 4.2 cm.

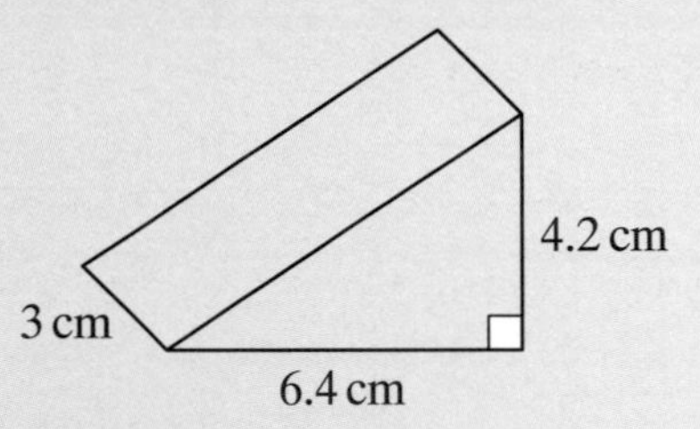

Find its volume.

5 Steve is doing some calculations.
For each part, say whether he is calculating a perimeter, an area or a volume:

a length × width

b 2 × (length + width)

c length × width × height

d 2 × length × width

9 Equations and inequalities

OBJECTIVES

F **Examiners would normally expect students who get an F grade to be able to:**

Solve equations such as $3x = 12$ or $x + 5 = 9$

E **Examiners would normally expect students who get an E grade also to be able to:**

Solve equations such as $\frac{x}{2} = 7$ or $3x - 1 = 9$

D **Examiners would normally expect students who get a D grade also to be able to:**

Solve equations such as $3x - 4 = 5 + x$ or $2(5x + 1) = 28$

C **Examiners would normally expect students who get a C grade also to be able to:**

Solve equations such as $3x - 12 = 2(x - 5)$, $\frac{7-x}{3} = 2$ or $\frac{2x}{3} - \frac{x}{4} = 5$

Solve inequalities such as $3x < 9$ and $12 \leqslant 3n < 20$

Solve linear inequalities such as $4x - 3 < 10$ and $4x < 2x + 7$

Represent sets of solutions on the number line

What you should already know ...

- Collect like terms
- Substitution
- Multiply out brackets (by a number only, which may be negative)
- Cancel fractions
- Inequality signs
- Solve equations

VOCABULARY

Term – a number, variable or the product of a number and a variable(s) such as $3, x$ or $3x$

Linear expression – a combination of terms where the highest power of the variable is 1

Equation – a statement showing that two expressions are equal, for example, $2y - 17 = 15$

Linear equation – an equation where the highest power of the variable is 1; for example, $3x + 2 = 7$ is a linear equation but $3x^2 + 2 = 7$ is not

Unknown – the letter in an equation that represents a quantity which is 'unknown', for example:

$3y = 6$

y is the unknown

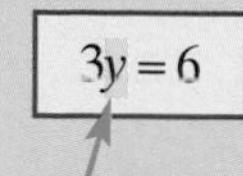

z is the unknown

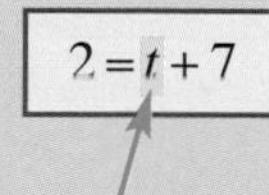

t is the unknown

Solution – the value of the unknown quantity, for example, if the equation is $3y = 6$, the solution is $y = 2$

Coefficient – the number (with its sign) in front of the letter representing the unknown, for example:

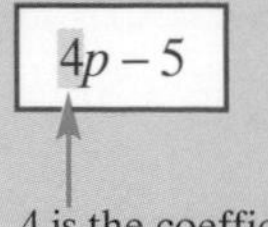

$2 - 3p^2$

-3 is the coefficient of p^2

Operation – a rule for combining two numbers or variables, such as add, subtract, multiply or divide

Inverse operation – the operation undoes or reverses a previous operation, for example subtract is the inverse of add:

$15 + 8 = 23$ Add 8

$23 - 8 = 15$ Subtract 8 to return to the starting number 15

Integer – any positive or negative whole number or zero, for example, $-2, -1, 0, 1, 2 \ldots$

Brackets – these show that the terms inside should be treated alike, for example,

$2(3x + 5) = 2 \times 3x + 2 \times 5 = 6x + 10$

Inequality – statements such as $a \neq b$, $a \leqslant b$ or $a > b$ are inequalities

Inequality signs – $<$ means less than, $\leqslant$ means less than or equal to, $>$ means greater than, $\geqslant$ means greater than or equal to

Number line – a line where numbers are represented by points upon it; simple inequalities can be shown on a number line

Learn 1 Equations where the unknown appears once only

Examples:

Solve these equations:

a $4x = 28$

b $x + 7 = 3$

c $\dfrac{x}{5} = 2$

d $2x - 7 = 11$

a $4x = 28$ Remember that $4x$ means $4 \times x$

$\dfrac{4x}{4} = \dfrac{28}{4}$ Divide both sides by 4

$x = 7$

b $x + 7 = 3$

$x + 7 - 7 = 3 - 7$ Subtract 7 from both sides

$x = -4$

c $\dfrac{x}{5} = 2$

$\dfrac{x}{5} \times 5 = 2 \times 5$ Multiply both sides by 5

$x = 10$

d $2x - 7 = 11$

$2x - 7 + 7 = 11 + 7$ Add 7 to both sides

$2x = 18$

$\dfrac{2x}{2} = \dfrac{18}{2}$ Divide both sides by 2

$x = 9$

Think about the operations (+, –, ×, ÷) that have been applied to the unknown.

Reverse these operations, doing the same to both sides of the equation.

Apply 1 

Solve these equations:

1 $3x = 21$

2 $5y = 14$

3 $8z = 4.8$

4 $6a = 9$

5 $3b = 10$

6 $4c = -8$

7 $x - 2 = 9$

8 $y + 8 = 3$

9 $z + 1.6 = 5.2$

10 $a + 4 = 1$

11 $b - 3 = -5$

<u>12</u> $c + 5.8 = 3.4$

13 $\frac{x}{3} = 4$

14 $\frac{y}{2} = 8$

15 $\frac{z}{5} = 0.7$

16 $\frac{a}{12} = 0.25$

17 $\frac{b}{3} = 0$

18 $\frac{c}{4} = -1$

19 $3x - 1 = 14$

20 $5y + 3 = 28$

<u>21</u> $6z + 5 = 27$

22 $2a + 5 = 1$

23 $3b + 16 = 4$

<u>24</u> $4c - 11 = 6$

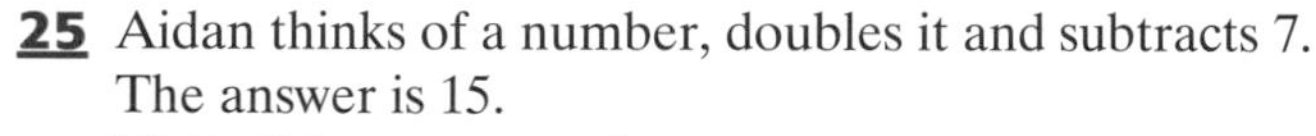

<u>25</u> Aidan thinks of a number, doubles it and subtracts 7.
The answer is 15.
Write this as an equation.
Solve the equation to find Aidan's number.

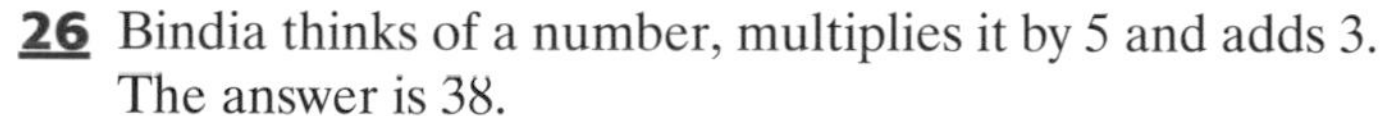

<u>26</u> Bindia thinks of a number, multiplies it by 5 and adds 3.
The answer is 38.
Write this as an equation.
Solve the equation to find Bindia's number.

<u>27</u>

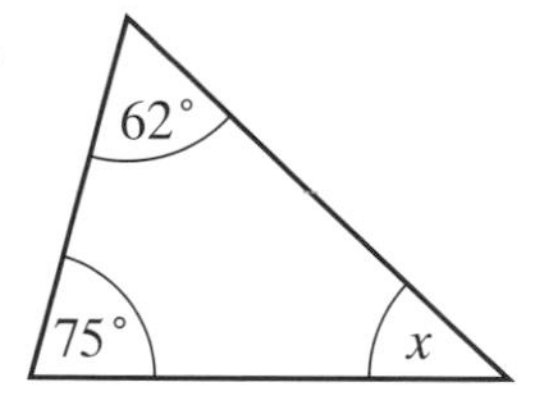

Use the sum of the angles in a triangle to write down an equation in x.
Solve your equation to find the value of x.

<u>28</u>

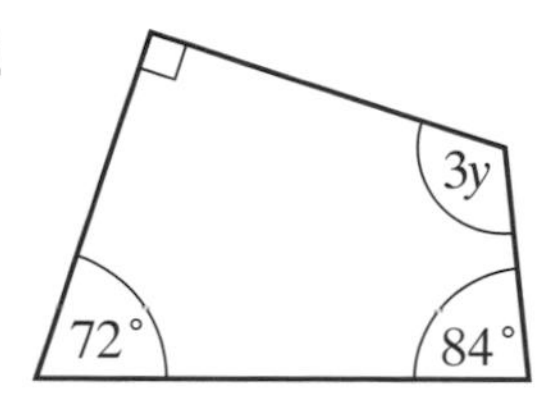

Use the sum of the angles in a quadrilateral to write down an equation in y.
Solve your equation to find the value of y.

Explore

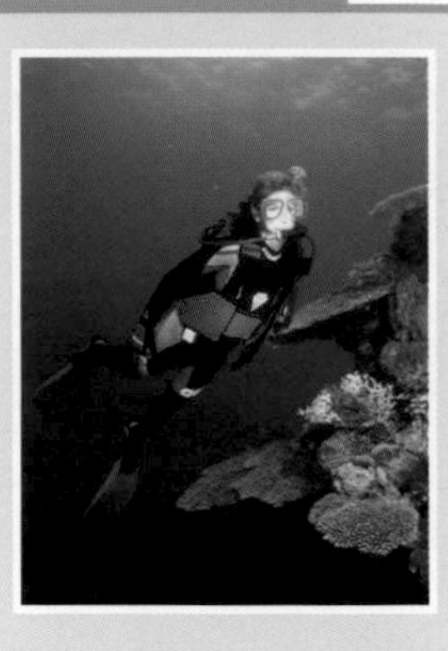

- Make up five different equations that have the solution $x = 8$
- Use a different style for each equation:
  - one which requires division to solve it
  - one which requires multiplication to solve it
  - one which requires addition to solve it
  - one which requires subtraction to solve it
  - one which requires a combination of reverse operations to solve it
- Explain how you did this

Investigate further

Learn 2 Equations where the unknown appears on both sides

Examples:

Solve these equations:

a $4x + 1 = 9 - x$ **b** $3y + 5 = 5y - 4$

a

$4x + 1 = 9 - x$

$4x + x + 1 = 9 - x + x$ Add x to both sides

$5x + 1 = 9$

$5x + 1 - 1 = 9 - 1$ Subtract 1 from both sides

$5x = 8$

$\frac{5x}{5} = \frac{8}{5}$ Divide both sides by 5

$x = 1.6$

b

$3y + 5 = 5y - 4$

$3y - 3y + 5 = 5y - 3y - 4$ Subtract $3y$ from both sides

If you took $5y$ from both sides, you would get $-2y$ on the left-hand side

$5 = 2y - 4$

$5 + 4 = 2y - 4 + 4$ Add 4 to both sides

$9 = 2y$

$\frac{9}{2} = \frac{2y}{2}$ Divide both sides by 2

$4.5 = y$

$y = 4.5$ Writing the equation with y as the subject

Collect together on one side all the terms that contain the unknown.
Collect together on the other side all the terms that do not contain the unknown.
Remember that a sign belongs to the term *after* it.

Apply 2 

Solve these equations:

1 $4x - 5 = 2x + 7$

2 $5y - 1 = y + 9$

3 $7z - 3 = 11 + 3z$

4 $t + 8 = 2 - 3t$

5 $3p + 2 = 5 - p$

6 $5 + 2q = 12 - 3q$

7 $8 + a = 5 - 2a$

8 $3b - 10 = 5b - 1$

9 $3c - 5 = 6 - c$

10 $2 + 5d = 7d + 3$

11 $2 - 3e = 8 - 6e$

12 $4f + 11 = 5 + f$

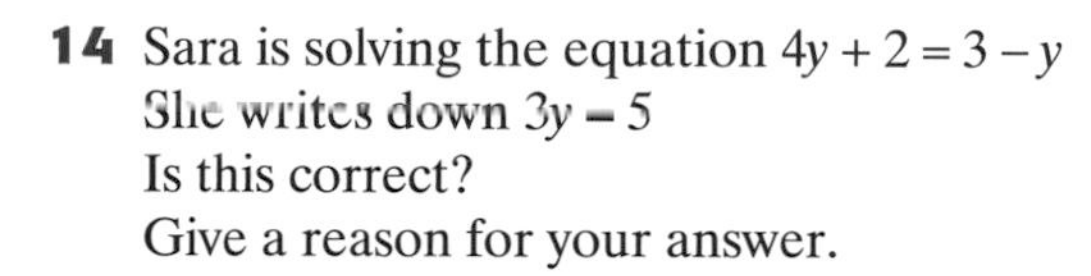

13 Andy is solving the equation $5x - 3 = 2 - 3x$
He writes down $2x - 3 = 2$
Is this correct?
Give a reason for your answer.

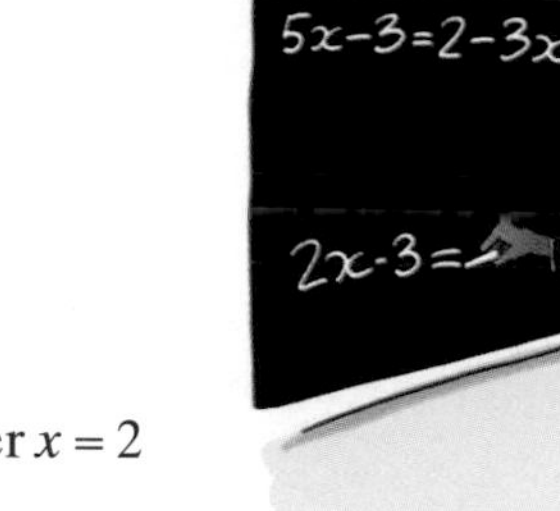

14 Sara is solving the equation $4y + 2 = 3 - y$
She writes down $3y = 5$
Is this correct?
Give a reason for your answer.

15 Tom solves the equation $6x - 7 = 9 + 2x$ and gets the answer $x = 2$
Can you find Tom's mistake?

16 Jan solves the equation $7y + 4 = 9 - 3y$ and gets the answer $y = 2$
Can you find Jan's mistake?

17 $4z - 1 = \bullet - z$
The answer to this equation is $z = 6$
What is the number under the ink blob?

18 $3a + \bullet = 1 - a$
The answer to this equation is $a = -1$
What is the number under the ink blob?

19 If $b = 4$, find the value of $2b - 5$
Hence explain why $b = 4$ is not the solution of the equation $2b - 5 = 9 - 3b$

20 If $c = -3$, find the value of $8 - 2c$
Hence explain why $c = -3$ is not the solution of the equation $3c + 11 = 8 - 2c$

Learn 3 Equations with brackets

Examples:

Solve these equations:

a $3(2x - 3) = 18$ **b** $5 - 3(y + 1) = 7 - 4y$

a $3(2x - 3) = 18$ Multiply out the brackets first, then follow the rules for solving equations

$6x - 9 = 18$ Remember to multiply both terms in the brackets by 3

$6x = 18 + 9$ Add 9 to both sides

$6x = 27$

$x = 4.5$ Divide both sides by 6

Alternative method:

$3(2x - 3) = 18$

$2x - 3 = 6$ Divide both sides by 3

$2x = 9$ Add 3 to both sides

$x = 4.5$ Divide both sides by 2

This alternative method cannot be used for all equations with brackets, as the next example shows

b $5 - 3(y + 1) = 7 - 4y$ Multiply out the brackets first, then follow the rules for solving equations

$5 - 3y - 3 = 7 - 4y$ Multiplying by -3 changes the sign in the brackets

$2 - 3y = 7 - 4y$ The numbers on the left-hand side have been collected

$-3y = 5 - 4y$ Subtract 2 from both sides

$-3y + 4y = 5$ Add $4y$ to both sides

$y = 5$

Apply 3 

Solve these equations:

1 $5(x-3)=20$

2 $4(y+2)=3y+9$

3 $8=2(z-1)$

4 $5(p+1)=3p+11$

5 $3(q-11)=7-q$

6 $2(3t-10)=13$

7 $5a-2=2(a-4)$

8 $3(3b-4)=2+7b$

9 $3+8c=6(c-1)$

10 $14d-1=3(2d+1)$

11 $2(5-2e)=8-3e$

12 $7-f=3(5-f)$

13 $3(5+2x)=x-5$

14 $2(y-3)+4(2y-7)=6$

15 $8-2(z+3)=5-3z$

16 $19=4-2(t-1)$

17 $6(2p-5)-4(p-2)=14$

18 $5(q-7)-3(2q-4)+25=0$

19 Sadhia thinks of a number, adds 5 and then doubles the result.
Her answer is 64.
Write this as an equation.
Solve the equation to find Sadhia's number.

20 Todd thinks of a number, subtracts 8 and then multiplies the result by 3.
His answer is 42.
Write this as an equation.
Solve the equation to find Todd's number.

Learn 4 Equations with fractions

Examples:

Solve these equations:

a $\frac{x}{4}-6=3$ **b** $\frac{2x}{3}-\frac{x}{4}=5$ **c** $\frac{3x+5}{2}=7$

a $\frac{x}{4}-6=3$

$\frac{x}{4}=3+6$ Add 6 to both sides

$\frac{x}{4}=9$

$x=9\times 4$ Multiply both sides by 4

$x=36$

b $\frac{2x}{3}-\frac{x}{4}=5$ The lowest common denominator is 12

${}^{4}\cancel{12}\times\frac{2x}{\cancel{3}_1}-{}^{3}\cancel{12}\times\frac{x}{\cancel{4}_1}=5\times 12$ Multiply **each term** by 12

$4\times 2x-3\times x=5\times 12$ Cancel (this should clear all the fractions)

$8x-3x=60$

$5x=60$

$x=12$

c $\dfrac{3x+5}{2} = 7$ ← This is the same as $\frac{1}{2}(3x+5) = 7$

$^1\not{2} \times \dfrac{3x+5}{\not{2}_1} = 7 \times 2$ Multiply **both** sides by 2 to clear the fraction

$3x + 5 = 14$

$3x = 14 - 5$ Subtract 5 from both sides

$3x = 9$

$x = 3$ Divide both sides by 3

Remove the fraction by multiplying both sides by the denominator.
If there is more than one fraction, multiply by the lowest common denominator.
There are 'invisible brackets' around the terms on top of an algebraic fraction.

Apply 4

Solve these equations:

1 $\dfrac{x}{7} = 5$

2 $\dfrac{x}{4} = 3$

3 $\dfrac{2x}{3} = 5$

4 $\dfrac{x}{2} + 7 = 9$

5 $\dfrac{y}{3} - 1 = 5$

6 $6 = 2 + \dfrac{z}{5}$

7 $3 + \dfrac{a}{2} = 7$

8 $4 - \dfrac{b}{3} = 1$

9 $\dfrac{c}{6} + 4 = 3$

10 $\dfrac{3x+2}{5} = 4$

11 $\dfrac{2y-3}{4} = 3$

12 $1 = \dfrac{9-z}{3}$

13 $\frac{1}{4}(3p + 5) = 2$

14 $\frac{1}{2}(5q + 3) = 14$

15 $\frac{1}{3}(2t - 7) = 4$

<u>16</u> $\frac{1}{2}(5a - 2) = a - 4$

<u>17</u> $\frac{1}{8}(b - 1) = 10 - b$

<u>18</u> $2c - 11 = \frac{1}{3}(2 - c)$

<u>19</u> $\dfrac{x}{3} + \dfrac{x}{4} = 7$

<u>20</u> $\dfrac{y}{2} - \dfrac{y}{5} = 6$

<u>21</u> Jayne says the answer to the equation $\dfrac{x+5}{2} = 4 - x$ is $x = -1$

Use substitution to check whether Jayne is correct.

<u>22</u> Tom and Jared are solving the equation $\dfrac{3y-7}{4} = y + 2$

Tom gets the answer $y = 1$ and Jared gets $y = -15$
Check their answers to see if either of them is correct.

<u>23</u> Explain why you cannot solve the equation $\dfrac{4z-1}{2} = 1 + 2z$

Learn 5 Inequalities and the number line

Examples:

a Draw number lines to show these inequalities.

i $x < 2$ **ii** $x \geqslant -3$ **iii** $-6 < x \leqslant -3$ **iv** $x \leqslant -2$ or $x > 1$

i

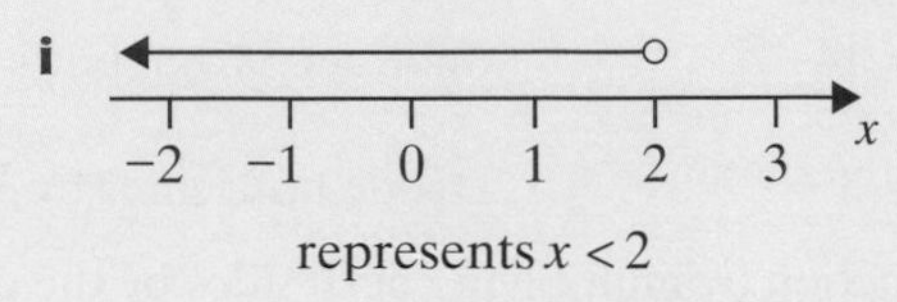

represents $x < 2$

An **open** circle shows that x can be very close to 2 but not equal to 2

If x is an integer, it could be 1, 0, −1, −2, −3, ... (but not 2).
If x is any real number, it could be any number less than 2.

ii

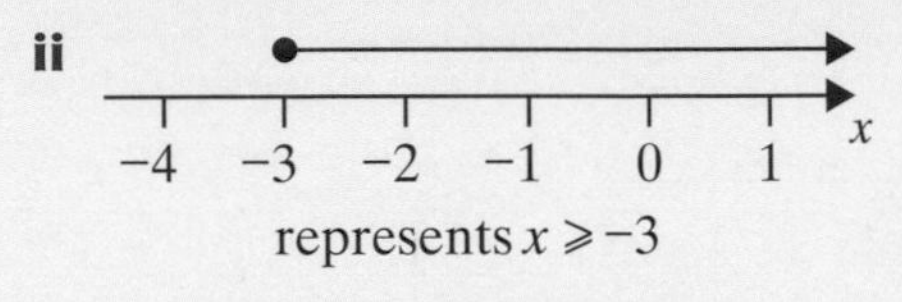

represents $x \geqslant -3$

A **closed** circle shows that x can equal −3

If x is an integer, it could be −3, −2, −1, 0, 1, 2, ...
If x is any real number, it could be any number greater than or equal to −3.

iii

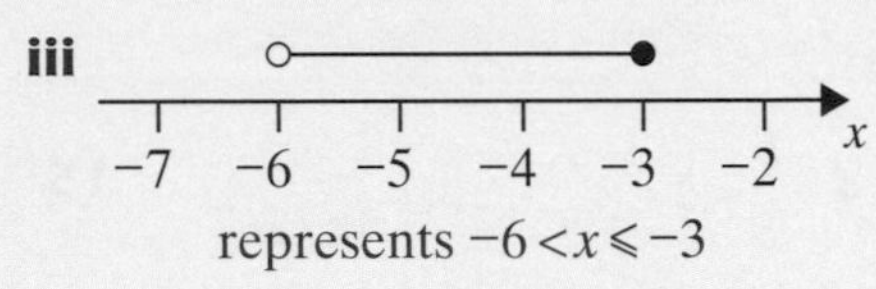

represents $-6 < x \leqslant -3$

If x is an integer, it could only be −5, −4 or −3 (but not −6).
If x is any real number, it could be any number greater than −6 **and** less than or equal to −3.

iv

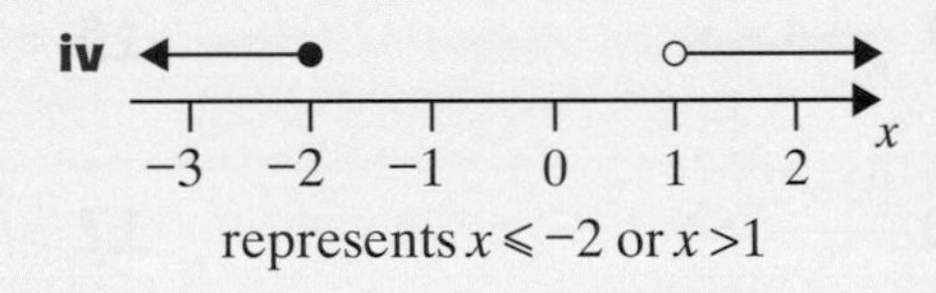

represents $x \leqslant -2$ or $x > 1$

If x is an integer, it could be −2, −3, −4 ... or 2, 3, 4, ... (but not −1, 0 or 1).
If x is any real number, it could be any number less than or equal to −2 **or** any number greater than 1.

b Solve $-5 < 3x + 4 \leqslant 10$

You can solve inequalities using the same methods as for solving equations.

First, split the inequality into two parts.

$-5 < 3x + 4$ and $3x + 4 \leqslant 10$

$-9 < 3x$ and $3x \leqslant 6$ Subtract 4 from both sides

$-3 < x$ and $x \leqslant 2$ Divide both sides by 3

Finally, put the inequalities back together: so $-3 < x \leqslant 2$

If the question had said that x is an integer, the possible values are −2, −1, 0, 1, 2

Add or subtract the same number from both sides and the inequality is still valid

Multiply or divide both sides by the same **positive** number and the inequality stays valid

However, if you multiply or divide both sides by a **negative** number, the inequality is reversed

c Find the largest integer that satisfies $x + 13 > 5x - 3$.

$x + 13 > 5x - 3$

$13 > 4x - 3$ Subtract x from both sides

$16 > 4x$ Add 3 to both sides

$4 > x$ Divide both sides by 4

$x < 4$ ← x is less than 4, so x can be any integer less than 4, but not 4

The largest integer that satisfies the inequality is 3.

Apply 5

1 Show each of the following inequalities on a number line:

a $x < -2$

b $x < -3$

c $x \geqslant -1$

d $x \leqslant 4$

e $-2 \leqslant x \leqslant 3$

f $-4 < x < 0$

g $x < -2$ or $x > 1$

h $x \leqslant 2$ or $x > 5$

i $-2 < x < 2$

2 Write the inequalities shown on these diagrams.

a

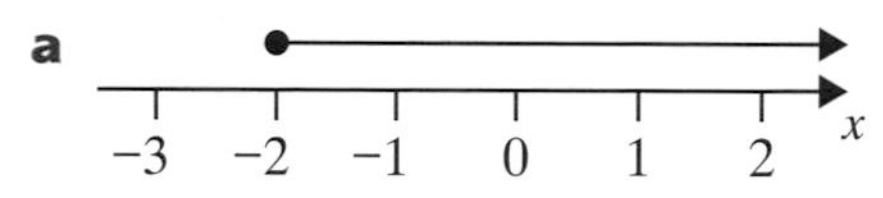

b

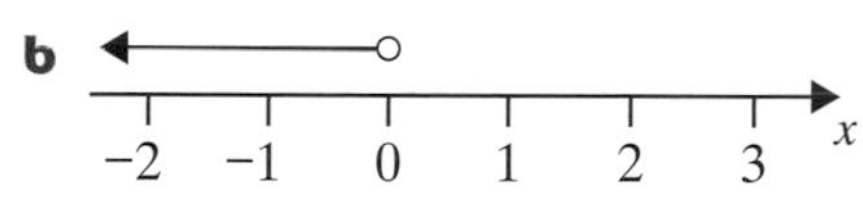

c

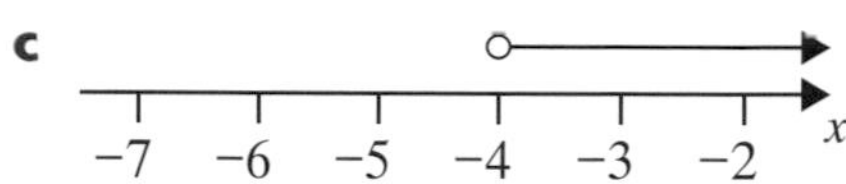

d

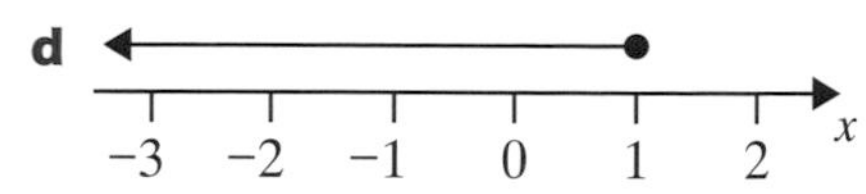

e

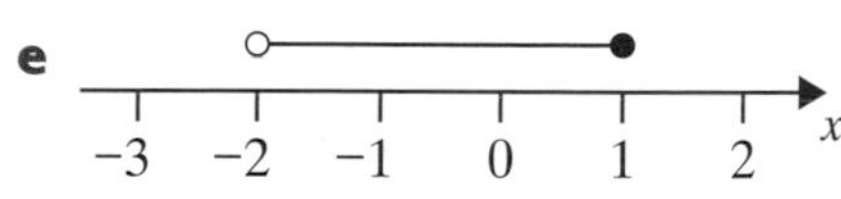

f

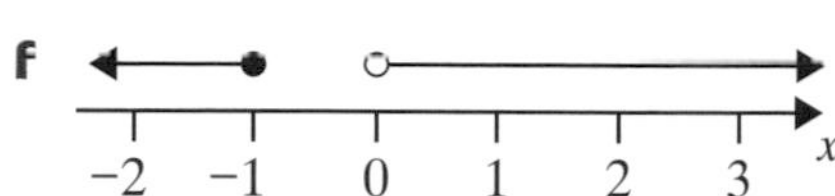

3 Oliver says that this diagram shows the inequality $-4 < x < 2$.
Julia says that he is wrong, it shows $-4 \leqslant x \leqslant 2$.
Do you agree with Oliver or Julia?
Give a reason for your answer.

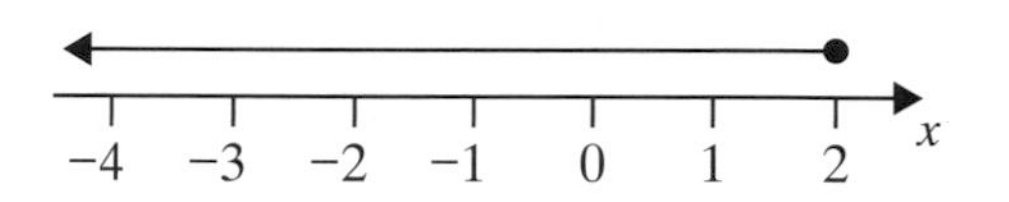

4 List all the integer values of n such that:

a $-8 \leqslant 4n < 15$

b $-3 < 2n \leqslant 12$

c $-5 \leqslant 2n - 1 < 6$

5 Solve these linear inequalities.

a $4x - 1 \geqslant 3$

b $2x + 16 \leqslant 29$

c $5x + 10 < 0$

d $x < 4x - 9$

e $3x - 7 > 8x + 8$

f $4(x + 3) \leqslant 3(x - 2)$

g $4 > 7 - x$

h $12 + 2x < 6 - x$

i $6 - 4x \geqslant 3 - x$

6 Given that $5y > 2$ and $\frac{1}{2}y \leqslant 3\frac{1}{2}$ and that y is an integer, find the possible values of y.

7 Find the smallest integer that satisfies:

a $0 > 4 - x$

b $3(2x - 10) \geqslant 2$

c $2x + 8 \geqslant 0$

d $6 \leqslant 5(2x + 7)$

8 Find the largest integer that satisfies:

a $9 - 2x \geqslant 5$

b $4(3x + 9) < 50$

c $6 - 3x \geqslant 10$

d $7(2x - 5) < 4x + 5$

9 Find all the possible pairs of positive integers, x and y, such that $2x + 3y \leqslant 9$.

10 Solve these linear inequalities.

a $8 \geqslant 12 - \frac{x}{4}$

b $\frac{2x - 3}{5} < 6$

c $\frac{3x + 8}{4} > 1$

d $\frac{x}{3} - \frac{x}{4} \leqslant -2$

e $\frac{x}{8} + 5 \geqslant 4 - \frac{x}{4}$

f $\frac{3 - 2x}{4} < x + 3$

11 Get Real!
Tolu is saving up for a new bike that costs £99.
She has £20 in her account at the moment.
Write an inequality and solve it to find the least number of pounds that Tolu must save every month for the next five months if she is to have enough money for the bike.

12 Two integers x and y are such that $2 \leqslant x < 7$ and $-4 \leqslant y \leqslant 1$.

a What is the largest value of y^2?

b What is the smallest value of xy?

c If $y^2 = 4$, what is the value of y?

13 Get Real!
Class 11Y want to know how old their teacher is.
Mrs Hirst gives them a clue.
She says: 'In 10 years time I will be **less than** double the age I was 10 years ago'.
Can you write an inequality and solve it to find the youngest age that Mrs Hirst could be?

Explore

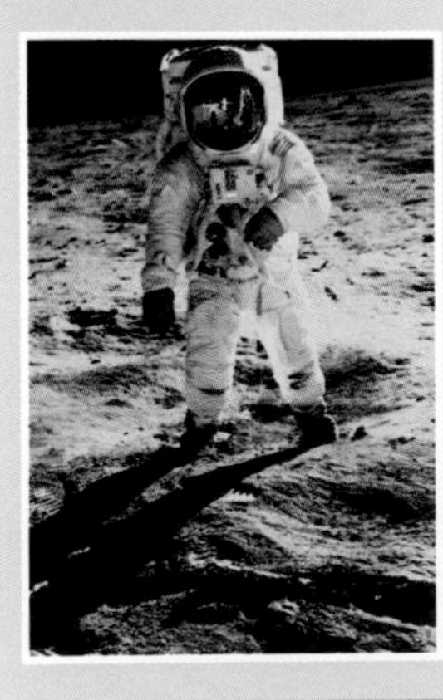

- Think of two numbers that satisfy the inequality $x^2 < 9$
- Find the two solutions of the equation $x^2 = 9$
- Show the set of solutions of $x^2 < 9$ on a number line
- Investigate which values of x satisfy the inequality $x^2 > 9$
- Where does this set of solutions fit on the number line?
- Now look at the solutions of $x^2 = 4$

Investigate further

Equations and inequalities

ASSESS

The following exercise tests your understanding of this chapter, with the questions appearing in order of increasing difficulty.

1 Solve these equations:

a $2x = 8$ **b** $3a = 21$ **c** $4b = 18$ **d** $5z = -30$ **e** $m + 5 = 8$ **f** $n - 4 = 2$ **g** $7 = 1 + d$ **h** $p + 9 = 2$

2 Solve these equations:

a $\frac{c}{3} = 4$ **b** $\frac{1}{4}k = 9$ **c** $\frac{1}{3}x = -5$ **d** $2v + 1 = 7$ **e** $4f - 6 = 2$ **f** $2q + 5 = 1$ **g** $4 - 3w = 1$ **h** $\frac{1}{4}g + 7 = 2$

3 Solve these equations:

a $5x + 1 = 15 - 2x$

b $4x + 8 = 6x - 5$

c $4w + 7 = 37 - 2w$

d $\frac{1}{2}z - 1 = 11 - \frac{1}{4}z$

4 Solve these equations:

a $2(3p - 1) = 28$ **b** $3(a - 1) = 2(a + 1)$ **c** $\frac{2s + 14}{5} = 1$

5 Write down an equation and use it to solve the following problems.

a Demelza's dad is three times as old as Demelza. The sum of their ages is 52. How old is Demelza?

b Three consecutive odd numbers add up to 93. Find these numbers.

c Jack and Jill went up the hill to fetch a pail of water. The water weighed 15 kg more than the pail. The total weight was 18 kg. How heavy was the pail?

d Tweedledum and Tweedledee bought a cake and cut it into 3 pieces, one piece for each of them and one piece for Alice. Tweedledum's piece was 50 g heavier than Tweedledee's piece. Tweedledee's piece was 30 g heavier than Alice's piece. The total mass was 710 g. What was the mass of Alice's piece?

6 Show each of the following inequalities on a number line.

a $x > -2$

b $x < 4$

c $4 < x \leqslant 9$

d $-3 \leqslant x \leqslant 7$

7 List all the **integer** solutions to the inequalities:

a $2 < 3x < 14$

b $-7 \leqslant 4a \leqslant 5$

c $15 \geqslant 3z \geqslant -4$

8 Solve these inequalities:

a $3a < 12$

b $4b \geqslant -20$

c $4g - 2 > -3$

d $3x - 2 > 4x + 1$

Try some real past exam questions to test your knowledge:

9 Solve these equations:

a $4x + 7 = 3$

b $3y - 11 = 9 - y$

Spec B, Mod Paper 1, Nov 04

10 Solve the equation:

$\frac{1}{2}x - 5 = \frac{1}{4}x + 3$

Spec A, Int Paper 1, June 05

10 Reflections and rotations

OBJECTIVES

G

Examiners would normally expect students who get a G grade to be able to:

Draw a line of symmetry on a 2-D shape

Draw the reflection of a shape in a mirror line

F

Examiners would normally expect students who get an F grade also to be able to:

Draw **all** the lines of symmetry on a 2-D shape

Draw the line of reflection for two shapes

Give the order of rotation symmetry of a 2-D shape

Name, draw or complete 2-D shapes from information about their symmetry

E

Examiners would normally expect students who get an E grade also to be able to:

Reflect shapes in the axes of a graph

D

Examiners would normally expect students who get a D grade also to be able to:

Reflect shapes in lines parallel to the axes, such as $x = 2$ and $y = -1$

Rotate shapes about the origin

Describe fully reflections in a line and rotations about the origin

Identify reflection symmetry in 3-D solids

C

Examiners would normally expect students who get a C grade also to be able to:

Reflect shapes in lines such as $y = x$ and $y = -x$

Rotate shapes about any point

Describe fully reflections in any line and rotations about any point

Find the centre of a rotation and describe it fully

Combine reflections and rotations

What you should already know ...

- Coordinates and equations of lines, such as $x = 3$, $y = -2$, $y = x$, $y = -x$
- Names of 2-D and 3-D shapes

VOCABULARY

Congruent – exactly the same size and shape; one of the shapes might be rotated or flipped over

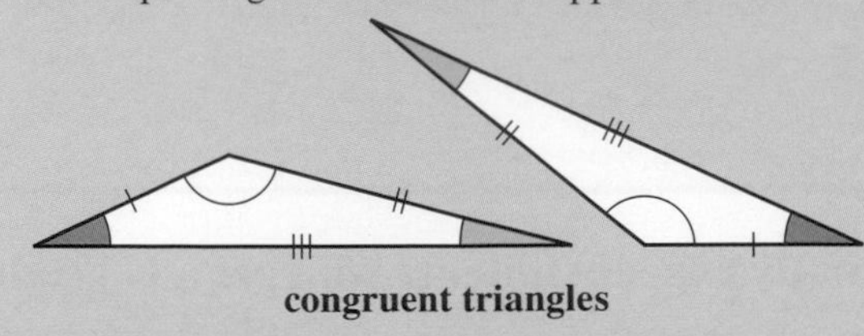

congruent triangles

Symmetry (reflection) – a shape has (reflection) symmetry if a reflection through a line passing through its centre produces an identical-looking shape. The shape is said to be symmetrical

Symmetrical – a shape that has symmetry

Line of symmetry – a shape has reflection symmetry about a line through its centre if reflecting it in that line gives an identical-looking shape

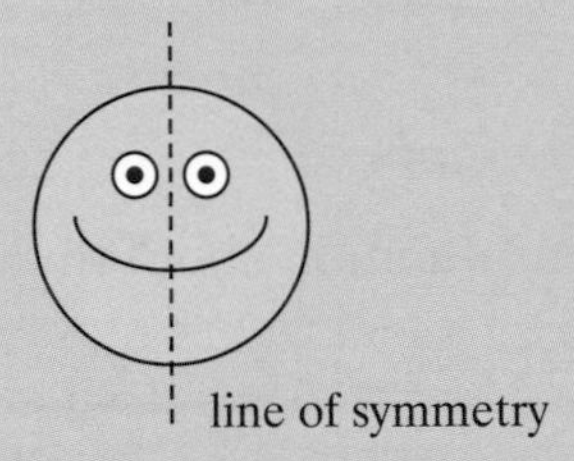

line of symmetry

Symmetry (rotation) – a shape has (rotation) symmetry if a rotation about its centre through an angle greater than 0° and less than 360° gives an identical-looking shape

Order of rotation symmetry – the number of ways a shape would fit on top of itself as it is rotated through 360°

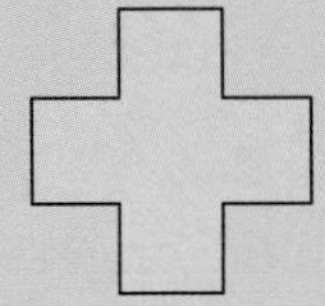

rotation symmetry order 4

(Shapes that are not symmetrical have rotation symmetry of order 1 because a rotation of 360° always produces an identical-looking shape)

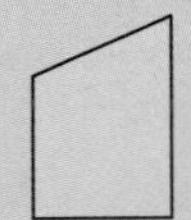

rotation symmetry order 1 (i.e. not symmetrical)

Transformation – reflections, rotations, translations and enlargements are examples of transformations as they transform one shape onto another

Reflection – a transformation involving a mirror line (or axis of symmetry), in which the line from the shape to its image is perpendicular to the mirror line. To describe a reflection fully, you must describe the position or give the equation of its mirror line, for example, the triangle A is reflected in the mirror line $y = 1$ to give the image B

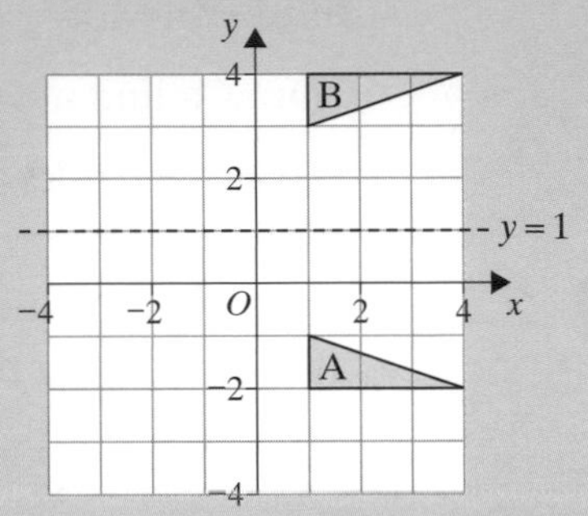

Axis of symmetry – the mirror line in a reflection

Image – the shape after it undergoes a transformation, for example, reflection or rotation

Rotation – a transformation in which the shape is turned about a fixed point called the centre of rotation. To describe a rotation fully, you must give the centre, angle and direction (a *positive angle* is *anticlockwise* and a *negative angle* is *clockwise*), for example, the triangle A is rotated about the origin through 90° anticlockwise to give the image C

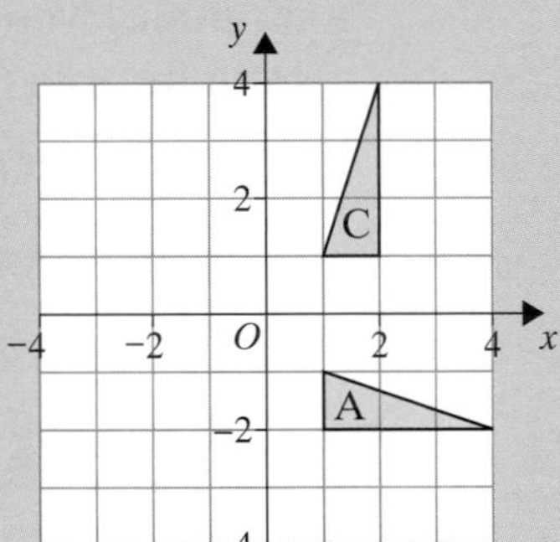

Centre of rotation – the fixed point around which the object is rotated

Learn 1 Reflections

Example: Show the image of the object ABCD in the given mirror line.

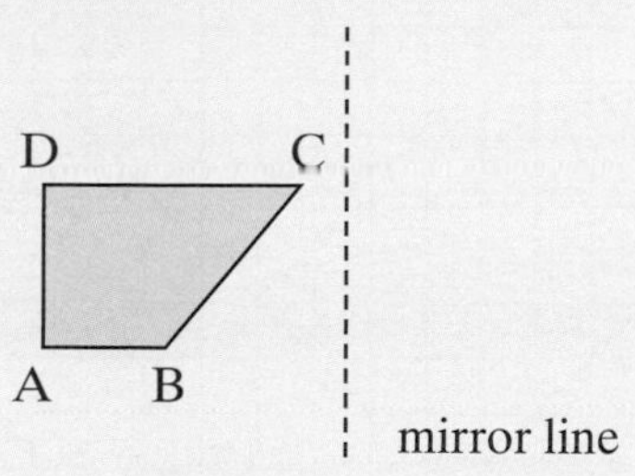

The diagram below shows the image of the object ABCD in the given mirror line.

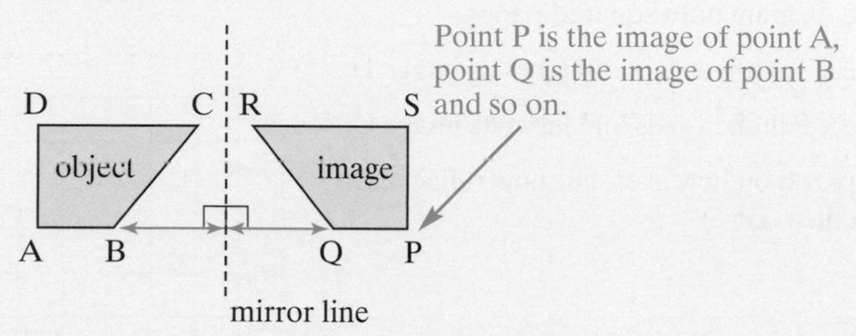

For each point on the object, there is a point on the image an **equal distance from the mirror line, but on the other side of it.** The mirror line is a line of symmetry. The object and image are congruent (the same size and shape).

Apply 1

1 Copy each shape onto squared paper. Draw its reflection in the mirror line.

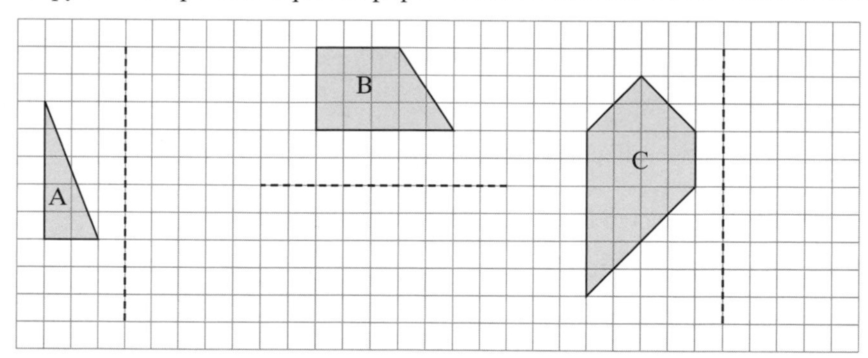

2 Copy each shape onto squared paper. Draw its reflection in the mirror line.

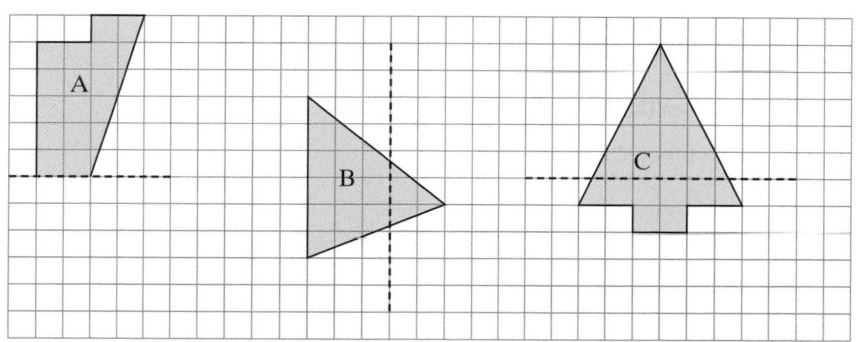

3 Copy each shape onto squared paper. Draw its reflection in the mirror line.

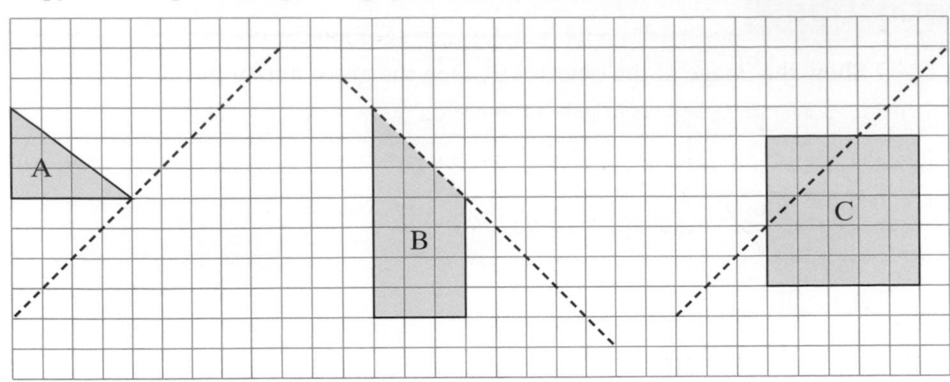

4 **a** Copy the diagram onto squared paper.

b **i** Reflect T in the x-axis and label its image U.

ii Reflect P in the x-axis and label its image Q.

c Repeat part **b** on new axes, but now reflect each shape in the y-axis.

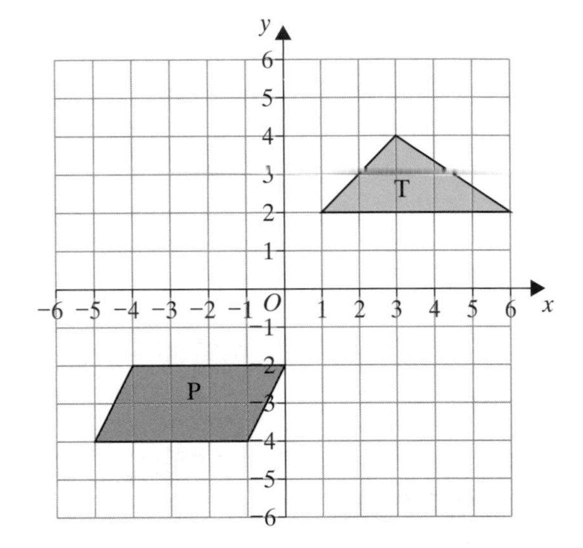

5 **a** Use axes of x and y from -6 to 6 on squared paper.

i Draw the triangle, T, by joining the points $(-3, -4)$, $(-5, 0)$ and $(-1, 0)$.

ii Draw the trapezium, P, by joining $(2, 0)$, $(5, 0)$, $(5, 3)$ and $(0, 3)$.

iii Reflect each shape in the y-axis.

b Repeat part **a** on new axes, but now reflect each shape in the x-axis.

6 Neil says that when point P(2, 3) is reflected in the x-axis the image is Q$(-2, 3)$.

a Draw a sketch to show that this is not true.

b What is the image of P when it is reflected in the x-axis?

c What did Neil do wrong?

7 Use axes of x and y from -8 to 8.

a Draw a rhombus, R, with vertices $(0, 1)$, $(2, 4)$, $(0, 7)$ and $(-2, 4)$.

b Reflect R in the line $x = 3$ and label the image S.

c Reflect R in the line $x = -3$ and label the image T.

8 Use axes of x and y from -6 to 6.

a Draw a pentagon with vertices at $(-2, -1)$, $(2, -1)$, $(2, 1)$, $(0, 2)$ and $(-2, 1)$. Label the pentagon P.

b Draw the reflection of P in the line $y = 2$. Label the image Q.

c Draw the reflection of P in the line $y = -2$. Label the image R.

9 When Naomi was asked to reflect the shape S in the line $x = 1$, she drew this diagram.

a What did she do wrong?

b Draw a diagram to show the correct image.

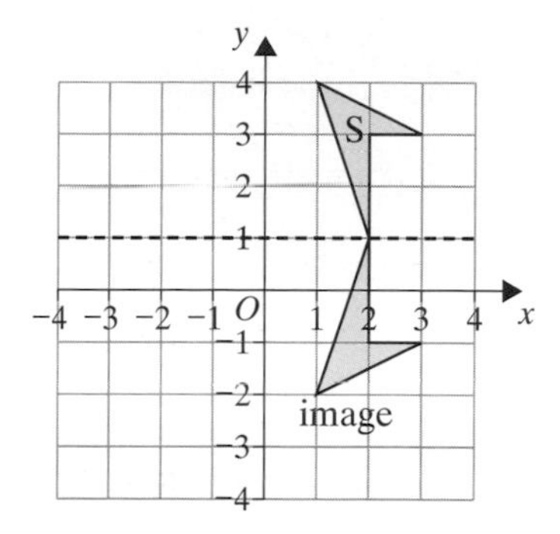

10 Get Real!

A Rangoli is a floor decoration used to welcome visitors to an Indian home.
This diagram shows half of a Rangoli pattern. Copy it onto isometric dotty paper and reflect the shapes in the line AB to show the rest of the pattern.

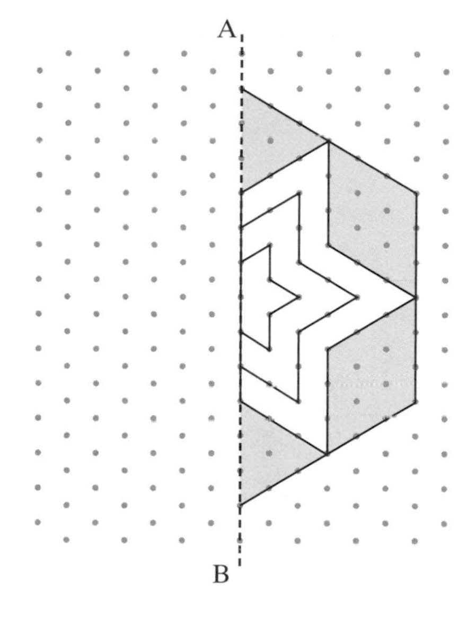

11 a Draw the mirror line $y = x$ on axes of x and y from -6 to 6.

b i Draw the triangle, T, by joining the points $(1, 6)$, $(3, 6)$ and $(3, 3)$.

ii Reflect T in the line $y = x$ and label its image U.

c i Draw the kite, K, by joining $(0, -2)$, $(2, -3)$, $(0, -6)$ and $(-2, -3)$.

ii Reflect K in the line $y = x$ and label it L.

12 Use axes of x and y from -8 to 8.

a Draw:

i triangle T with vertices at $(-4, 7)$, $(1, 6)$ and $(-2, 3)$

ii quadrilateral Q with vertices at $(-6, -7)$, $(2, -6)$, $(1, -2)$ and $(-3, -3)$.

b Draw the image of each shape after reflection in the line $y = -x$.

c Each shape and its image are congruent. Mark the sides and angles to show which are equal to each other.

Explore

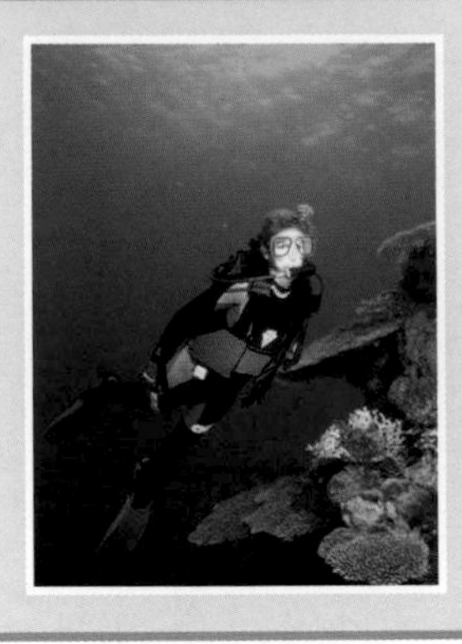

- Copy and complete the table to show the coordinates of the images of the points when they are reflected in the x-axis
- What do you notice?

Point	Image
(3, 2)	
(−2, 4)	
(1, −3)	
(−1, −2)	
(0, 5)	
(0, −1)	

Investigate further

Learn 2 Lines of symmetry

Examples:

For each of the following, write down the number of lines of symmetry.

a

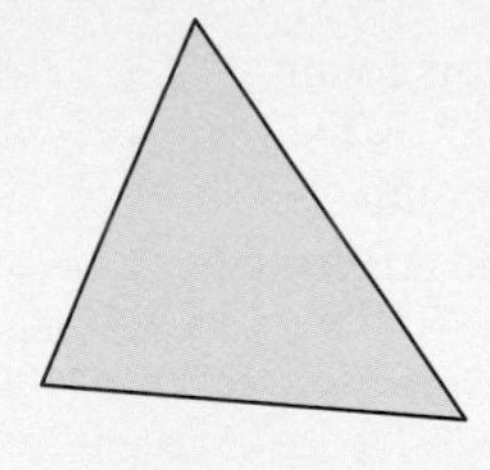

b

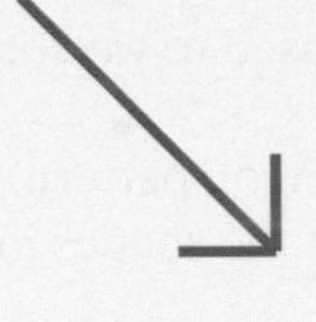

c

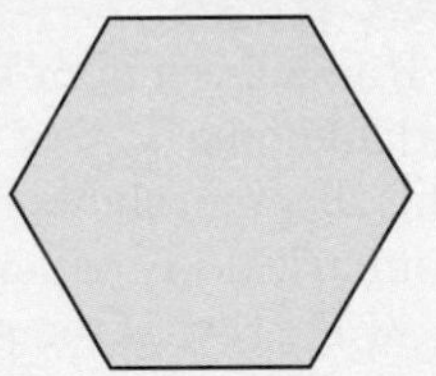

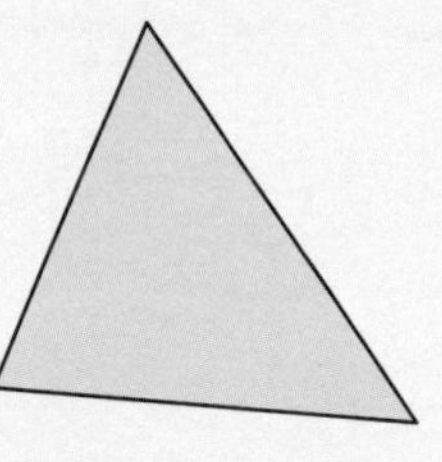

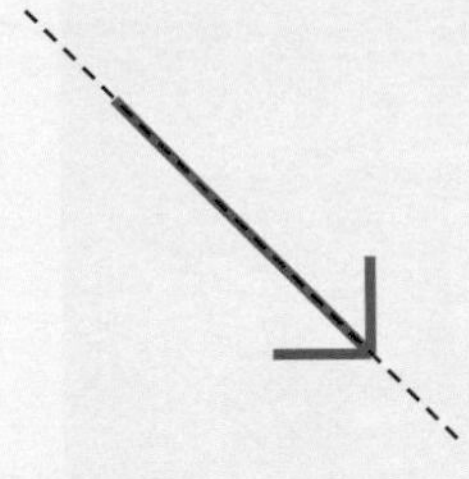

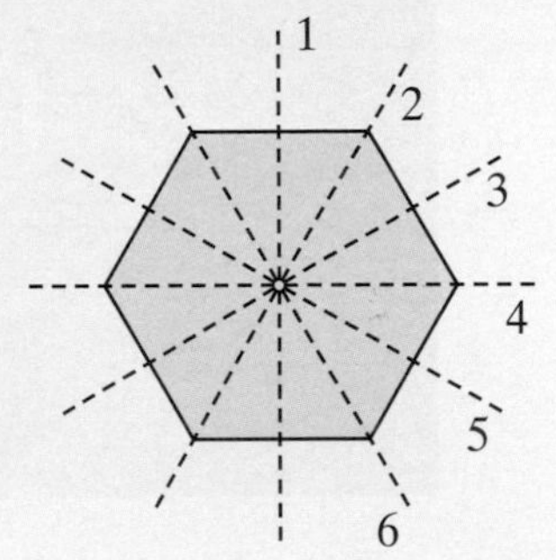

a This triangle has **no lines of symmetry**.

b This arrow has **one line of symmetry**.

c A regular hexagon has **six lines of symmetry**.

You can check lines of symmetry using a mirror or tracing paper.

Some 3-D shapes have reflection symmetry. There are many ways that you can cut a cylinder into matching halves. Two ways are shown here. In this case the mirror is a plane, rather than a line. Each point in one half of the cylinder has an image in the other half that is an equal distance from the plane.

Note that the vertical plane in the second diagram could be through any diameter of the top.

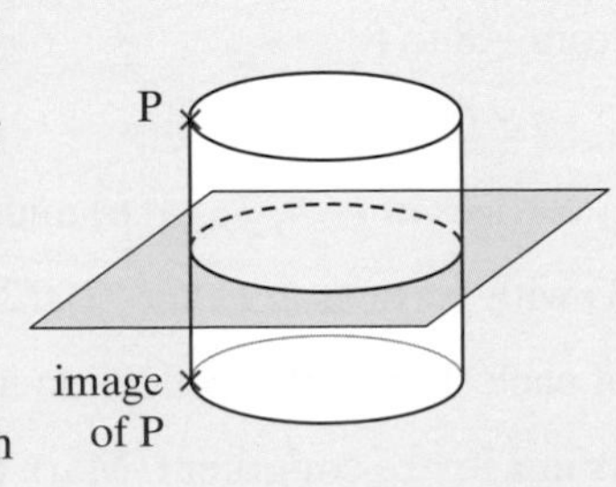

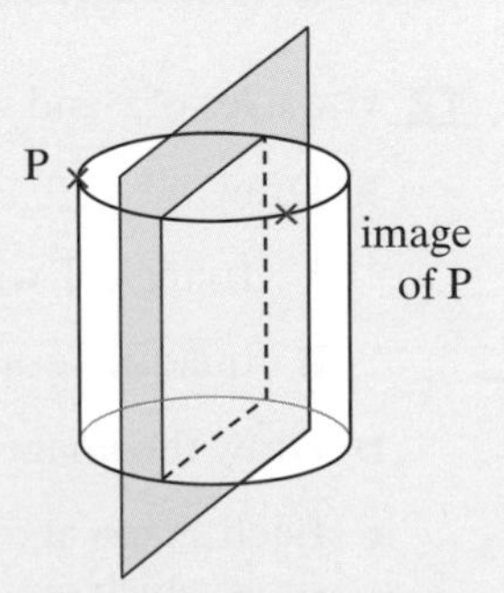

Apply 2

You can use tracing paper to help you with this exercise.

1 Each of these shapes has one line of symmetry. Copy each shape onto squared paper and draw its line of symmetry.

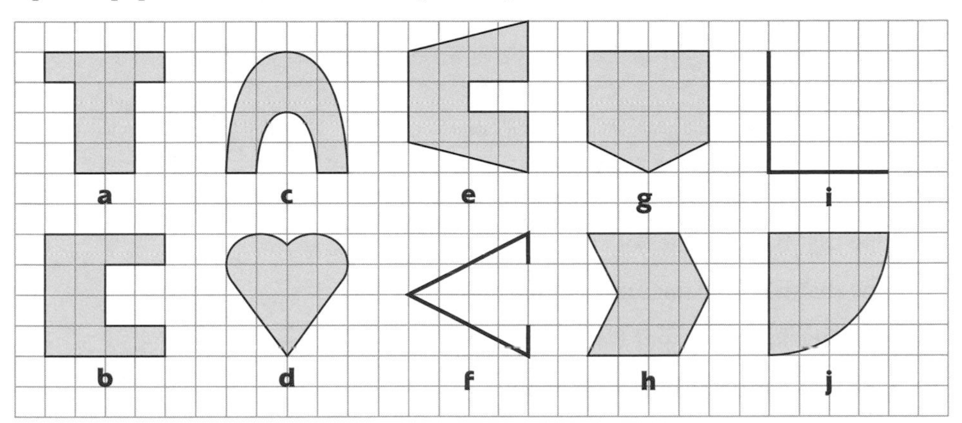

2 Copy these shapes onto squared paper. Draw **all** their lines of symmetry.

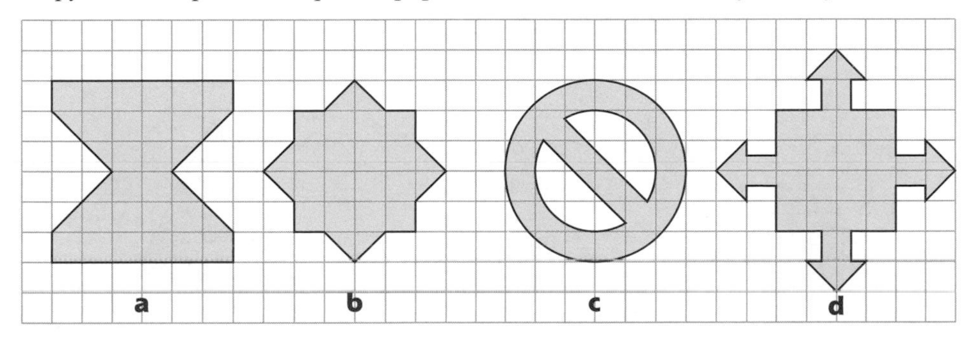

3 Get Real!
How many lines of symmetry does each of these road signs have?

No entry

No stopping

Keep left

Ice

4 How many lines of symmetry does each letter in the word MATHS have?

5 Copy and complete this table.

Shape	Number of lines of symmetry
Rectangle	
Square	
Parallelogram	
Rhombus	
Kite	

6 Copy and complete each shape so that the dotted line is a line of symmetry.

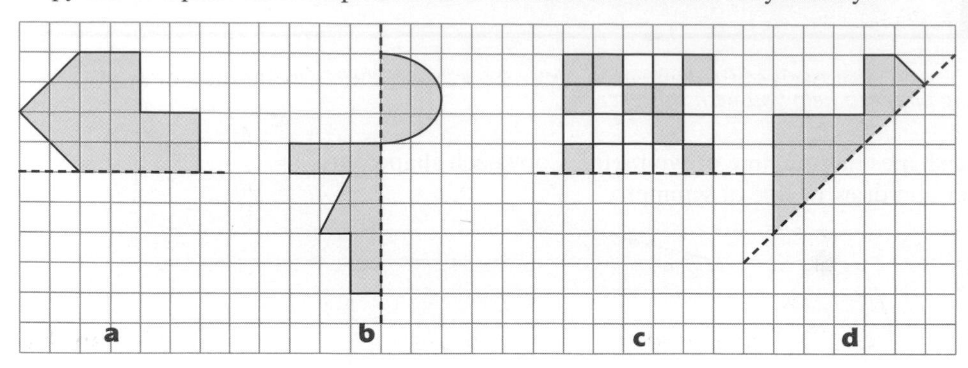

7 **a** Use squared paper for this part.

- **i** Draw any shape that has two lines of symmetry.
- **ii** Draw any shape that has four lines of symmetry.

b Use isometric or isometric dotty paper for this part.

- **i** Draw any shape that has three lines of symmetry.
- **ii** Draw any shape that has six lines of symmetry.

8 Which of these 3-D solids have reflection symmetry?
In each case say how many different ways you can cut the solids into matching halves.

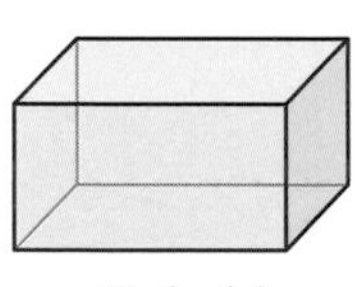
Cuboid

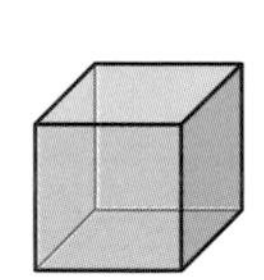
Cube

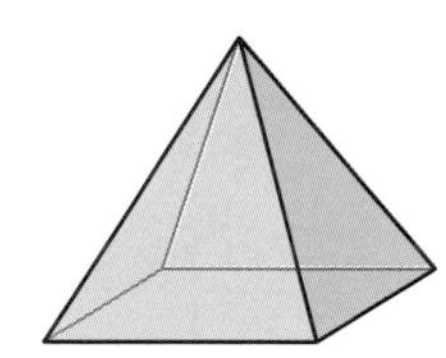
Square-based pyramid

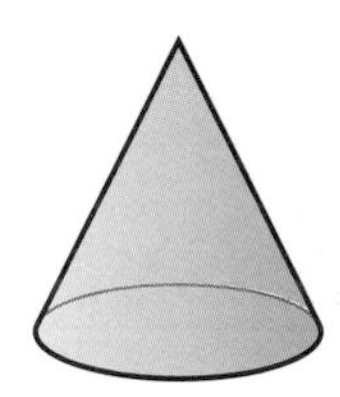
Cone

9 These solids are made from cubes.
How many planes of symmetry does each solid have?

a
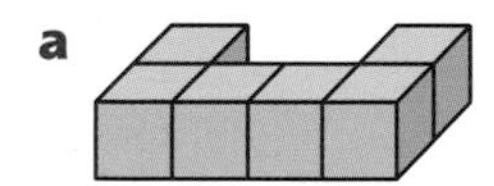

b

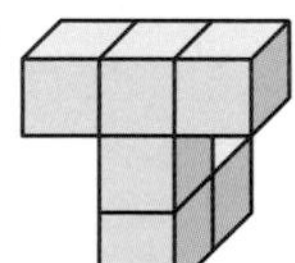

c

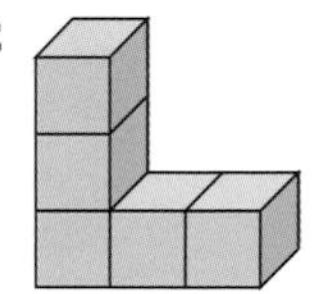

Explore

How many lines of symmetry does each of the following shapes have?

- Equilateral triangle
- Square
- Regular pentagon
- Regular hexagon

Investigate further

Learn 3 Rotation symmetry

Examples: For each of the following, write down the order of rotation symmetry.

a

b

c

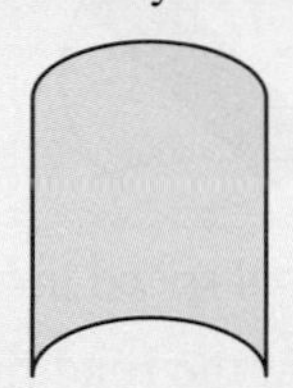

a This double-headed arrow has **rotation symmetry of order 2**.

It would look the same if it was turned upside down.

b An equilateral triangle has **rotation symmetry of order 3**.

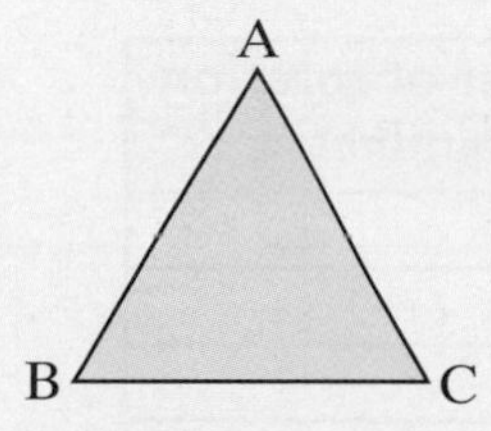

The triangle could be turned so that the vertex now at A could also be at B or C.

c This shape has **rotation symmetry of order 1**.

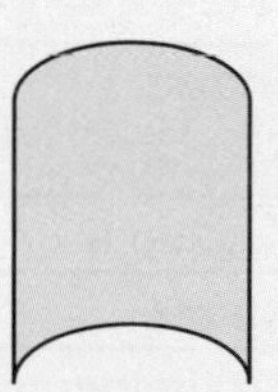

You cannot turn it to another position so that it looks the same as this.

Apply 3

You can use tracing paper to help you with this exercise.

1 Write down the order of rotation symmetry of each shape.
Use tracing paper to help if you wish.

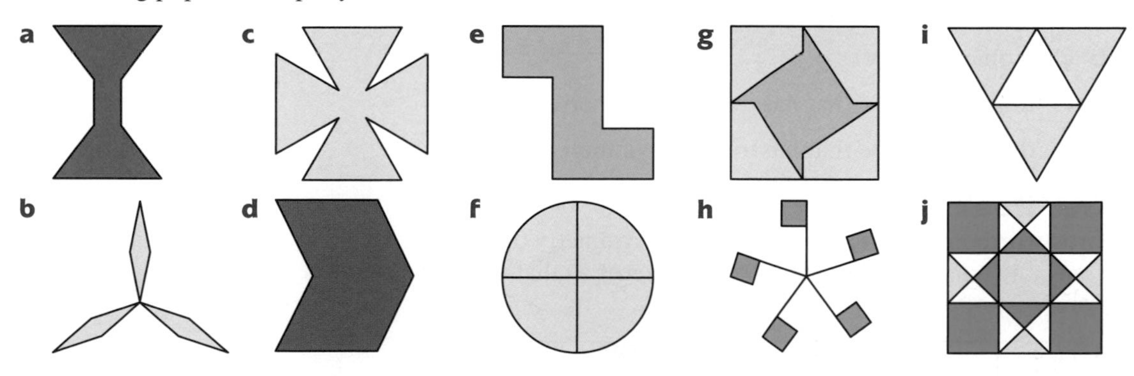

2 Find the order of rotation symmetry of each shape in question **2** of Apply **2**.

3 Get Real!

a What order of rotation symmetry does each of these road signs have?

National speed limit

Level crossing

Mini-roundabout

b Look at the road signs in question **3** of Apply **2**.
For each sign write down the order of rotation symmetry.

4 What order of rotation symmetry does each letter in the word MATHS have?

5 Copy and complete this table.

Shape	Order of rotation symmetry
Rectangle	
Square	
Parallelogram	
Rhombus	
Kite	

6 Describe carefully the line and rotation symmetry in each of the following logos.

a 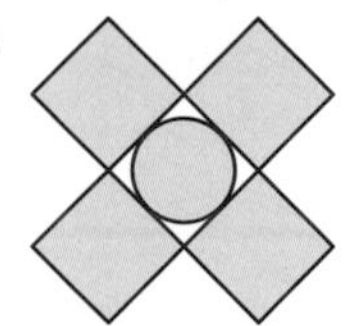

b 

c 

d 

e 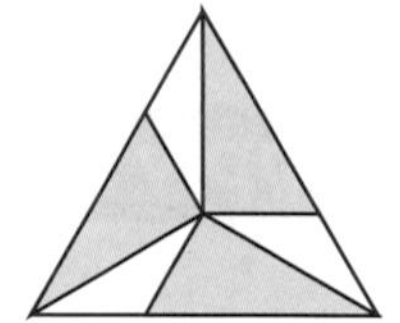

7 **a** On squared paper:

i draw any shape that has rotation symmetry of order 2

ii draw any shape that has rotation symmetry of order 4.

b On isometric paper:

i draw any shape that has rotation symmetry of order 3

ii draw any shape that has rotation symmetry of order 6.

8 Get Real!

Crossword puzzles usually have rotation symmetry of order 2.
Copy and complete these crossword patterns so that they have rotation symmetry of order 2.

a

b 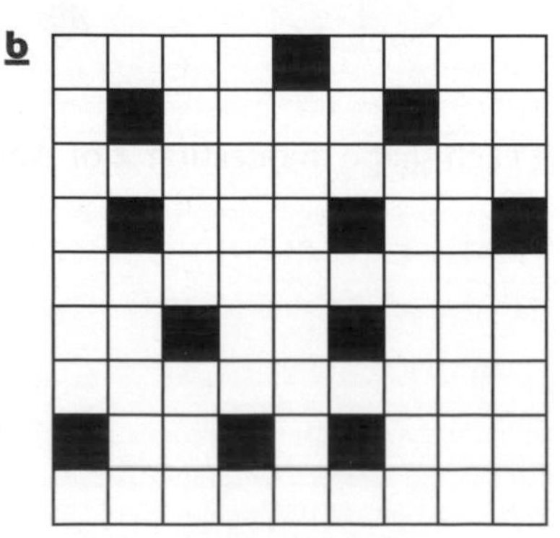

9 **a** Which special quadrilaterals have two lines of symmetry and rotation symmetry of order 2?

b Draw a hexagon that has two lines of symmetry and rotation symmetry of order 2.

Explore

Copy and complete this table

Shape	Order of rotation symmetry
Equilateral triangle	
Square	
Regular pentagon	
Regular hexagon	

Investigate further

Learn 4 Rotations

Examples:

a Show the image of the following object after a rotation of 180° about the point R.

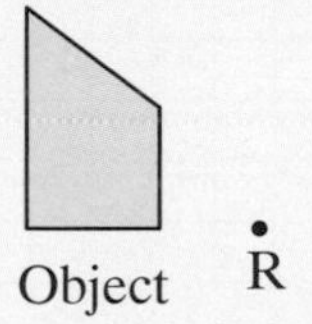

b Show the image of the following object after a rotation of 120° anticlockwise about the point R.

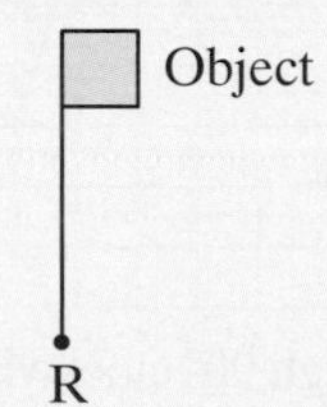

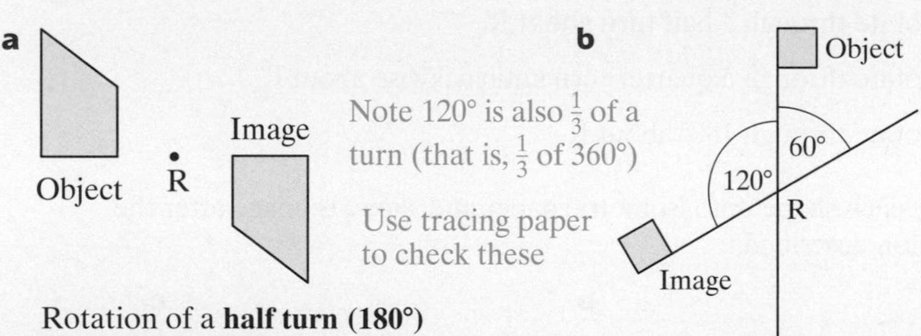

Rotation of a **half turn (180°) about R** (could be clockwise or anticlockwise).

Rotation of **120° anticlockwise** (or **240° clockwise**) **about R**.

In a rotation everything is turned through an angle about a point called the centre of rotation. The angle of rotation can be given in degrees or as a fraction of a turn and can be clockwise (–) or anticlockwise (+). When an object is rotated it gives an image of the same size and shape (that is, the object and image are congruent).

Apply 4

You can use tracing paper to help you with this exercise.

1 Copy each shape and show its image after the rotation described.

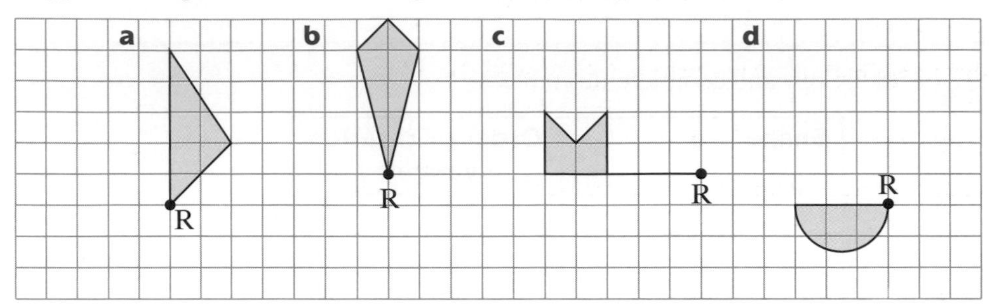

a Rotate through 90° anticlockwise about R.

b Rotate through 180° about R.

c Rotate through a quarter turn clockwise about R.

d Rotate through a half turn about R.

2 Copy each shape and show its image after the rotation described.

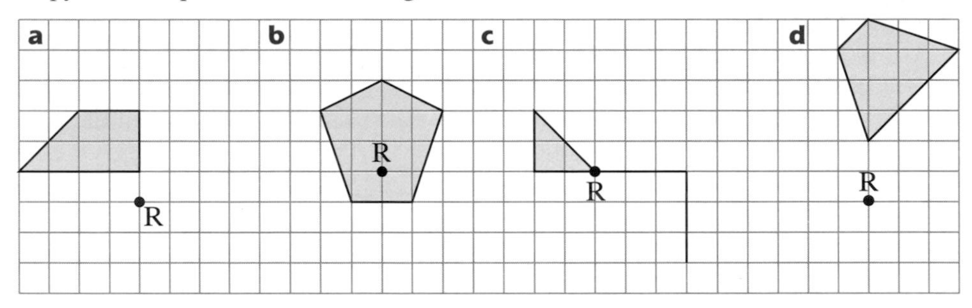

a Rotate through 90° clockwise about R.

b Rotate through a half turn about R.

c Rotate through a quarter turn anticlockwise about R.

d Rotate through 180° about R.

3 Copy each shape onto isometric paper and show its image after the rotation described.

a **b** **c**

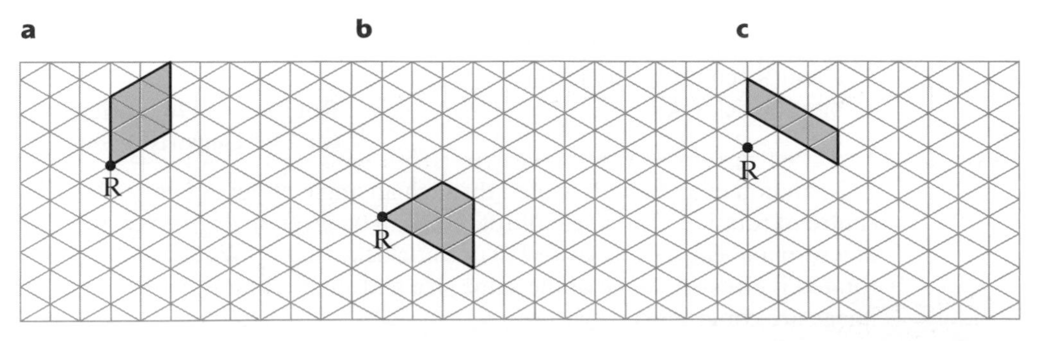

a Rotate through 60° clockwise about R.

b Rotate through 120° anticlockwise about R.

c Rotate through $\frac{1}{3}$ of a turn clockwise about R.

4 a On squared paper draw axes of x and y from -6 to 6.
Join the points (1, 1), (3, 1) and (1, 5) to form a triangle and label it A.

b Rotate A through 180° about the origin O and label the image B.

5 a Draw axes of x and y from -6 to 6 and a trapezium with vertices $(-6, 1)$, $(-2, 1)$, $(-3, 4)$ and $(-4, 4)$ on squared paper.
Label the trapezium T.

b Rotate T through a half turn about the origin O and label the image U.

6 a Copy the diagram onto squared paper.

b Draw the image of the T-shape after a rotation of 90° clockwise about the origin O. Label it A.

c Draw the image of the T-shape after a rotation of 180° about the origin O. Label it B.

d Draw the image of the T-shape after a rotation of 90° anticlockwise about the origin O. Label it C.

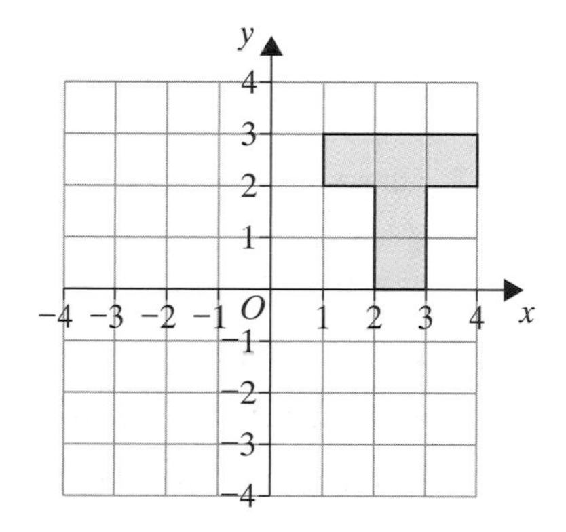

7 a Using axes x and y from -6 to 6, draw the triangle, T, with vertices (3, 4), (5, 0) and (1, 0).

b Draw the image of T after a rotation of 90° anticlockwise about O and label it U.

8 a Using axes of x and y from -6 to 6, join the points $(-3, 0)$, $(-1, -2)$, $(-3, -6)$ and $(-5, -2)$ to form a kite, K.

b Draw the image of the kite after a quarter-turn rotation anticlockwise about O. Label it L.

9 Angela says that when point P(2, -3) is rotated 90° clockwise about the origin the image is Q(3, 2).

a Draw a sketch to show that this is not true.

b What is the image of P when it is rotated 90° clockwise about the origin?

c What did Angela do wrong?

10 Get Real!
The diagram shows the side view of a cat flap.
From this position it can rotate up to 120° clockwise or 120° anticlockwise about the top, R.
Copy the cat flap and show its positions when it is fully open in each direction.

R

11 On squared paper draw axes of x and y from -8 to 8.

a Draw a trapezium with vertices (2, 1), (4, 0), (4, 3) and (2, 2).
Label the trapezium T.

b Draw the image of T after a half turn about the point (3, 4).
Label the image U.

c Draw the image of T after a half turn about the point (3, -2).
Label the image V.

d Draw the image of T after a half turn about the point (-2, -2).
Label the image W.

e What do you notice about the three images?

12 On squared paper draw axes of x and y from -6 to 6.

a Draw a triangle with vertices (-2, 4), (-2, 2) and (-6, 2).
Label the triangle A.

b Draw the image of A after a rotation of 90° anticlockwise about the point (0, 4). Label it B.

c Draw the image of A after a rotation of 90° anticlockwise about the point (-2, 0). Label it C.

d Mark the sides and angles in A, B and C to show which are equal.

e What can you say about the coordinates of the images?

Explore

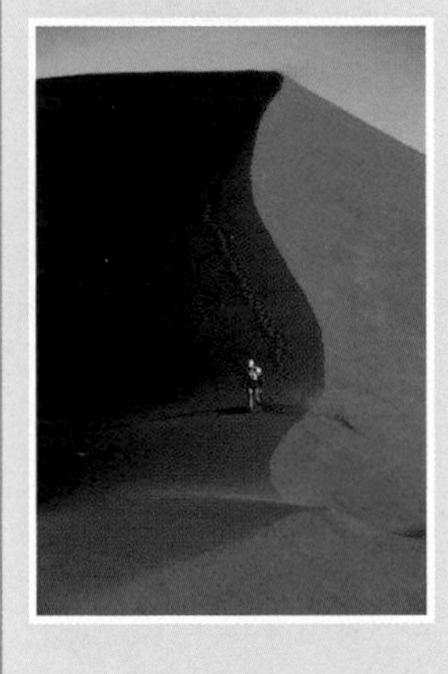

Copy and complete the table to show the coordinates of the images of the given points when they are rotated through 180° about the origin, O

Point	Image
(3, 2)	
(−2, 4)	
(1, −3)	
(−1, −2)	
(0, 5)	
(0, −1)	

What do you notice?

Investigate further

Learn 5 Describing reflections and rotations

Examples:

a Describe fully the transformation that maps (moves)

i T onto U **ii** T onto V.

b Describe fully the transformation that maps (moves)

i A onto B **ii** A onto C.

a

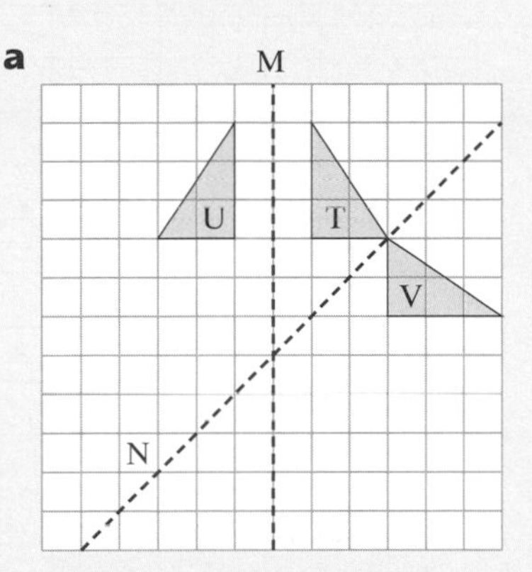

i The transformation that **maps** T onto U is a **reflection in the mirror line M.**

ii The transformation that maps T onto V is a **reflection in the mirror line N.**

If it is difficult to find the mirror line, mark points halfway between each vertex and its image. Joining these gives the mirror line

b

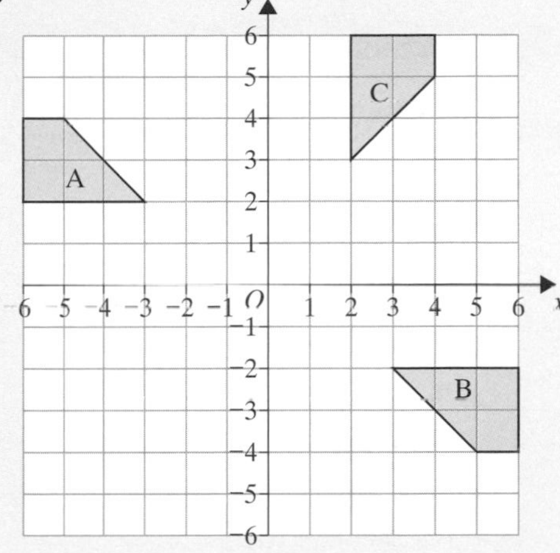

i The transformation that maps A onto B is a **rotation of 180° about the origin *O*.**

ii The transformation that maps A onto C is a **rotation of 90° clockwise (–) about the origin *O*.**

You must say anticlockwise (+) or clockwise (–) for every angle except 180°

You can use tracing paper to check these

To describe a **reflection** fully you must show the position or give the **equation of the mirror line.**

To describe a **rotation** fully you must give the **centre** and the **angle** and say whether it is **clockwise** or **anticlockwise.**

Apply 5

1 Copy these objects and their images onto squared paper.
In each case draw the mirror line.

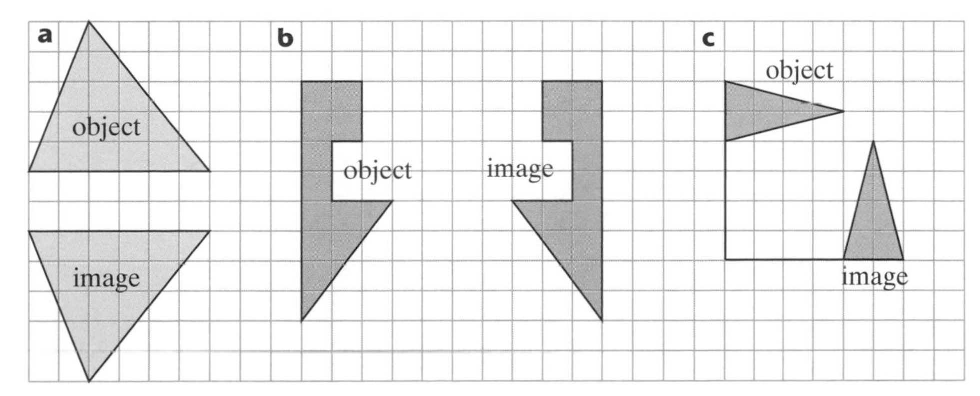

2 Write the equation of the mirror line that maps:

a A onto B

b A onto C

c B onto D

d F onto G

e A onto E

f B onto H

g C onto F

h G onto D

i C onto D

j D onto I

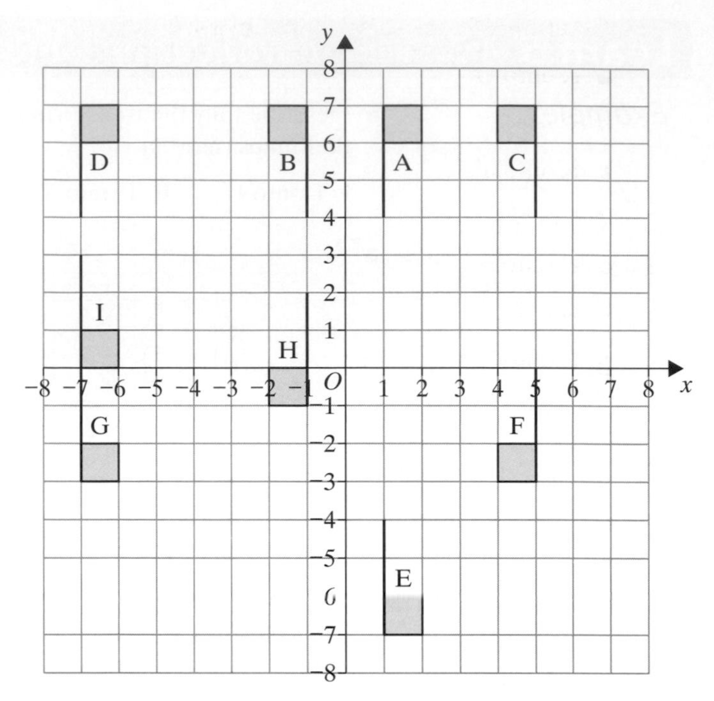

3 **a** Using axes of x and y from -8 to 8, draw and label trapezium A with vertices (3, 1), (7, 1), (6, 3) and (4, 3).

b **i** Draw and label trapezium B with vertices (1, 3), (1, 7), (3, 6) and (3, 4).

ii Describe fully the reflection that maps A onto B.

c **i** Draw and label trapezium C with vertices $(-1, -3)$, $(-1, -7)$, $(-3, -6)$ and $(-3, -4)$.

ii Describe fully the reflection that maps A onto C.

4 For each part the diagram shows an object and its image after a rotation about R.

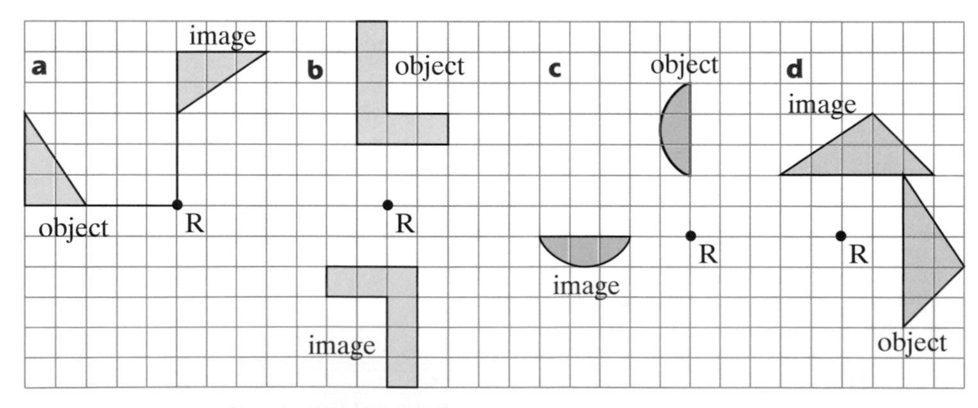

In each case describe the rotation **fully**.

5 Describe fully the rotation that maps:

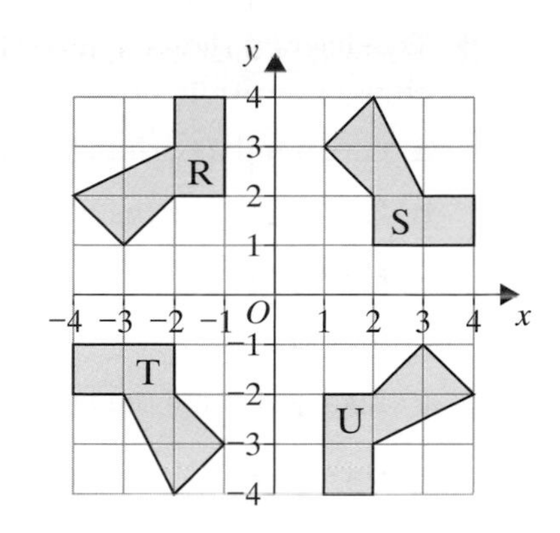

a R onto U
b U onto R
c R onto S
d S onto R
e U onto S
f U onto T
g R onto T
h T onto S

6 Describe fully the rotation that maps:

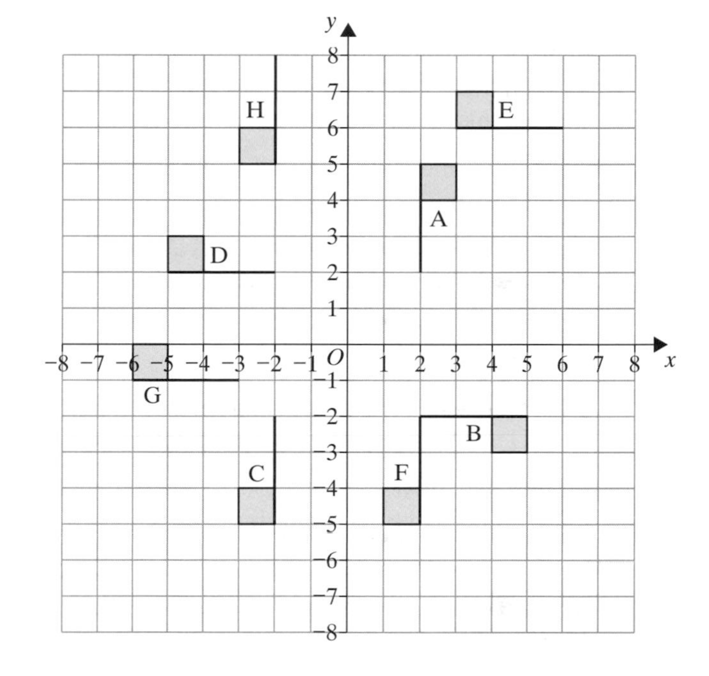

a A onto B
b A onto C
c A onto D
d D onto B
e B onto F
f A onto E
g C onto G
h H onto A
i B onto E
j D onto C

7 Describe fully the transformation that maps:

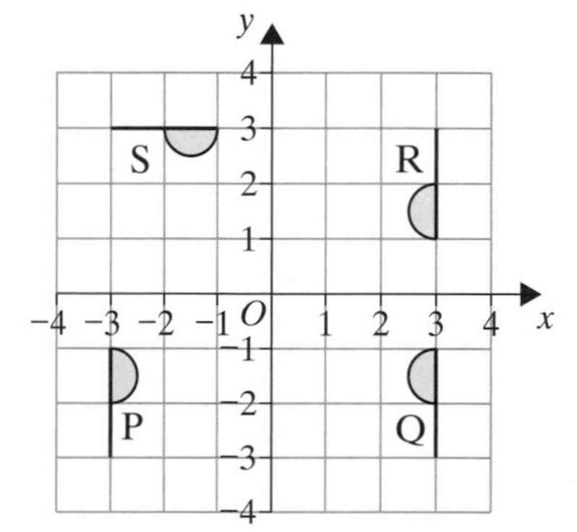

a P onto Q
b P onto R
c P onto S
d Q onto R
e Q onto S
f R onto S
g S onto P

8 **Get Real!**

a Describe fully the transformation that maps the minute hand on a clock from its position at one o'clock to its position at half past one.

b Describe fully the transformation that maps the hour hand on a clock from its position at one o'clock to its position at five o'clock.

9 The diagram shows a quadrilateral ABCD and its image PQRS after a transformation.

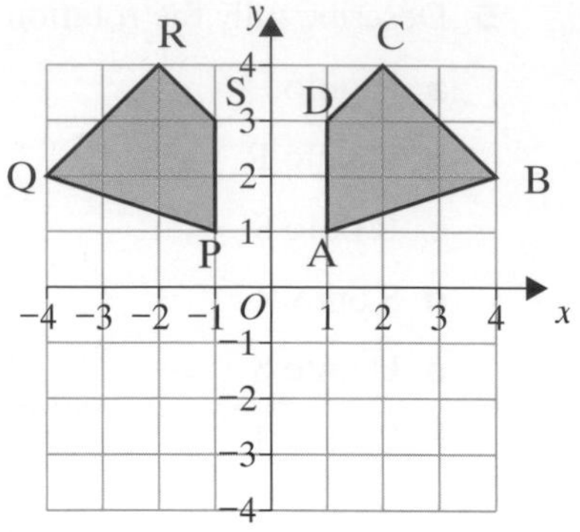

a Give a full description of the transformation.

b Find and name the length that is equal to:

i AB **iii** BD

ii CD **iv** AC

c Find and name the angle that is equal to:

i ∠CDA **ii** ∠CAD

d Find and name a triangle that is congruent to:

i DAB **ii** CAD

10 a Using axes of x and y from -6 to 6, draw the square with vertices (2, 2), (5, 2), (5, 5) and (2, 5).
Label this square S.

b i Draw and label square T with vertices (2, −2), (5, −2), (5, −5) and (2, −5).

ii Find as many transformations as you can that map S onto T.
Describe each transformation fully.

c i Draw and label square U with vertices (−2, −2), (−5, −2), (−5, −5) and (−2, −5).

ii Find as many transformations as you can that map S onto U.
Describe each transformation fully.

d Find as many transformations as you can that map U onto T.
Describe each transformation fully.

Explore

- What do you notice about the coordinates of the pairs of object and image points given in the table?
- Draw the triangle ABC and its image DEF
- Describe fully the transformation that maps ABC onto DEF
- On new axes draw the triangle PQR and its image STU
- Describe fully the transformation that maps PQR onto STU

Object	Image
A(1, 2)	D(2, −1)
B(5, 4)	E(4, −5)
C(3, −1)	F(−1, −3)
P(−5, 2)	S(2, 5)
Q(1, −1)	T(−1, −1)
R(−2, −1)	U(−1, 2)

Investigate further

Learn 6 Combining reflections and rotations

Example:

The shape A is reflected in the line $x = 2$ to give shape B.
The shape B is rotated through 180° about the point (2, 0) to give shape C.

What single transformation maps shape A onto shape C?

When shape A is reflected in the line $x = 2$, the image is shape B.

When shape B is then rotated through 180° about the point (2, 0), the image is shape C.

The single transformation that maps shape A onto shape C is a reflection in the x-axis.

The x-axis is the same as the line $y = 0$

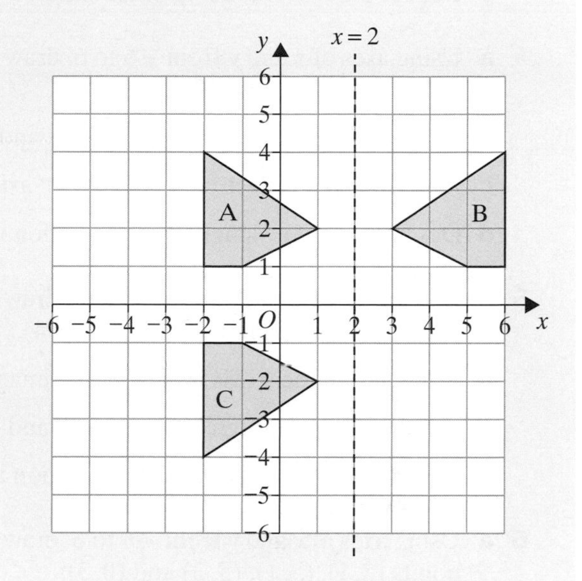

Apply 6

1 Use squared paper for this question.

a Draw eight parallel mirror lines.

b Draw a parallelogram in the centre as shown.

c Reflect the parallelogram in each mirror line that it touches (that is, in mirror lines M and N).

d Reflect the images in each mirror line the images touch.

e Continue until there is a parallelogram in each space between the lines.

M N

2 **a** On squared paper, draw two perpendicular mirror lines M and N that intersect at R and an L-shape, A, like this.

b Reflect A in mirror line M. Label the image B.

c Now reflect B in mirror line N. Label the image C.

d Describe fully the transformation that would map A onto C.

e Repeat parts **a** to **d** using other shapes.

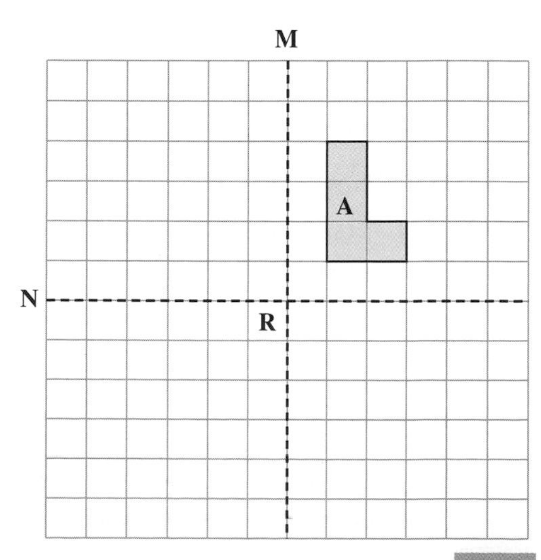

3 **a** Draw another copy of the diagram in question **2**.

b Rotate A through 90° clockwise about R. Label the image D.

c Now rotate D through 180° about R. Label the image E.

d Describe fully the transformation that would map A onto E.

e Repeat parts **a** to **d** using other shapes.

4 **a** Using axes of x and y from -6 to 6, draw the triangle, T, with vertices at (3, 4), (1, 2) and (6, 2).

b Draw the reflection of T in the x-axis and label it U.

c Now draw the reflection of U in the y-axis and label it V.

d Describe fully the single transformation that would map T onto V.

5 **a** Using axes of x and y from -6 to 6, draw the quadrilateral, Q, with vertices at (3, 6), (1, 6), (3, 1) and (5, 4).

b Reflect Q in the y-axis and label the image R.

c Now rotate R through 180° about O and label the image S.

d Describe fully the single transformation that would map Q onto S.

6 **a** Using axes of x and y from -6 to 6, draw the trapezium, A, by joining points (2, 1), (5, 1), (5, 3) and (0, 3).

b Rotate A through 90° clockwise about O and label the image B.

c Now draw the reflection of B in the y-axis and label it C.

d What single transformation would move A onto C?

7 Use axes of x and y from -8 to 8.

a Draw the pentagon, P, with vertices at $(-1, 4)$, (4, 4), (3, 5), (3, 6) and (1, 6).

b Draw the reflection of P in the line $x = -2$ and label the image Q.

c Now rotate Q through 180° about $(-2, 3)$ and label the image R.

d Describe fully the single transformation that would map P onto R.

8 Use axes of x and y from -8 to 8.

a Join the points (0, 4), (2, 6), (5, 6) and (4, 3) to give a quadrilateral, Q.

b **i** Reflect Q in the line $y = 3$ and label the image R.

ii Mark the sides and angles in Q and R to show which are equal.

c **i** Now reflect R in the line $y = -x$ and label the image S.

ii Mark the sides and angles in S to show which are equal to those in Q and R.

d Describe fully the single transformation that would map Q onto S.

9 Gary says, 'If you rotate a shape 90° clockwise about O and then reflect it in the line $x = -2$, you always get the same image that you would get if you reflected the shape in the line $x = -2$ and then rotated it 90° clockwise about O.' Use a triangle with vertices at $(-4, 2)$, $(-4, 5)$ and $(-2, 2)$ to check whether this statement is true.

10 **a** Using axes of x and y from -6 to 6, draw the kite with vertices K(3, 5), L(1, 4), M(3, 1) and N(5, 4).
Label each vertex.

b Reflect KLMN in the y-axis and label the image $K_1L_1M_1N_1$ where K_1 is the image of K, L_1 is the image of L and so on.

c Now rotate $K_1L_1M_1N_1$ through 180° about O and label the image $K_2L_2M_2N_2$ where K_2 is the image of K_1, L_2 is the image of L_1 and so on.

d Describe fully the single transformation that would map KLMN onto $K_2L_2M_2N_2$ with K mapped to K_2, L to L_2 etc.

e What other transformation would map KLMN onto $K_2L_2M_2N_2$ but in which L would not be mapped to L_2?

11 What single transformation is equivalent to a reflection in the line $y = x$ followed by a reflection in the line $y = -x$?
Draw a diagram to illustrate your answer.

Explore

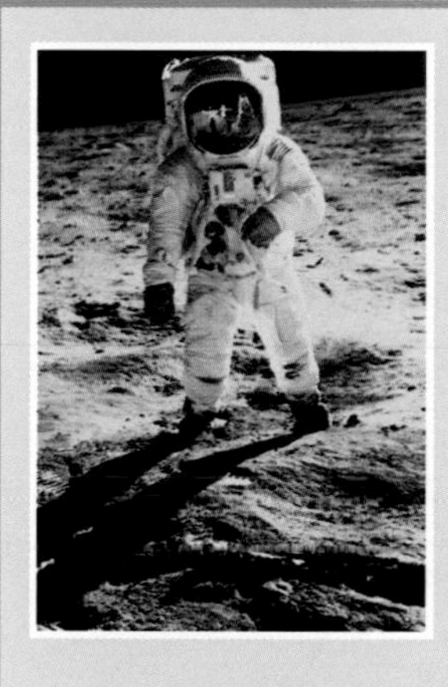

- Using axes of x and y from 0 to 12 draw triangle A as shown
- Draw the reflection of A in the line $x = 3$ and label it B
- Draw the reflection of B in the line $x = 6$ and label it C
- Describe fully the single transformation that maps A onto C
- Try this with other shapes
- Draw a new copy of the diagram and reflect A in the line $x = 3$ to give image B then reflect B in the line $x = 7$ to give C
- Describe again the single transformation that maps A onto C
- Try this with other shapes

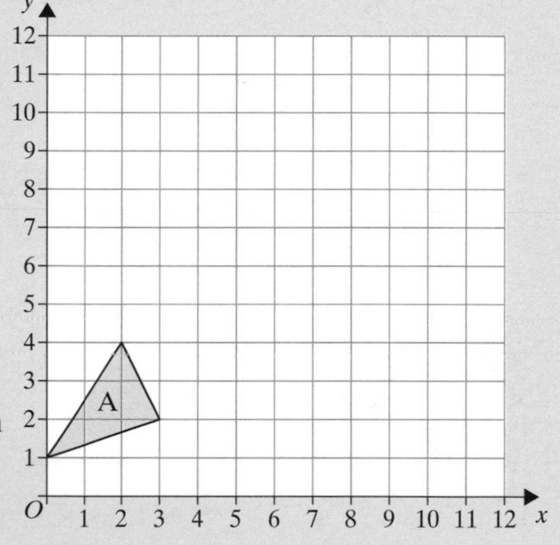

Investigate further

Reflections and rotations

ASSESS

The following exercise tests your understanding of this chapter, with the questions appearing in order of increasing difficulty.

1 Copy the given shapes and draw a line of symmetry on each one:

a 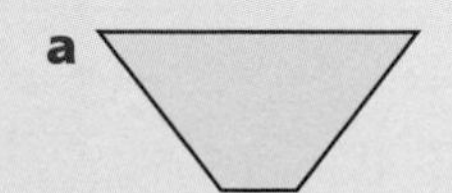

b 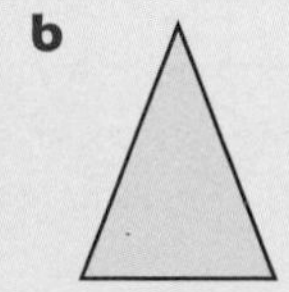

c 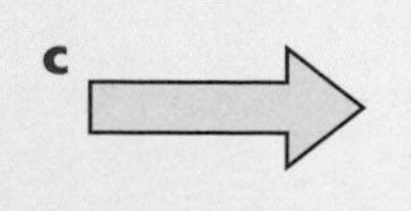

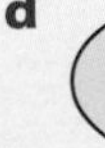

d 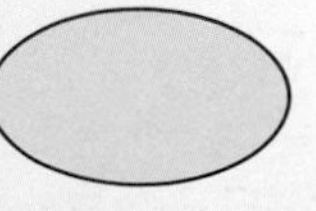

2 **a** Copy the given shapes and draw all the lines of symmetry on each one:

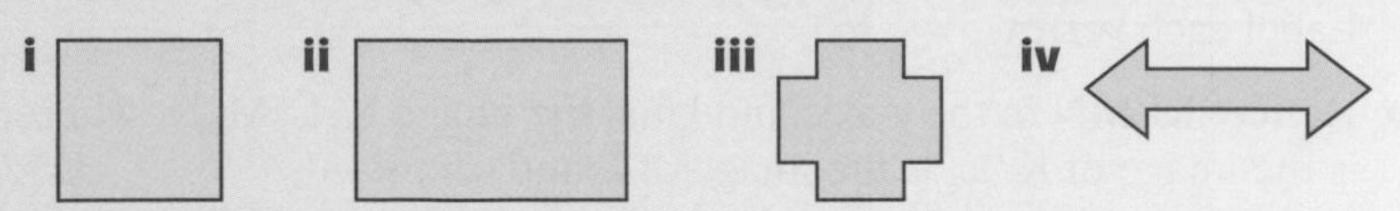

b Write down the order of rotation symmetry of the given shapes:

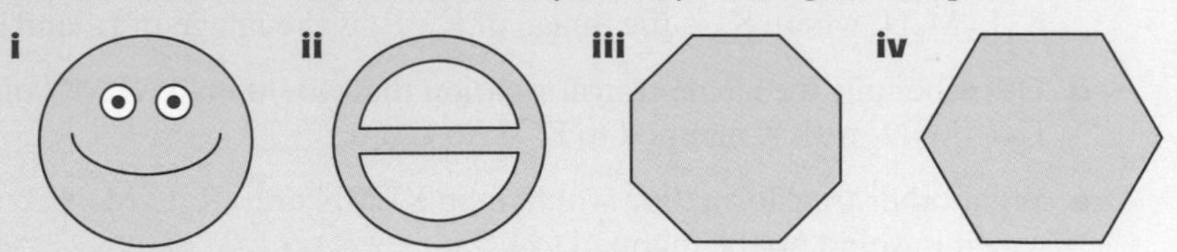

c Write down a capital letter from the alphabet that could have:

i two lines of symmetry

ii four lines of symmetry

iii an infinite number of lines of symmetry

iv rotation symmetry of order 2

v rotation symmetry of order 4

vi infinite rotation symmetry

vii rotation symmetry of order 2 but no line symmetry

3 **a** Copy each of the given shapes and draw its reflection in the mirror line shown.

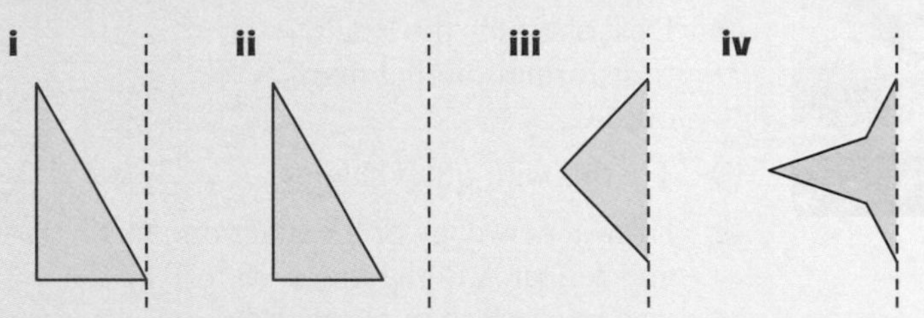

b Copy each of the given shapes and reflect them in both the x-axis and the y-axis.

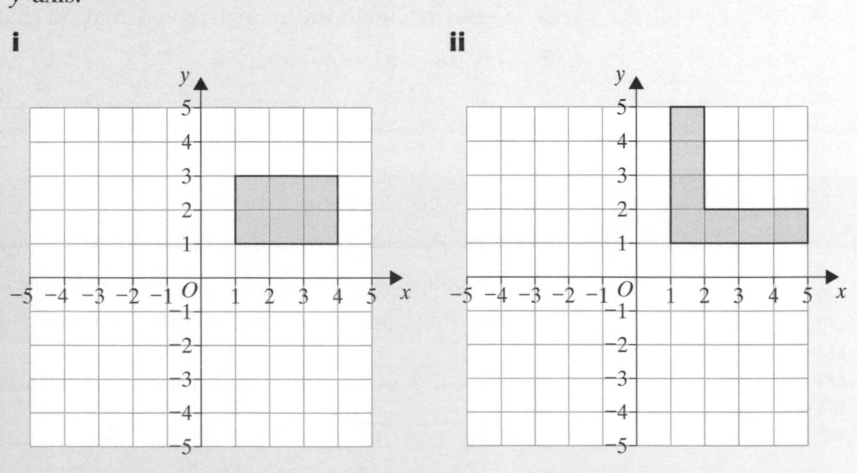

4 Reflect the object shown in the lines:

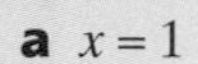

a $x = 1$

b $y = 2$

c $y = x$

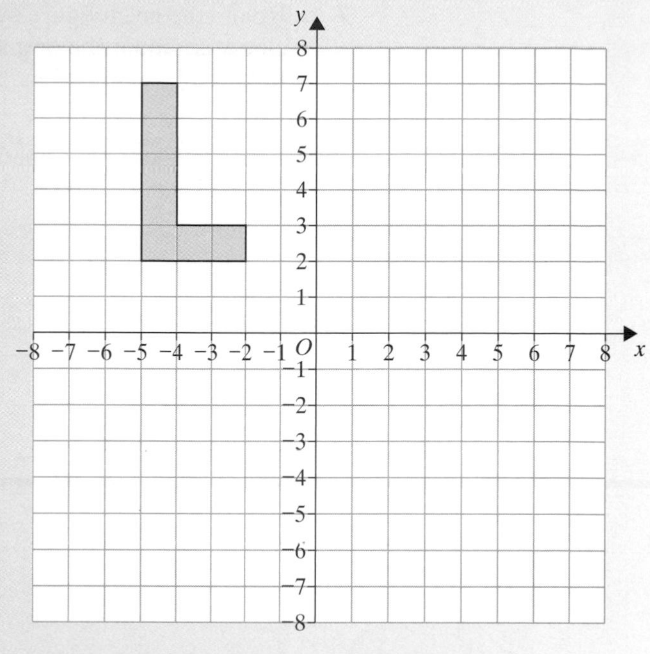

5 Copy the following diagrams and draw the image obtained by rotating the shape:

i 90° anticlockwise about the given point

ii 90° clockwise about the given point.

a

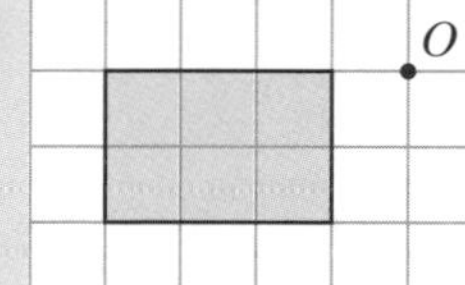

b

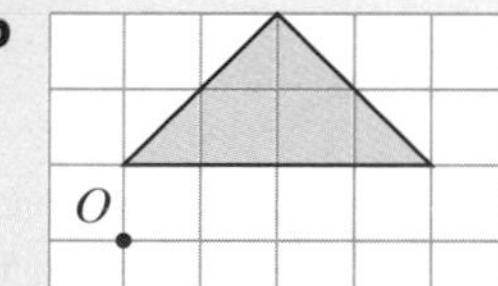

6 Find the images of the given shape after rotations about the origin of

a 90° anticlockwise

b 90° clockwise

c $\frac{3}{4}$ of a turn anticlockwise.

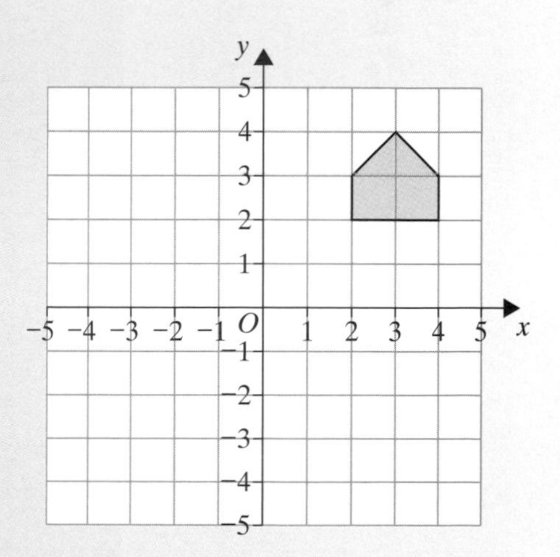

7 **a** Rotate the given figure by 90° clockwise about the origin followed by a reflection in the x-axis.

b Which single transformation is the equivalent of this?

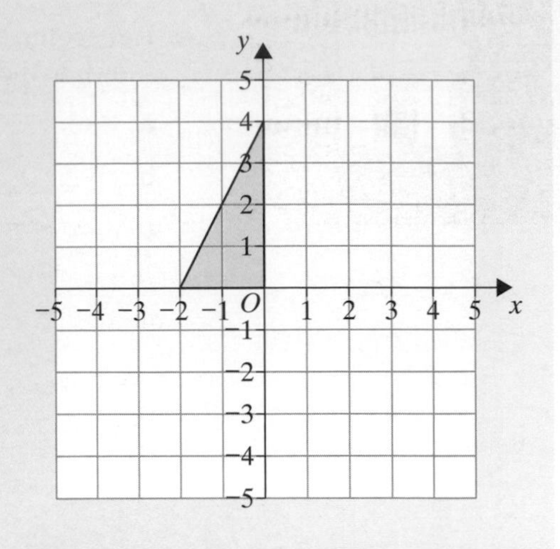

11 Trial and improvement

OBJECTIVES

Examiners would normally expect students who get a C grade to be able to:

Form and solve equations such as $x^3 + x = 12$ using trial and improvement methods

What you should already know ...

- Substitute into algebraic expressions
- Use the bracket and power buttons on your calculator

VOCABULARY

Trial and improvement – a method for solving algebraic equations by making an informed guess then refining this to get closer and closer to the solution

Decimal places – the digits to the right of a decimal point in a number, for example, in the number 23.657, the number 6 is the first decimal place (worth $\frac{6}{10}$), the number 5 is the second decimal place (worth $\frac{5}{100}$), and 7 is the third decimal place (worth $\frac{7}{1000}$); the number 23.657 has 3 decimal places

Square number – a square number is the outcome when a whole number is multiplied by itself; square numbers are 1, 4, 9, 16, 25, ...

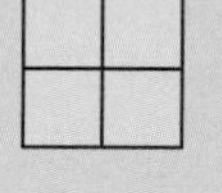
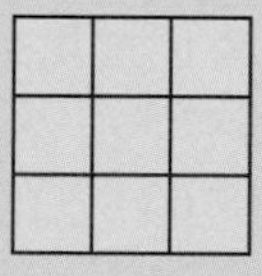
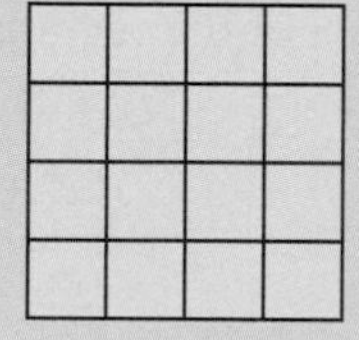

Square root – the square root of a number such as 16 is a number whose outcome is 16 when multiplied by itself

Cube number – a cube number is the outcome when a whole number is multiplied by itself then multiplied by itself again; cube numbers are 1, 8, 27, 64, 125, ...

Cube root – the cube root of a number such as 125 is a number whose outcome is 125 when multiplied by itself then multiplied by itself again

Learn 1 Trial and improvement

Examples:

a Use the trial and improvement method to solve $x^2 - x = 60$ for a positive value of x.
Give your answer to one decimal place.

Laying out the information in tabular form:

Trial value of x	x^2	$x^2 - x$	Comment
6	36	30	Too low
7	49	42	Too low
8	64	56	Too low
9	81	72	Too high
Now you know that the value lies between 8 and 9 Try 8.5			
8.5	72.25	63.75	Too high
8.4	70.56	62.16	Too high
8.3	68.89	60.59	Too high
8.2	67.42	59.04	Too low
Now you know that the value lies between 8.2 and 8.3 Try 8.25			
8.25	68.0625	59.8125	Too low
Now you know that the value lies between 8.25 and 8.3 Both 8.25 and 8.3 are the same as 8.3 to one decimal place			

You can see that the answer is closer to 8 than 9, so you might try 8.3, for example, rather than 8.5

The required value is 8.3 to one decimal place.

b The area of a rectangle is 63 cm^2.
One side is 4 cm longer than the other side as shown.

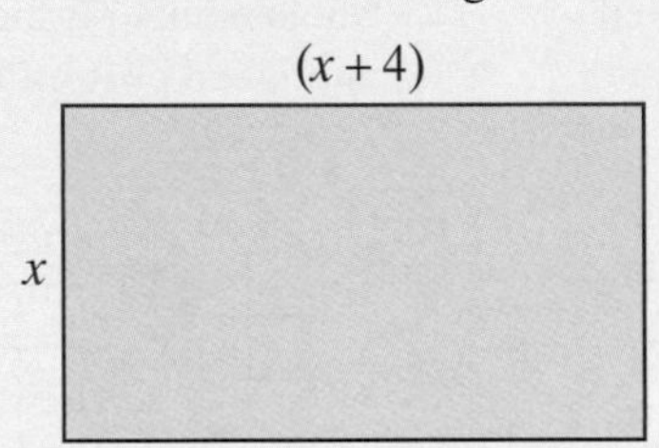

Use the trial and improvement method to find the value of x.

Give your answer to two decimal places.

Use the fact that the area of the rectangle = length × width:

$$(x + 4) \times x = 63$$
$$x^2 + 4x = 63$$

Laying out the information in tabular form:

Start with a good guess

You can see that the answer is closer to 6 than 7, so you might try 6.2, for example, rather than 6.5

Trial value of x	$x^2 + 4x$	Comment
6	$6^2 + (4 \times 6) = 60$	Too low
7	$7^2 + (4 \times 7) = 77$	Too high
So now you know that the value lies between 6 and 7 Try 6.5		
6.5	$6.5^2 + (4 \times 6.5) = 68.25$	Too high
6.3	$6.3^2 + (4 \times 6.3) = 64.89$	Too high
6.2	$6.2^2 + (4 \times 6.2) = 63.24$	Too high
6.1	$6.1^2 + (4 \times 6.1) = 61.61$	Too low
Now you know that the value lies between 6.1 and 6.2 Try 6.15		
6.15	$6.15^2 + (4 \times 6.15) = 62.4225$	Too low
6.18	$6.18^2 + (4 \times 6.18) = 62.9124$	Too low
6.19	$6.19^2 + (4 \times 6.19) = 63.0761$	Too high
Now you know that the value lies between 6.18 and 6.19 But you need to give the value to two decimal places Try 6.185		
6.185	$6.185^2 + (4 \times 6.185) = 62.994225$	Too low
Now you know that the value lies between 6.185 and 6.19 But both values are 6.19 to two decimal places		

The required value is $x = 6.19$ to two decimal places.

Apply 1

1 Match the following numbers to their squares.
The first one is done for you.

x	x^2
3.2	62.41
4	68.89
6	16
7.9	94.09
8.1	81
8.3	100
9	36
9.3	65.61
9.7	86.49
10	10.24

(8.1 → 65.61)

 **2** Decide which of the following statements are true and which are false.

a 4.2^2 is between 16 and 25.

b 7.11^2 is between 64 and 81.

c 10.8^2 is between 64 and 81.

d 8.6^2 is between 64 and 81.

e 0.3^2 is between 0 and 1.

f 1.99^2 is between 1 and 4.

g 14.3^2 is between 144 and 196.

3 Use the trial and improvement method to solve the equation $x^2 + 4 = 55$ for a positive value of x.
Give your answer to one decimal place.
The table has been started for you. Copy and complete it.

Don't just guess the next value: 7.5 is exactly halfway between 7 and 8 so makes the most sense

Trial value of x	x^2	$x^2 + 4$	Comment
3	9	13	Too low
4	16	20	Too low
5	25	29	Too low
6	36	40	Too low
7	49	53	Too low
8	64	68	Too high
Now you know that the value lies between 7 and 8.			
7.5	56.25	60.25	Too high
?			
?			

4 Use the trial and improvement method to solve the equation $x^2 + 11 = 134$ for a positive value of x.
Give your answer to one decimal place.
The table has been started for you. Copy and complete it.

Trial value of x	x^2	$x^2 + 11$	Comment
9	81	92	Too low
10	100	111	Too low

5 **Get Real!**

Angus the farmer knows the area of his square fields but not their lengths.
Use the trial and improvement method to find the lengths of the fields.
Give your answers to two decimal places.

a

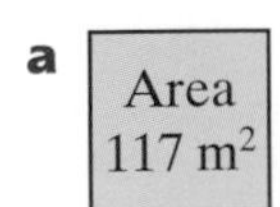

b Area $500\,\text{m}^2$

c Area $350\,\text{m}^2$

6 Find, using the trial and improvement method, exact positive solutions of these equations.
Remember to show all your working.

a $x^2 + 2x = 63$

b $x^2 - 2x = 675$

c $x^2 + 5x = 336$

d $x^2 - 7x = 368$

e $x^3 + x = 520$

f $x^5 = 32\,768$

g $x - x^3 = -336$

h $2x - x^3 = -711$

7 Get Real!
The area of Sunita's lawn is 50 square metres.
The length of the lawn is one metre longer than the width.
Use the trial and improvement method to work out the length and width of the lawn.
Give your answers to one decimal place.

8 Use the trial and improvement method to find the lengths of these rectangles correct to one decimal place.
Remember to show all your working.

a $(x + 3)$ by x: Area = 40 cm^2

b $(x + 1)$ by x: Area = 18 cm^2

c

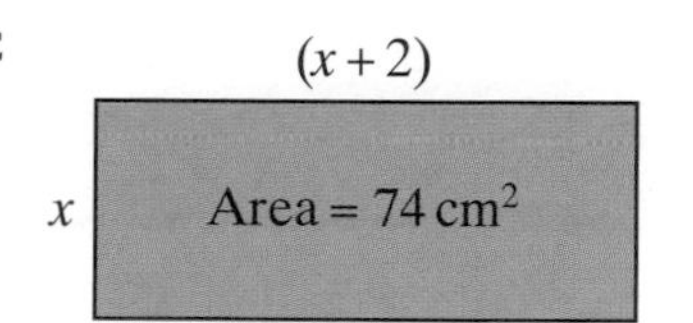

d

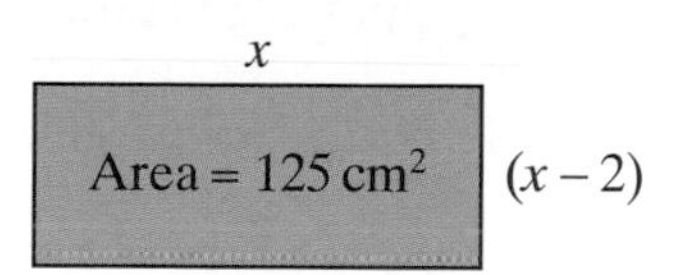

9 Use the trial and improvement method to find one solution to each of these equations.

Give your answers to two decimal places.
Remember to show all your working.

a $x^2 - 2x = 11.4$ if the solution lies between 4 and 5

b $x^3 + x = 616$ if the solution lies between 8 and 9

10 Use the trial and improvement method to find one solution to each of these equations.

Give your answers to two decimal places.
Remember to show all your working.

a $x^3 = 20$ if x lies between 2 and 3

b $x^2 - 2x + 5 = 23.9$ if x lies between 5 and 6

c $x^3 + 50 = 0$ if x lies between −4 and −3

d $x^3 + 19 = 0$ if x lies between −2 and −3

e $x^3 - 4x + 2 = 0$ if x lies between −2 and −3

f $x^3 + 3x = -176$ if x lies between −5 and −6

Explore

- This shape consists of a central square and four equal arms

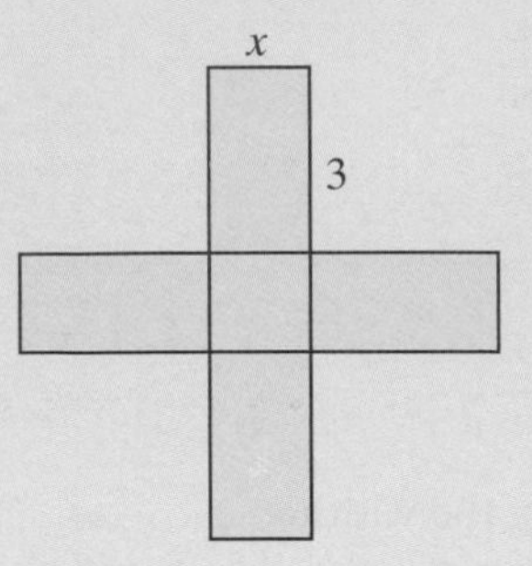

- The area of the shape is 69 cm^2
- Form an equation and use the trial and improvement method to find the value of x to two decimal places
- What happens if you change the value of the area?

Investigate further

Trial and improvement

ASSESS

The following exercise tests your understanding of this chapter, with the questions appearing in order of increasing difficulty.

1 Without using a calculator, decide which of these statements is true and which is false.

a 3.7^2 is between 9 and 16

b 4.6^3 is between 16 and 25

c $\sqrt{55}$ is less than 7

d $\sqrt[3]{55}$ is greater than 3 but less than 4.

2 a Surindra is using a calculator to find the positive square root of 115 by the trial and improvement method. Some of her working is shown below.

x	x^2	Comment
10	100	Too low
11	121	Too high
10.5		

Copy the table and continue it to find $\sqrt{115}$ to two decimal places.

b Malachi is using a calculator to find the cube root of 115 by the trial and improvement method. The start of his working is shown below.

x	x^3	Comment
4	64	Too low
5		
4.5		

Copy the table and continue it to find $\sqrt[3]{115}$ to two decimal places.

3 a The equation $x^2 - 5x = 4$ has one solution between $x = 5$ and $x = 6$.

x	$x^2 - 5x$	Comment
5	$25 - 25 = 0$	Too low
6	$36 - 30 = 6$	Too high
5.5	$30.25 - 27.5 = 2.75$	Too low
5.75		

Copy the table and continue it to find this solution to two decimal places.

b The other solution to the same equation lies between 0 and −1.
Use the method shown above to find this solution to two decimal places.

4 A solution to the equation $x^3 + 2x = 150$ lies between 5 and 6.
Use the trial and improvement method to find this solution to one decimal place.

5 Josie and Kim are using the trial and improvement method to find a solution to the equation $x^3 - 3x^2 = 10$ to two decimal places.
Josie's answer is 3.72 and Kim's answer is 3.73
Which of them is correct? Give reasons for your answer.

Try some real past exam questions to test your knowledge:

6 Gary is using the trial and improvement method to find a solution to the equation $x^3 - 5x = 56$

This table shows his first two trials.

x	$x^3 - 5x$	Comment
4	44	Too small
5	100	Too big

Continue the table to find a solution to the equation.
Give your answer to one decimal place.

Spec B, Int Paper 2, Nov 03

7 Dario is using the trial and improvement method to find a solution to the equation

$$x + \frac{1}{x} = 5$$

The table shows his first trial.

x	$x + \frac{1}{x}$	Comment
4	4.25	Too low

Continue the table to find a solution to the equation.
Give your answer to one decimal place.

Spec B, Int Paper 2, June 04

12 Translation and enlargement

OBJECTIVES

Examiners would normally expect students who get an F grade to be able to:

Give a scale factor of an enlarged shape

Examiners would normally expect students who get an E grade also to be able to:

Enlarge a shape by a positive scale factor

Find the measurements of the dimensions of an enlarged shape

Use map scales to find distance

Examiners would normally expect students who get a D grade also to be able to:

Enlarge a shape by a positive scale factor from a given centre

Translate a shape using a description such as 4 units right and 3 units down

Compare the area of an enlarged shape with the original shape

Examiners would normally expect students who get a C grade also to be able to:

Enlarge a shape by a fractional scale factor

Translate a shape by a vector such as $\begin{pmatrix} 4 \\ -3 \end{pmatrix}$

Transform shapes by a combination of translation, rotation, and reflection

What you should already know ...

- Plot positive and negative coordinates
- Add and subtract negative numbers
- Reflect shapes in a line of symmetry or mirror line
- Rotate shapes around a given centre
- Understand and use units of length
- Work out the area of rectangles and triangles
- Multiplication of integers and decimals
- Understand the concept of ratio
- Find the HCF of two numbers

VOCABULARY

Object – the shape before it undergoes a transformation, for example, translation or enlargement

Image – the shape after it undergoes a transformation, for example, translation or enlargement

Mapping – a transformation or enlargement is often referred to as a mapping with points on the object mapped onto points on the image

Congruent – exactly the same size and shape; one of the shapes might be rotated or flipped over

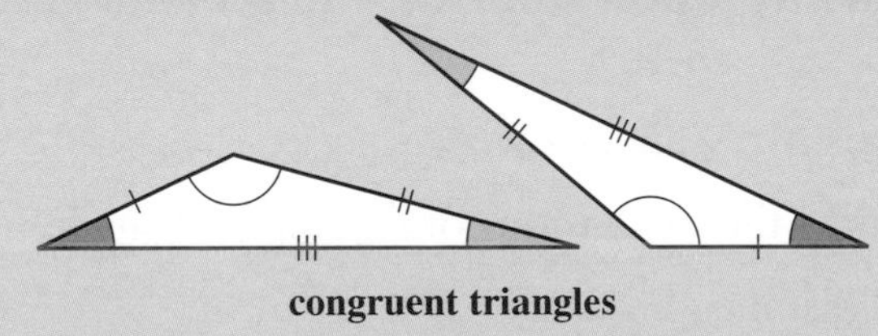

congruent triangles

Similar – figures that are the same shape but different sizes so that one shape may be an enlargement of the other

Scale factor – the ratio of corresponding sides usually expressed numerically so that:

$$\text{Scale factor} = \frac{\text{length of line on the enlargement}}{\text{length of line on the original}}$$

Transformation – reflections, rotations, translations and enlargements are examples of transformations as they transform one shape onto another

Enlargement – an enlargement changes the size of an object (unless the scale factor is 1); but not its shape; it is defined by giving the centre of enlargement and the scale factor; the object and the image are similar

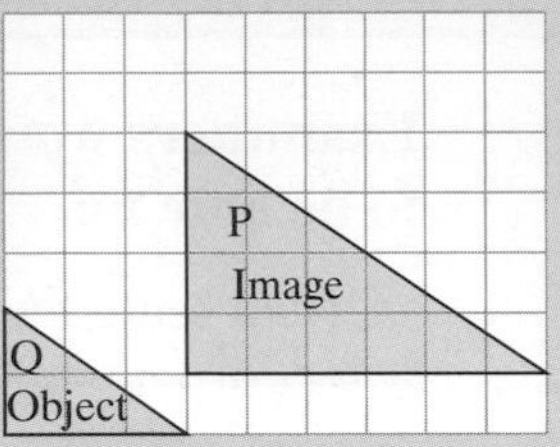

Triangle P is an enlargement of triangle Q
All the lines have doubled in size
The scale factor of the enlargement is 2

Translation – a transformation where every point moves the same distance in the same direction so that the object and the image are congruent

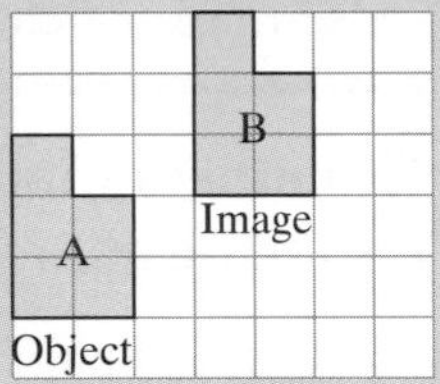

Shape A has been mapped onto Shape B by a translation of 3 units to the right and 2 units up

The vector for this would be $\begin{pmatrix} 3 \\ 2 \end{pmatrix}$

A translation is defined by the distance and the direction (vector)

Vector – used to describe translations

It is written in the form:

$$\begin{pmatrix} \text{units moved in the positive } x\text{-direction} \\ \text{units moved in the positive } y\text{-direction} \end{pmatrix}$$

Learn 1 Introduction to enlargement and scale factor

Examples: These two boats are drawn on centimetre grids.

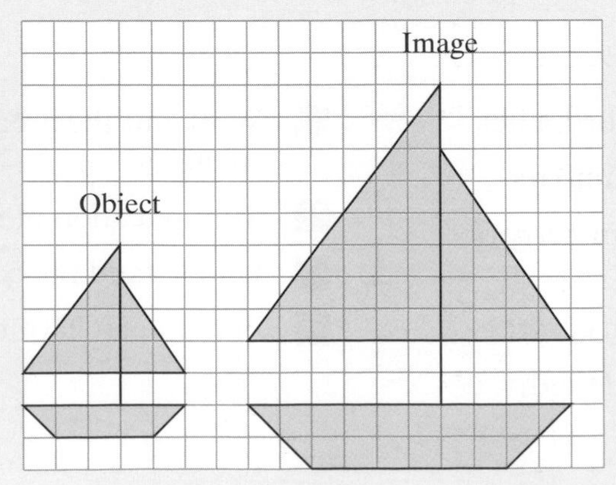

Notice that the two boats are similar
They are the same shape
All corresponding angles are equal
All corresponding lengths are in the same ratio

	Answers
a How long is the boat in the original drawing (Object)?	5 cm
b How long is the boat in the enlarged drawing (Image)?	10 cm
c What is the height of the mast in the original drawing?	5 cm
d What is the height of the mast in the enlarged drawing?	10 cm
e What is the width of the large sail in the original drawing?	3 cm
f What is the width of the large sail in the enlarged drawing?	6 cm
g How many times larger is the enlarged shape than the original shape?	2

The scale factor of the enlargement is 2

To enlarge a shape, multiply all of its lengths by the scale factor of the enlargement.

Length of line on enlargement = scale factor × length of line on the original

$$\text{Scale factor} = \frac{\text{length of line on the enlargement}}{\text{length of corresponding line on the original}}$$

Apply 1

1 **a** Enlarge these shapes by making every line twice as long.

i

ii

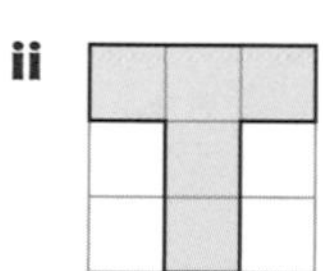

iii

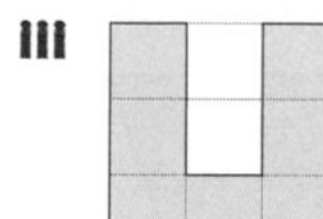

b Work out the areas of the original and the enlarged shapes.

c What do you notice about the areas of the shapes?

2 **a** Enlarge these shapes by making every line 3 times as long.

i

ii

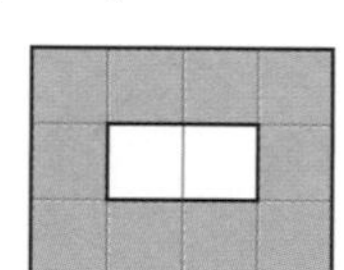

b Work out the areas of the original and the enlarged shapes.

c What do you notice about the areas of the shapes?

3 In these diagrams A has been enlarged to B.
What is the scale factor of each enlargement?

a

b

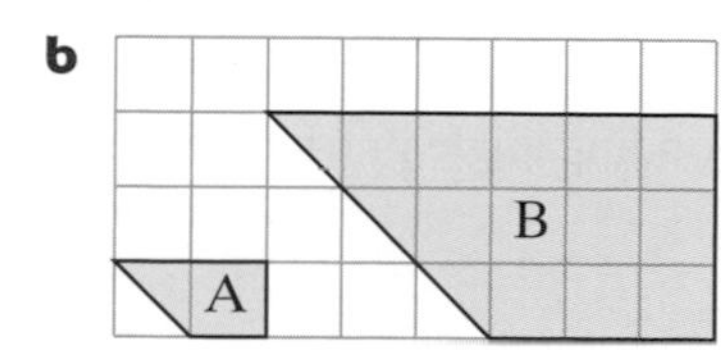

c

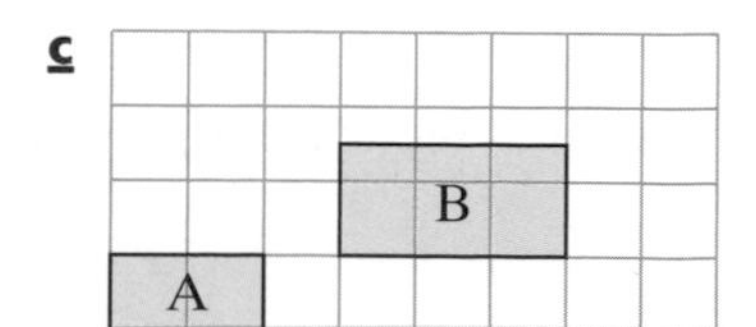

4

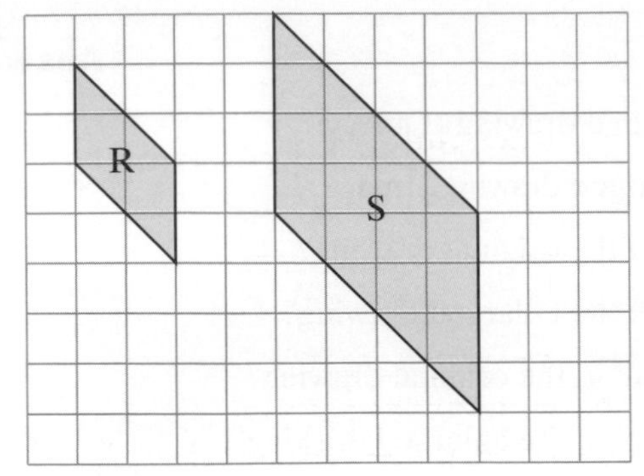

a What is the scale factor of the enlargement which transforms figure R to figure S?

b What is the scale factor of the enlargement which transforms figure S to figure R?

5 A trapezium ABCD is an enlargement of the trapezium WXYZ.

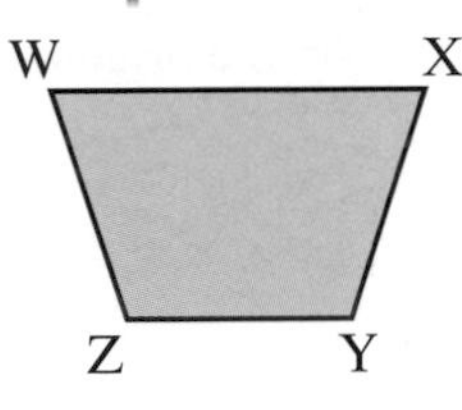

Not drawn accurately

WX = 5 cm
XY = 3 cm
YZ = 2 cm
ZW = 3 cm

The scale factor of the enlargement is 2.5

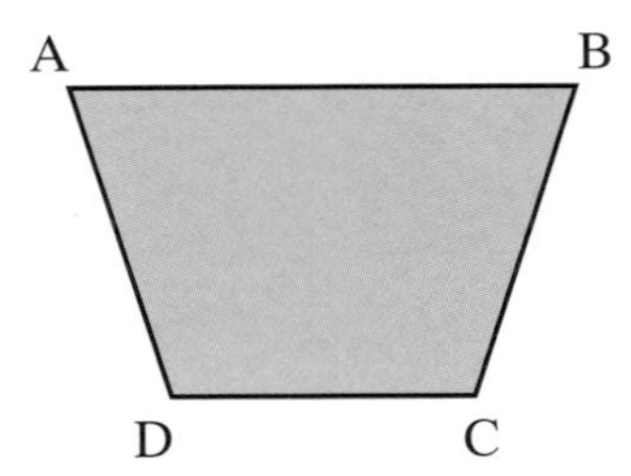

Find the lengths of the sides:

a AB

b BC

c CD

d DA

6 Triangle A is enlarged with scale factor 3 into triangle B.
One side of triangle B has length 4.5 cm.

a What is the length of the corresponding side of triangle A?

One angle of triangle B is 60°.

b What is the size of the corresponding angle in triangle A?

Learn 2 Centre of enlargement

Examples:

a Enlarge rectangle R by scale factor 2 with centre (−6, 1).

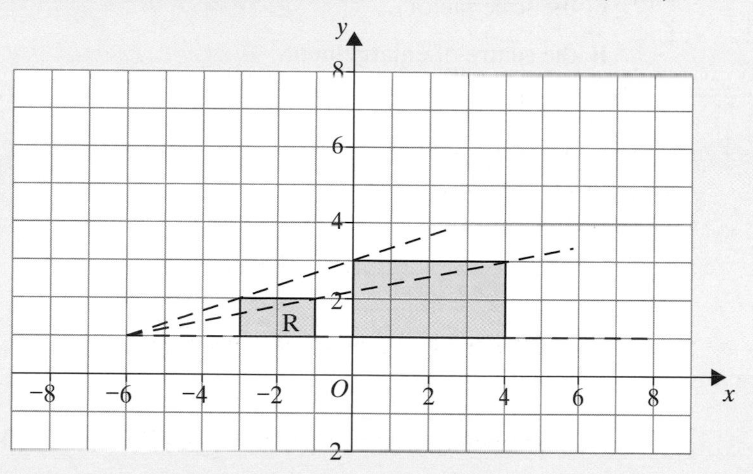

1 Plot the centre
2 Draw 'rays' from the centre to each vertex of the object
3 Measure the distance from the centre to each vertex of the object
4 Multiply each of these lengths by the scale factor to find the distance from the centre to the corresponding vertex of the image

b Triangle XYZ has been enlarged to the triangle QRS.

i What is the scale factor of the enlargement?

ii What is the centre of the enlargement?

Join QX and SZ; the point where these lines meet is the centre of enlargement, C

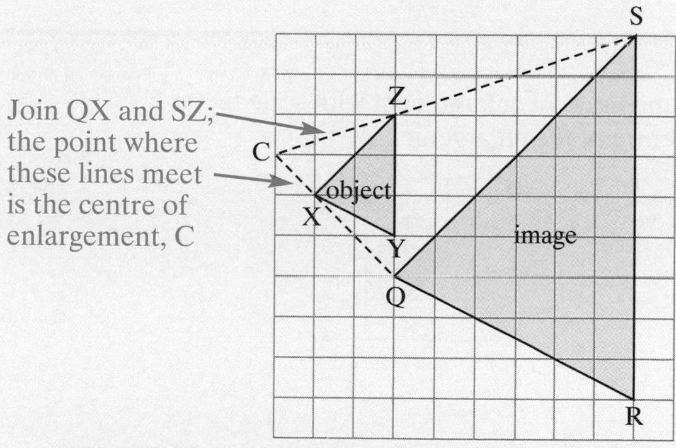

The distance from C to Q is 3 times the distance from C to X

The scale factor of enlargement is 3

i The length of YZ is 3 units. The length of RS is 9 units.

$$\text{Scale factor} = \frac{\text{enlarged length}}{\text{corresponding original length}} = \frac{9}{3} = 3$$

Scale factor = 3

ii Point C is the centre of the enlargement.

To define an enlargement, give the centre of the enlargement and the scale factor

c Trapezium B is an enlargement of trapezium A.
Find:

i the scale factor

ii the centre of enlargement.

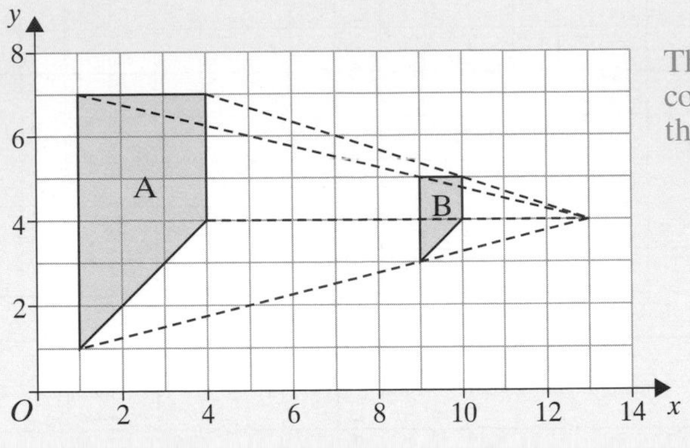

The point where 'rays' through corresponding vertices meet is the centre of enlargement

i The scale factor $= \frac{2}{6} = \frac{1}{3}$

Scale factor $= \dfrac{\text{enlarged length}}{\text{corresponding original length}}$

ii The centre of enlargement = (13, 4)

An enlargement by a scale factor between zero and one will make the shape smaller

Apply 2

1 Copy each diagram and draw an enlargement with scale factor 2.
Use point C as the centre of the enlargement.

a

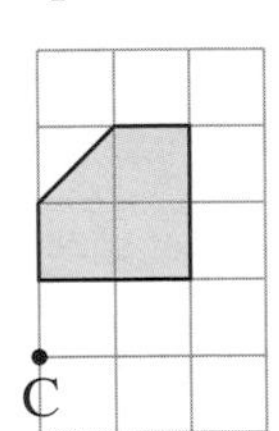

b

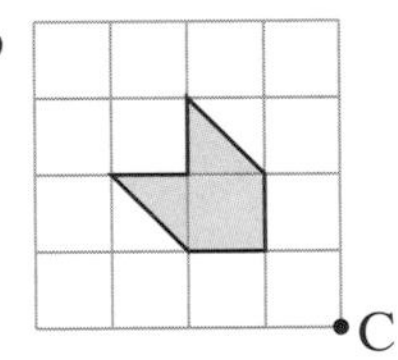

2 For each of these, copy the diagram and draw an enlargement with scale factor 3. Use point C as the centre of enlargement.

a

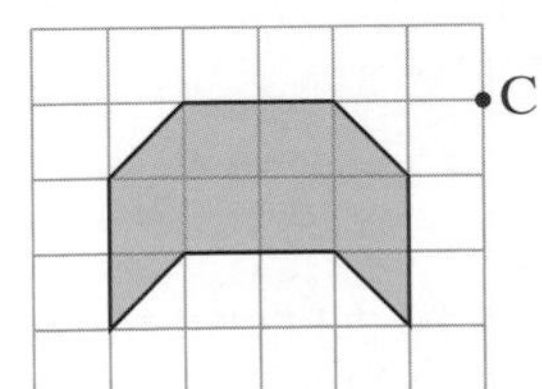

b

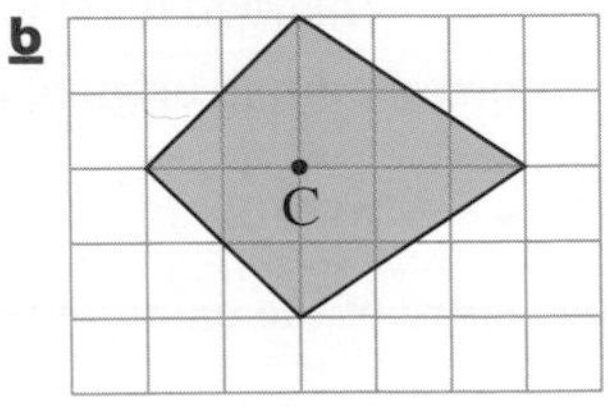

3 In each of these diagrams, shape X has been enlarged to make shape Y.
In each case find the scale factor and the centre of the enlargement.

a

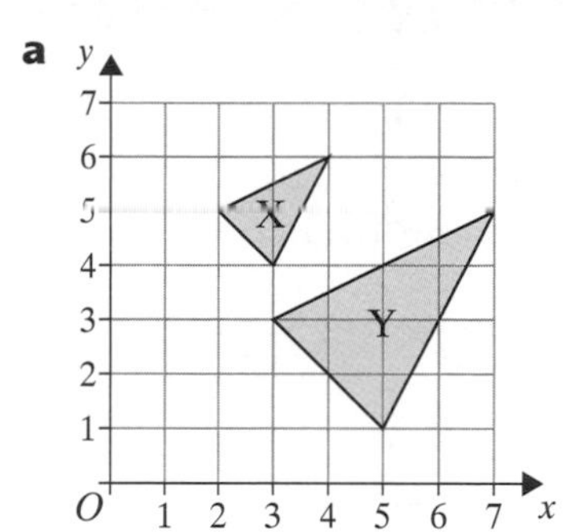

c

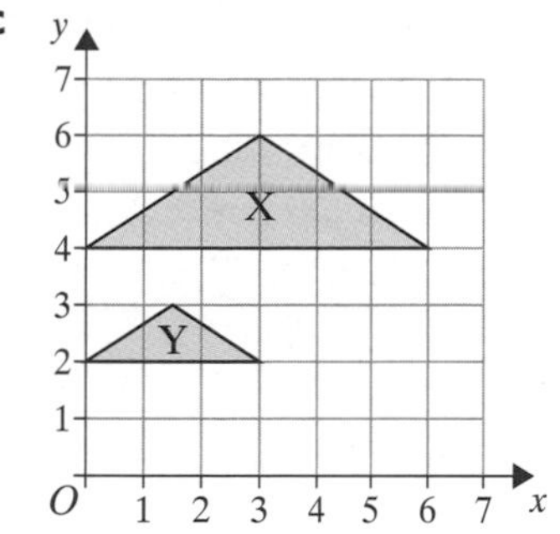

b

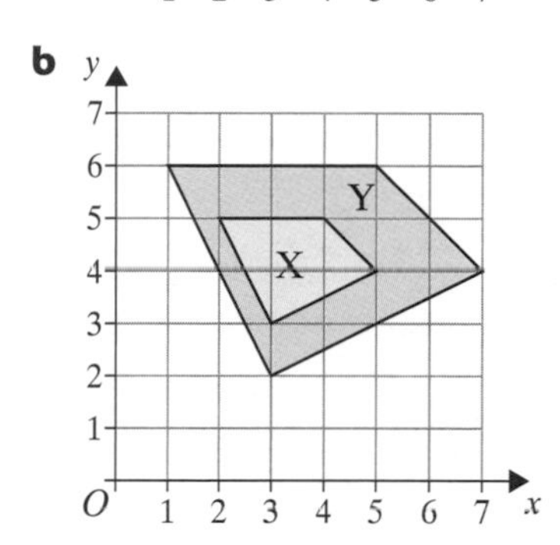

d

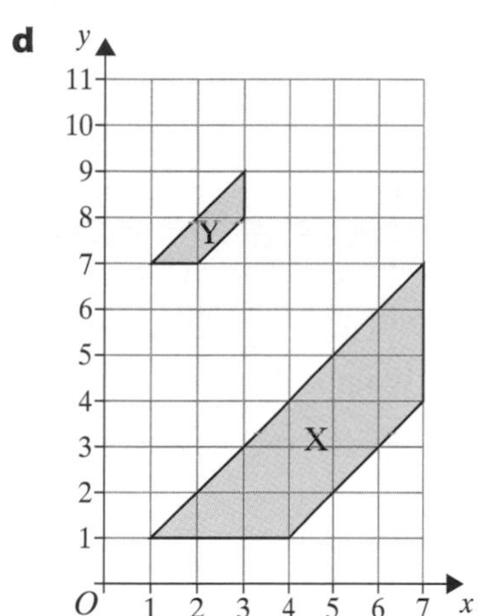

4 Draw a pair of axes with values from 0 to 7.
Plot and label triangle A with vertices at (5, 5), (5, 7) and (4, 7).

a Draw the image of triangle A after enlargement with scale factor 2, centre (7, 7). Label the image B.

b What are the coordinates of the vertices of B?

5 Draw a pair of axes with values from −7 to 7.
Plot and label triangle C with vertices at (−6, −5), (−3, −5) and (−3, −4).

a Draw the image of triangle C after enlargement with scale factor 3, centre (−6, −7). Label the image D.

b What are the coordinates of the vertices of D?

6 Draw and label the x-axis from 0 to 12 and the y-axis from 0 to 10.
Plot and label pentagon P with vertices at (7, 1), (11, 1), (10, 5), (11, 7) and (7, 7).

a Draw the image of pentagon P after enlargement scale factor $\frac{1}{2}$, centre (1, 9). Label the image Q.

b What are the coordinates of the vertices of Q?

7 Draw and label the x-axis from 0 to 18 and the y-axis from 0 to 20.
Draw a rectangle using the coordinates (4, 6), (12, 14), (8, 18) and (0, 10).
Enlarge this rectangle using scale factor $\frac{1}{4}$ and centre of enlargement (8, 14).

Explore

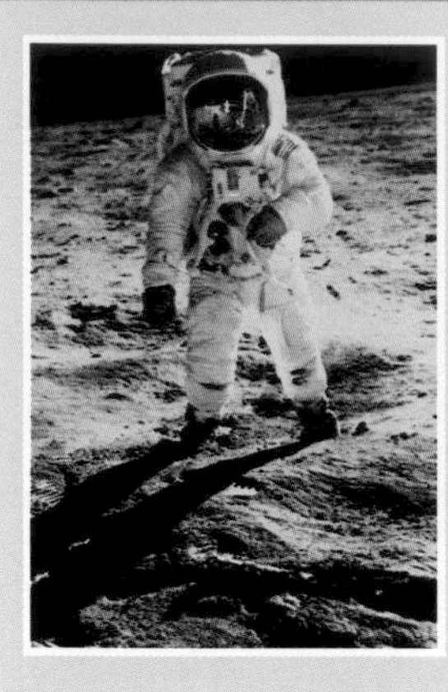

- Draw a shape of your choice
- Enlarge the shape, using centres of enlargement both inside and outside the shape
- What happens when you use different scale factors (whole numbers or ...)?

Investigate further

Learn 3 Enlargement and ratio

Examples:

Rectangle B is enlarged to give rectangle A.

a What is the ratio of the lengths and widths of their sides?

b What is the ratio of the perimeter of B to the perimeter of A?

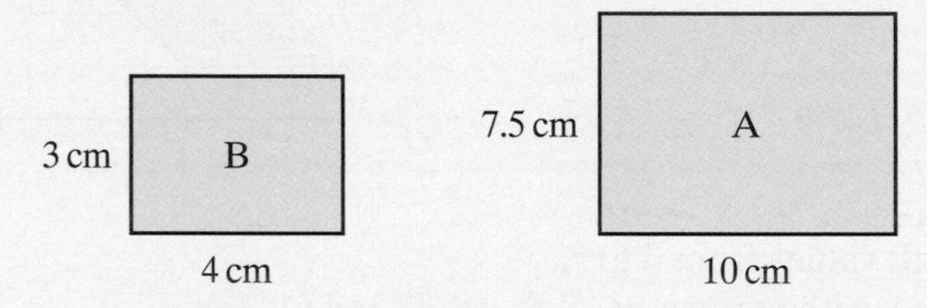

a In rectangle B the length is 4 cm. In rectangle A the length is 10 cm.
The ratio of the lengths is 4 : 10. This can be simplified to 2 : 5.

In rectangle B the width is 3 cm.
In rectangle A the width is 7.5 cm.
The ratio of their widths is 3 : 7.5.
This can be simplified to 6 : 15 = 2 : 5.

The ratios of all of the sides should be the same because A is an enlargement of B, scale factor 2.5
The two shapes are similar

b The perimeter of rectangle B is 2(3 + 4) = 14 cm.
The perimeter of rectangle A is 2(7.5 + 10) = 35 cm.
The ratio of the perimeter of B to A = 14 : 35.
This can be simplified to become 2 : 5.

Perimeter is a length also, so the ratio should be the same as in part **a**

Apply 3

1 Rectangle D is an enlargement of rectangle C.

a What is the ratio of the lengths of these rectangles (in its simplest form)?

b What is the ratio of the widths of these rectangles (in its simplest form)?

c What is the ratio of the perimeter of these rectangles (in its simplest form)?

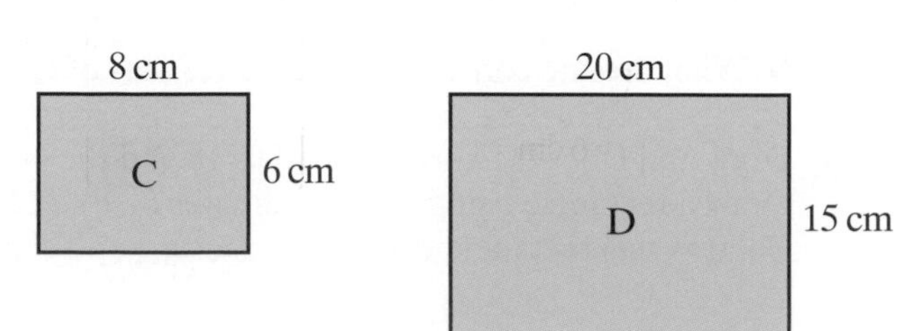

2 Triangle F is an enlargement of the isosceles triangle E.

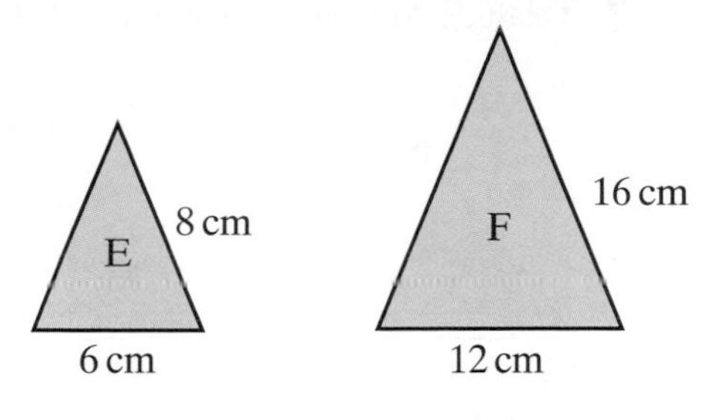

a What is the ratio of the base of these triangles (in its simplest form)?

b What is the ratio of the sides of these triangles (in its simplest form)?

c What is the ratio of the perimeter of these triangles (in its simplest form)?

3 James says that triangle B is an enlargement of triangle A.
Is he correct?
Give a reason for your answer.

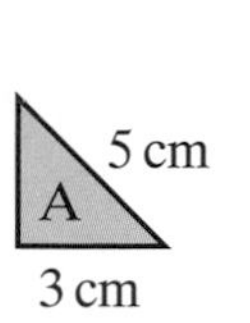

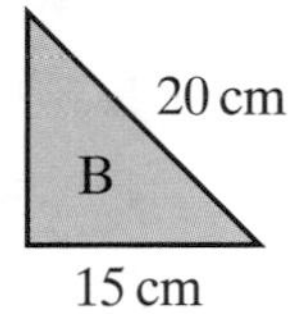

4 Triangle Y is an enlargement of triangle X.

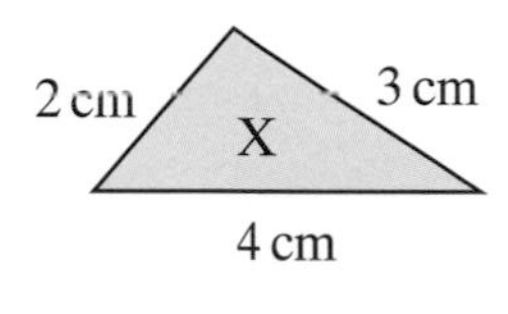

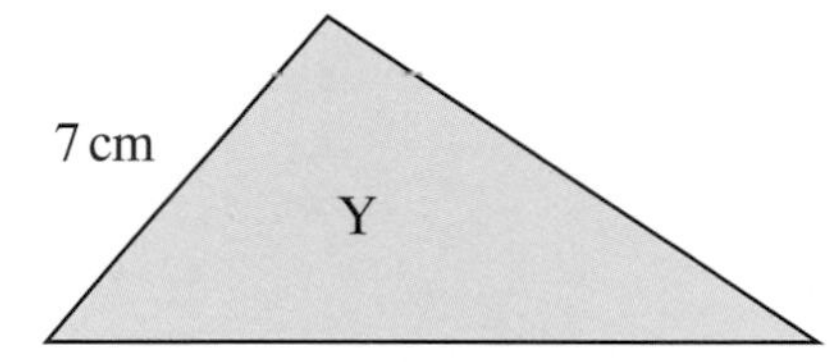

a Work out the ratio of the corresponding lengths.

b Calculate the lengths of the unknown sides.

c What is the ratio of the perimeters of these triangles (in its simplest form)?

5 Trapezium S is an enlargement of trapezium T.

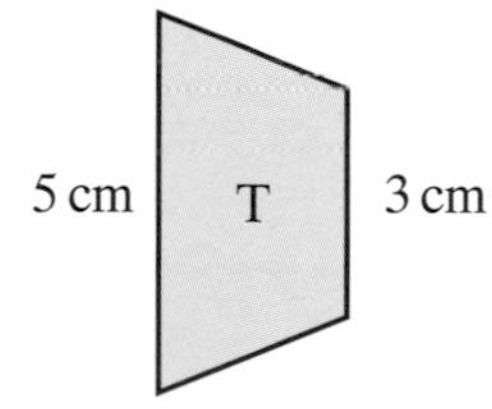

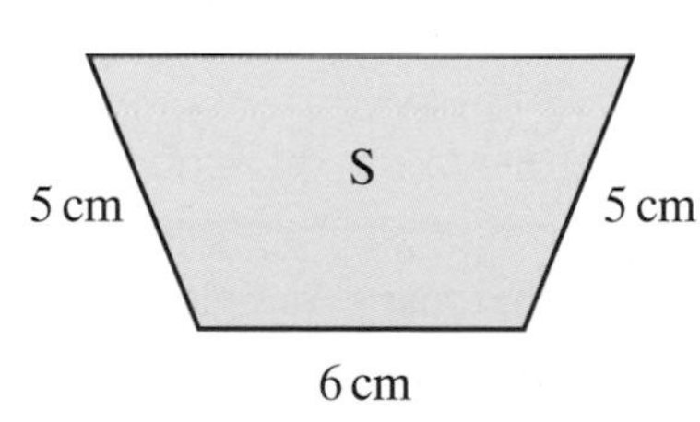

a Work out the ratio of corresponding lengths.

b Calculate the lengths of the unknown sides.

c What is the ratio of the perimeters of these trapeziums?

6 These pairs of solids are similar. State the ratio of the corresponding lengths. Calculate the lengths of w, x and y.

a

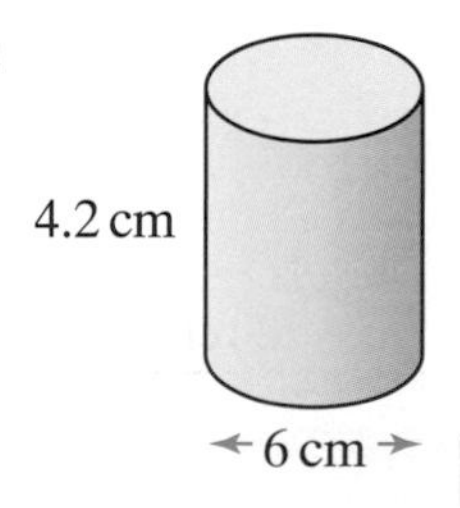

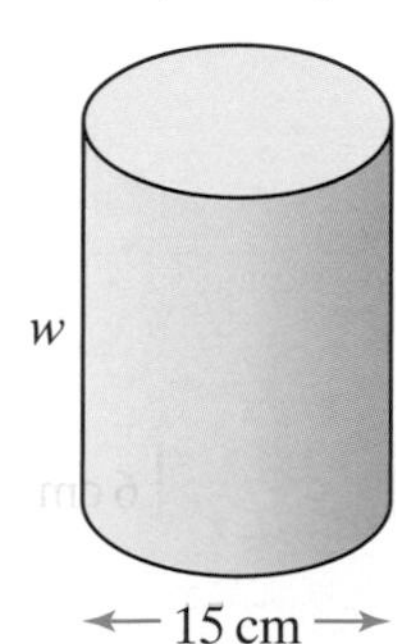

b

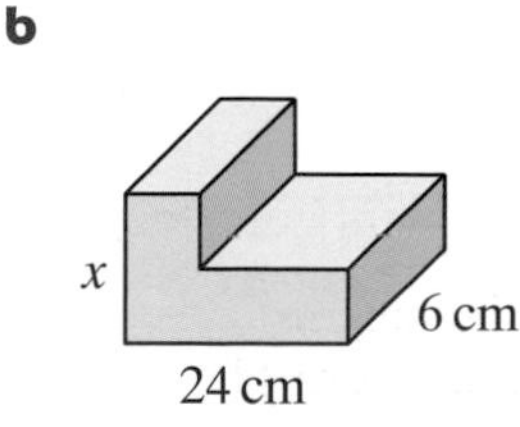

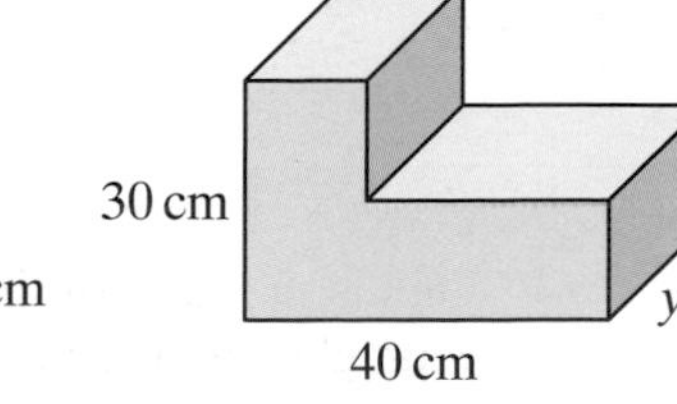

7 Triangle M is an enlargement of triangle N.

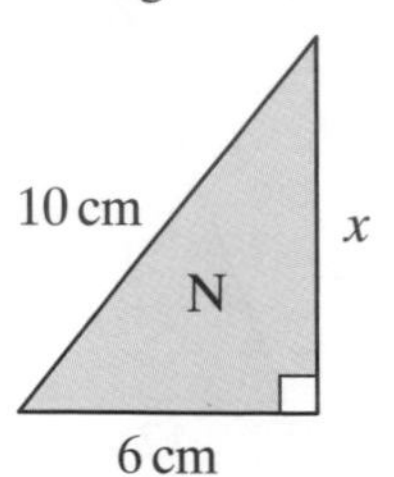

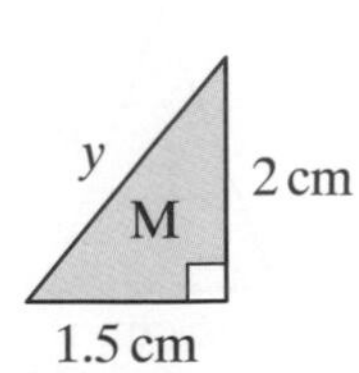

a What is the ratio of the corresponding sides?

b What is the length of the sides x and y?

c What is the ratio of the areas of these triangles in its simplest form?

Explore

What is the ratio between the lengths of A4 and A3 paper?

Investigate further

Learn 4 Translation

Examples:

Describe the translation that maps triangle P onto triangle Q:

a in words

b as a vector.

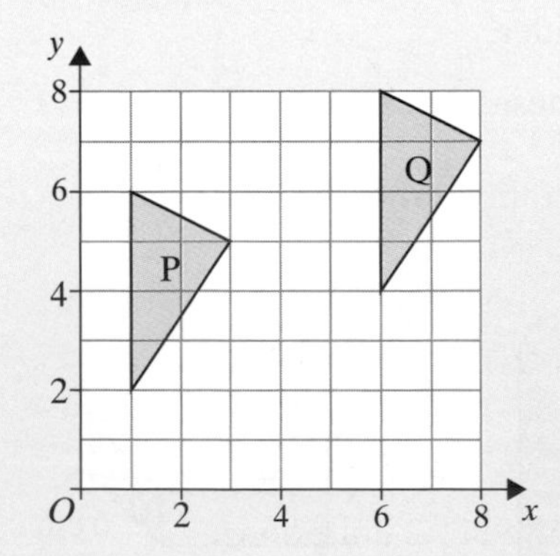

The translation that maps P onto Q is:

a 5 units to the right and 2 units up.

b $\begin{pmatrix} 5 \\ 2 \end{pmatrix}$

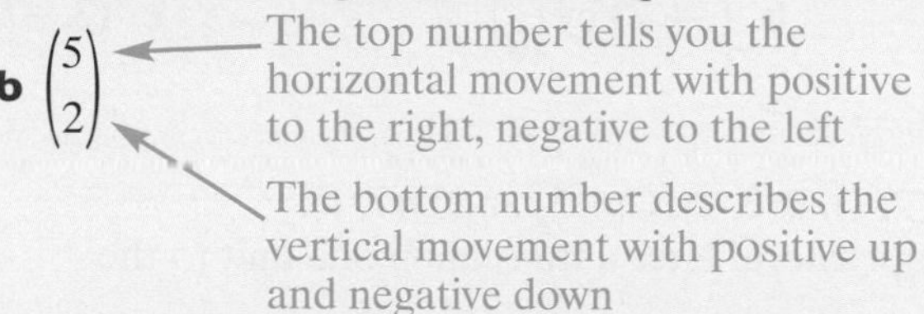

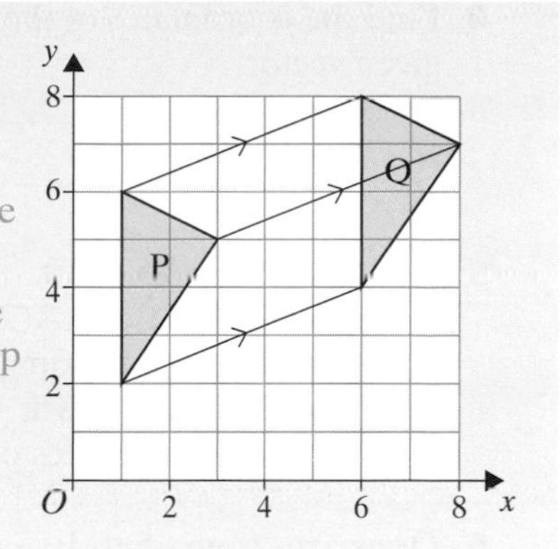

Apply 4

1 Describe the translation that maps triangle X onto triangle Y:

a in words

b as a vector.

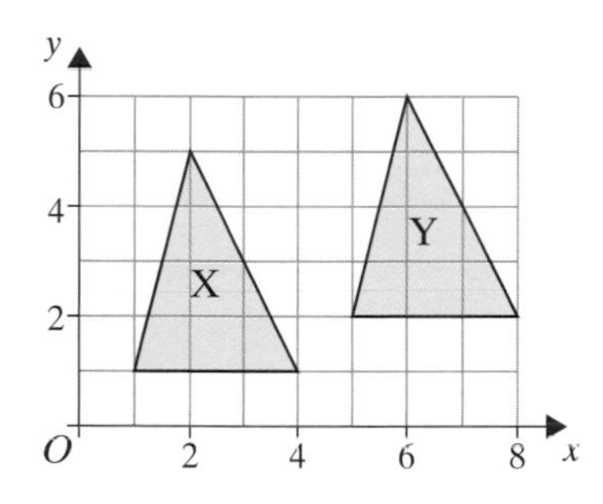

2 a The grey triangle is translated using the vectors

$\begin{pmatrix} 3 \\ 2 \end{pmatrix}$ $\begin{pmatrix} -5 \\ 0 \end{pmatrix}$ $\begin{pmatrix} 6 \\ -4 \end{pmatrix}$ $\begin{pmatrix} 5 \\ 4 \end{pmatrix}$

Write the letters of the images to spell a word.

b Write the vectors that translate the grey triangle onto the letters of the word FRIEND.

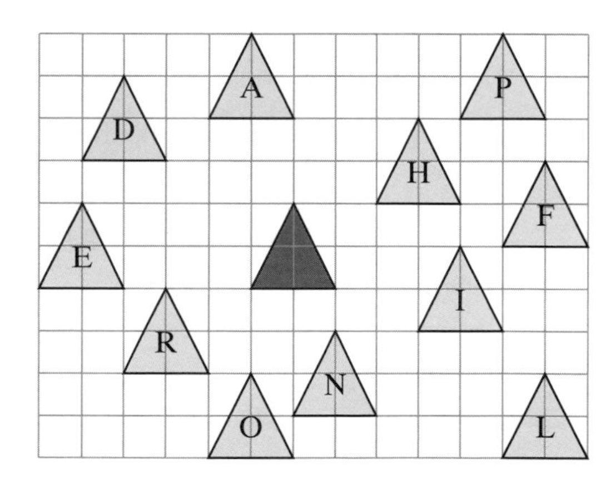

3 Find the coordinates of the images of these points after the translations described.

a (3, 4) translation 2 units to the right followed by 3 units up.

b (2, 6) translation 2 units to the left followed by 2 units down.

c (−2, −3) translation 4 units to the right followed by 6 units up.

d (4, −1) translation 5 units to the left followed by 3 units down.

4 Find the coordinates of the image of each point after a translation by the given vector.

a (2, 7) by $\begin{pmatrix} 3 \\ 4 \end{pmatrix}$ **b** (−4, 2) by $\begin{pmatrix} 1 \\ 0 \end{pmatrix}$ **c** (3, −7) by $\begin{pmatrix} -3 \\ -2 \end{pmatrix}$

5 Draw axes from −5 to 10.
Draw the quadrilateral A(2, 1), B(3, 4), C(6, 3) and D(7, 1).
Draw the image of the quadrilateral after a translation one unit to the right and four units down.
What are the coordinates of the vertices of the image of this quadrilateral?

6 Draw axes from −5 to 10.
Draw the triangle R(2, 2), S(2, 5) and T(4, 5).
Draw the image of the triangle after a translation three units to the left and six units down.
What are the coordinates of the image of this triangle?

7 What are the coordinates of the image of the point Z(3, 2) after a translation of:

a $\begin{pmatrix} 3 \\ 1 \end{pmatrix}$ **b** $\begin{pmatrix} 0 \\ -1 \end{pmatrix}$ **c** $\begin{pmatrix} -3 \\ -4 \end{pmatrix}$

8 Triangle PQR is translated to triangle P′Q′R′.
P and Q have coordinates (3, 7) and (5, 8) respectively.
P′ and R′ have coordinates (5, 10) and (4, 1).

Find R and Q′.

9 A translation maps the point L(3, 6) onto the point L′(−3, 2).
What is the image of the point M(2, 3) under this translation?

10 Triangle A is formed by joining the three points (−2, −2), (−1, −4) and (−1, −1).
Draw this triangle on a pair of axes.

a Show its image, triangle B, after the translation $\begin{pmatrix} 2 \\ -2 \end{pmatrix}$.

Triangle C is found by translating B by the vector $\begin{pmatrix} -2 \\ 2 \end{pmatrix}$.

b What do you notice?

Explore

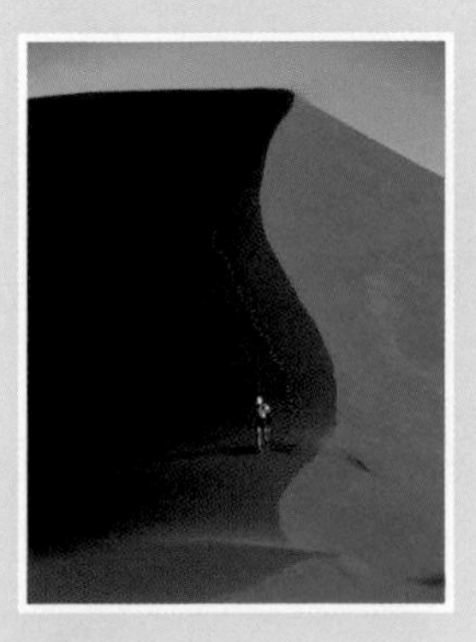

- Use a 10 × 10 grid and the triangle X(0, 0), Y(1, 0) and Z(0, 1)
- How many different translations are there that use integer values only? The translated shape must be on the grid

Investigate further

Learn 5 Combining rotation, reflection and translation

A combination of the same transformation or a combination of different transformations can sometimes be described as a single transformation.

Examples:

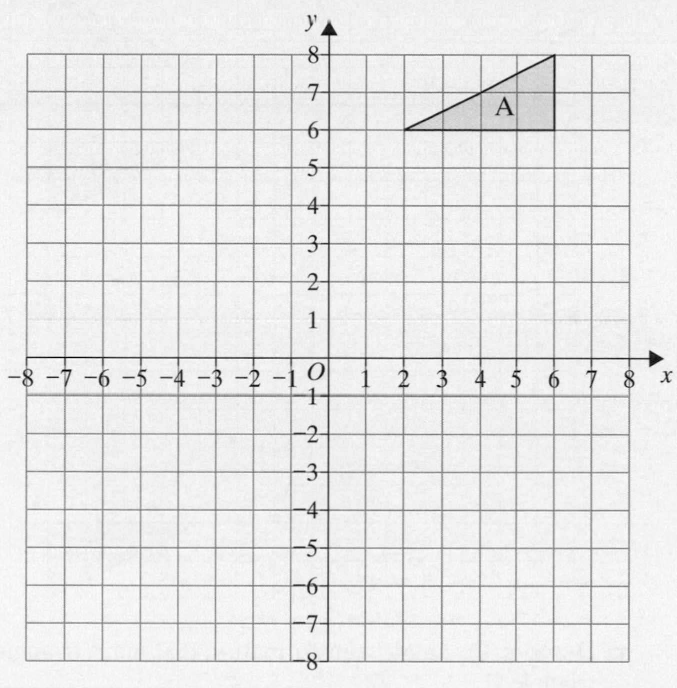

a Triangle A is reflected in the line $y = 5$. Draw this triangle and label it B.

a

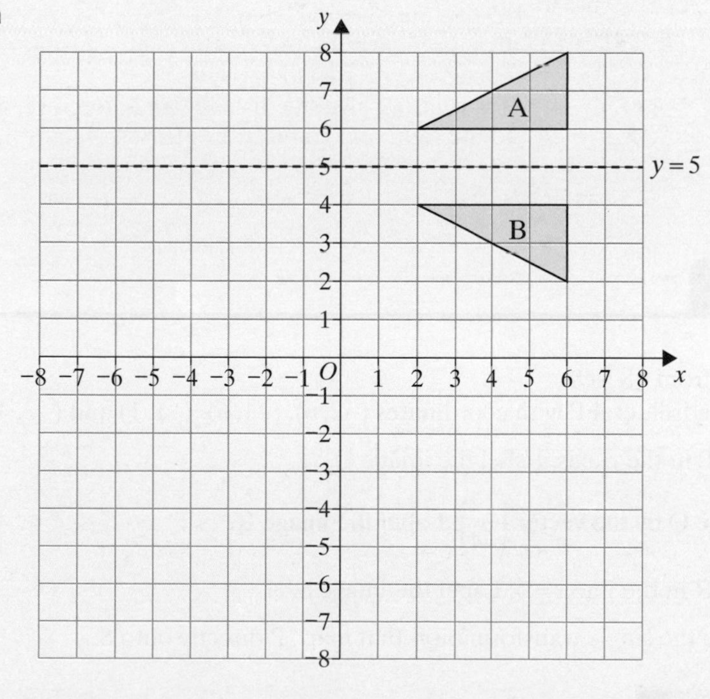

b Triangle B is then translated by $\begin{pmatrix} 0 \\ -10 \end{pmatrix}$ to C. Draw this triangle and label it C.

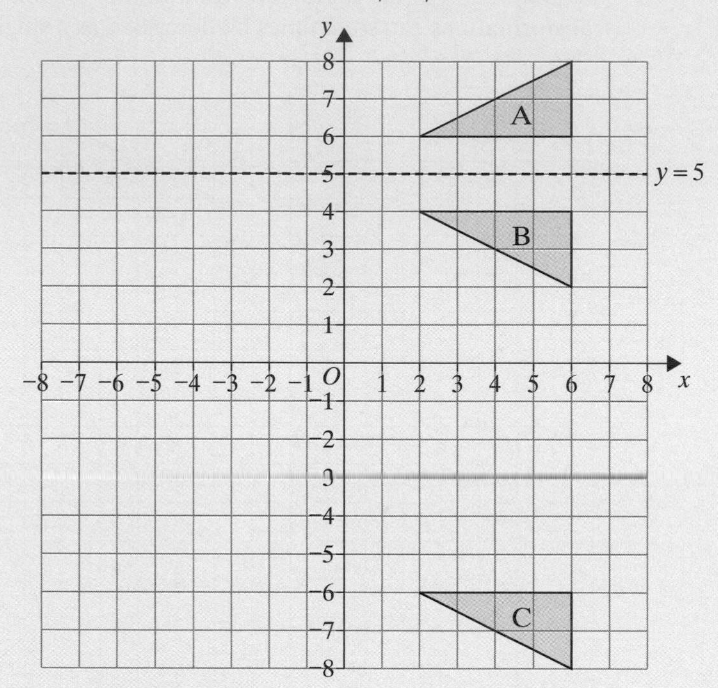

c Describe the single transformation that maps triangle A directly onto triangle C.

Reflection in x-axis ($y = 0$)

Reflection: give the equation of the line of reflection
Rotation: give the centre, the angle of rotation and the direction (clockwise or anticlockwise)
Translation: give the distance and direction, for example, 2 units to the right and 3 units down (or use vector notation)

Apply 5

1 Draw axes from −8 to 8.
Plot the quadrilateral P with coordinates (−2, 6), (−1, 6), (−1, 1) and (−2, 4).

a Reflect P in the y-axis. Label the image Q.

b Translate Q by the vector $\begin{pmatrix} 2 \\ -4 \end{pmatrix}$. Label the image R.

c Reflect R in the line $x = 4$. Label the image S.

d Describe the single transformation that maps P directly onto S.

2 Draw axes from −8 to 8. Plot triangle A using the points (6, 0), (4, 3) and (5, 5).

a Reflect triangle A in the line $x = 4$. Label the image B.

b Reflect triangle B in the x-axis. Label the image C.

c Reflect triangle C in the y-axis. Label the image D.

d Reflect triangle D in the x-axis. Label the image E.

e Describe the single transformation that maps triangle A directly onto triangle E.

3 Draw axes from −8 to 8. Triangle X has vertices at (−4, −4), (−4, −8) and (−2, −8). It is rotated anticlockwise 90° about the point (−2, −2) onto triangle Y.

Triangle Y is then translated by $\begin{pmatrix} 2 \\ 6 \end{pmatrix}$ onto triangle Z.

Describe fully the single transformation that maps triangle X directly onto triangle Z.

4 Describe fully three different transformations that could move the square labelled L to the square labelled M.

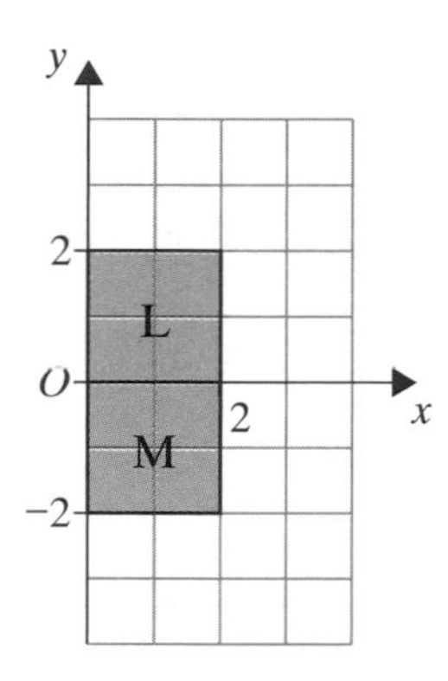

Learn 6 Maps and scales

Examples:

a A plan has a scale of 5 cm to represent 1 m. What is this scale in ratio form?

5 cm represents 100 cm — Make sure the units are the same

1 cm represents 20 cm — Simplify by dividing by the HCF of 5 and 100

The ratio is 1 : 20.

b A map has a scale of 1 : 5000. What is the actual distance in metres between two points which are 6.5 cm apart on the map?

1 cm represents 5000 cm

6.5 cm represents 6.5 × 5000 = 32 500 cm = 325 m — Remember to divide by 100 to change cm into m

The points are 325 m apart.

c Two towns are 10.5 km apart. On a map the distance between them is 15 cm. What is the scale of the map?

10.5 km = 10 500 m = 1 050 000 cm — The units must be the same

15 cm represents 1 050 000 cm

15 : 1 050 000 — Simplify by dividing by the HCF of 15 and 1 050 000

1 : 70 000

The scale is 1 : 70 000.

Apply 6

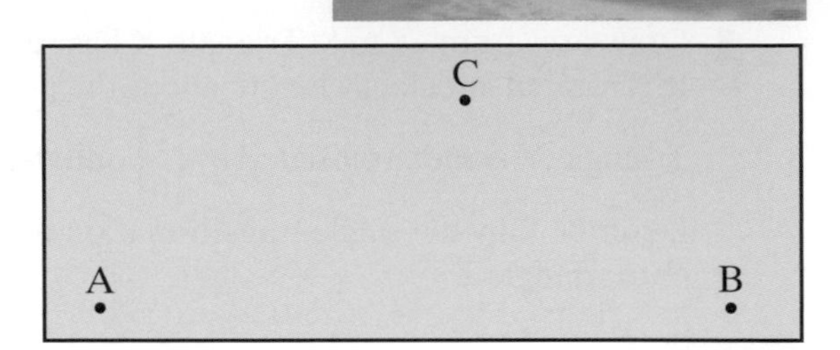

1 A model helicopter has a rotor span of 25 cm. The real length is 8 m. What is the scale of the model?

2 On a plan the length of a sports hall is 4 cm. The real length is 32 m. What is the scale of the plan?

3 The position of three towns is shown on the map. The scale of the map is 1 cm to 20 km. Use the map to estimate the distances between:

a town A and town B

b town A and town C

c town B and town C.

C

A B

4 Jennifer has measured the length of the classroom as 6 m. She plans to use a scale of 1 : 50. She says that the length on her scale drawing will be 30 cm. Is she correct? Give a reason for your answer.

5 Find the distance, in kilometres, which 1 cm represents on a map with a scale of:

a 1 : 100 000

b 1 : 200 000

c 1 : 800 000

d 1 : 25 000 000

6 A map has a scale of 1 : 800 000. Find the actual distance, in kilometres, which these lengths on a map represent.

a 1 cm

b 5 cm

c 7 cm

d 1.5 cm

e 2.3 cm

7 How many kilometres do these lengths represent on a map that has a scale of 1 : 250 000?

a 1 cm

b 4 cm

c 3.2 cm

d 12.8 cm

e 0.6 cm

8 A map has a scale of 1 : 6 000 000. Calculate the lengths on the map that represent these distances.

a 294 km

b 396 km

c 972 km

d 1230 km

9 A map has a scale of 1 : 22 000 000. Calculate the lengths on the map that represent these distances.

a 946 km

b 1408 km

c 1342 km

d 1672 km

10 Find the scale of the map on which:

a 1 cm represents 3 km
b 1 cm represents 250 km
c 1 cm represents 12.5 km
d 2 cm represents 8 km
e 3 cm represents 36 km
f 4 cm represents 10 km

11 Get Real!
On a map Plymouth and Southampton are 4 cm apart.
The actual distance between these two places is 200 km.
What is the scale of the map?

12 Get Real!
On a map Stockport and Sheffield are 3 cm apart.
The actual distance between these two places is 45 km.
What is the scale of the map?

Translation and enlargement

ASSESS

The following exercise tests your understanding of this chapter, with the questions appearing in order of increasing difficulty.

1 Look at the following shapes:

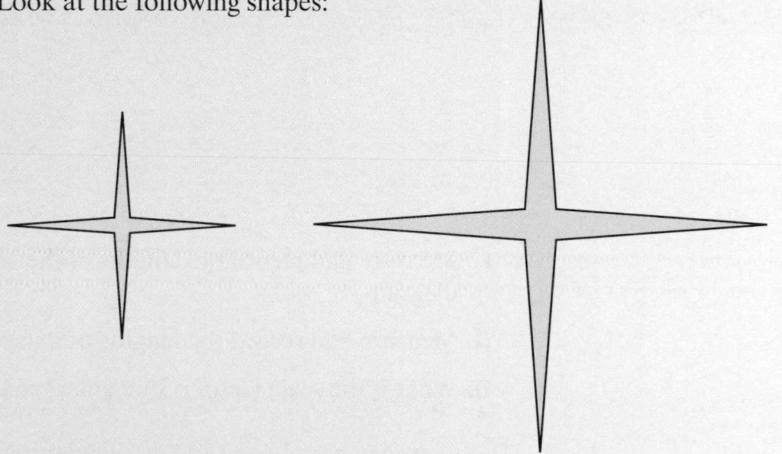

a **i** Measure and record the lengths of a sloping side in each diagram.
ii Measure and record the lengths of the total height in each diagram.
iii Measure and record the lengths of the total width in each diagram.
iv What is the scale factor of the enlargement?

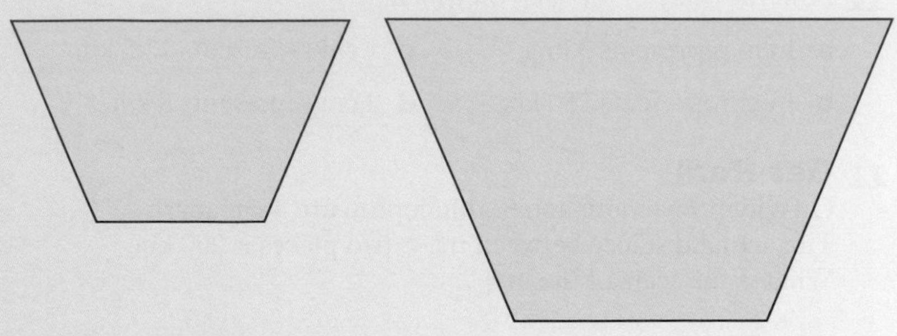

b **i** Measure and record the lengths of a sloping side in each diagram.

ii Measure and record the lengths of the perpendicular height in each diagram.

iii Measure and record the lengths of the parallel sides in each diagram.

iv What is the scale factor of the enlargement?

c **i** Measure and record the lengths of the sloping straight side in each diagram.

ii Measure and record the lengths of a maximum width in each diagram.

iii What is the scale factor of the enlargement?

2 Use squared paper to copy and draw enlargements of the following shapes with the given scale factors.

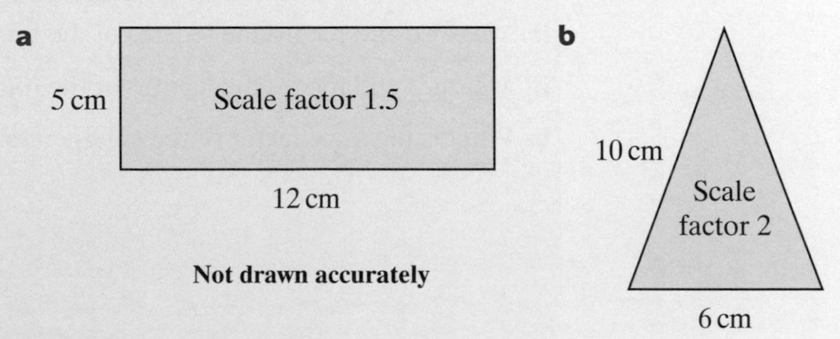

a **i** Measure the length of the diagonals of each rectangle.
What do you notice?

ii Work out the area of each rectangle.
What do you notice?

b i Measure the length of the perpendicular height of each triangle.
What do you notice?

ii Work out the area of each triangle.
What do you notice?

3 a On a European map the scale is 1 inch : 20 miles.
The Newhaven to Dieppe ferry route is 3.9 inches.
How far is Dieppe from Newhaven?

b A street map of London has a scale of 1 : 14 000.

i On the map Nelson's Column is 8.6 cm from Buckingham Palace.
How many metres apart are they on the ground?

ii Great Ormond Street Children's Hospital is 2.5 km from St Thomas's Hospital. How far apart are they on the map?

4 Use squared paper to copy and draw enlargements of these shapes with the given scale factors from the given centres of enlargement O.

a

O

Scale Factor 2

b

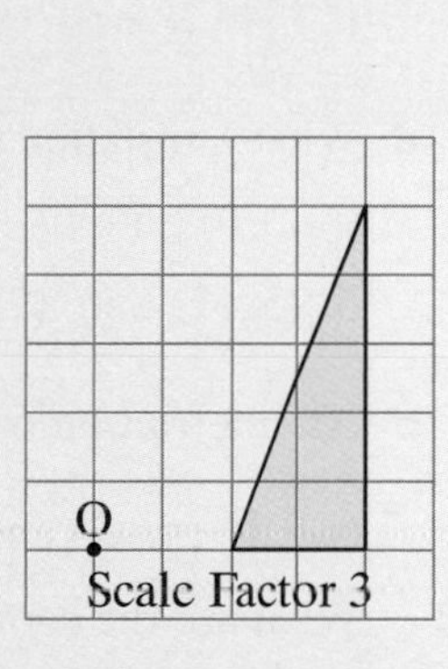

Scale Factor 3

c

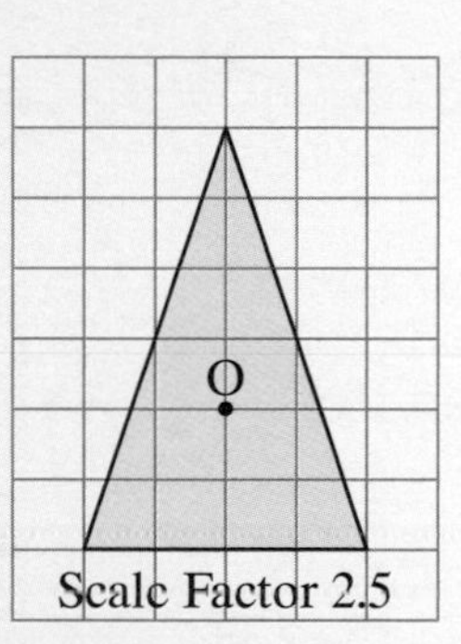

Scale Factor 2.5

5 Use squared paper to copy and draw enlargements of these shapes with the given scale factors from the given centres of enlargement O.

a

O

Scale Factor $\frac{1}{2}$

b

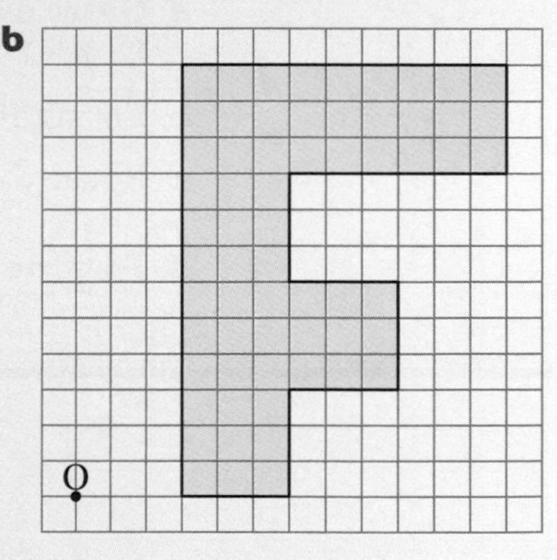

Scale Factor $\frac{1}{3}$

6 **a** A walker needs to get to the other side of a lake. To do this she must first walk 4 km due west and then 3 km due north. Draw her journey on squared paper and then measure and write down the actual distance she is from her original starting place.

b Write down the coordinates of the images of each of the following points after the given translations:

i Object (2, 7), Translation $\begin{pmatrix} 2 \\ 1 \end{pmatrix}$

ii Object (−3, 9), Translation $\begin{pmatrix} 4 \\ -3 \end{pmatrix}$

iii Object (2,−7), Translation $\begin{pmatrix} -5 \\ -1 \end{pmatrix}$

iv Object (−4, −6), Translation $\begin{pmatrix} -6 \\ 2 \end{pmatrix}$

c What is noticeable about the following translations?

i $\begin{pmatrix} 3 \\ 0 \end{pmatrix}$ **ii** $\begin{pmatrix} 0 \\ -2 \end{pmatrix}$

d What translation transforms:

i (3, 0) onto (7, −2)

ii (5, −2) onto (2, 4)

iii (−4, −7) onto (4, 7)

iv (−2, 3) onto (2, 3)?

e After a translation of $\begin{pmatrix} -3 \\ 4 \end{pmatrix}$, the coordinates of a rectangle are (4, 3), (−1, 3), (4, −2) and (−1, −2).

What are the coordinates of the object?

7 Draw and label axes for values of x between −16 to 16 and y between −8 to 16.

a On the diagram draw a letter F by plotting and joining the points (2, 1), (2, 3), (2, 5), (3, 3) and (4, 5).

b Enlarge the letter F by scale factor 3, centre (0, 0).

c Rotate your enlargement by 90° anticlockwise about (0, 0).

d Translate this rotation by the vector $\begin{pmatrix} 10 \\ -13 \end{pmatrix}$.

13 Measures

OBJECTIVES

G **Examiners would normally expect students who get a G grade to be able to:**

Decide which metric unit to use for everyday measurements

F **Examiners would normally expect students who get an F grade also to be able to:**

Convert between imperial and metric units

Know rough metric equivalents of pounds, feet, miles, pints and gallons

Make sensible estimates of a range of measures in everyday settings

E **Examiners would normally expect students who get an E grade also to be able to:**

Solve simple speed problems

C **Examiners would normally expect students who get a C grade also to be able to:**

Solve more difficult speed problems

Understand and use compound measures such as speed and density

Recognise accuracy in measurements given to the nearest whole unit

What you should already know ...

- Calculate areas and volumes
- The circumference and area of circles
- Rounding numbers

VOCABULARY

Unit – a standard used in measuring, for example, a metre is a unit of length

Metric units – these are related by multiples of 10 and include:

- metres (m), millimetres (mm), centimetres (cm) and kilometres (km) for lengths
- grams (g), milligrams (mg), kilograms (kg) and tonnes (t) for mass
- litres (ℓ), millilitres (mℓ) and centilitres (cℓ) for capacity

Imperial units – these are units of measurement historically used in the United Kingdom and other English-speaking countries; they are now largely replaced by metric units. Imperial units include:

- inches (in), feet (ft), yards (yd) and miles for lengths
- ounces (oz), pounds (lb), stones and tons for mass
- pints (pt) and gallons (gal) for capacity

Conversion factor – the number by which you multiply or divide to change measurements from one unit to another. The approximate conversion factors that you should know are:

Length	Mass	Capacity
1 foot ≈ 30 cm 5 miles ≈ 8 km	1 kg ≈ 2.2 pounds	1 gallon ≈ 4.5 litres 1 litre ≈ 1.75 pints

The table below gives conversion factors for metric units of length, mass and capacity

Metric system Learn these facts

Length	Mass	Capacity
1 cm = 10 mm	1 g = 1000 mg	1 ℓ = 100 cℓ or 1000 mℓ
1 m = 100 cm or 1000 mm	1 kg = 1000 g	
1 km = 1000 m	1 t = 1000 kg	

Compound measure – a measure formed from two or more other measures, for example,
speed ($= \frac{\text{distance}}{\text{time}}$), density ($= \frac{\text{mass}}{\text{volume}}$),
population density ($= \frac{\text{population}}{\text{area}}$)

Speed – the rate of change of distance with respect to time. To calculate the average speed, divide the total distance moved by the total time taken. It is usually given in metres per second (m/s) or kilometres per hour (km/h) or miles per hour (mph)

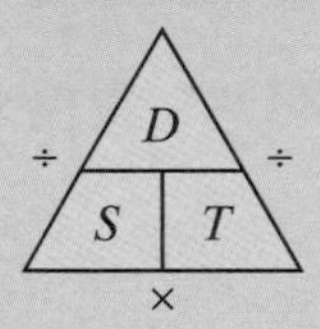

In the triangle, cover the item you want, then the rest tells you what to do

$\text{Speed} = \frac{\text{distance}}{\text{time}}$ $\frac{\text{metres}}{\text{seconds}}$ gives m/s

$\text{Distance} = \text{speed} \times \text{time}$

$\text{Time} = \frac{\text{distance}}{\text{speed}}$

Density – to calculate density, divide the mass of the object by the volume of the object. It is usually given in grams per cubic centimetre (g/cm^3) or kilograms per cubic metre (kg/m^3)

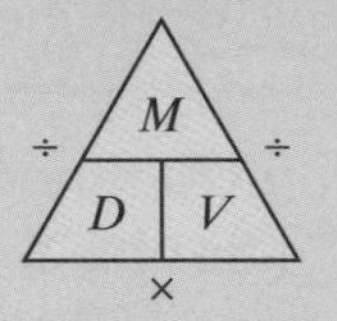

In the triangle, cover the item you want, then the rest tells you what to do

$\text{Density} = \frac{\text{mass}}{\text{volume}}$ $\frac{\text{kilograms}}{\text{cubic metres}}$ gives kg/m^3

$\text{Mass} = \text{density} \times \text{volume}$

$\text{Volume} = \frac{\text{mass}}{\text{density}}$

Lower bound – this is the minimum possible value of a measurement, for example, if a length is measured as 37 cm correct to the nearest centimetre, the lower bound of the length is 36.5 cm

Upper bound – this is the maximum possible value of a measurement, for example, if a length is measured as 37 cm correct to the nearest centimetre, the upper bound of the length is 37.5 cm

Learn 1 Converting measurements

Examples:

Convert:

a 8 ounces to grams (use 1 ounce = 28.3 g)

b 14 feet to metres (use 1 metre = 3.28 feet).

To change from one unit to another, **multiply or divide** by the conversion factor.

a When the **units** are **smaller**, there are **more** of them.
Each ounce is equivalent to 28.3 g
so 8 ounces = 8 × 28.3 = 226.4 g.
To 3 significant figures (3 s.f.) this is 226 g.

Grams are lighter than ounces, so there are more grams

It is sensible to round to 3 s.f. because the conversion factor, 28.3, has 3 s.f.

b When the **units** are **larger**, there are **fewer** of them.
For each metre you need 3.28 feet
so 14 feet = 14 ÷ 3.28 = 4.268 ... m.
This rounds to 4.27 m (to 3 s.f.).

Metres are longer than feet, so there are fewer metres

Note that sometimes you will be given the conversion factors to use, especially for imperial units.

Apply 1

1 Use the metric conversion table on page 170 to answer the following questions.

a Convert to metres:

i 7500 mm **ii** 225 cm **iii** 0.36 km

b Convert to kilograms:

i 1.3 t **ii** 450 g **iii** 800 mg

c Convert to litres:

i 75 cℓ **ii** 2450 mℓ **iii** 330 mℓ

2 a In the imperial system 1 gallon = 8 pints.
Write in pints:

i 5 gallons **ii** $2\frac{1}{2}$ gallons

b In the imperial system 1 pound = 16 ounces and 1 stone = 14 pounds.
Write in pounds:

i 4 stones **ii** 80 ounces **iii** $10\frac{1}{2}$ stones

c In the imperial system 1 foot = 12 inches, 1 yard = 3 feet and 1 mile = 1760 yards.
Write in feet:

i 60 inches **ii** 5 yards **iii** half a mile

3 Get Real!

The table gives the dimensions of some furniture in centimetres.
Copy the table, but give the dimensions in metres.

Item	Width	Height	Depth
2-door wardrobe	86 cm	176 cm	52 cm
3-door wardrobe	127 cm	177.8 cm	52 cm
dressing table	137.2 cm	67.7 cm	38.5 cm
chest of drawers (6)	80 cm	96.7 cm	38.5 cm
chest of drawers (3)	42.7 cm	53.2 cm	31.2 cm

4 Copy and complete the following, giving your answers to 2 significant figures.
The conversion factors are given in brackets.

a 7 inches = ... cm (1 inch = 2.54 cm)

b 12 feet = ... m (1 m = 3.28 feet)

c 58 miles = ... km (1 mile = 1.61 km)

d 112 pounds = ... kg (1 kg = 2.2 pounds)

e 10 pints − ... ℓ (1 ℓ − 1.76 pints)

f 32 gallons = ... ℓ (1 gallon = 4.55 ℓ)

5 Say which of the following are wrong and why.

a 17 m = 0.17 cm

b 1.2 cm = 12 mm

c 3.5 kg = 3500 g

d 75 mg = 0.75 g

e $2\frac{1}{4}$ ℓ = 2250 mℓ

f 25 mℓ = 250 cℓ

6 Get Real!

The Smiths have weighed their cases before going on holiday.
Convert each mass to kilograms to 1 d.p.

HINT lb means 'pounds'
Use 1 kg = 2.2 lb

7 Get Real!

The dimensions of a computer workstation are:
height 790 mm width 900 mm depth 430 mm

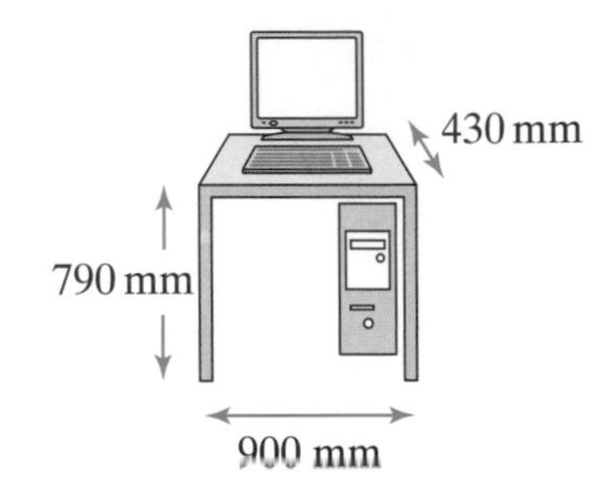

a Convert each of these dimensions to

i cm **ii** m **iii** inches to 1 d.p. (use 1 inch = 2.54 cm).

b How many computer workstations would fit along a wall that is

i 5 m long **ii** 9 feet 6 inches long (1 foot = 12 inches)?

8 Get Real!

Here is an old recipe that makes 4 pints of fruit punch.
A caterer wants to make enough for 120 drinks, each of 300 mℓ, for a school party.
Work out how much he needs of each ingredient.
Give answers to the nearest litre, 0.1 kg, tablespoon or teaspoon.
(Use 1 litre = 1.76 pints, 1 ounce = 28.3 g.)

Spicy Fruit Punch

1 pint orange juice
2 pints ginger ale
$\frac{1}{2}$ pint pineapple juice
$\frac{1}{4}$ pint water
6 ounces sugar
2 tablespoons lemon juice
1 teaspoon mixed spice

9 A group of friends have measured their heights and weights in imperial units.
Copy and complete the table to give each measurement in metric units correct to the nearest centimetre or kilogram.

	Height		Weight	
	Imperial	Metric	Imperial	Metric
Jenny	5 ft 2 in	cm	7 st 8 lb	kg
Sufia	5 ft 6 in	cm	8 st 3 lb	kg
Liam	5 ft 8 in	cm	9 st 11 lb	kg
Max	6 ft 1 in	cm	10 st 4 lb	kg

Notes: ft = feet in = inches 1 ft = 12 in 1 inch = 2.54 cm
st = stones lb = pounds 1 st = 14 lb 1 kg = 2.2 lb

Learn 2 Rough equivalents of imperial units

Examples:

a A room is approximately 10 feet wide.
Roughly what is the width of the room in metres?

b A girl says she weighs about $8\frac{1}{2}$ stones.
Approximately what is this in kilograms?
(1 stone = 14 pounds.)

a Using 1 foot $\approx$ 30 cm gives 10 feet $\approx$ 300 cm = 3 m.
The room is about 3 m wide.

b Given that 1 stone = 14 pounds, 8 stones = 8 × 14 = 112 pounds
and $\frac{1}{2}$ stone = 7 pounds, so $8\frac{1}{2}$ stones = 112 + 7 = 119 pounds.
Using 1 kg $\approx$ 2.2 pounds, 119 pounds $\approx$ 119 ÷ 2.2 $\approx$ 54 kg.
Her weight is about 54 kg.

The calculation is only approximate, so don't give too many figures in the answer

If you are asked for an approximate answer or a rough estimate, use the approximate conversion factors given below. You need to know all of these, **except** 1 inch $\approx 2\frac{1}{2}$ cm.

$\approx$ means 'approximately equal to'

Length	Mass	Capacity
1 foot $\approx$ 30 cm 5 miles $\approx$ 8 km 1 inch $\approx 2\frac{1}{2}$ cm	1 kg $\approx$ 2.2 pounds	1 gallon $\approx$ 4.5 litres 1 litre $\approx$ 1.75 pints

Learn this useful rhyme: 'A litre of water's a pint and three quarters'

Apply 2

1 **a** Estimate each of these lengths in centimetres:

i 3 feet **ii** 5 feet **iii** $1\frac{1}{2}$ feet **iv** 2 inches **v** 7 inches

b Estimate each of these lengths in metres:

i 20 feet **ii** 32 feet **iii** 150 feet **iv** $7\frac{1}{2}$ feet **v** 20 inches

2 **a** Estimate each of these masses in pounds:

i 5 kg **ii** 2.5 kg **iii** 6.75 kg **iv** 800 g

b Estimate each of these masses in kilograms:

i 5 pounds **ii** 35 pounds **iii** $12\frac{1}{2}$ pounds **iv** 160 pounds

3 Estimate each of these capacities in litres:

a 4 pints **b** 9 gallons **c** 25 gallons **d** $3\frac{1}{2}$ gallons **e** $3\frac{1}{2}$ pints

4 **Get Real!**
Ken lives 150 miles from Dover. He wants to convert this distance to kilometres for his French pen pal who is coming to visit.
Roughly how far is it in kilometres?

 **5** Sally says that 10 pounds is about the same as 22 kg.

a Explain why this is wrong.

b Give the correct weight to the nearest 0.1 kg.

 **6 Get Real!**

A vintage car has a 12-gallon petrol tank.
Roughly how many litres will it hold?

 **7 Get Real!**

Imagine you are on holiday in Italy.
You want to buy the items on the shopping list.

Shopping list

5 lb potatoes	3 lb onions
$\frac{1}{2}$ lb tomatoes	$\frac{1}{4}$ lb mushrooms
2 pints milk	$\frac{1}{2}$ pint cream

Re-write the list, giving approximate metric amounts to the nearest 0.1 kilogram or 0.1 litre.

8 Get Real!

The chart gives the distances in kilometres between some of the capital cities in Europe.

Amsterdam								
650	Berlin							
197	764	Brussels						
1093	1696	941	Edinburgh					
2331	2869	3141	2795	Lisbon				
480	1074	333	608	2187	London			
1790	2364	1600	2254	651	1646	Madrid		
510	1051	320	1007	1821	399	1280	Paris	
1691	1502	1520	2404	2653	1796	2002	1389	Rome

Find the distance in miles, to the nearest 5 miles, for each of the following journeys:

a London to Amsterdam

b Brussels to Paris

c London to Paris

d Edinburgh to Lisbon

e London to Rome

f Berlin to Madrid

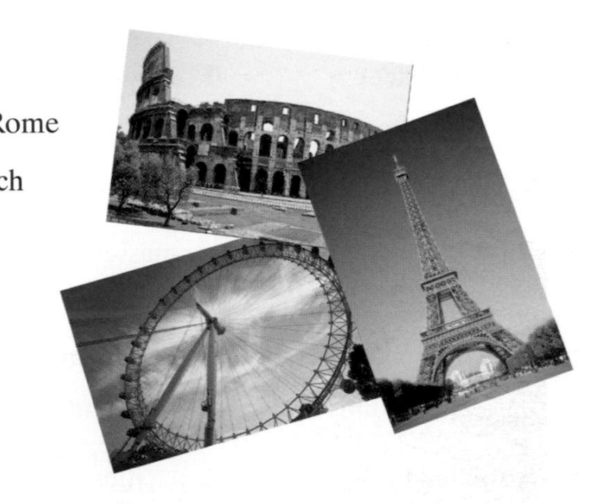

9 Get Real!

A drawing in an estate agent's leaflet gives the dimensions of the rooms in a flat in feet (ft).

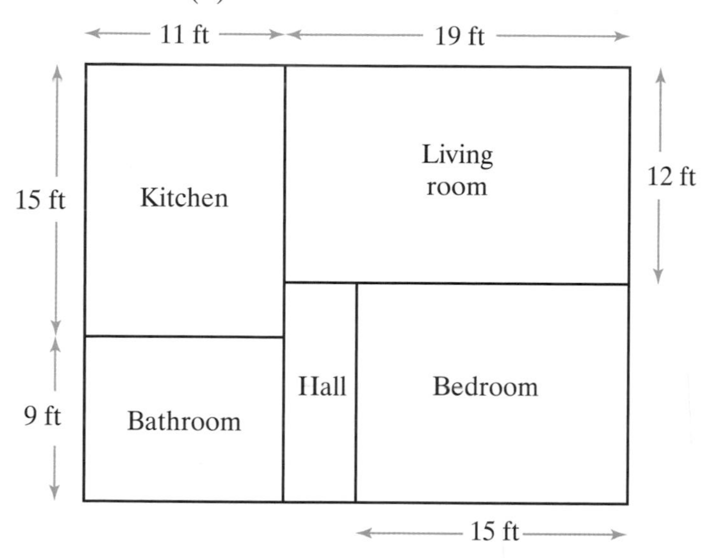

Draw a table that gives the approximate length and width of each room in metres.

10 Get Real!

An old cookbook contains these recipes.

Blackcurrant jam
4 lb blackcurrants
5 lb sugar
2 pints water

Apple crumble
$1\frac{1}{2}$ lb cooking apples
8 oz self-raising flour
4 oz margarine
6 oz sugar

HINT oz means ounces and 1 lb = 16 oz

Write the ingredients in metric units, giving masses to the nearest 25 g and the amount of water to the nearest 50 mℓ.

Explore

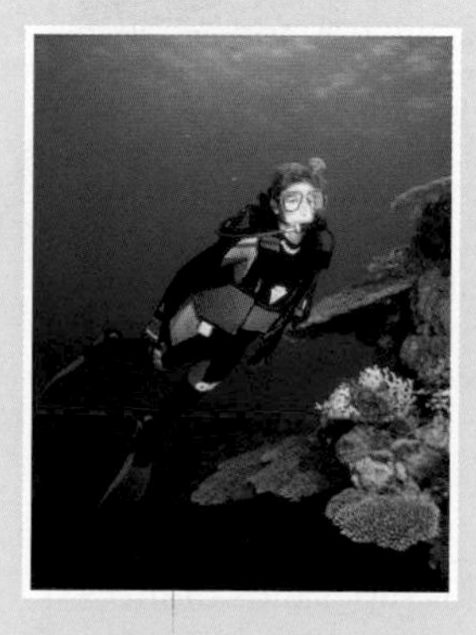

- Use these conversion factors to complete the table (an inch is roughly 2.5 cm, but 2.54 cm gives a more accurate conversion)

Length (in)		0	1	2	3	4
Length (cm)	Rough		2.5			
	Accurate		2.54			
Difference (cm)			0.04			

- What happens to the differences as the lengths increase?

Investigate further

Learn 3 Estimating measures in everyday settings

Examples:

a Write down which metric unit of length you would use to measure:

i the length of a carpet

ii the weight of a baby

iii the capacity of a water tank.

b Write down which imperial unit you would use to measure each item above.

a i metres

ii kilograms

iii litres

b i feet (or possibly yards where 3 feet = 1 yard)

ii pounds

iii gallons

When estimating the length, mass or capacity of something, it is useful to compare it with something similar that you know.

	Size	Units	Useful examples
Length	**Small**	millimetres centimetres inches	Width of a larger full-stop ≈ 1 mm Width of your little finger ≈ 1 cm Distance from the tip to the first joint of your thumb ≈ 1 inch A ruler is usually marked up to 30 cm, 300 mm or 12 inches
	Medium	metres feet	A long stride (or the distance from the tip of your nose to the tip of your longest finger when your arm is outstretched) ≈ 1 m The length of a man's foot ≈ 1 foot
	Big	kilometres miles	The distance from London to Birmingham is roughly 100 miles (or 160 km)
Mass	**Small**	milligrams grams ounces	A pill may contain 100 mg of a drug A pack of butter usually weighs 250 g or 8 ounces (oz)
	Medium	kilograms pounds	Sugar is usually sold in 1 kg packs A jar of jam weighs about 1 pound
	Big	tonnes	Small cars often weigh about 1 tonne
Capacity	**Small**	millilitres	A teaspoon ≈ 5 mℓ A tablespoon ≈ 15 mℓ A can of drink generally contains ≈ 330 mℓ
	Medium	litres pints	Cartons of fruit juice usually hold 1ℓ This is a bit less than 2 pints A typical watering can holds 9 litres or 2 gallons
	Big	litres gallons	A swimming pool holds thousands of litres or thousands of gallons

Apply 3

1 **a** Write down which metric unit of length you would use to measure each of these items:

i length of a bedroom
ii length of a needle
iii length of a river
iv width of a stream
v height of a mountain
vi distance from Leeds to York

b Write down which imperial unit of length you would use to measure each item above.
Choose from **inches**, **feet** or **miles**.

2 Which metric unit would you use to weigh each item listed below?

a Man
b Letter
c Parcel of books
d Delivery van
e Tin of salmon

3 Which metric unit would you use to measure the capacity of each container listed below?

a Large bottle of pop
b Carton of cream
c Bottle of shampoo
d Paddling pool

4 Choose the most likely measurement of each item:

Width of a garden gate 84 m 84 cm 84 mm	Length of a nail 2.5 mm 2.5 cm 2.5 m	Height of a dining chair 4300 mm 430 mm 43 mm
Width of a rugby pitch 0.7 km 70 m 700 cm	Distance between motorway services 350 m 35000 mm 35 km	Diameter of a tennis ball 65 mm 65 cm 0.65 m

5 Choose the most likely mass of each item:

Bar of chocolate 20 g 200 g 2000 g CHOCO	Sack of potatoes 25 g 25 kg 25 t POTATOES	Elephant 60 000 g 6000 kg 60 t
Newborn baby 3500 mg 350 g 3.5 kg	Sheet of paper 50 g 50 mg 5 g	Apple 10 g 0.1 kg 0.01 t

6 Choose the most likely capacity of each item:

Saucepan 3 mℓ 3 cℓ 3 ℓ	Teacup 250 mℓ 250 cℓ 250 ℓ	Washbasin 250 mℓ 15 cℓ 15 ℓ
Bottle of suntan oil 20 mℓ 2 cℓ 0.2 ℓ	Paddling pool 2000 mℓ 2000 cℓ 2000 ℓ	Teaspoon 5 mℓ 50 cℓ 0.5 ℓ

7 **a** Estimate the length, width and height of your classroom.

b Estimate the length and width of each door and window in your classroom.

8 Write down the names of five towns that are near where you live and estimate their distances from your school or college.

9 The following estimates are incorrect. Say why and in each case give a better estimate.

a My cat weighs 4.5 g.

b Your teacher is $5\frac{1}{2}$ m tall.

c I put about 20 ℓ of milk in a cup of tea.

d A feather weighs about 10 kg.

e A tablecloth measures 180 mm by 160 mm

f A can of cola holds 330 cℓ.

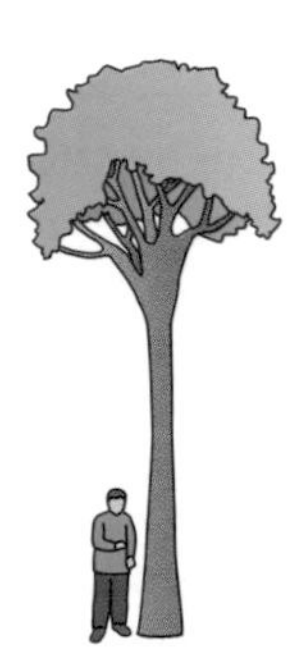

<u>10</u> The diagram shows a man standing next to a tree.
Estimate the height of the tree.

<u>11</u> **a** Name three things that you think could measure about 15 cm.

b Name three items that you think could weigh about 400 g.

c Name three containers that you think could hold about 200 mℓ.

<u>12</u> The diagram shows a car in front of a building.

a Estimate the length of the building.

b Estimate the height of the building.

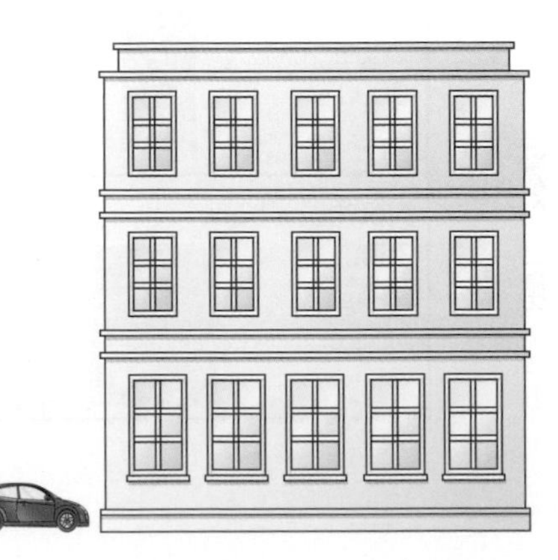

Learn 4 Compound measures

Examples:

a A swimmer crosses a river that is 120 metres wide in 8 minutes.

Find her average speed in **i** metres per second

ii kilometres per hour.

b The density of gold is 19.3 g/cm^3.
A gold bar is 20 cm long, 8 cm wide and 8 cm high.
How much does it weigh to the nearest kilogram?

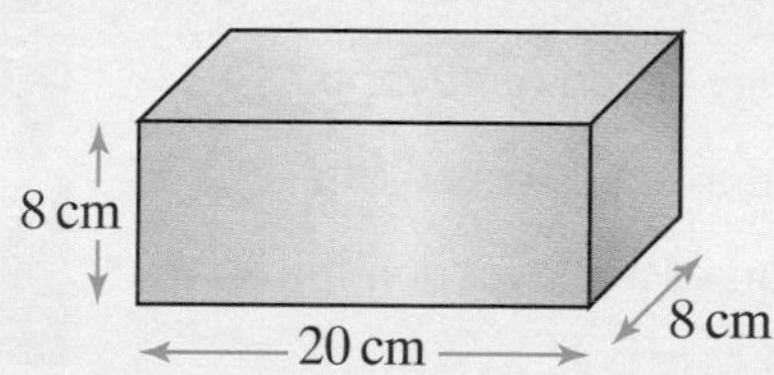

a Speed is the rate at which something moves.

$\text{Speed} = \frac{\text{distance}}{\text{time}}$ Distance = speed × time $\text{Time} = \frac{\text{distance}}{\text{speed}}$

$\frac{\text{metres}}{\text{seconds}}$ gives m/s

Change the distance and time units to those required

Ask yourself if the answer is sensible

i 8 minutes = 8 × 60 = 480 seconds

$\text{Speed} = \frac{\text{distance}}{\text{time}} = \frac{120}{480} = \frac{1}{4}$ or 0.25 m/s

Change the units one at a time

seconds $\xrightarrow{\times 60}$ minutes $\xrightarrow{\times 60}$ hours

ii 0.25 m/s = 0.25 × 60 = 15 metres/min
15 metres/min = 15 × 60 metres/hour
= 900 metres/hour
= 0.9 kilometres/hour

Is this answer sensible?

b Density is mass per unit volume.

$\text{Density} = \frac{\text{mass}}{\text{volume}}$ Mass = density × volume $\text{Volume} = \frac{\text{mass}}{\text{density}}$

$\frac{\text{kilograms}}{\text{cubic metres}}$ gives kg/m^3

Volume = 20 × 8 × 8 = 1280 cm^3
Mass = density × volume = 19.3 × 1280
= 24 704 g = 25 kg to the nearest kilogram

Apply 4

 **1** Find the average speed:

a a sprinter ran 100 m in 12 seconds

b a cheetah took 6 seconds to run 150 metres

c the steam engine, Mallard, took 2 hours to travel 253 miles.

2 Find the distance travelled:

a A car travels for 5 hours at an average speed of 65 mph.

b A ball rolls for 12 seconds at an average speed of 2.5 m/s.

c A snail crawls for 15 seconds at a steady speed of 0.2 cm/s.

d Ben cycled for 20 minutes at a steady speed of 18 mph.

3 How long does each of these journeys take?

a Kath ran the 400 metre race at an average speed of 8 m/s.

b A coach travels 120 miles at an average speed of 40 mph.

c A walker travels a distance of 3 km at a steady speed of 2 m/s.

d Chocolates move along a 15 m section of a production line at a steady speed of 2.5 cm/s.

4 Copy and complete the table.

Distance travelled	Time taken	Average speed
200 metres	16 seconds	
350 kilometres	4 hours	
	1 minute	4.5 m/s
	$2\frac{1}{2}$ hours	56 mph
150 metres		12 m/s
8 metres		5 cm/s

5 Get Real!

The table gives some of the results from men's events in the 2004 Olympic Games.

Winner	Distance	Time
Justin Gatlin	100 m	9.85s
Shawn Crawford	200 m	19.79s
Jeremy Wariner	400 m	44.00s
Yuriy Borzakovskiy	800 m	1min 44.45s
Hicham El Guerrouj	1500 m	3min 34.18s
Hicham El Guerrouj	5000 m	13min 14.39s
Kenenisa Bekele	10000 m	27min 05.10s

Find the average speed of each runner in:

a metres per second

b kilometres per hour.

6 Five students are told that an animal travels half a kilometre in 25 minutes. The answers they give for the average speed are:

Natasha 12.5 m/s Sue 20 m/s Sam 50 m/s Iftikhar 33 cm/s Mike 3 m/s

a Who gave the correct answer?

b Explain what you think each of the other students did wrong.

7 The average speed for a journey was 2 m/s.
Give a possible distance and time:

a in metres and seconds **b** in kilometres and minutes.

8 A driver plans to start a journey of 150 miles at 10 a.m.

a He expects to travel at an average speed of 60 mph.
What time does he expect to arrive?

b If his car's rate of petrol consumption is 7.5 miles per litre, how much petrol will he use on this journey?

9 **Get Real!**

a i Find the distance you travel to school.

ii Time how long this journey takes on a number of occasions.

b Work out your average speed on each journey and your overall average speed.

10 A delivery van sets off at 7:45 a.m. to deliver a package to a customer who lives 20 miles from the store. It arrives at the customer's house at 8:10 a.m.

a Calculate the van's average speed in miles per hour.

The van's average speed on the return journey in rush hour traffic is only 25 miles per hour.

b How long does the return journey take?

c Calculate the van's average speed for the whole journey.

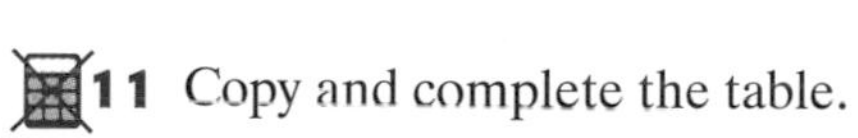

11 Copy and complete the table.

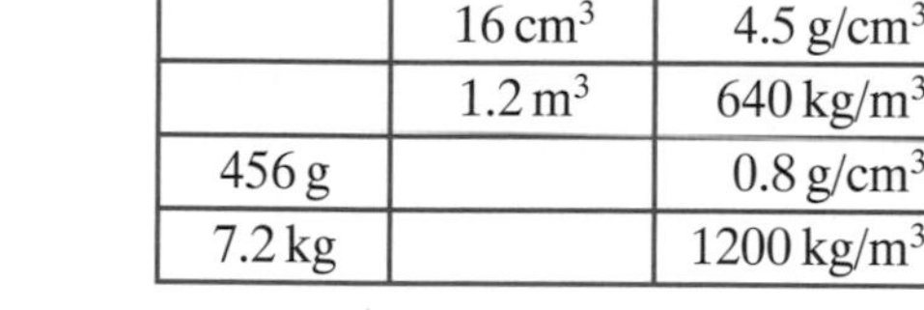

Mass	Volume	Density
248 g	200 cm^3	
4.5 kg	0.6 m^3	
	16 cm^3	4.5 g/cm^3
	1.2 m^3	640 kg/m^3
456 g		0.8 g/cm^3
7.2 kg		1200 kg/m^3

12 A log of wood weighs 712 kg and its volume is 1.25 m^3.

Find its density in **a** kg/m^3 **b** g/cm^3

13 The density of cork is 0.24 g/cm^3.
Find the volume of a cork of mass 65 grams.

14 The density of petrol is 737 kg/m^3.
What is the mass of the petrol in a 100 litre tank when it is full?
Use the fact that 1 litre = 1000 cm^3.

15 The density of copper is 8.9 g/cm^3 and the density of tin is 7.3 g/cm^3.
400 g of copper and 600 g of tin are melted together to form an alloy.
Find:

a the volume of the copper

b the volume of the tin

c the density of the alloy.

16 Find the value of the missing quantities in each part:

a

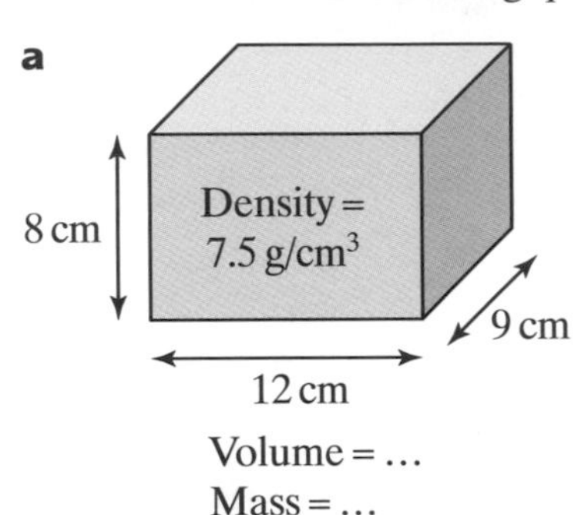

Volume = …
Mass = …

b

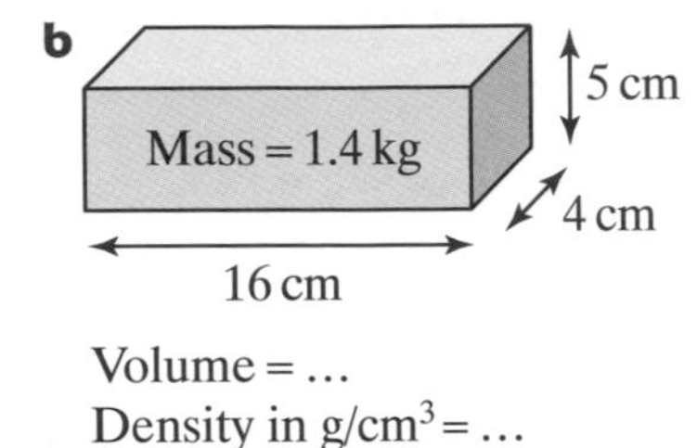

Volume = …
Density in g/cm^3 = …

c

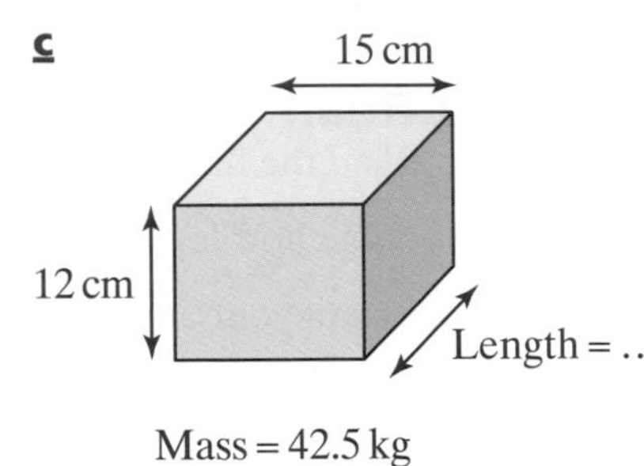

Mass = 42.5 kg
Density = 8.5 g/cm^3

17 Population density is the number of people per square kilometre who live in an area.

a A town covers an area of 16 square kilometres and has a population of 35 360.
What is its population density?

b It is planned to extend the town by building a new housing estate on adjacent land of area 0.25 square kilometres. The plan assumes the population density of the estate will be 4500 people per square kilometre.
According to the plan, how many people will live on the estate?

Explore

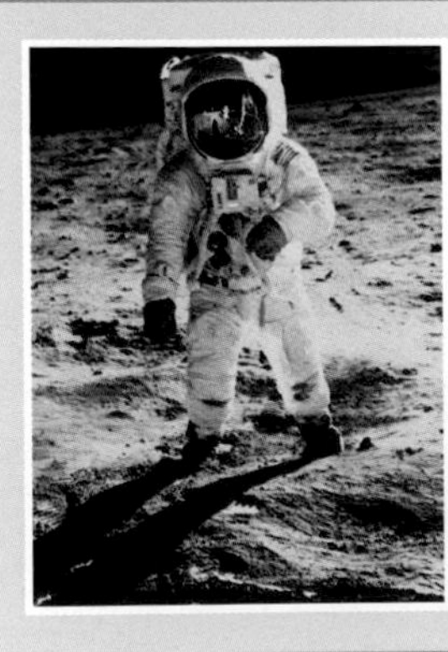

- Find the diameter, thickness and weight of a 10p coin
  You will get a more accurate result if you weigh a number of coins
- Calculate the density of the coin

Investigate further

Learn 5 Accuracy in measurements

Example: If the length of a pencil is given as 7 cm to the nearest cm, what is the minimum and maximum possible length?

Measurements given to **the nearest whole unit** may be **up to half a unit larger or smaller**, so the minimum length of the pencil is 6.5 cm and the maximum length of the pencil is 7.5 cm.

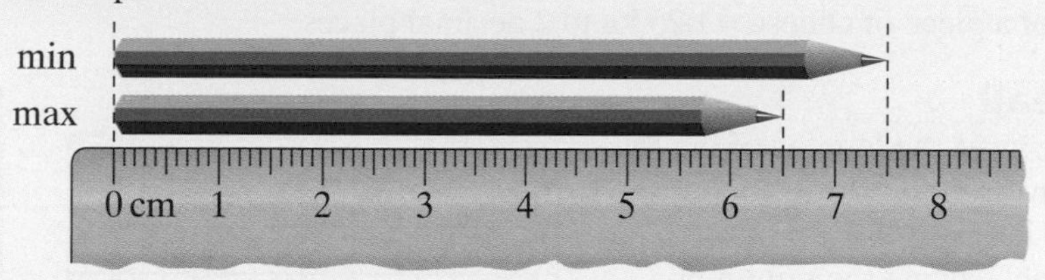

In general: $6.5 \leqslant$ length of pencil < 7.5 since a length of 7.5 cm would normally be rounded to 8 cm.

The minimum value, sometimes called the **lower bound** (or limit) is 6.5 cm.

The maximum value, sometimes called the **upper bound** (or limit) is 7.5 cm.

When a measure is expressed to a given unit, the maximum error is half of this unit.

Accuracy	Maximum possible error
Nearest 100	50
Nearest 10	5
Nearest whole number	0.5
To 1 decimal place (i.e. nearest 0.1)	0.05
To 2 decimal places (i.e. nearest 0.01)	0.005

Apply 5

1 In each part, write the lower and upper limits for the measurement.

a The height of a tree is 14 m to the nearest metre.

b The distance to the next service station is 32 km to the nearest kilometre.

c The mass of a letter is 32 g to the nearest gram.

d A girl's height is 154 cm to the nearest centimetre.

e The midday temperature at a resort is 29°C to the nearest degree.

f The estimated time for a job is 40 minutes to the nearest minute.

2 Give the upper and lower bounds of the following measurements.

a Mass of a package = 3.4 kg to 1 decimal place

b The time taken by a runner for a 100 m sprint was 10.27 seconds to 2 decimal places

c Capacity of a can = 330 mℓ to the nearest 10 mℓ

d Length of a flight = 2400 miles to the nearest 100 miles

3 Give the range of possible values of the following measurements in the form $a \leqslant$ measurement $< b$.

a The length of a swimming pool is 25 metres to the nearest metre.

b The temperature in a furnace is 500°C to the nearest 10°C.

c Water in a paddling pool = 1600 ℓ to the nearest 100 ℓ

d Patient's temperature = 38.6°C to 1 decimal place

e Mass of a piece of cheese = 1.25 kg to 2 decimal places

4 Get Real!
The table gives the lengths of the five longest rivers in the world to the nearest 5 km.
Draw a table giving the minimum and maximum values of each length.

River	Length (km)
Nile	6825
Amazon	6435
Chang Jiang	6380
Mississippi	5970
Yenisey-Angara	5535

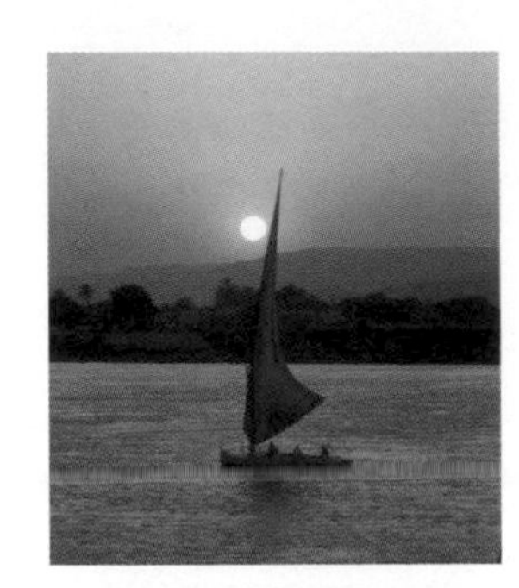

5 A book says that the mass of a package is 8.2 kilograms to 2 significant figures.
Chris writes, 'If the mass is x kg then $8.15 \leqslant x < 8.249$.'
Adam writes, 'If the weight is x kg then $8.15 \leqslant x < 8.25$.'
Who is correct? Give a reason for your answer.

6 The capacity of a swimming pool is given as 12 000 litres.

a State whether the capacity has been given to the nearest litre, 10 litres, 100 litres or 1000 litres if the least possible capacity is:

i 11 500ℓ **ii** 11 950 ℓ **iii** 11 999.5 ℓ **iv** 11 995 ℓ.

b What is the least possible capacity if 12 000 litres is correct to the nearest tenth of a litre?

7 A plan gives dimensions for a landscaped garden.
What can you say about the accuracy of the diameter of a pond if it is given on the plan as:

a 3 m **b** 3.0 m **c** 3.00 m?

Explore

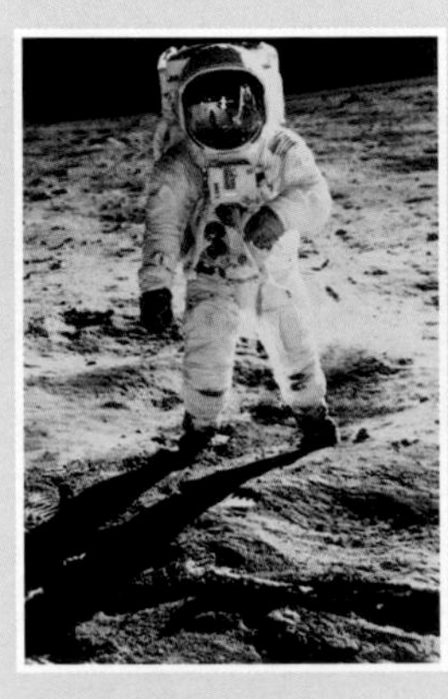

◎ Fence panels are 2 metres long to the nearest centimetre
Copy the table and complete the first row to give the lower and upper bounds of the length of the fence panel and the maximum error in the nominal length

Number of fence panels	Length	Lower bound of length	Upper bound of length	Maximum error in length
1	2m			
2	4m			
3	6m			
4	8m			

◎ Complete the rest of the table for other numbers of fence panels

Investigate further

Measures

ASSESS

The following exercise tests your understanding of this chapter, with the questions appearing in order of increasing difficulty.

1 millimetres centimetres metres kilometres
grams kilograms tonnes
millilitres litres

Write down which of the above units you would use to measure each of the following:

a The distance from Birmingham to Bury
b The length of a football pitch
c The thickness of a CD case
d Your weight
e The length of a comb
f The weight of a jelly-baby sweet
g The amount in a dose of liquid cough medicine
h The amount of beer drunk in a rugby club after a match
i The amount of wine in one glass
j The length of David Beckham's right foot
k The weight of a Ferrari Testarossa
l The weight of a full suitcase on the way to Tenerife
m The thickness of a gold necklace
n The distance run in a marathon
o The weight of a lipstick
p The capacity of a swimming pool
q The weight of a jumbo jet
r The distance from your classroom to the school gate

2 **a** How many millimetres are there in $4\frac{3}{4}$ metres?
b How many metres are there in 3200 centimetres?
c How many milligrams are there in 4570 grams?
d How many kilometres are there in 765 metres?
e How many centimetres are there in 2 feet?
f How many litres are there in 10 gallons?
g How many pounds are there in 5 kg?
h How many metres are there in 10 miles?
i How many miles are there in 800 km?
j How many pints are there in 5 litres?
k How many millilitres are there in 4 gallons?

3 **a** David and Debbie went from London to Cape Town on holiday.
The distance is about 7000 miles. Roughly how many kilometres is this?

b Akira weighs 55 kg. Roughly how many pounds is this?

c At a wedding reception eight bottles of champagne were used to toast the bride and groom. Each bottle contains 750 mℓ.
Roughly how many pints of champagne were used altogether?

d At the same reception a 'Chocolate Fountain' was eaten.
It contained 4 kg of chocolate and there was none left over.
On average, each guest ate 1 ounce of the Chocolate Fountain.
Estimate how many guests were at the reception. (16 ounces = 1pound)

e Tessa throws the javelin 55 metres. Roughly how many feet is this?

4 **a** A cyclist travels 45 miles in $2\frac{1}{2}$ hours.
What is her average speed?

b A racing car travels 7000 metres in 4 minutes.
What is its average speed in km/h?

c A Virgin Voyager travels for $4\frac{1}{3}$ hours at an average speed of 72 km/h.
How far does it go?

d An Airbus travels for $3\frac{1}{2}$ hours at an average speed of 520 mph.
How far has it flown?

e A motorcyclist travels 140 miles at 30 mph.
How long does his journey take?

f A coach travels 21 miles at 45 mph.
How long, in minutes, does the journey take?

5 **a** An object has a mass of 12 grams and a volume of 18 cm^3.
What is its density?

b A gold bar has a mass of 30 g and a density of 19.3 g/cm^3.
What is its volume?

c A metal bar measuring 50 cm by 20 cm by 5 cm has a density of 15 g/cm^3.
What is the mass of the bar?

6 Write down the lower and upper limits of the following measurements.

a The Sun is 93 000 000 miles from the Earth to the nearest million miles.

b Euan's arm is 1 m long to the nearest 10 cm.

c A parcel weighs 940 g to the nearest 5 g.

d The 1500 m race was won in a time of 246 s to the nearest second.

e The number of matches in a box is 100 to the nearest 10 matches.

f Jane is 35 years old to the nearest year.

g A stick has a length of 56.4 cm to the nearest mm.

14 Real-life graphs

OBJECTIVES

F **Examiners would normally expect students who get an F grade to be able to:**

Plot points of a conversion graph and read off positive values

E **Examiners would normally expect students who get an E grade also to be able to:**

Read from a conversion graph for negative values

Interpret distance–time graphs

D **Examiners would normally expect students who get a D grade also to be able to:**

Calculate simple average speeds from distance–time graphs

C **Examiners would normally expect students who get a C grade also to be able to:**

Calculate complex average speeds from distance–time graphs

What you should already know ...

- Plot coordinates
- Plot and interpret a line graph
- Solve problems involving proportional reasoning

VOCABULARY

Conversion graph – a graph used to convert one unit into another unit, for example, pounds to kilograms, litres to pints

Speed – the rate of change of distance with respect to time. To calculate the average speed, divide the total distance moved by the total time taken. It is usually given in metres per second (m/s) or kilometres per hour (km/h) or miles per hour (mph)

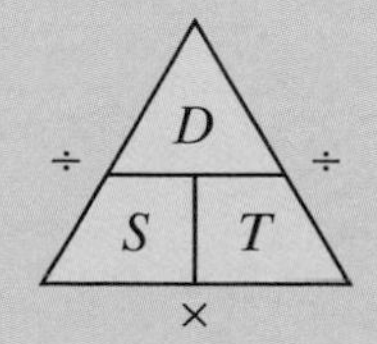

In the triangle, cover the item you want, then the rest tells you what to do

$$\text{Speed} = \frac{\text{distance}}{\text{time}}$$ $\frac{\text{metres}}{\text{seconds}}$ gives m/s

Distance = speed × time

$$\text{Time} = \frac{\text{distance}}{\text{speed}}$$

Gradient – a measure of how steep a line is

$$\text{Gradient} = \frac{\text{change in vertical distance}}{\text{change in horizontal distance}} = \frac{y}{x}$$

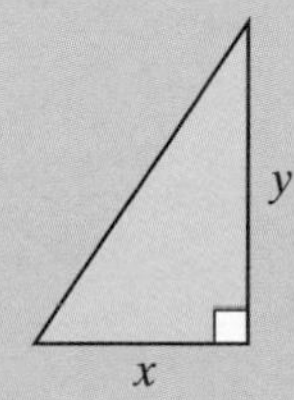

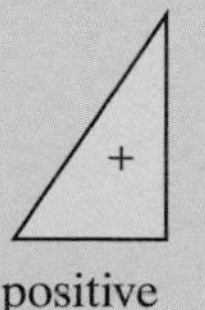

positive gradient negative gradient

Learn 1 Conversion graphs and other real-life linear graphs

Example: Paul knows that £5 = 8.5 US dollars ($)

a Draw a conversion graph to convert pounds (£) to dollars ($).

b Use the graph to convert **i** £5.50 to dollars **ii** $7 to pounds.

a Plot the points (5, 8.5) and (0, 0).

Two points are needed to draw a conversion graph but three points provide a good check

Here the point (5, 8.5) is included as well as (0, 0) as £0 = $0

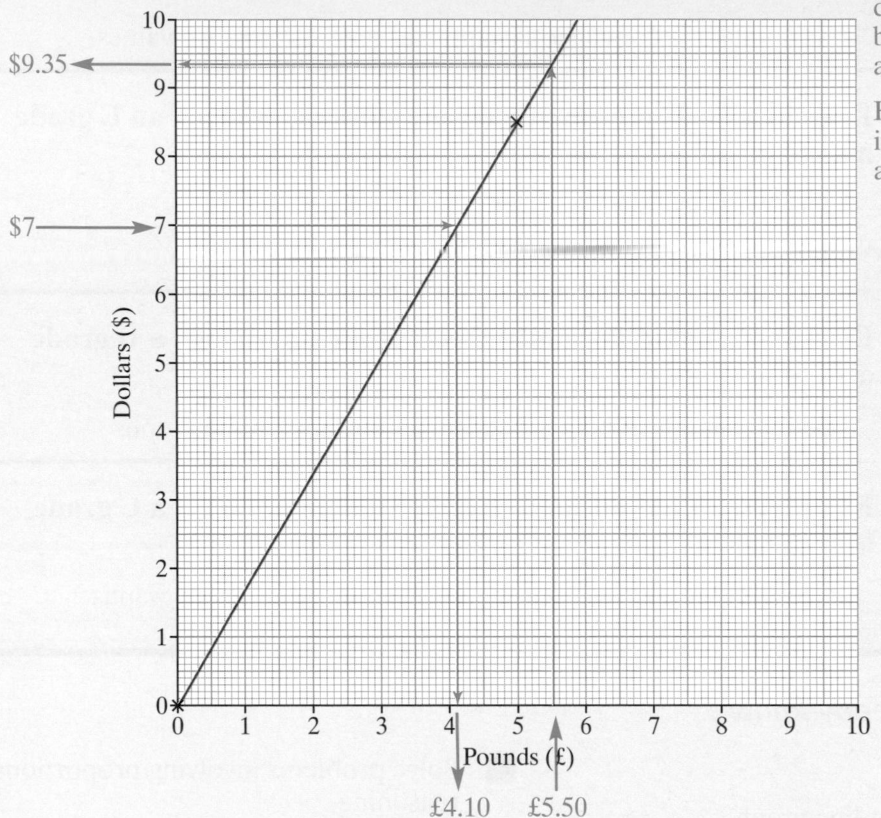

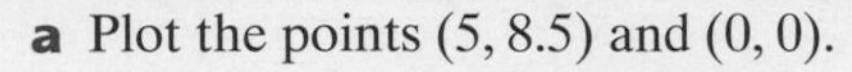

b i Draw a line from £5.50 on the pounds axis to the graph and a line from the graph to the dollars axis:
£5.50 = $9.35

ii Draw a line from $7 on the dollars axis to the graph and a line from the graph to the pounds axis:
$7 = £4.10

Apply 1

1 a Copy the set of axes shown below, using each large square to represent 1 unit.

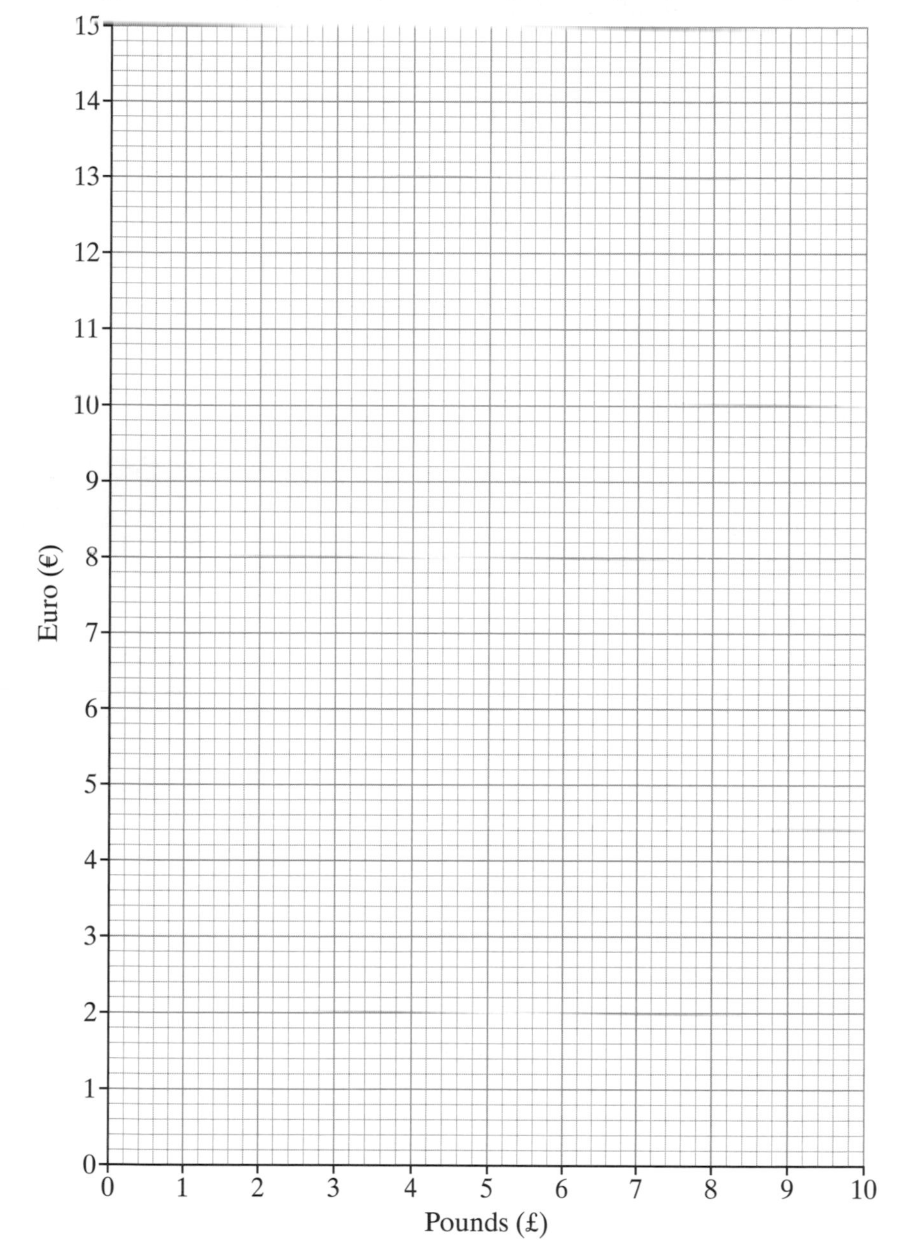

b Margaret knows that £10 = €14
Use this information to draw a conversion graph for pounds to euro.

c Use your conversion graph to convert the following amounts into euro:

i £3 **ii** £5 **iii** £2.50 **iv** 50p **v** £9.60

d Use your conversion graph to convert the following amounts into pounds:

i €1 **ii** €2 **iii** €4.50 **iv** €10 **v** €0.10

e Steve is travelling to France by Eurostar. At Waterloo station, he buys a cup of coffee for £1.75. At the Gare du Nord station, he buys a cup of coffee for €2.10. Is a cup of coffee cheaper in England or France? Give a reason for your answer.

2 **a** Copy the set of axes shown, using each large square to represent 5 units.

b Abdula knows that 20 miles is equivalent to 32 kilometres.
Use this information to draw a conversion graph for miles to kilometres.

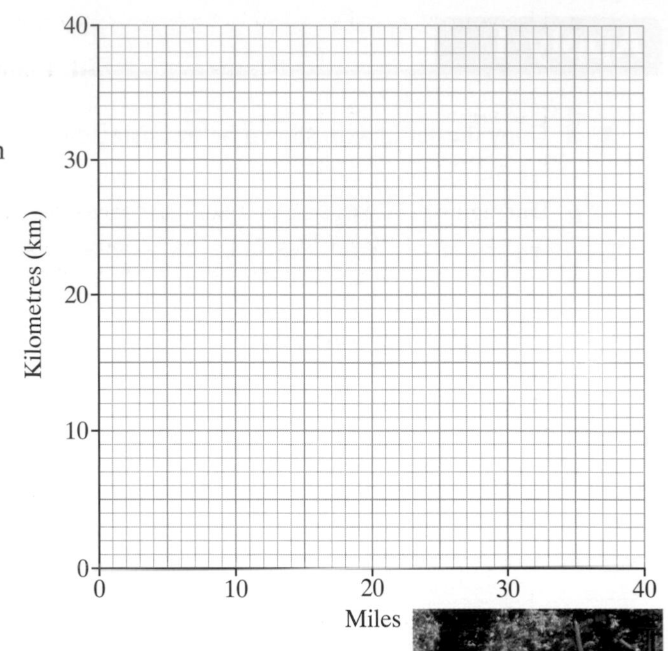

c Use your conversion graph to convert the following distances into kilometres:

i 7 miles
ii 2 miles
iii 15 miles
iv 25 miles

d Use your conversion graph to convert the following distances into miles:

i 10 km
ii 2 km
iii 30 km

e Marika is going to raise money for a charity by entering either a 20 km fun run or a half marathon (13 miles). Which one is the shorter to run?

3 **a** Copy the set of axes shown below using each large square to represent 1 unit.

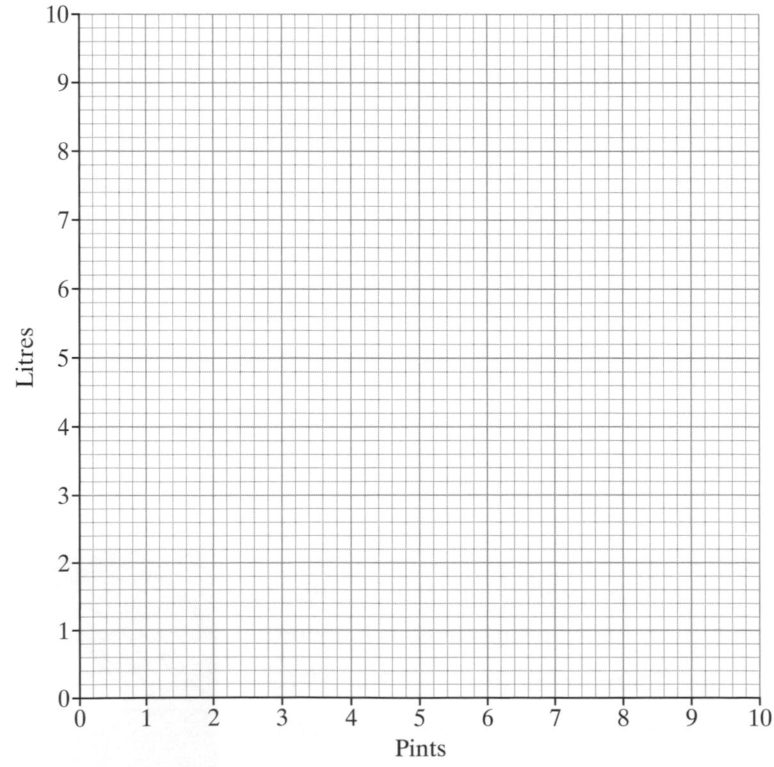

b Nasreem notices on the label of her milk carton that 4 pints is equivalent to 2.3 litres. Use this information to draw a conversion graph for pints to litres.

c Use your conversion graph to convert the following volumes into litres:

i 10 pints **ii** 6 pints **iii** 1 pint

d Use your conversion graph to convert the following volumes into pints:

i 5 litres **ii** 1 litre

e During an oil crisis in the 1980s, motorists had to pay £5 for a gallon of petrol. Is £1 for 1 litre of petrol a better or a worse deal? Give a reason for your answer. (1 gallon = 8 pints)

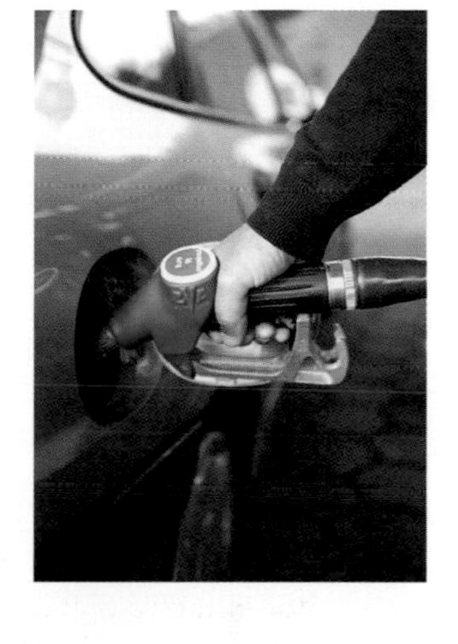

4 Matt and Erin are using this conversion graph to convert kilograms to pounds.
Their daughter Lily weighed 3.3 kg when she was born.

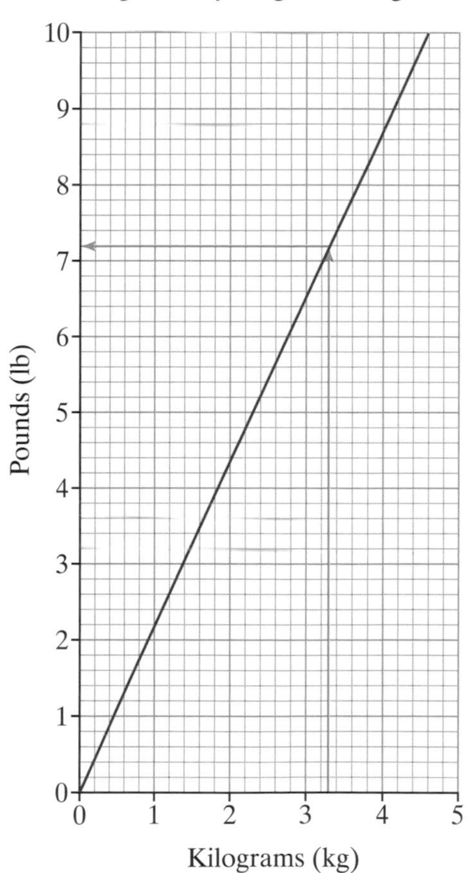

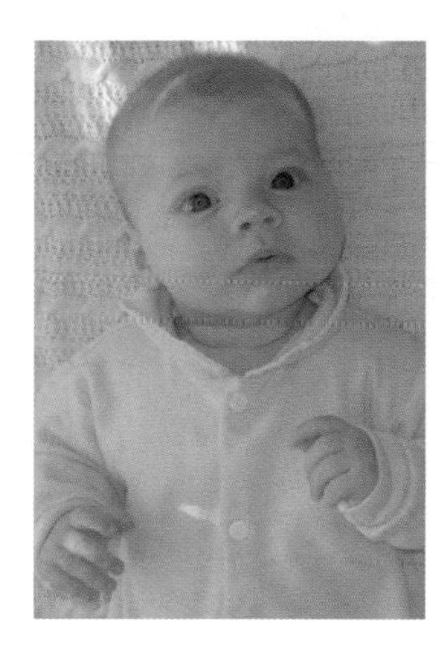

Using the graph, Matt and Erin convert this to 7.1 lb.
Are they correct? Give a reason for your answer.

5 **a** Copy the set of axes shown below, using each large square to represent 20 units.

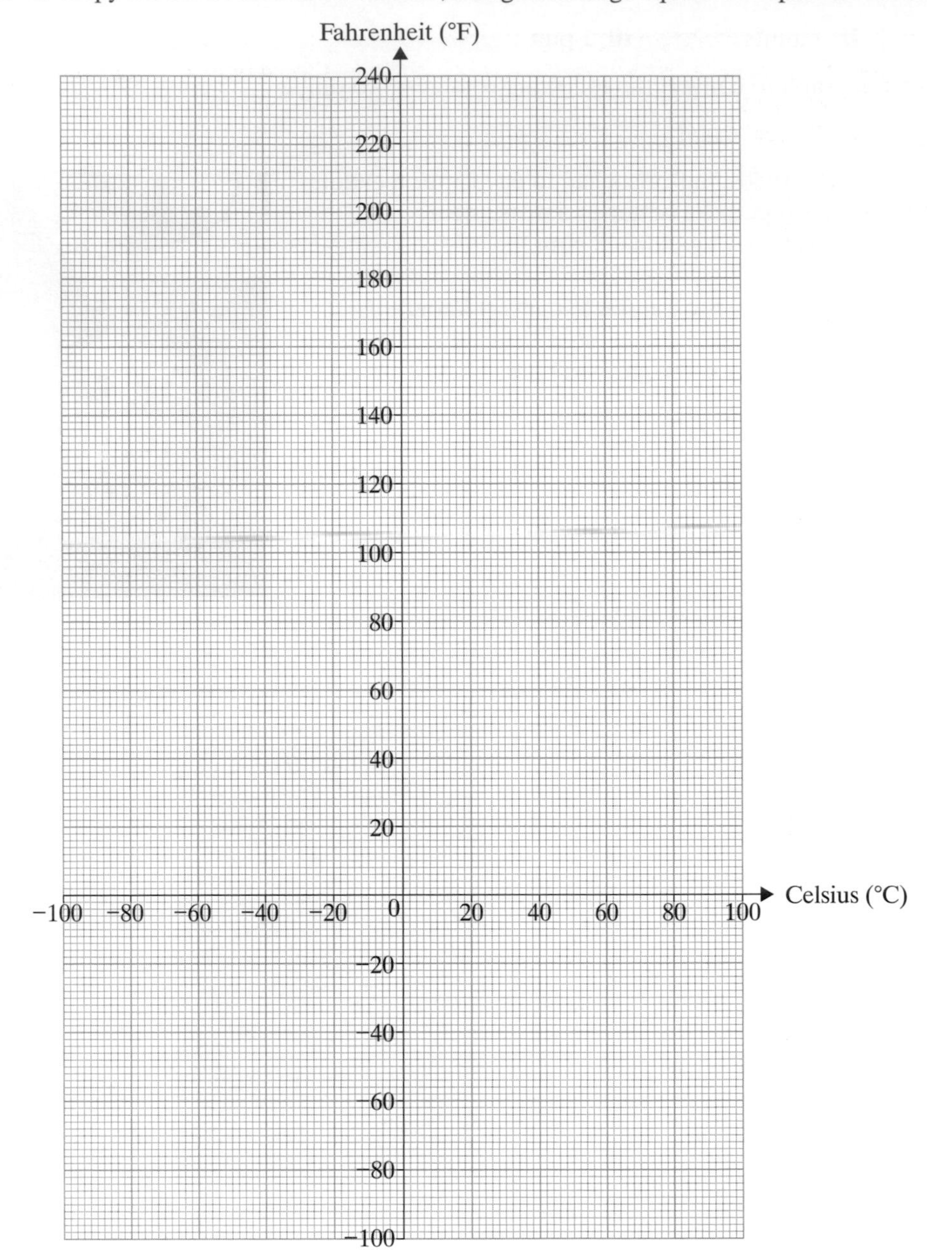

b 60°C is equivalent to 140°F and −40°C is equivalent to −40°F.
Use this information to draw a conversion graph for Celsius to Fahrenheit.

c Use your conversion graph to convert the following temperatures to °F:

i 40°C **ii** 10°C **iii** −60°C **iv** −18°C

d Water freezes at 0°C and boils at 100°C. Convert these temperatures to °F.

e Use your conversion graph to convert the following temperatures to °C:

i 40°F **ii** 10°F **iii** −80°F **iv** −30°F

f The normal body temperature of a human is 37°C. Use the conversion graph to find this temperature in degrees Fahrenheit.

g The hottest day ever in Britain was on Sunday 10th August 2004 in Gravesend, Kent. The record temperature was 100.6°F. Convert this temperature to degrees Celsius.

Explore

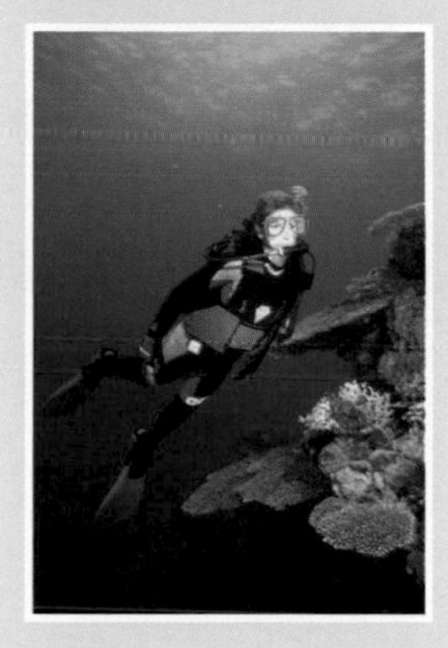

- Copy the set of axes using the same scale as shown below, using each large square to represent half a year on the horizontal axis and 10 centimetres on the vertical axis

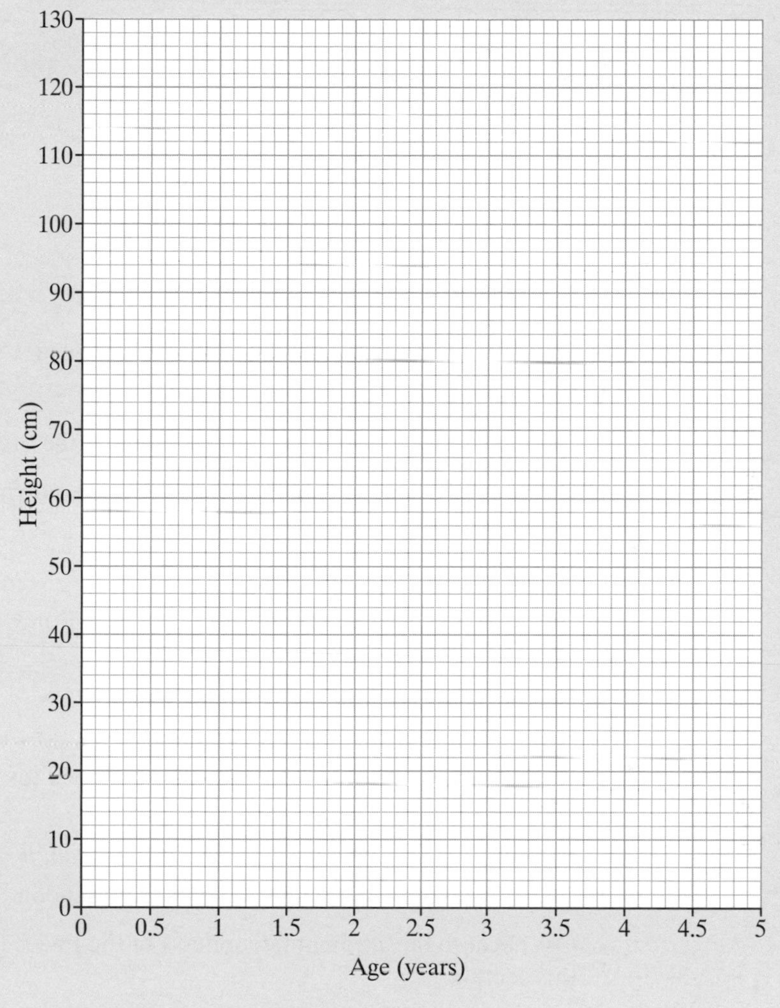

- The average 3-year-old boy is 95 cm tall
  Plot this point on the graph
- The average 5-year-old boy is 110 cm tall
  Plot this point on the graph
- Join the two points with a straight line
- Use the graph to find the following:

  a the approximate height of a boy aged 2.5 years

  b the approximate age of a boy under 99 cm tall
- Could the graph be used to estimate the height of:

  a a newborn baby

  b a 19-year-old male?

Investigate further

Learn 2 Distance–time graphs

Example: The distance–time graph below shows the journey of a school bus dropping students off at two villages.

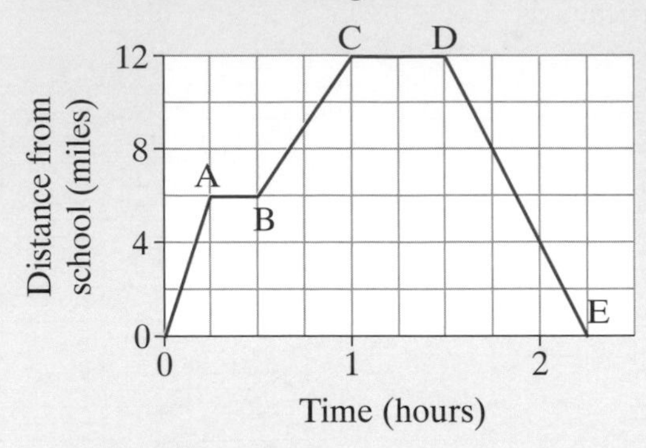

a Describe the journey, giving reasons for the shape of the graph.

b How far are the villages from school?

c During which section is the bus travelling at the fastest speed?

d Calculate the speed of the bus, in miles per hour (mph), during these sections:

i OA **ii** CD **iii** DE

e Calculate the average speed for the whole journey.

a The bus leaves school at a constant speed.
After $\frac{1}{4}$ hour, the bus stops for $\frac{1}{4}$ hour.
It then carries on the journey at a slower speed for $\frac{1}{2}$ hour before stopping again after 12 miles.
It stops for $\frac{1}{2}$ hour and then travels back to school at a constant speed.

b The villages are 6 miles and 12 miles from school.

c OA because the gradient (steepness) of the line is the greatest during this section.

d i Distance = 6 miles, time = $\frac{1}{4}$ hour, so in 1 hour the bus would travel 4×6 miles = 24 miles.
Hence, speed = 24 miles per hour = 24 mph

Remember: 'miles per hour' means 'how many miles you would travel in 1 hour'

ii Distance = 0 miles, time = $\frac{1}{2}$ hour, so speed = 0 mph

iii Distance = 12 miles, time = $\frac{3}{4}$ hour, so:

$$\text{Speed} = \frac{\text{distance}}{\text{time}} = \frac{12}{\frac{3}{4}} = \frac{24}{1.5} = \frac{48}{3} = 16 \text{ mph}$$

(×2 applied to numerator and denominator at each step)

Creating equivalent fractions

Alternatively, 12 miles in $\frac{3}{4}$ hour is the same as 4 miles in $\frac{1}{4}$ hour or 16 miles in 1 hour

e Total distance = 24 miles, total time = $2\frac{1}{4}$ hours, so

$$\text{Average speed} = \frac{\text{distance}}{\text{time}} = \frac{24}{2\frac{1}{4}} = \frac{48}{4\frac{1}{2}} = \frac{96}{9} = 10.\dot{6} = 11 \text{ mph}$$

Apply 2 

1 The graph below shows the journeys of three people – Sam, Joshua and Hannah.

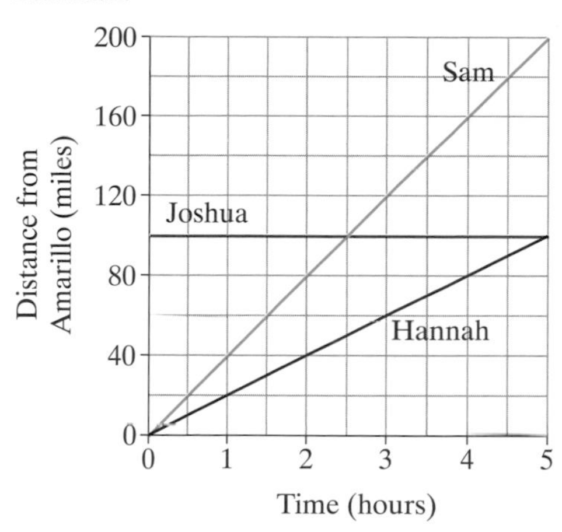

a Who does the graph suggest is travelling by:

i car **ii** bicycle?

b Who is the furthest distance from Amarillo after:

i 2 hours **ii** 5 hours?

c How far is Joshua from Amarillo after:

i 2 hours **ii** 4 hours?

d What can you say about Joshua's journey?

e Who travels the fastest?

f Calculate Hannah's speed in miles per hour (mph).

2 Miranda goes out jogging every day.
Her journey is shown:

'Miranda runs very fast for the first 2 minutes and then jogs at a constant speed for a short while. She then runs even faster until she is 1500 m from home. She jogs at a constant speed for 4 minutes before running backwards to her home.'
Do you agree with this description of Miranda's journey? Give reasons for your answer.

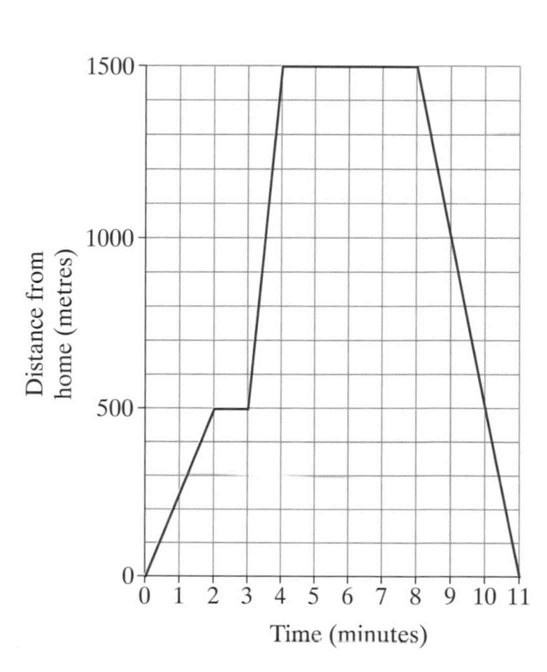

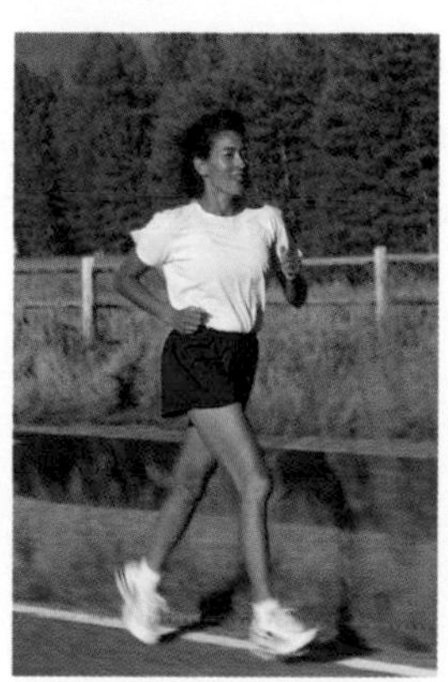

3 The graph shows a journey by a car from Swinton.

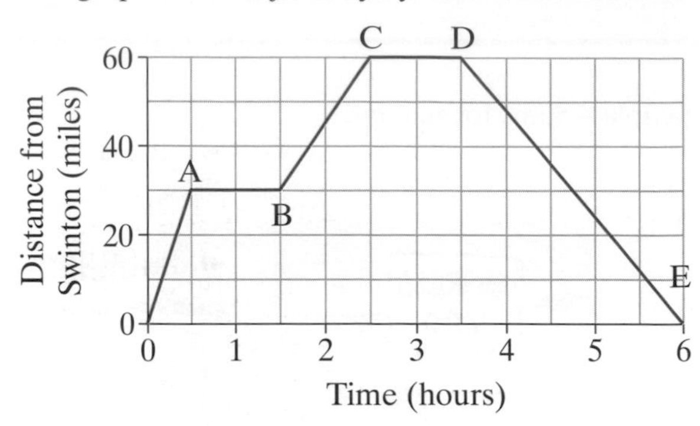

a Jess works out the speed of the car during section OA as follows:

$$\text{Speed} = \frac{\text{distance}}{\text{time}} = \frac{30}{\frac{1}{2}} = 15 \text{ miles per hour}$$

Do you agree? Give a reason for your answer.

b Jess also calculates the average speed for the whole journey as follows:

$$\text{Average speed} = \frac{\text{total distance}}{\text{total time}} = \frac{60}{6} = 10 \text{ miles per hour}$$

Do you agree? Give a reason for your answer.

4 Jim the courier is delivering parcels from his depot in Nelson.
He leaves the depot at 9.45 a.m.

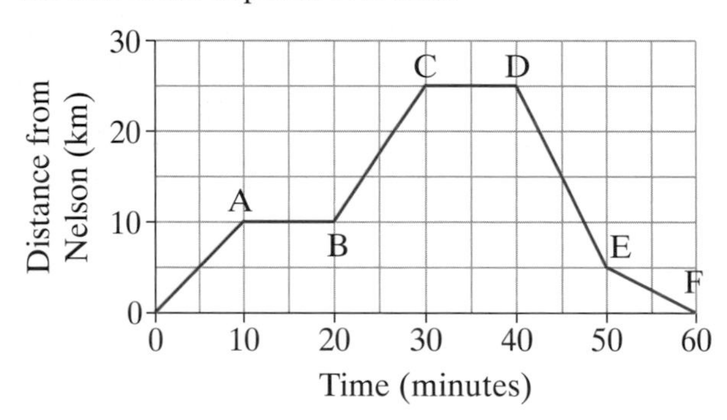

a How many times does Jim stop to deliver parcels?

b At what time does Jim arrive at his first delivery?

c How far does Jim travel in the first 10 minutes?

d Calculate Jim's speed during section BC in kilometres per hour (km/h).

e Calculate Jim's average speed over the first half hour of the journey.

f At what time does Jim start to return to the depot?

g During which section is Jim travelling the fastest?

h Calculate Jim's average speed for the whole journey.

5 Two snails, Sandra and Brian, are racing along a 30 cm ruler. Write a commentary for the race. You should include key distances, times and speeds in your commentary.

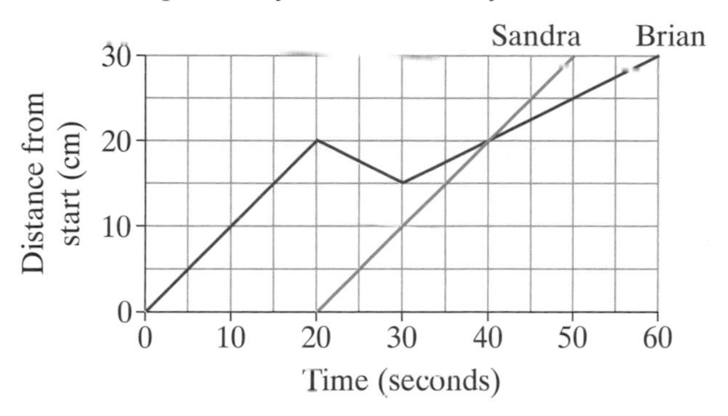

6 Pat walks to her local park for 15 minutes at 4 km/h. She sits on a bench for 10 minutes and then walks back home at a speed of 3 km/h. Draw a distance–time graph representing Pat's journey.

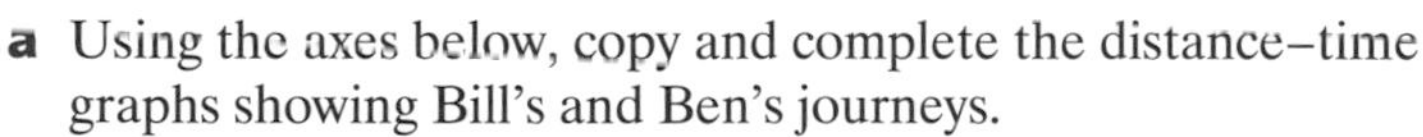

7 Bill and Ben both leave their homes at 10:30 a.m. Bill is travelling the 155 miles from his home in Cheltenham to Manchester. He travels along the M5 at 60 mph for 50 minutes and then is held up in a traffic jam for half an hour as the M5 joins the M6. He then travels the remainder of the journey at 70 mph
Ben leaves Manchester at the same time that Bill leaves Cheltenham. He travels to Cheltenham at a constant speed of 50 mph.

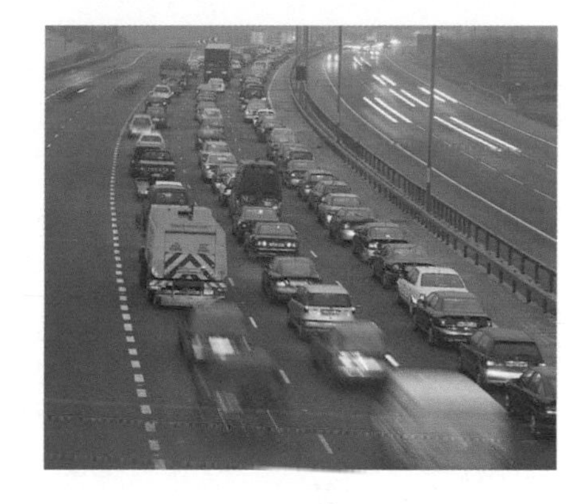

a Using the axes below, copy and complete the distance–time graphs showing Bill's and Ben's journeys.

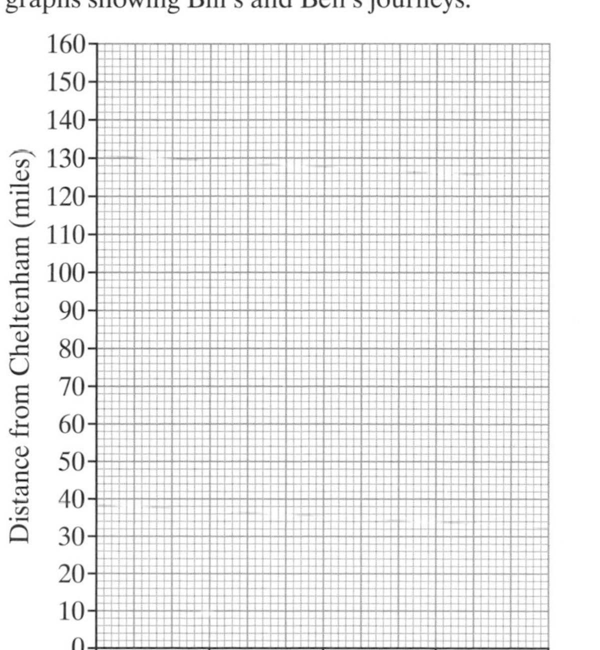

b At what times do Bill and Ben arrive at their destinations?

c At what time are Bill and Ben the same distance from Cheltenham?

Real-life graphs

ASSESS

The following exercise tests your understanding of this chapter, with the questions appearing in order of increasing difficulty.

1 Cooking a turkey needs 15 minutes for each pound weight plus a further 20 minutes.

a Copy and complete the following table:

Weight (lb)	0	4	8	12	16	20	24
Cooking time (min)	20	80		200			380

b Draw a graph of the above information, joining all the points with a straight line.

c Use your graph to answer the following questions:

i Liz cooks a 22 lb turkey. How long does it need to cook?

ii Shaheed cooks a 10 lb turkey. How long does it need to cook?

iii Jamie's turkey cooks for 245 minutes. How heavy is it?

iv Delia's turkey cooks for 305 minutes. How heavy is it?

v Sunita says that a 20 lb turkey will take twice as long to cook as a 10 lb turkey.
Is she correct?

2 The graph opposite shows the conversion from degrees Celsius to degrees Fahrenheit.

Use the graph to answer the following questions:

a Convert the following Celsius temperatures to Fahrenheit:

i 0° **ii** 50° **iii** 100° **iv** −20° **v** −100°

b Convert the following Fahrenheit temperatures to Celsius:

i 50° **ii** 104° **iii** 176° **iv** −76° **v** −130°

c The 'normal' temperature of the human body is said to be 98.6°F.
What is this temperature in °C?

d Alcohol boils at 85°C. What is this in °F?

e One temperature has the same value both in Fahrenheit and in Celsius.
What is it?

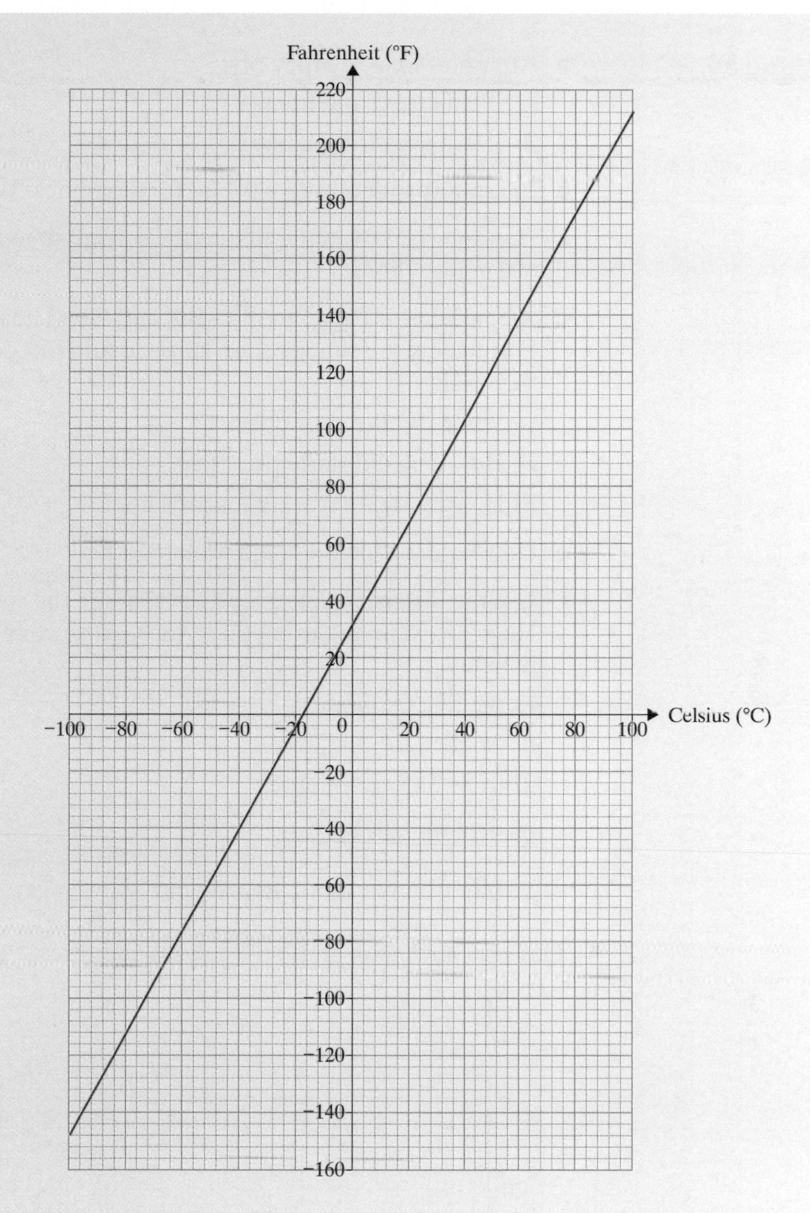

3 Bill walks his dog, Hugo. The distance they are from home is shown in the graph below.

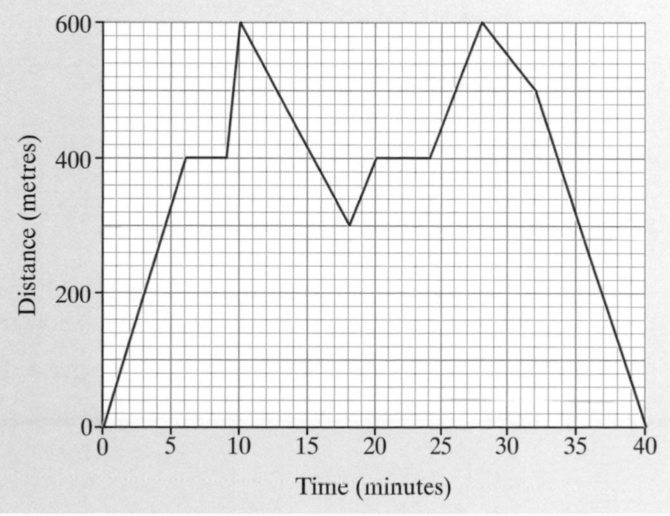

a After how long are they first:

i 200 metres from home

ii 550 metres from home?

b Bill stands and talks to a friend for 4 minutes. When did this happen?

c How long is it from the first to the second time that Bill and Hugo are 600 m from home?

d At one stage it began to rain. Bill and Hugo headed for home. Fortunately the shower soon stopped and Bill decided to resume their walk.

i When did the shower start?

ii How long did the shower last?

iii How far did they walk towards home?

e How far did Bill and Hugo walk together?

f Use the distance–time graph to calculate the speed at which Bill and Hugo are walking during these times (give your answers in metres per minute):

i the first 6 minutes

ii from 10 to 18 minutes

iii from 24 to 28 minutes

iv the last 8 minutes.

Try a real past exam question to test your knowledge:

4 This is a conversion graph for gallons and litres.

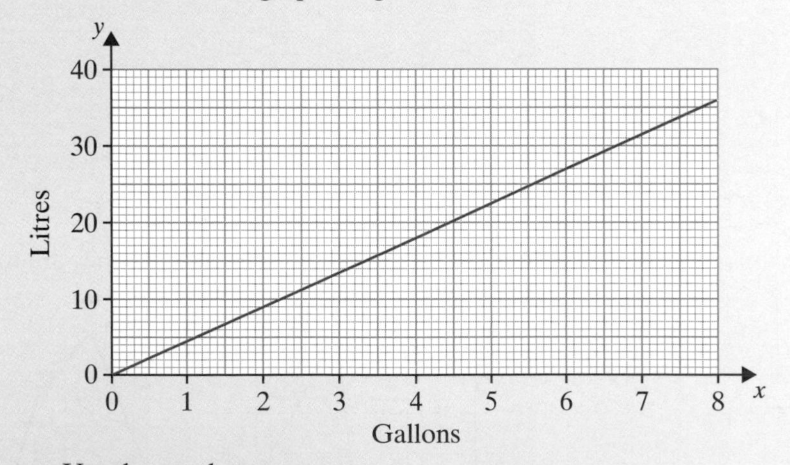

a Use the graph to convert:

i 4 gallons to litres,

ii 30 litres to gallons.

b 50 gallons is approximately 225 litres.
Explain how you can use the graph to show this.

Spec A, Foundation Paper 1, June 05

15 Formulae

OBJECTIVES

G **Examiners would normally expect students who get a G grade to be able to:**

Use a formula written in words, such as cost = 20 × distance travelled

F **Examiners would normally expect students who get an F grade also to be able to:**

Use a simple formula such as $P = 2l + 2w$

Substitute positive numbers into a simple formula

E **Examiners would normally expect students who get an E grade also to be able to:**

Write an expression from a problem

Substitute negative numbers into a simple formula

Use formulae from mathematics and other subjects

D **Examiners would normally expect students who get a D grade also to be able to:**

Substitute numbers into more complicated formulae such as

$$C = \frac{(A + 1)D}{9}$$

C **Examiners would normally expect students who get a C grade also to be able to:**

Find a solution to a problem by forming an equation and solving it

Rearrange linear formulae such as $p = 3q + 5$

What you should already know ...

- The four rules applied to negative numbers
- Calculate the squares, cubes and other powers of numbers
- Simplify expressions by collecting like terms
- Solve linear equations
- Know facts about angle properties, area, volume

VOCABULARY

Expression – a mathematical statement written in symbols, for example, $3x + 1$ or $x^2 + 2x$

Equation – a statement showing that two expressions are equal, for example, $2y - 7 = 15$

Formula – an equation showing the relationship between two or more variables, for example, $E = mc^2$

Identity – two expressions linked by the ≡ sign are true for all values of the variable, for example, $3x + 3 \equiv 3(x + 1)$

Sum – the result of adding together two (or more) numbers, variables, terms or expressions

Product – the result of multiplying together two (or more) numbers, variables, terms or expressions

Subject of a formula – in the formula $P = 2(l + w)$, P is the subject of the formula

Learn 1 Writing formulae using letters and symbols

Examples:

a A rough rule for changing inches into centimetres is to multiply the number of inches by 2.5
Write a formula for this rule.

b Write an expression for each of the following:
- **i** Think of a number. Double it. Add 4.
- **ii** Think of a number. Add 1. Multiply by 3.
- **iii** Think of a number. Square it. Add 2.
- **iv** Think of a number. Double it. Add 3. Find the square root of the answer.

c I think of a number, multiply it by 6 and add 1. The answer is 43.
What was the number?

This part of the question tells you that you need to write an equation

d Andrew needs to work out the sizes of the angles in this diagram.
Form an equation to help him work out the answer.
What size are the three angles?

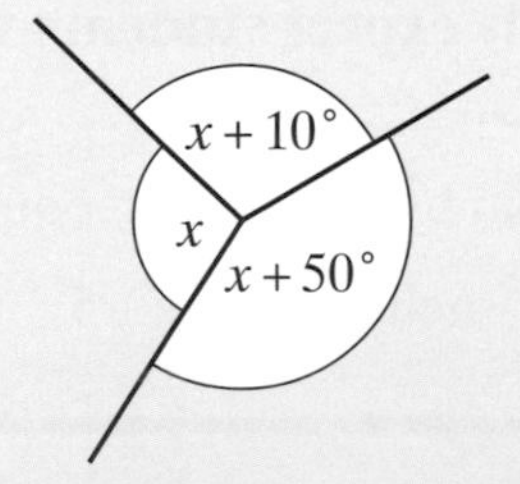

Not drawn accurately

a In words, the formula would be
'number of centimetres = number of inches × 2.5'.
Using symbols, the formula would be

$c = i \times 2.5$ ← c stands for number of centimetres, i stands for number of inches

or $c = 2.5i$ ← The number is usually written before the letter

b i Think of a number x
Double it $2x$
Add 4 $2x + 4$

ii Think of a number x
Add 1 $x + 1$
Multiply by 3 $3(x + 1)$

The answers are all expressions

iii Think of a number x
Square it x^2
Add 2 $x^2 + 2$

Remember: expressions do not have an equals sign but equations do

iv Think of a number x
Double it $2x$
Add 3 $2x + 3$
Square root $\sqrt{2x + 3}$

c First, write an equation:

Let the unknown number be x
Multiply by 6 $6x$
Add 1 $6x + 1$
The answer is 43 $6x + 1 = 43$ ← Create an equation

Next, solve the equation:

$6x + 1 - 1 = 43 - 1$ ← Subtract 1 from both sides

$6x = 42$

$6x \div 6 = 42 \div 6$ ← Divide both sides by 6

$x = 7$

d $x + x + 10° + x + 50° = 360°$ ← Use the fact that angles at a point add up to 360° to help you form an equation

$3x + 60° = 360°$ ← Simplify by adding like terms together

Solve the equation:

$3x + 60° - 60° = 360° - 60°$ ← Subtract 60 from both sides

$3x = 300°$

$3x \div 3 = 300° \div 3$ ← Divide both sides by 3

$x = 100°$

The three angles are $x = 100°$, $x + 10° = 110°$ and $x + 50° = 150°$

Check to see that your answers add up to 360°

Apply 1

1 Get Real!

A cookery book gives this rule for roast beef:

60 minutes per kilogram plus another 25 minutes

Write an expression for this rule.
Use k to stand for the number of kilograms.

2 Get Real!
The rule for finding out how far away thunderstorms are is:
'Count the number of seconds between the lightning and the thunder. Divide the answer by 5. The answer gives the distance in miles.'
Write a formula for this rule.
Use s to stand for the number of seconds and d to stand for the distance in miles.

3 Get Real!
You can find out how many amps an electrical appliance will use by using this rule:

Number of amps equals the number of watts divided by 240.

Write a formula for this rule.
Use a to stand for the number of amps and w to stand for the number of watts.

4 Write an expression for each of these statements.

a Five times the number x.

b Seven times the number y then take away 5.

c The sum of three times the number x and double the number y.

d The product of the numbers a, b and c.

e Four times the number z plus six times the number x.

f Half the number x subtracted from eight times the number y.

5 Write an expression for each of these statements.
Let the unknown number be x.

a Think of a number. Add 4.

b Think of a number. Double it.

c Think of a number. Subtract 3.

d Think of a number. Multiply by 4. Add 1.

e Think of a number. Add 2. Multiply by 5.

6 a Nick says that the answer to question **5b** is $x2$. Is he correct? Give a reason for your answer.

b Becky says that the answer to question **5b** is x^2. Is she correct? Give a reason for your answer.

7 There are six buttons on a blouse. Write an expression to show how many buttons there would be on y blouses.

8 Write an expression for each of these statements.
Let the unknown number be y.

a Think of a number. Square it.

b Think of a number. Square it and add 3.

c Think of a number. Double it. Add 4.

d Think of a number. Subtract 2. Multiply the result by 2.

e Think of a number. Double it. Add 6. Divide the result by 2.

9 Write down an expression for the total cost of

a x lollies at 60p each and y lollies at 80p each

b c cakes at 75p each and b biscuits at 16p each.

10 Write an equation and solve it to find the unknown number in each of these questions.
Let the unknown number be z.

a Think of a number. Add 3. The answer is 29.

b Think of a number. Multiply it by 3. Add 5. The answer is 17.

c Think of a number. Multiply it by 4. Subtract 5. The answer is 23.

d Think of a number. Add 3. Multiply by 4. The answer is 12.

e Think of a number. Subtract 7. Multiply by 2. The answer is 6.

11 **Get Real!**
A cookery book gives this rule to roast a chicken:

Allow 20 minutes per pound plus 20 minutes.

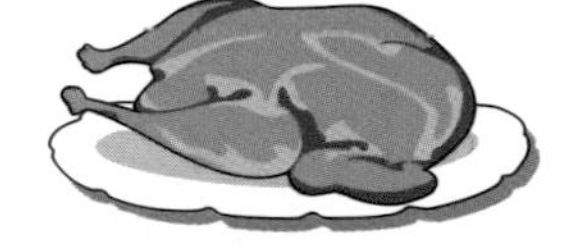

a Write a formula for this rule. Use c to represent the time needed to roast a chicken and p to represent the number of pounds.

b How long would a 3-pound chicken take to roast?

12 Write down an equation for each of these diagrams.
Solve it to find the value of x.

a

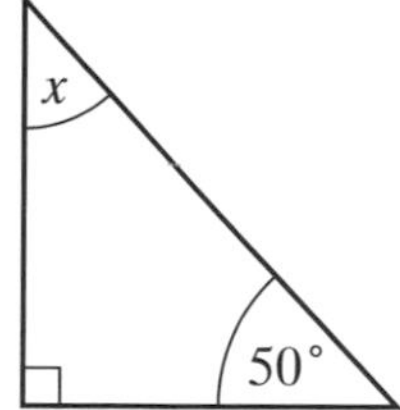

c

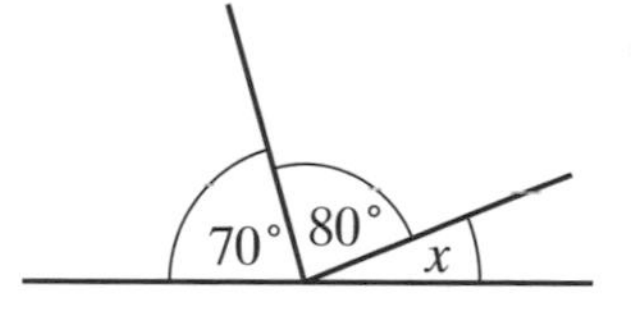

Not drawn accurately

b

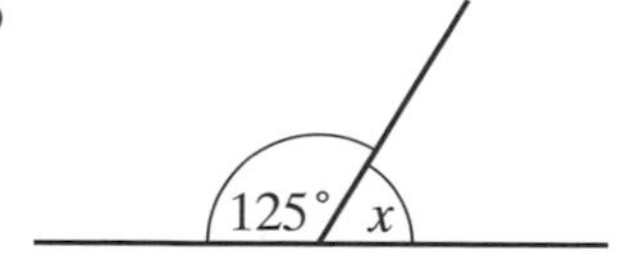

d

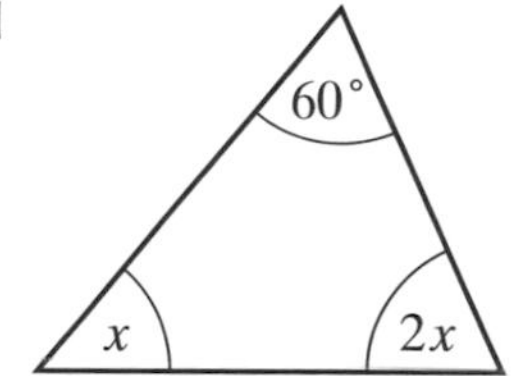

13 Write down an equation for each of these diagrams.
Solve it to find the value of x.

a

x $x + 20°$ $2x$

b

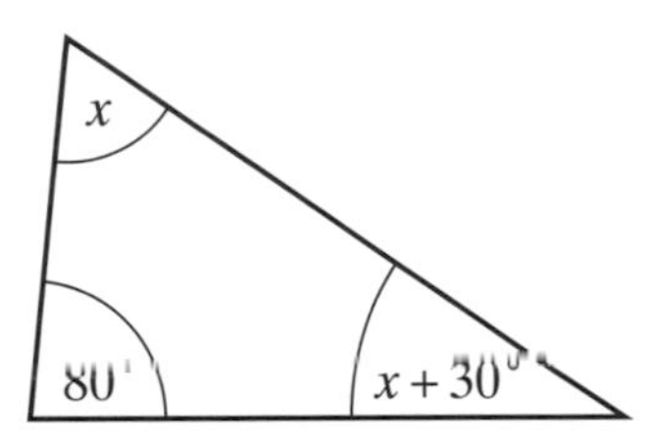

c

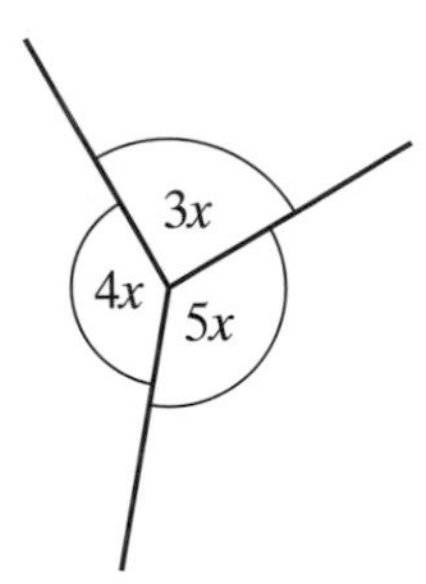

d

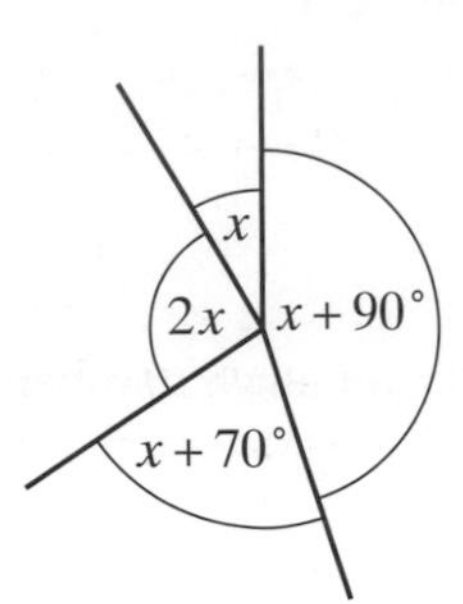

e

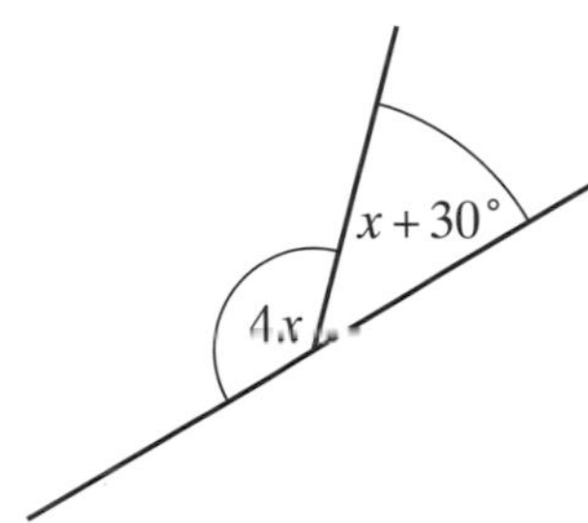

f

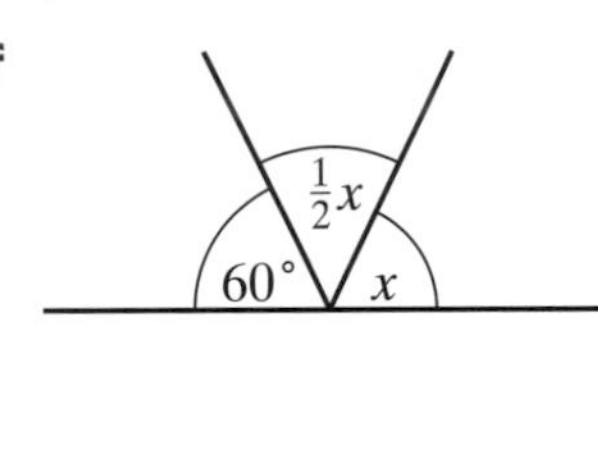

g

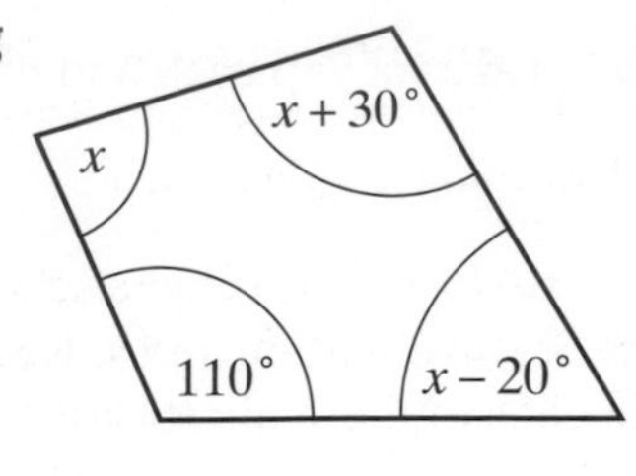

h

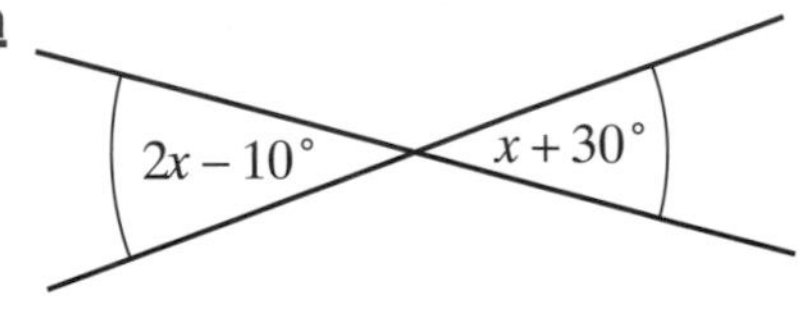

Not drawn accurately

14 Find the area of each of these shapes.
The perimeter of each shape is given in cm.

a

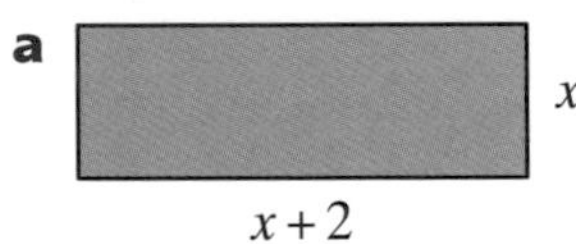

$x + 2$

Perimeter = 8

b

$x + 4$

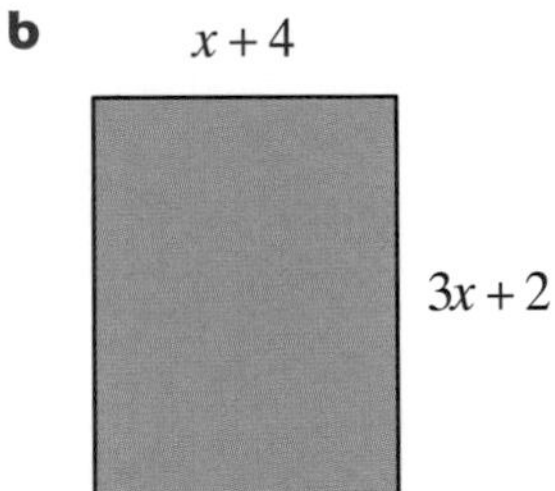

$3x + 2$

Perimeter = 28

15 The length of a rectangle is three times its width. The width is w cm.

a Write down an expression for the length.

The perimeter of the rectangle is 120 cm.

b Find the area of the rectangle.

16 The length of a rectangle is 5 cm more than its width. The width is b cm.

a Write down an expression for the length.

The perimeter of the rectangle is 90 cm.

b Find the area of the rectangle.

17 A square has sides of length $x + 2$. The perimeter is 32 cm.
Find the area of the square.

18 Katie goes on holiday with $\$x$. Sam has three times as much as Katie.
James has \$5 more than Katie. Altogether they have \$250.
How much do they each have?

19 Miss Anstey organised a school trip to London on 4 coaches.
Each coach has the same number of seats and there are 5 spare seats.
On the way back 2 coaches broke down. The students and teachers continued their journey either in one of the two full coaches or by train. 101 students and teachers had to catch the train.
Find the number of seats on each coach.

Learn 2 Substitution

Examples:

a If $a = 3$, $b = 2$ and $c - -1$, find the value of:

i $a + b + c$ **ii** ab **iii** $2a - 3b + 4c$ **iv** $\frac{2a}{3b}$

b In science, Philip is using the formula $s = ut + \frac{1}{2}at^2$
Hc needs to find s when:

i $u = 4$, $t = 3$ and $a = 5$ **ii** $u = 3$, $t = 4$ and $a = 0.5$

a i $a + b + c$
$= 3 + 2 + -1$ ← Replace the letters with the values given
$= 4$

ii ab ← Remember that ab means $a \times b$
$= 3 \times 2$
$= 6$

iii $2a - 3b + 4c$
$= (2 \times 3) - (3 \times 2) + (4 \times -1)$
$= 6 - 6 - 4$ ← $+ -4 = -4$
$= -4$

Remember to show all stages of your working; you gain method marks for this in an examination

iv $\frac{2a}{3b}$
$= \frac{(2 \times 3)}{(3 \times 2)}$
$= \frac{6}{6}$
$= 1$

b i $s = ut + \frac{1}{2}at^2$
$s = (4 \times 3) + (\frac{1}{2} \times 5 \times 3^2)$
$s = 12 + 22.5$
$s = 34.5$

First, write down the formula to use; next, substitute the given values into the formula

$\frac{1}{2}at^2$ means $\frac{1}{2} \times a \times t^2$ so only the t is squared

ii $s = ut + \frac{1}{2}at^2$
$s = (3 \times 4) + (\frac{1}{2} \times 0.5 \times 4^2)$
$s = 12 + 4$
$s = 16$

Apply 2

1 Find the value of each of these expressions when $x = 10$

a $x + 4$ **b** $3x$ **c** $x - 8$ **d** $2x + 5$ **e** $12 - x$ **f** $40 - 4x$ **g** $5x - 32$ **h** $2(x + 2)$

2 Find the value of each of these expressions when $x = 3$

a $4x$ **b** $3x - 2$ **c** $2x + 2$ **d** $5 - x$ **e** $6 + 2x$ **f** $4x - 9$ **g** x^2 **h** $x^2 + 4x - 6$

3 Find the value of each of these expressions when $y = -2$

a $5 + y$ **b** $3y$ **c** $4y + 6$ **d** $6 - y$ **e** $8 - 2y$ **f** $5y - 1$ **g** $y^2 - 3$ **h** $2y^2 + 3y$

4 If $x = 2$, $y = 3$, $s = 4$ and $t = 0.5$, find the value of:

a $2x$ **b** $6y$ **c** $3y + x$ **d** $3s - t$ **e** $\frac{s}{2} + 1$ **f** $\frac{4s}{x}$ **g** xy **h** $2s - t + y$ **i** xys **j** $st \div x$ **k** $xy - st$ **l** $\frac{st}{xy}$

5 If $p = 3$, $q = 6$ and $r = -7$, find the value of:

a $q + 3$ **b** $12 - r$ **c** $3r$ **d** pqr **e** $\frac{12}{p}$ **f** $p + 3q + 4r$ **g** $\frac{pq}{9}$ **h** $p + 4q - 2r$ **i** $\frac{r}{7}$ **j** $pr - qp$ **k** $r + 3pq$ **l** $\frac{q}{r}$

6 If $a = 2$, $b = -3$ and $c = 5$, find the value of:

a a^2 **b** b^2 **c** abc **d** $a^2 + b$ **e** $b^2 - c$ **f** $3c^2$ **g** $2c - 4a$ **h** a^2c **i** abc^2 **j** $a^2 + b^2$ **k** $b^2 - a^3$ **l** $a^4 - b^3$

7 Emily has worked out the answer to question **6f** as 225. Is she correct? Give a reason for your answer.

8 $ab + c = 10$
Find five different sets of numbers that make this statement correct. Use negative numbers, fractions, ...

9 Antony is using the formula $C = \frac{(A + 1)D}{9}$
He knows that $D = 30$ and $A = -7$
Work out the value of C.

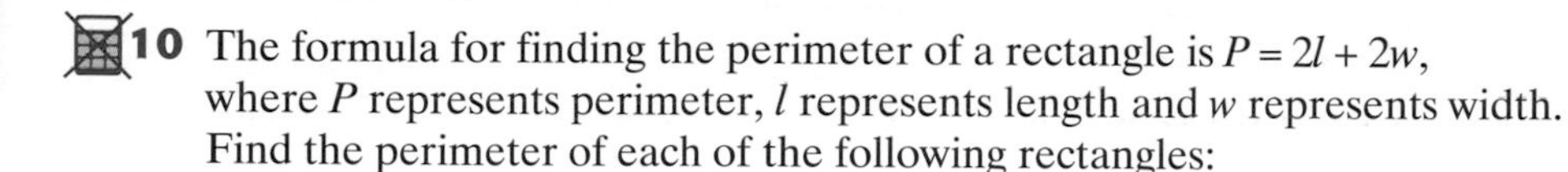

10 The formula for finding the perimeter of a rectangle is $P = 2l + 2w$, where P represents perimeter, l represents length and w represents width. Find the perimeter of each of the following rectangles:

- **a** Length = 12 cm; Width = 5 cm
- **b** Length = 52 cm; Width = 22 cm
- **c** Length = 3.5 cm; Width = 2.5 cm

11 Get Real!

The cost in pounds (C) of framing a picture depends on its length in cm (l) and its width in cm (w).
The formula used by Framing For You is $C = 3l + 2w$
Find the cost of framing a picture that is:

- **a** 50 cm long and 20 cm wide
- **b** 30 cm long and 20 cm wide
- **c** 80 cm long and 60 cm wide.

12 Get Real!

Tasty Pizzas works out its delivery charge with the formula:
Delivery Charge = Number of pizzas × 50p + £1.25

- **a** Find the delivery charge for three pizzas.
- **b** Richard pays a delivery charge of £4.25
  How many pizzas did he order?

13 Get Real!

Mrs Sturgess is working out the wages at the factory. She uses the formula:

Wages equal hours worked multiplied by rate per hour.

- **a** Write this formula in algebra.
- **b** How much does Eileen earn if she is paid £7.50 an hour and she works for 35 hours?
- **c** How much does Freddy earn if he works for 25 hours and is paid £8.25 an hour?

14 Get Real!

In Childhood Studies, Tom is using the formula $h = \frac{28 - a}{2}$ to find out how many hours of sleep a child needs: h is the number of hours and a is the age, in years, of the child.
Copy and complete this table.

a	2	4	6	7
h	13			

15 In science, Andrew is using the formula $V = IR$
Find V if:

a $I = 3$ and $R = 52$ **b** $I = 2.5$ and $R = 63$

16 In maths, Sarah is using the formula $A = \pi r^2$
Find A if:

a $\pi = 3.14$ and $r = 15.5$ **b** $\pi = 3.14$ and $r = 8$

17 In geography, Keith is using the formula $C = \frac{5(F - 32)}{9}$

Find C if:

a $F = 68$ **b** $F = 32$

18 In science, Briony is using the formula $s = ut + \frac{1}{2}at^2$
Find s if:

a $u = 3$, $t = 6.5$ and $a = 14.3$ **b** $u = 4$, $t = 3$ and $a = 8$

19 In maths, Didier is using the formula $V = \frac{4\pi r^3}{3}$

Find V if:

a $\pi = 3.14$ and $r = 12$ **b** $\pi = 3.14$ and $r = 2.5$

Explore

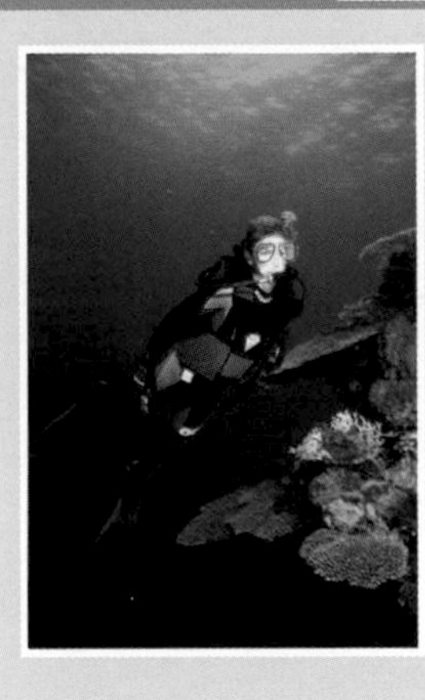

Andy's Uncle Jim is 70 today and he has decided to give Andy some money
Andy can choose one of four options:

- **Option 1:** £100 now, £90 next year, £80 the year after and so on
- **Option 2:** £10 now, £20 next year, £30 the year after and so on
- **Option 3:** £10 now, 1.5 times as much next year, 1.5 times as much the year after and so on
- **Option 4:** £1 now, £2 next year, £4 the year after, £8 the year after and so on

Each of the options will only operate whilst Uncle Jim is alive
Which option should Andy choose?

Investigate further

Learn 3 Changing the subject of a formula

Example: Make x the subject of the formula $y = 35x + 350$

$$y = 35x + 350$$

$$y - 350 = 35x + 350 - 350 \quad \leftarrow \text{Subtract 350 from both sides}$$

$$y - 350 = 35x$$

$$\frac{y - 350}{35} = \frac{35x}{35} \quad \leftarrow \text{Divide both sides by 35}$$

$$\frac{y - 350}{35} = x \quad \leftarrow x \text{ is now the subject of the formula}$$

$$x = \frac{y - 350}{35} \quad \leftarrow \text{Put } x \text{ on the left-hand side as the subject of the formula}$$

Apply 3

1 Rearrange the formula $M = l + 32$ to make l the subject.

2 Rearrange each of these formulae to make y the subject:

a $a + y = b$

b $y - c = d$

c $e + 2y = f$

d $g = h + 2y$

e $j = 3y - 2k$

f $2l + 5y = m$

3 Rearrange each of these formulae to make x the subject:

a $y = x - 22$

b $x + b = c$

c $y = bx$

d $a + x = c$

e $gx + p^2 = s^2$

f $ax - b^2 = c^2$

g $d = 7x - 50$

h $ex = f + g$

i $3x = h - i$

j $xy = j$

k $kx - l = m$

l $nx + p = q$

m $r + sx = t$

n $u + vx = 2w$

o $y - x = z$

4 The formula for finding the circumference of a circle is $C = \pi d$
Rearrange this formula to make d the subject.

5 Krishnan rearranges the formula $y = 4 - x$
He makes x the subject of the formula.
He thinks that the answer is $y - 4 = x$
Is he correct? Give a reason for your answer.

6 The formula for finding the speed of an object is $s = \frac{d}{t}$ where s stands for speed, d for distance and t for time. Rearrange the formula to make d the subject.

7 Katie rearranges the formula $y = \frac{4}{x}$ to make x the subject.

She gets the answer $x = \frac{y}{4}$

Is she correct? Give a reason for your answer.

8 Sarah is using the formula $v = u + at$ in science.
Rearrange the formula to make a the subject.

9 Sam is using the formula $v = Ri$ in science.
Rearrange the formula to make R the subject.

10 The formula for finding the volume of a cylinder is $V = \pi r^2 h$
where V = volume, r = radius and h = height. Write the formula for finding the height of the cylinder if you know the volume and the radius.

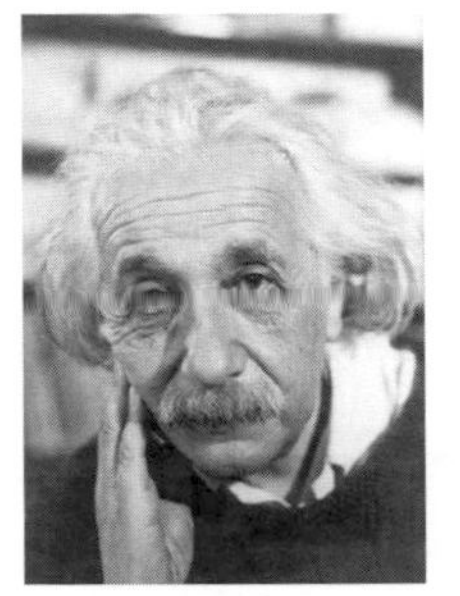

11 Einstein's famous formula is $E = mc^2$
Rearrange this formula to make m the subject.

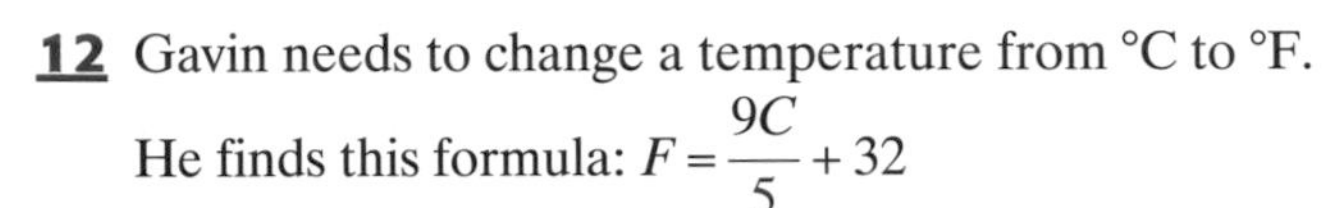

12 Gavin needs to change a temperature from °C to °F.

He finds this formula: $F = \frac{9C}{5} + 32$

Rearrange this formula to make C the subject.
Remember, you must show your working.

Formulae

ASSESS

The following exercise tests your understanding of this chapter, with the questions appearing in order of increasing difficulty.

1 Write an expression for each of the following statements:

a Five times the number x plus two.

b Six times the number y minus eleven.

c Twice the number c added to four times the number d.

d The number v squared.

e Add the numbers w and t together. Multiply the result by seven.

2 **a** In the formula $D = 3a + 2$, find the value of D when $a = 3$

b In the formula $y = \sqrt{z}$, find the value of y when $z = 36$

c In the formula $S = 5x + 7y$, find the value of S when $x = 4$ and $y = 8$

3 **a** a is the cost of 1 kg of apples and b is the cost of 1 kg of bananas. Write down the cost of 5 kg of apples and 3 kg of bananas.

b Write down the cost of 5 CDs at £c each together with 2 DVDs at £d each.

c Ally is y years old now. Write down:

i her dad's age if he is four times as old as Ally is now

ii Ally's age in nine years' time

iii her dad's age five years ago.

4 **a** The formula for the perimeter of a rectangle of length l and width w is $P = 2l + 2w$.
Find the perimeter of a rectangle:

i 5 cm long and 9 cm wide

ii 6 yards long and 2.5 yards wide.

b The formula for the area of a square is $A = l^2$, where l is the length of a side.
Find the area of a square of length:

i 3 m **ii** 20 in **iii** 4.5 ft

c The formula for the sum of the interior angles of a polygon of n sides is $S = 180(n - 2)°$

Find S when:

i $n = 3$ (What is such a polygon called?)

ii $n = 8$. (What is such a polygon called?)

iii $n = 102$

d Use the formula $p = m^2 + 5n$ to find:

i p when $m = 4$ and $n = -6$

ii p when $m = -7$ and $n = 8$

iii p when $m = -10$ and $n = -9$

5 The ferry fare to cross a lake is £7 for adults and £4 for children.

a Write down a formula to calculate the total ferry cost, £F, for A adults and C children.

b Use your formula to find:

i the total cost for 4 adults and 3 children

ii the total cost for 2 adults and 5 children.

6 Niamh measures the depth of a well, D m, by dropping a stone down it and timing the splash. The formula she uses is $D = 5t^2$

a Use this formula to find the depth of the well if she hears the splash after:

i 3 seconds **ii** 12 seconds.

Another well is 180 m deep.

b How long will it take to hear the splash if Niamh drops a stone down this well?

7 **a** Make y the subject of the equation $4x + 7y = 8$

b The change in momentum M of a body of mass m when it moves from a speed u to a new speed v is given by the formula $M = mv - mu$
Make v the subject.

16 Construction

OBJECTIVES

G

Examiners would normally expect students who get a G grade to be able to:

Measure a line accurately to the nearest millimetre

Recognise the net of a simple solid such as a cuboid

F

Examiners would normally expect students who get an F grade also to be able to:

Measure or draw an angle accurately to the nearest degree

Draw the net of a simple solid such as a cuboid

E

Examiners would normally expect students who get an E grade also to be able to:

Draw a triangle given three sides, or two angles and a side, or two sides and the included angle

D

Examiners would normally expect students who get a D grade also to be able to:

Draw a quadrilateral such as a kite or a parallelogram with given measurements

Understand that giving the lengths of two sides and a non-included angle may not produce a unique triangle

Construct and recognise the nets of 3-D solids such as pyramids and triangular prisms

C

Examiners would normally expect students who get a C grade also to be able to:

Construct the perpendicular bisector of a line

Construct the perpendicular from a point to a line

Construct the perpendicular from a point on a line

Construct angles of 60° and 90°

Construct the bisector of an angle

What you should already know ...

- Use a protractor and a pair of compasses
- Different types of angles and triangles
- Use bearings
- Round decimals to the nearest whole number
- Use a scale on a map and scale drawing

VOCABULARY

Bearing – an angle measured clockwise from North; all bearings should be written as three figure numbers, for example, 125° or 045°

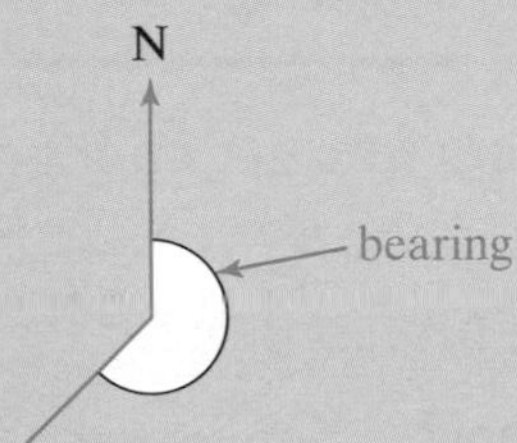

Equilateral triangle – a triangle with 3 equal sides and 3 equal angles – each angle is 60°

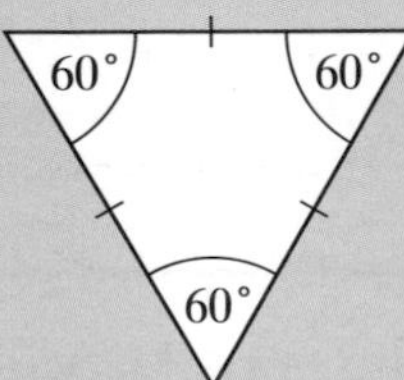

Isosceles triangle – a triangle with 2 equal sides and 2 equal angles; the equal angles are called **base angles**

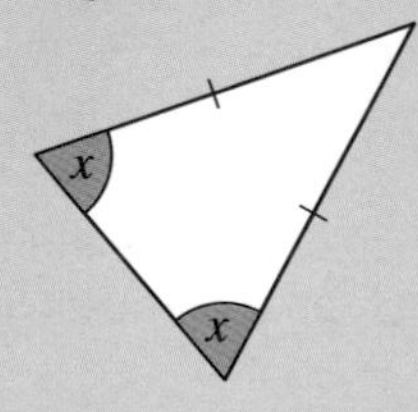

Right-angled triangle – a triangle with one angle of 90°

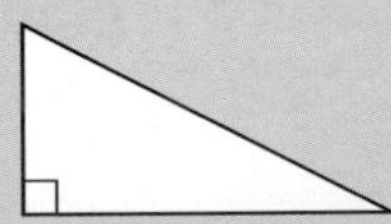

Congruent – exactly the same size and shape; one of the shapes might be rotated or flipped over

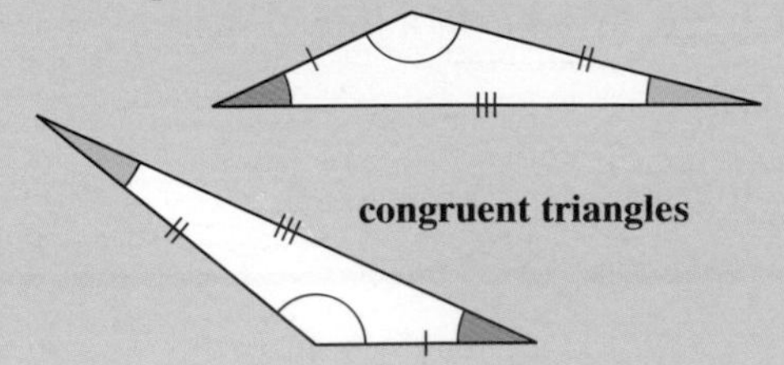

Perpendicular lines – two lines at right angles to each other

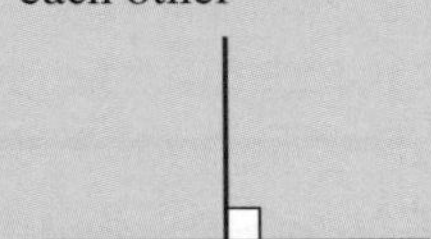

Parallel lines – two lines that never meet and are always the same distance apart

Arc (of a circle) – part of the circumference of a circle; a minor arc is less than half the circumference and a major arc is greater than half the circumference

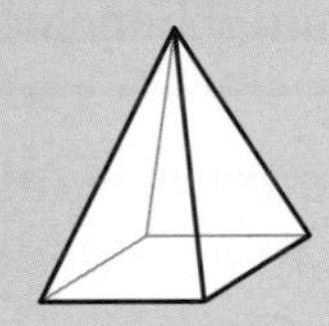

Square pyramid

Bisect – to divide into two equal parts

Midpoint – the middle point of a line

Perpendicular bisector – a line at right angles to a given line that also divides the given line into two equal parts

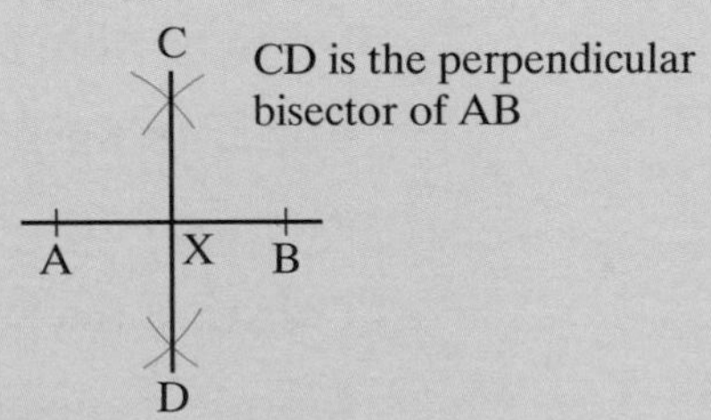

Angle bisector – a line that divides an angle into two equal parts

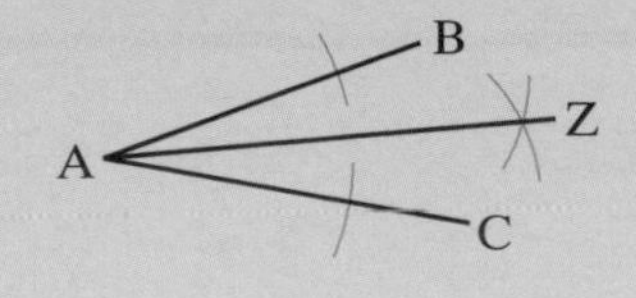

AZ is the angle bisector of angle BAC

Regular polygon – a polygon with all sides and all angles equal

Kite – a quadrilateral with two pairs of equal adjacent sides

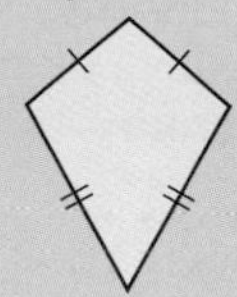

Trapezium (pl. **trapezia**) – a quadrilateral with one pair of parallel sides

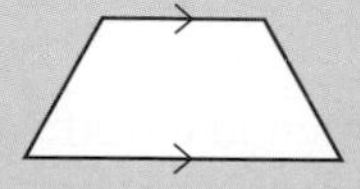

Parallelogram – a quadrilateral with opposite sides equal and parallel

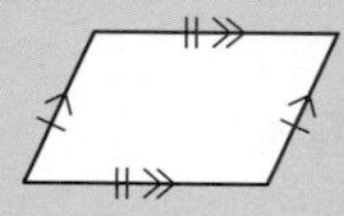

Rhombus – a quadrilateral with four equal sides and opposite sides parallel

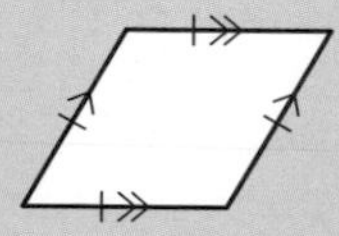

Solid – a three-dimensional shape

Cube – a solid with six identical square faces

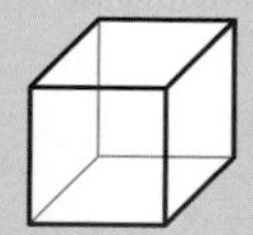

Cuboid – a solid with six rectangular faces (two or four of the faces can be squares)

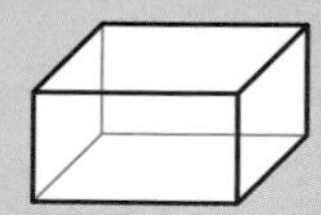

Face – one of the flat surfaces of a solid

Vertex (pl. **vertices**) – the point where two or more edges meet

Edge – a line segment that joins two vertices of a solid

Prism – a three-dimensional solid with two cross-sectional faces that are identical polygons, parallel to each other; all other faces are either parallelograms or rectangles

Prisms are named according to the cross-sectional face; for example,

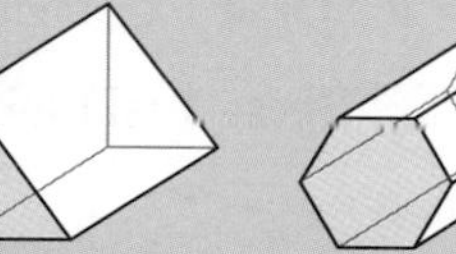

Triangular prism

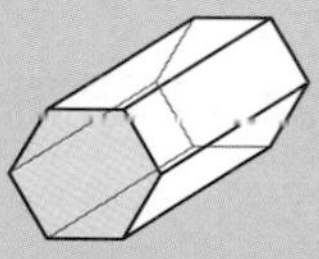

Hexagonal prism

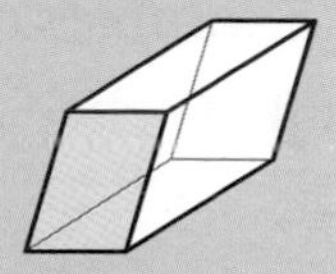

Parallelogram prism

Cylinder – a prism with a circle as a cross-sectional face

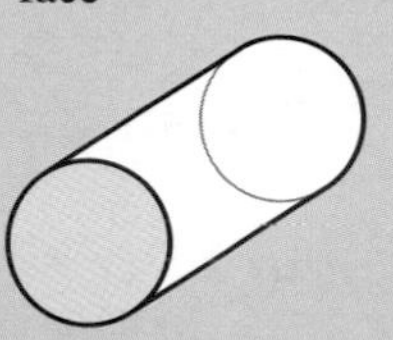

Pyramid – a solid with a polygon as the base and one other vertex; all the vertices of the base are joined to this vertex forming triangular faces. Pyramids are named according to their base, for example,

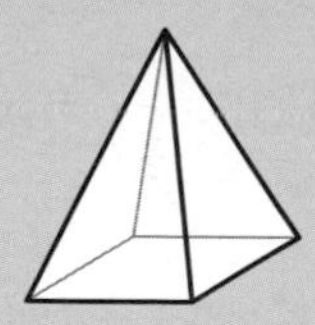

Square pyramid

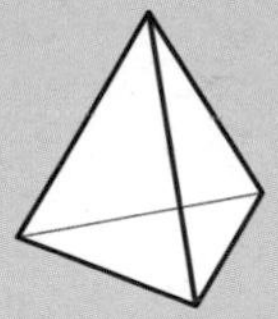

Triangular pyramid

Regular tetrahedron – a triangular pyramid with equilateral triangles as its faces

Cone – a pyramid with a circular base and a curved surface rising to a vertex

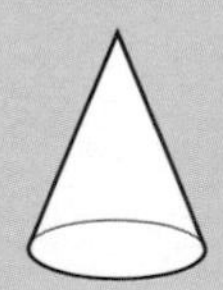

Net – a two-dimensional shape made of polygons that can be folded to make a three-dimensional solid, for example,

Net of a cuboid Net of a triangular prism

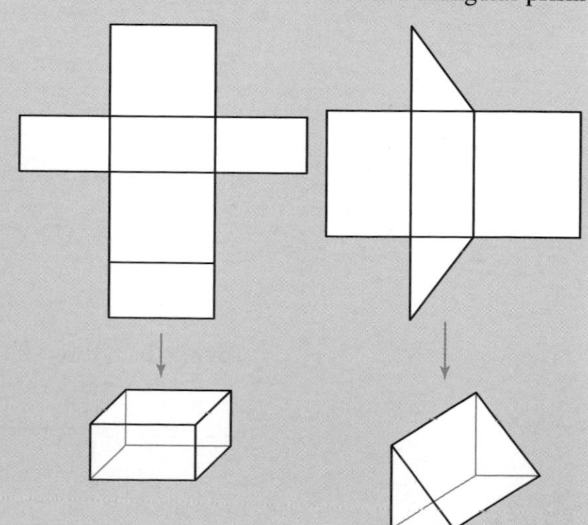

Learn 1 Measuring lines and angles and drawing 2-D shapes

Examples:

a Draw a triangle ABC with AB = 6 cm, angle A = 47° and angle B = 32°.

'Draw' means you can use any piece of equipment (for example, a protractor, a ruler) to produce an accurate diagram

Step 1 Draw a sketch. You don't need to use a ruler, just draw your sketch freehand.

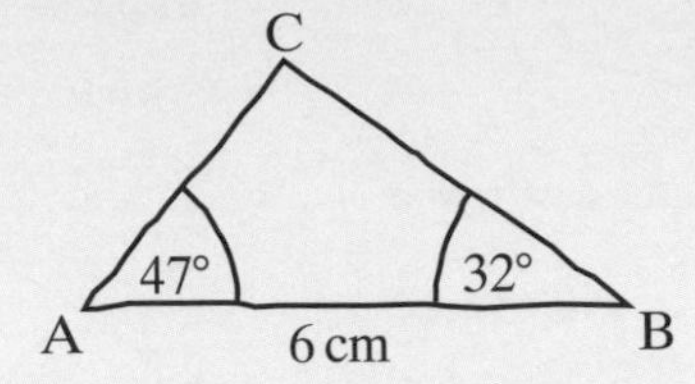

Step 2 Draw your first line a little longer than 6 cm, then mark off A and B. This will give you a cross on which to put the centre of your protractor.

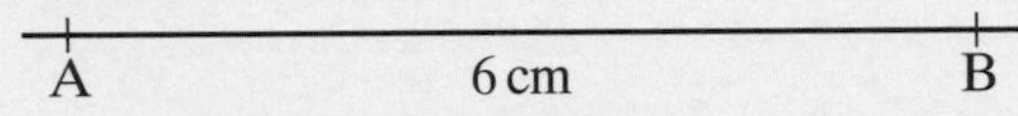

Step 3 Use your protractor to draw an angle of 47° at A.

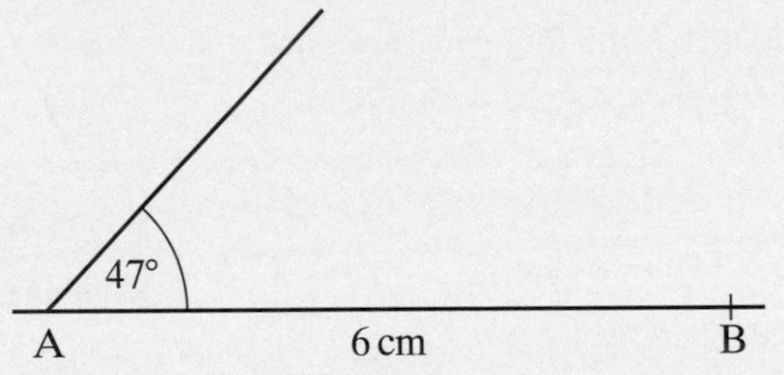

Step 4 Use your protractor to draw an angle of 32° at B.

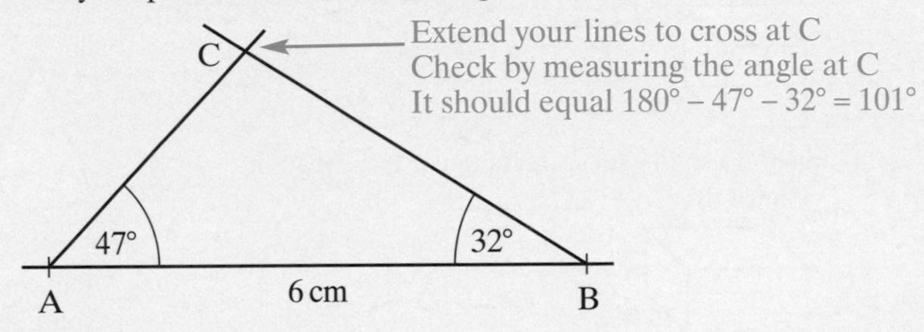

b Draw a triangle DEF with DE = 5 cm, DF = 3.2 cm and angle D = 122°.

Step 1 Draw a sketch.

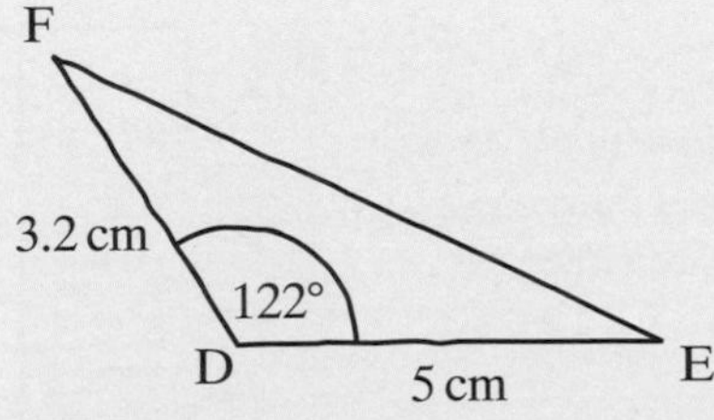

Step 2 Draw your first line a little longer than 5 cm, then mark off D and E.

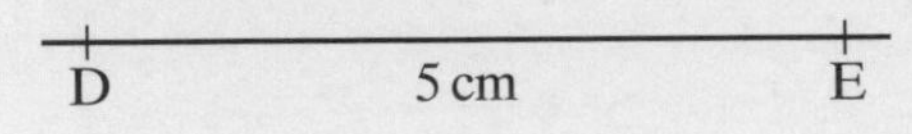

Step 3 Use your protractor to draw an angle of 122° at D.

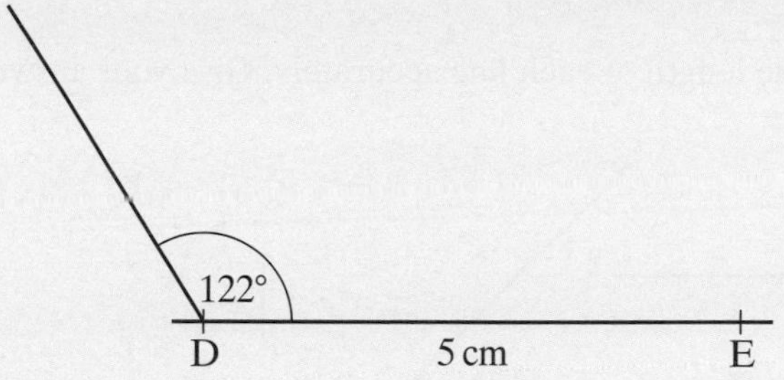

Step 4 Measure 3.2 cm along the line with your ruler (or use your compasses set to 3.2 cm, centre D) to mark the point F. Join FE.

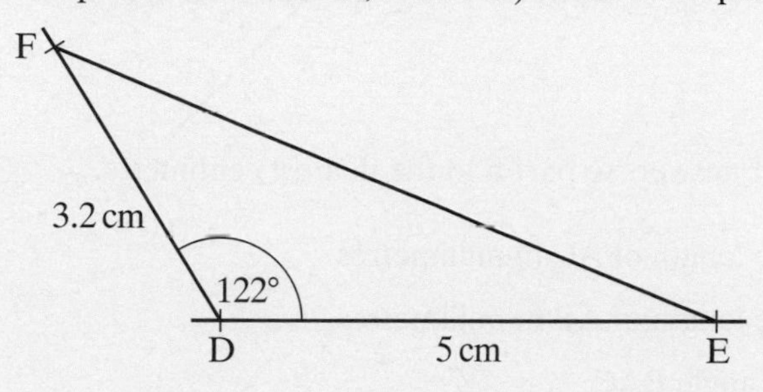

c Draw a triangle PQR with PQ = 3.4 cm, PR = 6.8 cm and QR = 4.3 cm.

Step 1 Draw a sketch. It is a good idea to have the longest side as the base of your triangle. That way you won't run out of space.

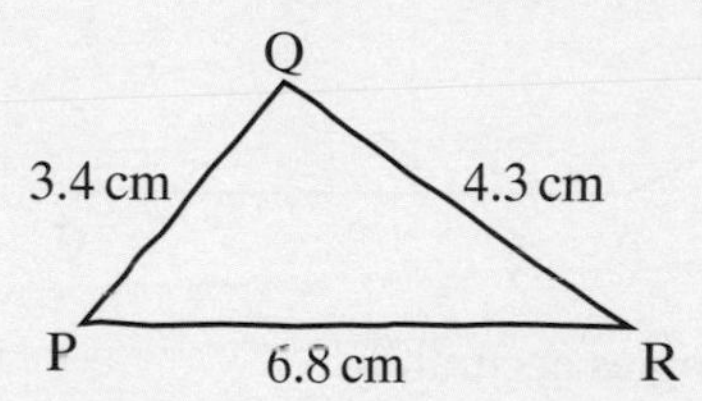

Step 2 Draw your first line a little longer than 6.8 cm, then mark off P and R.

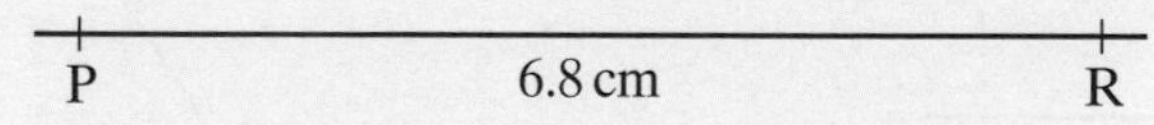

Step 3 Use your compasses with centre P, radius 3.4 cm to draw a big arc.

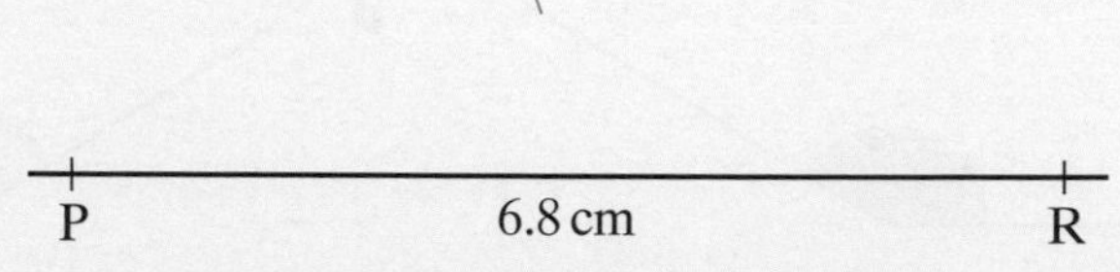

Step 4 Use your compasses with centre R, radius 4.3 cm to draw an arc that cuts your first arc at Q. Join PQ and QR.

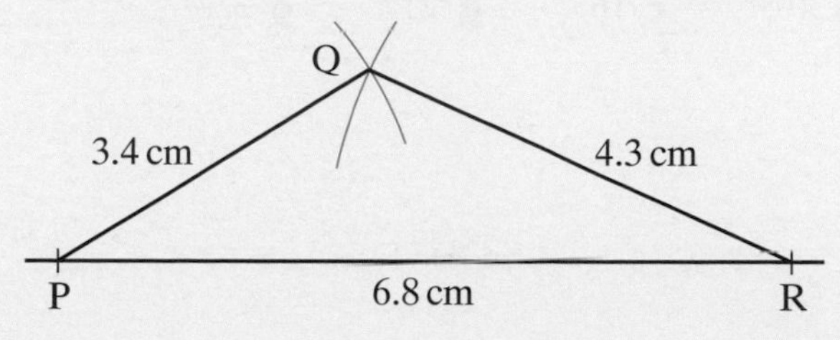

Apply 1

1 **a** Measure the length of each line accurately. Give your answers in centimetres.

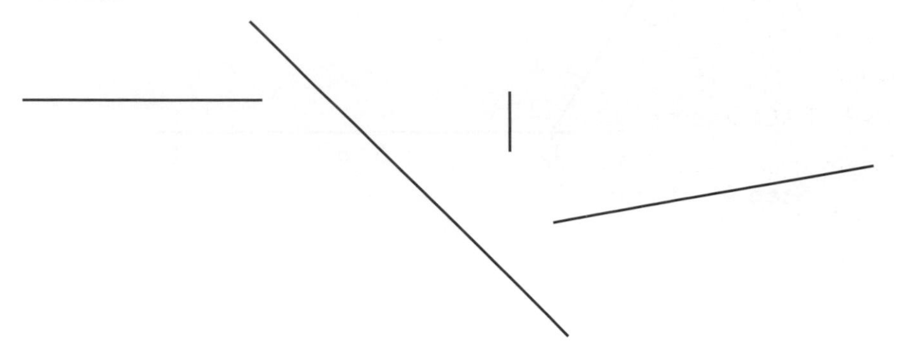

b Give your answers to part **a** to the nearest centimetre.

2 **a** Write the length of AC in millimetres.

b Write the length of AB in millimetres.

c Measure angle BAC.

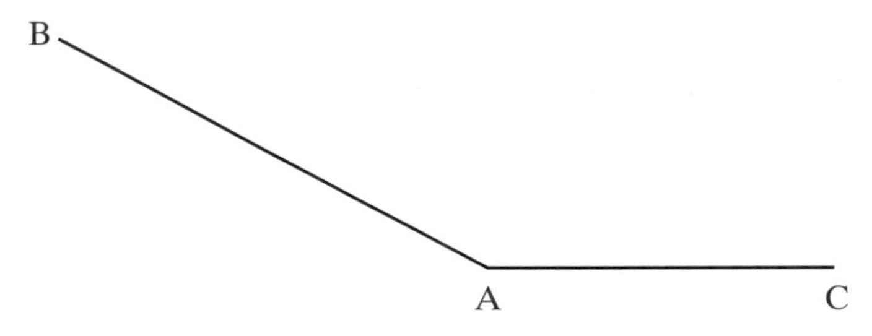

3 Measure each of the following angles accurately.

a

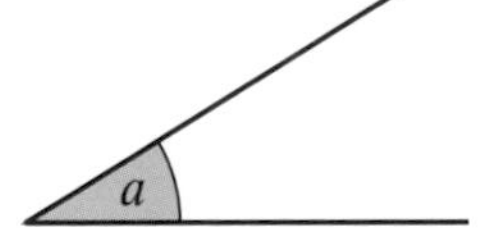

c

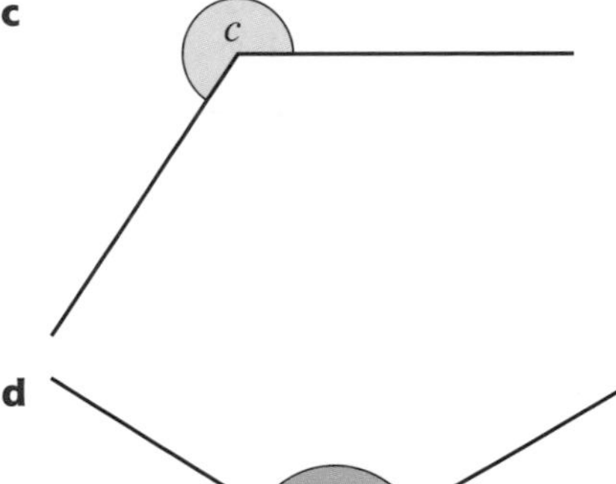

b

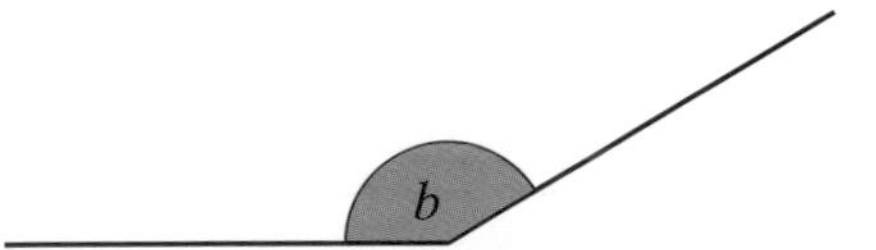

d

4 Draw and label these angles.

a 24° **b** 105° **c** 161° **d** 217° **e** 322°

5 Draw these triangles accurately, using the information that you are given.

a

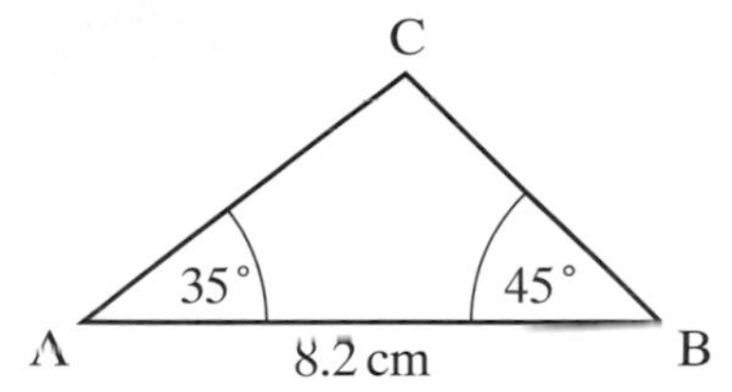

What is the length of AC on your diagram?

b Triangle PQR, with angle P = 72°, PQ = 3.7 cm, PR = 6.3 cm.

Measure and write down angle R.

c Triangle DEF, with DE = 7.1 cm, EF = 4.8 cm, DF = 6.4 cm.

Measure and write down angle F.

6 Make an accurate drawing of a rhombus in which all the sides are 6 cm and the length of the shorter diagonal is 7 cm.
What is the length of the longer diagonal?

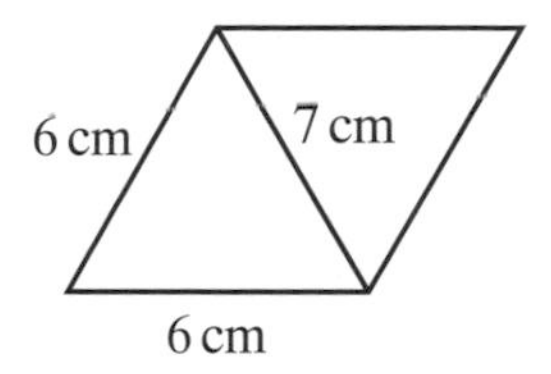

7 Make an accurate drawing of a rectangle in which the length of each of the longer sides is 8.4 cm and the length of each diagonal is 9.1 cm.
What is the length of each of the shorter sides?

8 Get Real!

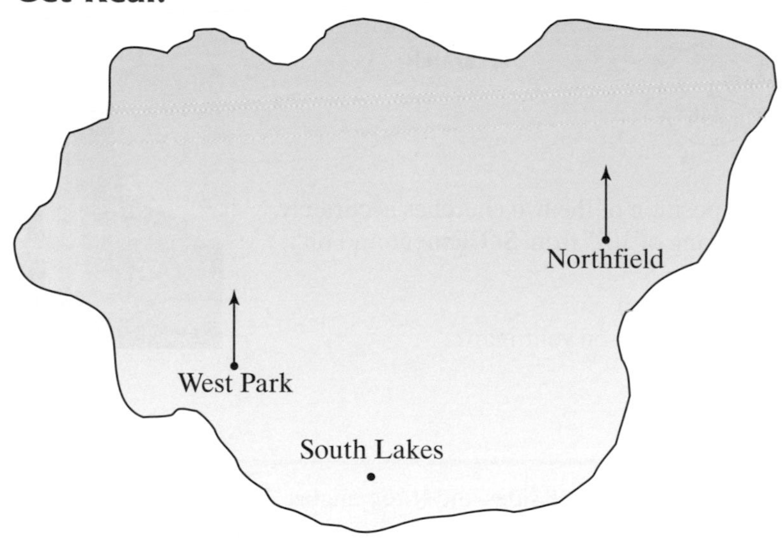

By measuring on the map, work out:

a the bearing of South Lakes from West Park

b the bearing of West Park from Northfield.

The map has a scale of 1 cm to 10 km.

c Find the distance from Northfield to South Lakes to the nearest kilometre.

9 Get Real!

Brian is making a model of his sailing boat. On the actual boat the mainsail is a right-angled triangle. The width of the mainsail is 3 m. Its hypotenuse is 8 m.
Using a scale of 1 : 100, draw accurately a scale plan for the mainsail of the model boat.
What is the height of the sail on the scale drawing?

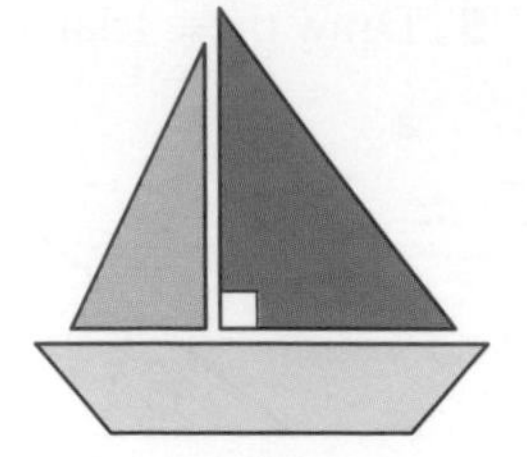

10 Get Real!

Bhavika is writing a book about making kites.
She makes a sketch of her own kite.
Copy the sketch. Calculate the missing angles and write them on your sketch.
Using a scale of 1 cm to represent 10 cm, make an accurate scale drawing of Bhavika's kite.
What is the width of your scale drawing?

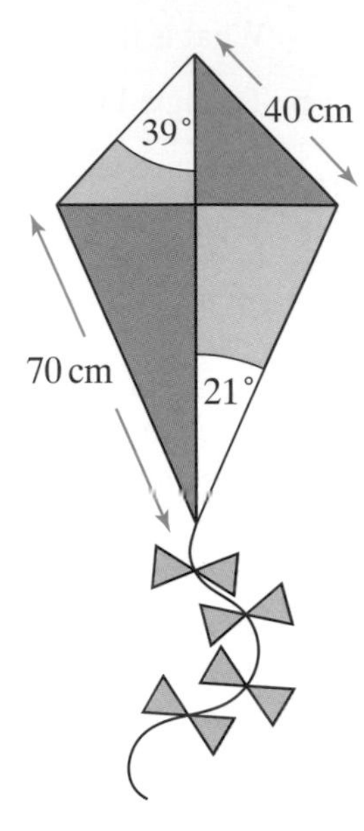

11 Get Real!

On a map the churches of St Clement and St Mary are 9.6 cm apart, with St Clement being due west of St Mary.

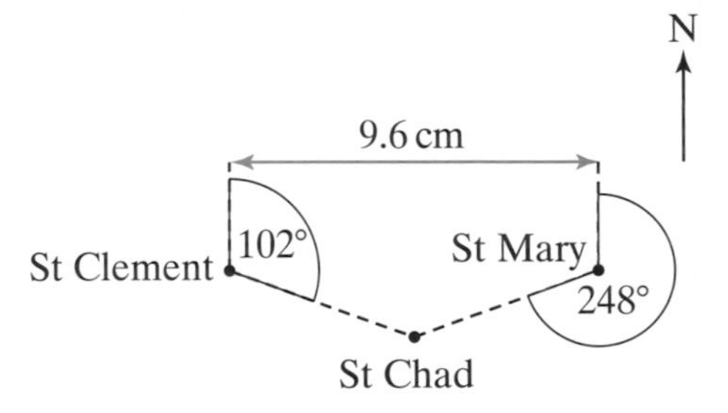

Not drawn accurately

Use this information to draw the position of the two churches accurately.
The church of St Chad is on a bearing of 102° from St Clement and on a bearing of 248° from St Mary.
Mark the position of St Chad on your map.
How far apart are St Chad and St Clement on your map?

Explore

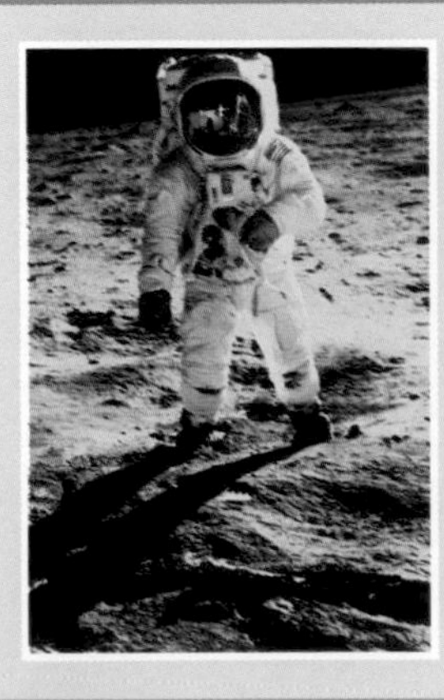

- Every triangle has three sides and three angles. You don't need to know all of these measurements to draw a unique triangle. Which measurements did you need in Apply **1** question **5**?
- Triangle ABC has angle A = 30° and AB = 5 cm. If BC = 4.5 cm, how many triangles can you construct that fit these measurements? What happens for different lengths of BC?
- If you want a company to make you a triangular sign, which measurements would you need to state to make sure that the sign is exactly the right size and shape?

Investigate further

Learn 2 Bisecting lines and angles

Examples:

a Construct the bisector of the line AB.

'Construct' means you must produce an accurate diagram using only your compasses and a straight edge (ruler)

Always leave in your construction arcs and remember:
NO ARCS – NO MARKS!

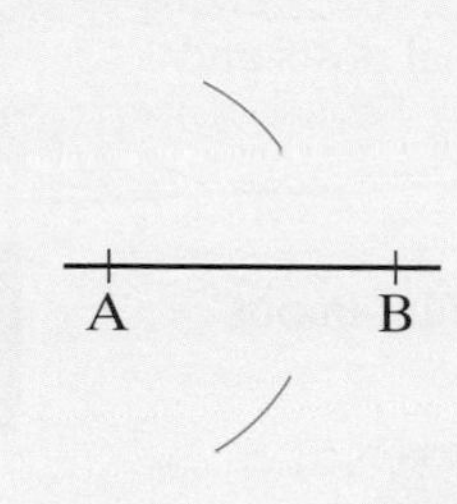

Set the radius of your compasses to more than half of AB.
Put the point on A and draw arcs above and below AB.

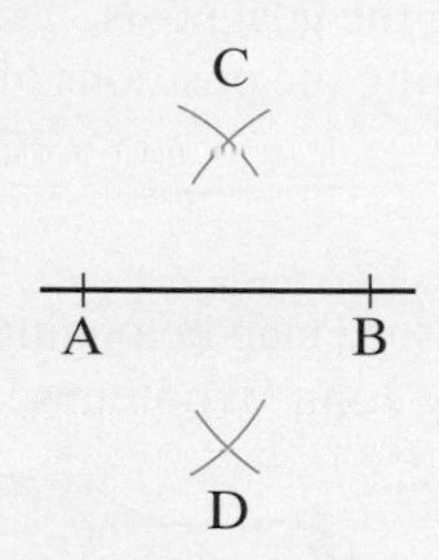

Keeping the same radius, put the point of your compasses on B and draw two new arcs to cut the first two at C and D.

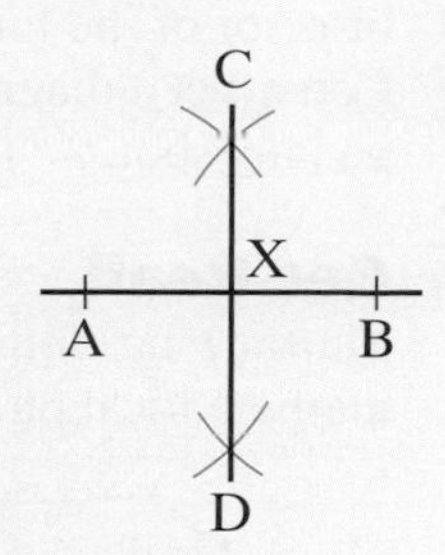

Join CD.
X is the midpoint of AB.
CD is not only the bisector of AB, it is the perpendicular bisector.

Use your ruler to check that AX = XB and use your protractor to check that angle CXB = 90° when you have finished your construction

b Construct the bisector of angle BAC.

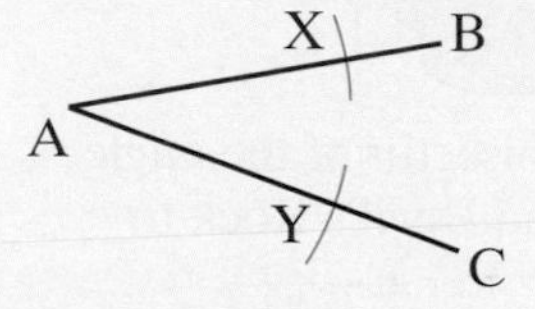

Set the radius of your compasses to less than the length of the shorter line.
Put the point on A and draw arcs to cut AB and AC at X and Y.

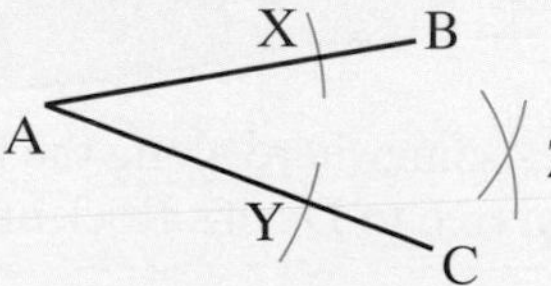

Put the point of your compasses on X and Y in turn and draw arcs that intersect at Z.

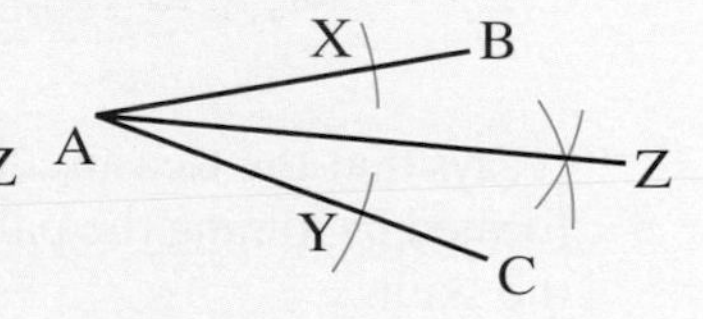

Join AZ.
AZ is the angle bisector of angle BAC.

Use your protractor to check that angle BAZ = angle ZAC when you have finished your construction

Apply 2 

1 **a** Draw the line AB = 9 cm.

b Construct its perpendicular bisector, CD, cutting AB at O.

c Construct the bisector of angle COB.

2 David bisects angle BAC by putting the point of his compasses on B and drawing an arc. Then he puts the point of his compasses on C and, with the same radius as before, he draws a second arc crossing the first. Finally he joins the point where the arcs cut to A.
What has he done wrong?

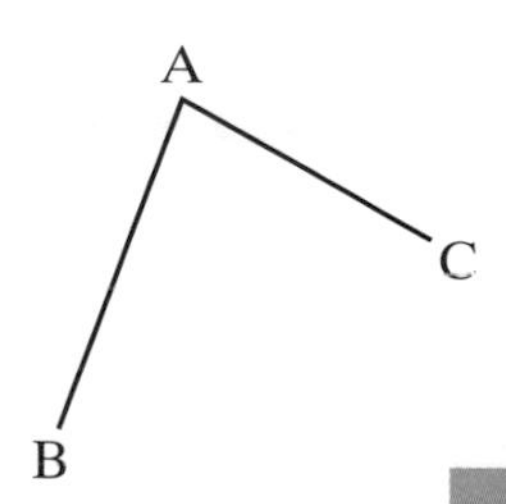

3 Get Real!

Adam is making a miniature table football game.
The distance between the goal posts is 8 cm.
He needs to mark the penalty spot 12 cm along the perpendicular bisector of the line joining the goal posts.
Construct a diagram showing the positions of the goal posts and penalty spot.

4 Get Real!

Matthew has found a treasure map belonging to the infamous mathematical pirate, Long John Hypotenuse.

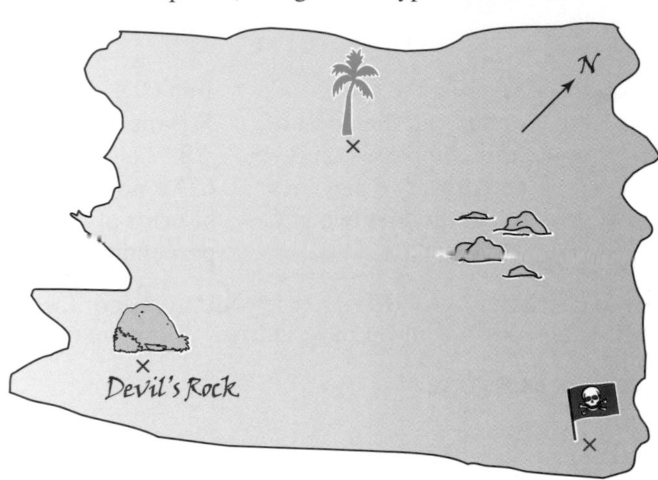

It says that the treasure lies somewhere along the bisector of the angle formed by joining the palm tree to Devil's Rock and Devil's Rock to the skull.

Roughly copy the positions of the palm tree, Devil's Rock and the skull. Join them up and construct the angle bisector at Devil's Rock to help Matthew find the path along which the treasure lies.

5 Ben says that he could use his compasses and straight edge to divide a straight line into three equal parts, four equal parts, five equal parts or any number of equal parts.

Zoë says there are only certain numbers of parts that can be found.
She has a rule for the number of equal parts.
What is Zoë's rule?

6 Get Real!

The council is designing a roundabout for a new housing estate.
At the moment two roads meet at 144°. The council plans to add a third road on the angle bisector of the two roads.

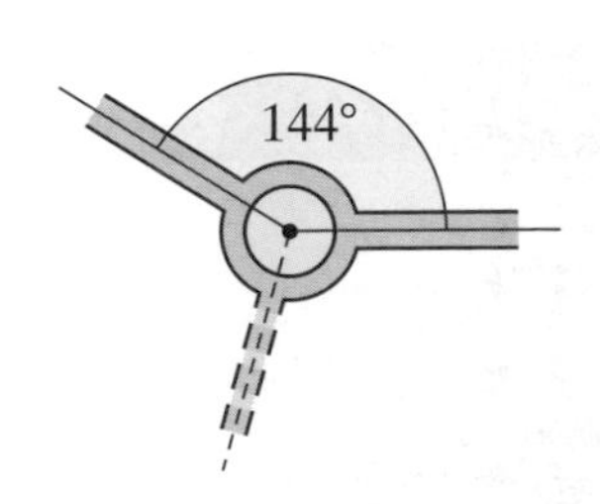

Draw accurately the angle between the first two roads and construct the angle bisector (shown dotted in the diagram) to make the third.

7 Kasim created this pattern, with a ruler and compasses, using the method he had learned for constructing perpendicular and angle bisectors.

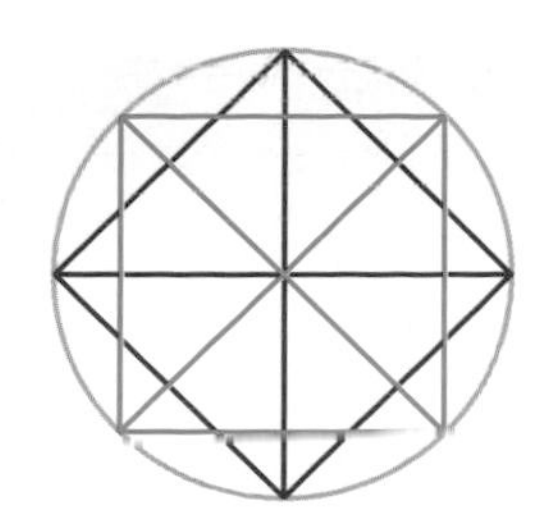

His instructions start:

- Draw a circle and mark in a diameter.
- Perpendicularly bisect the diameter.
- Join the four points at the edge of the circle.

Can you complete the instructions to add the blue lines to the diagram? Check that your instructions work by seeing what design you create when you follow them.

8 **Get Real!**

James is making a scale drawing of a snooker table.
He draws a 9 cm line and marks each end with a cross for the corner pockets. He constructs the perpendicular bisector of the line and marks a point 1.5 cm along the bisector.
This is the position of the black ball.

Construct the same diagram as James.

If James is using a scale of 1 : 20, using your diagram, work out the distance in real life between the black ball and a corner pocket.

Explore

- Draw a square accurately
- Construct the midpoints of each side
- Join them in order
- What shape have you made?
- Now repeat the steps starting with a rectangle, trapezium, kite, ...

Investigate further

Explore

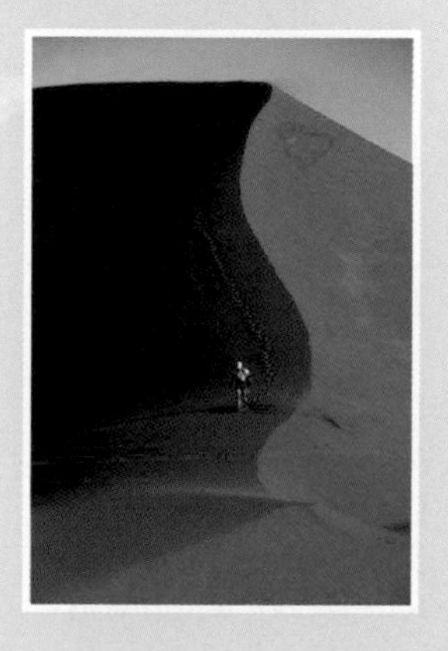

- Draw a triangle. Construct the perpendicular bisectors of the three sides. They should all meet at one point. This is called the **circumcentre** of the triangle
- Draw another triangle. Construct the angle bisectors of the three angles. They should all meet at one point. This is called the **incentre** of the triangle
- Draw another triangle. Construct the midpoints of the three sides. Join each angle to the midpoint of the opposite side. The three lines cross at one point called the **centroid** of the triangle
- One of the above centres is the centre of gravity of the triangle – that is, the triangle's point of balance. Experiment to find out which centre it is
- Use the Internet to find out about the other two centres
- Experiment with different types of triangle

Investigate further

Learn 3 Constructing angles of 90° and 60°

Examples:

a Construct a perpendicular from a point on a line.

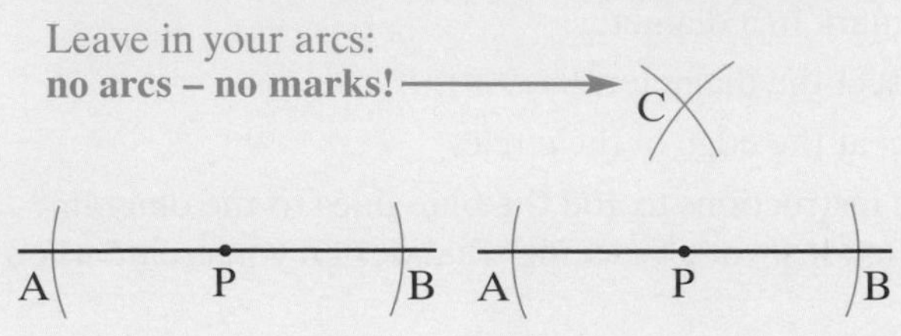

With the point of your compasses on P, draw two arcs to cut the line either side of P at A and B.

Make the radius of your compasses larger. Put the point on A and B in turn and draw arcs that intersect at C.

Join CP. CP is perpendicular to the original line with a 90° angle at P. Use your protractor to check that angle CPB = 90°.

b Drop a perpendicular from the point P onto the line AB.

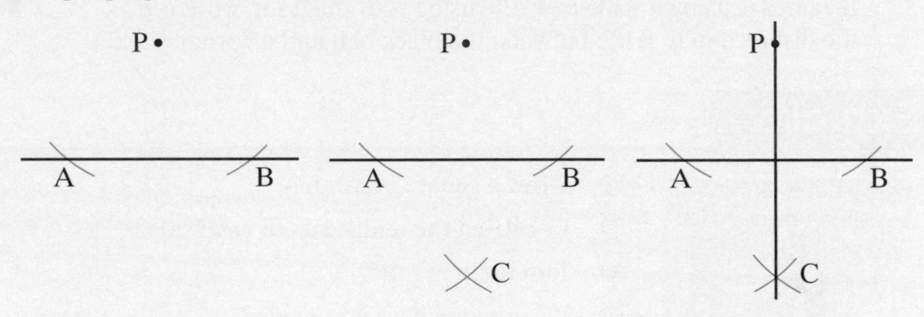

With the point of your compasses on P, draw two arcs that intersect the line at A and B.

Make the radius of your compasses larger. Put the point on A and B in turn and draw arcs that intersect at C.

Join PC. PC is the perpendicular to the original line from the point P.

c Construct an angle of 60°. If you also join RQ you have constructed an equilateral triangle!

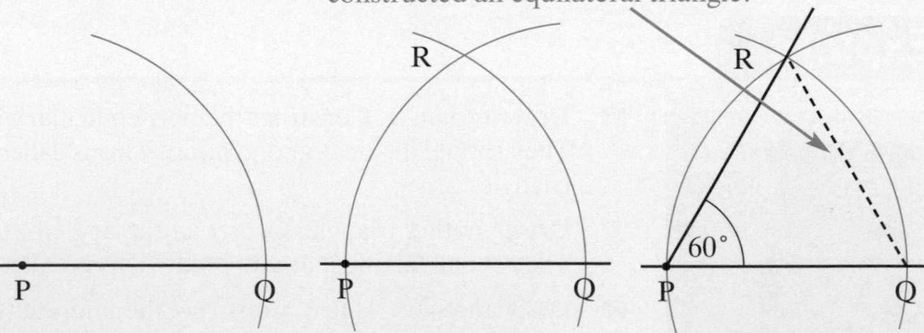

With the point of your compasses on P, draw a large arc that intersects the line at Q.

Keeping the radius of your compasses the same, put the point on Q and draw an arc that passes through P and cuts the first arc at R.

Join PR. Angle RPQ is 60°. Use your protractor to check that angle RPQ = 60° when you have finished your construction.

Apply 3

1 Using only a ruler and compasses:

a construct an angle of 60°

b bisect your angle to make two angles of 30°.

2 Construct a right-angled triangle with angles of 60° and 30°.

3 **a** Draw a line AB = 10 cm.

b Mark the point C on the line AB, with AC = 3.5 cm.

c Construct a line through C perpendicular to AB.

4 **a** Draw a line AC = 8 cm.

b Construct the equilateral triangle ABC.

c Drop a perpendicular from the point B onto the line AC.

d Measure the length of the perpendicular to the nearest millimetre.

5 Get Real!

A furniture designer is drawing a chair to go on the cover of his catalogues. The chair has the measurements shown in the diagram.

Construct the image accurately with a ruler and compasses, leaving in your construction arcs.

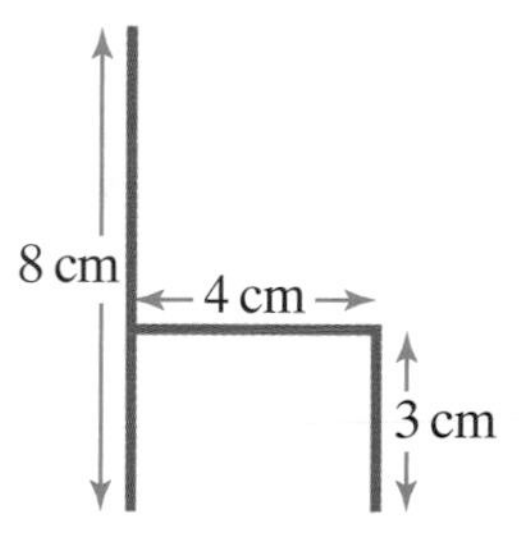

6 Get Real!

The diagram shows a roof truss design for a new house.

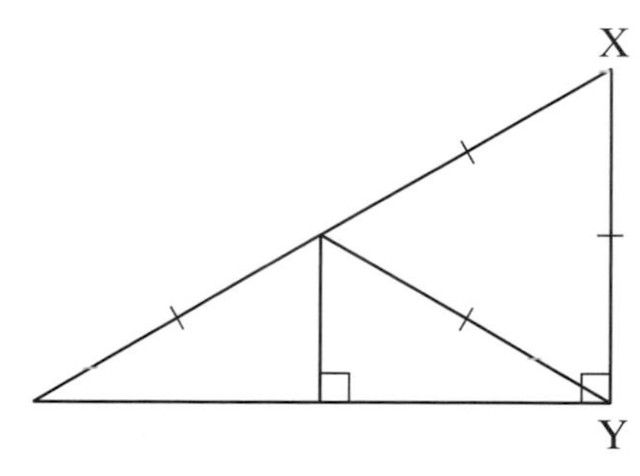

Construct the diagram, starting with the vertical line XY = 3 cm.
Measure the width of the base of the roof truss to the nearest millimetre.

7 Get Real!

Hasan has designed a bow logo for a fashion shop.

He draws a circle of radius 5 cm, centre O. Then he draws a diameter, AB.

Next he marks a point C on AB, 3 cm to the left of O.
He constructs the perpendicular to AB through C and extends it to D and E on the circumference of the circle.
He draws a line from E, through the centre O, to the edge of the circle, F, and another line from D, through O, to G.
Finally he joins GF.

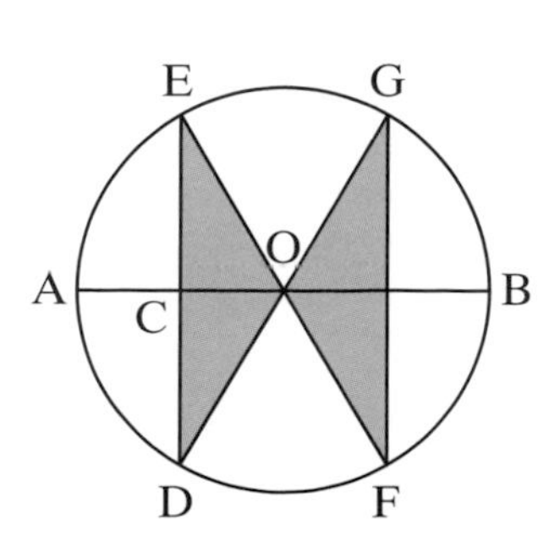

Follow Hasan's design instructions.
What is the height of the bow shape that you have constructed?

8 Get Real!

Sarah wants to know how far away a pylon is from the hedge at the boundary of her property.
She marks two points on the hedge, A and B, that are 75 metres apart.
Then she measures the angles at A and B between the hedge and the lines of sight of the pylon.
The angle at A is 60° and the angle at B is 30°.

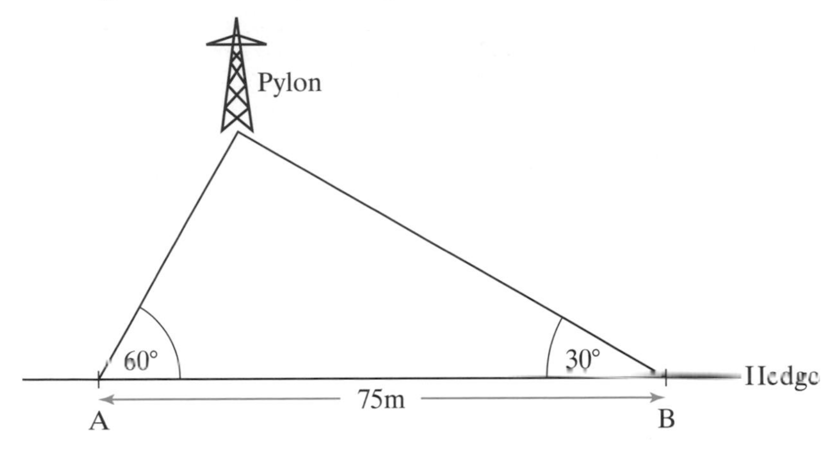

Using a ruler and compasses, construct a scale diagram of Sarah's results, using a scale of 1 : 1000.
By constructing the perpendicular from the pylon to the line AB, find the shortest distance of the pylon from Sarah's hedge.

Explore

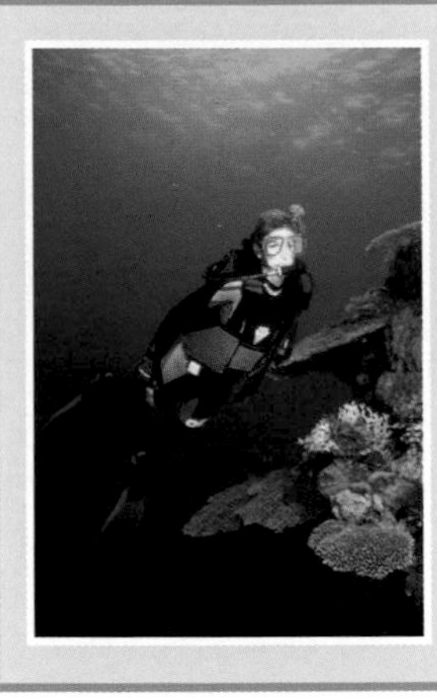

- You know how to construct angles of 90° and 60° using only a ruler and compasses
- Can you construct angles of 30° and 45°?
- What about 150° and 330°?
- What other angles can you construct accurately?

Investigate further

Explore

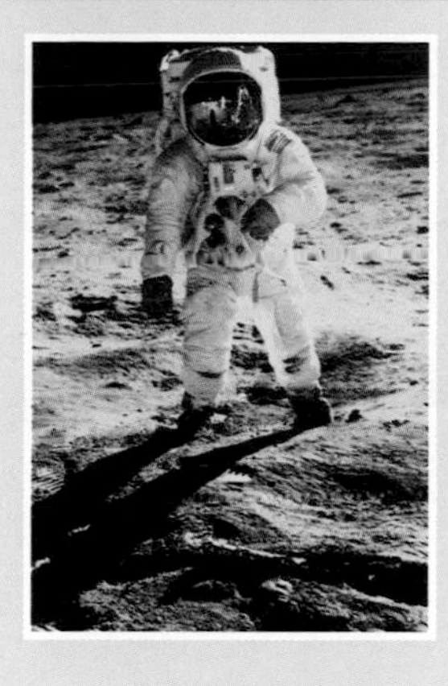

- A group of students are studying the patterns that are all around us, such as snowflakes, rose windows, clocks, fractals, etc.
- They start to create some of their own designs

 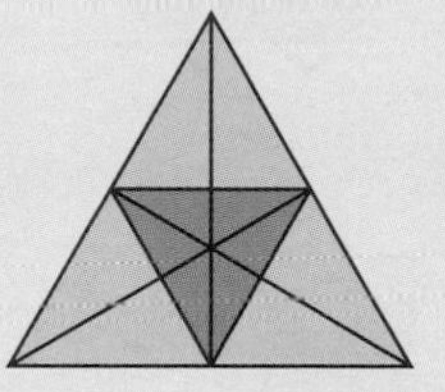 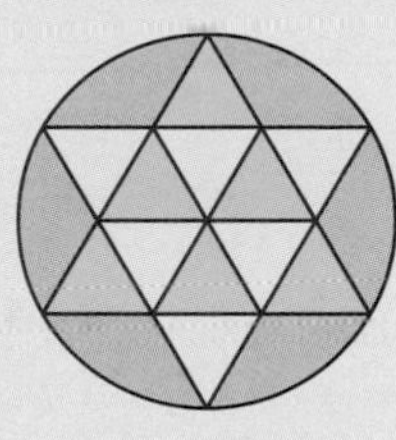 

- Can you use the constructions you have learned to recreate one of the patterns?
- What designs of your own can you create using only a ruler and compasses?

Investigate further

Learn 4 3-D solids and nets

Example:

Draw the nets of the following solids:

a Cuboid

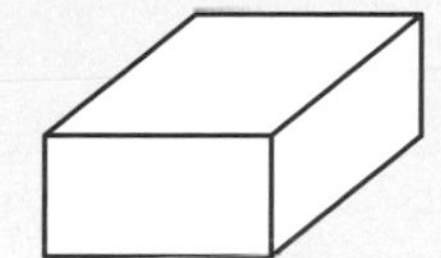

b Triangular prism

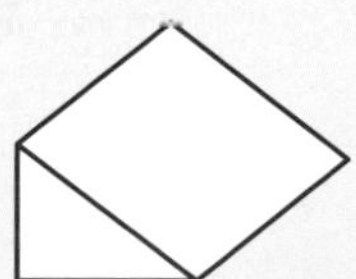

a

b

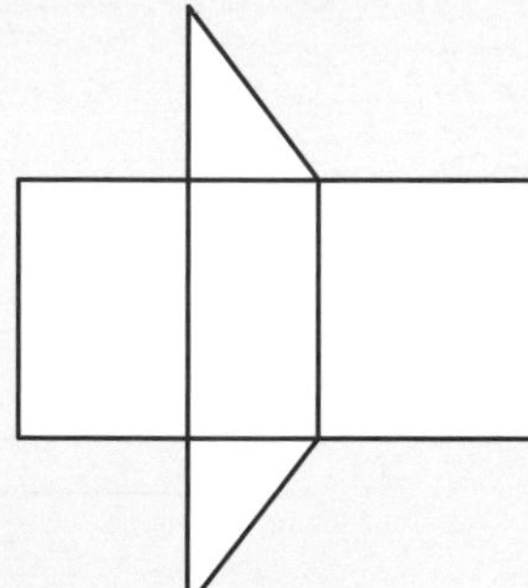

Apply 4 

1 Which of these are **not** nets of a cube? Give a reason for your answer.

a

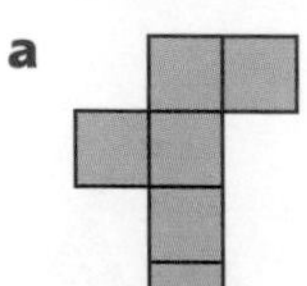

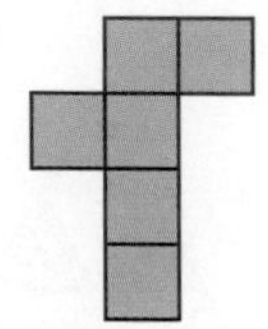

c

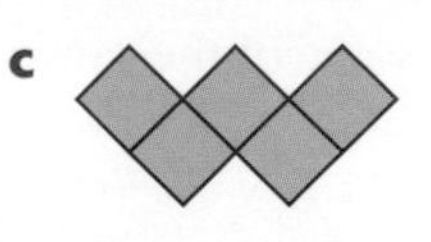

e

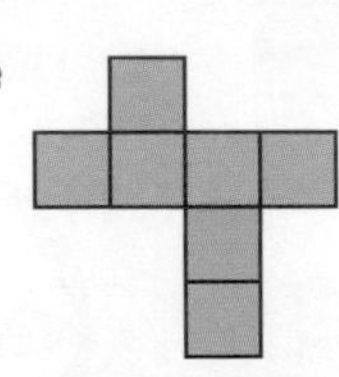

b **d**

f

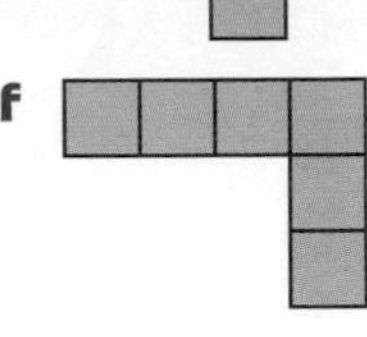

2 Which of these are **not** nets of a regular tetrahedron?
Give a reason for your answer.

a

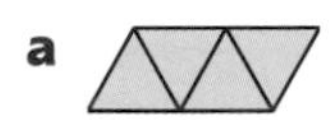

c

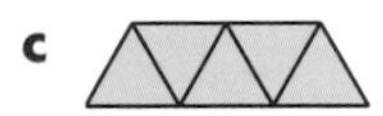

e

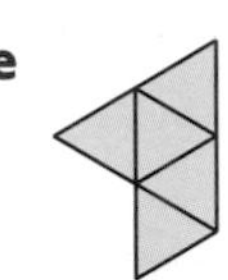

b

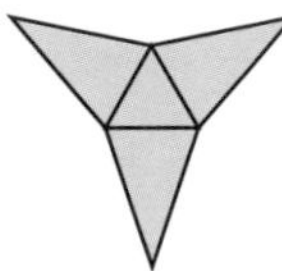

d

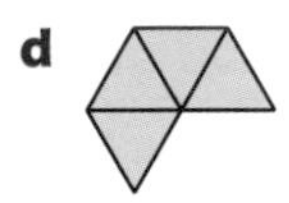

f

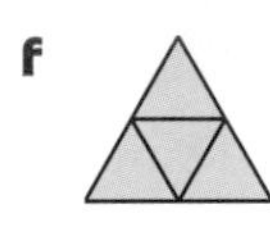

3 Which 3-D solids can be made from these nets?

a

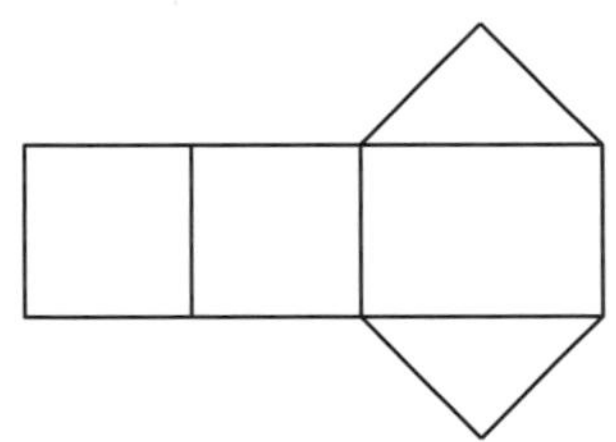

c

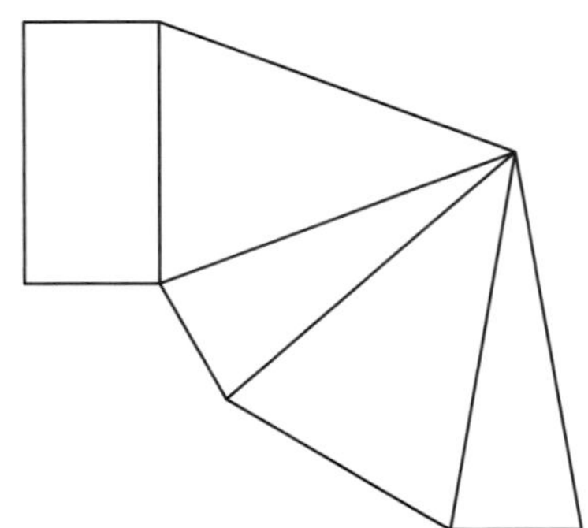

b

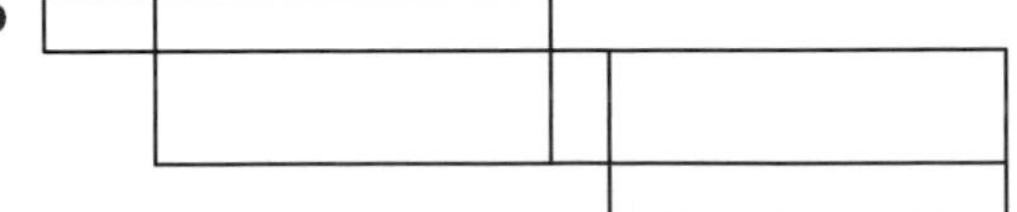

d

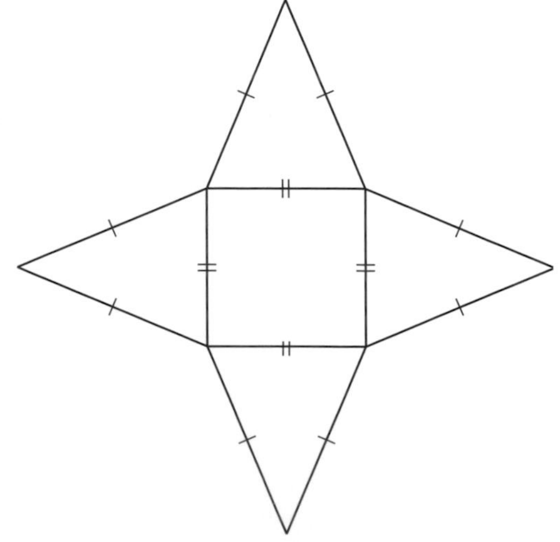

4 Get Real!

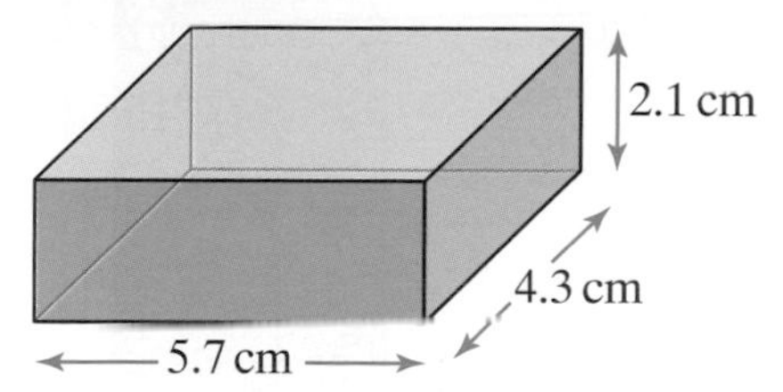

This is a picture of an open box for drawing pins.
Draw accurately the net of the box.
You may find it useful to use squared paper for your net.

5 Get Real!

Ciarán is making a gift box that looks like a gold bar.
The cross-section of the prism is an isosceles trapezium.

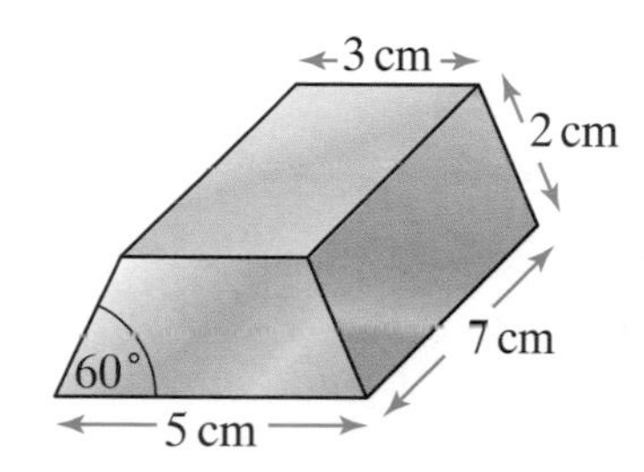

Draw a sketch of a possible net, marking in the missing angles of the trapezium.

Draw accurately the net of this gift box.
You may find it useful to use squared paper for your net.

6 This is a diagram of the net for a square-based pyramid.

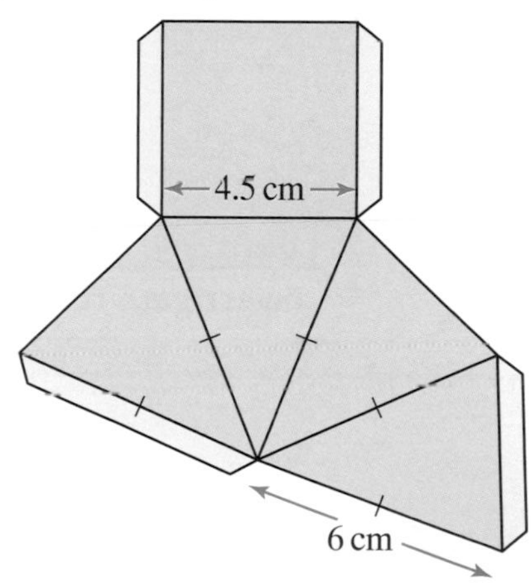

a Construct, using a ruler and compasses only, an accurate full-size net on cardboard.

b Score along the edges and make a model pyramid.

7 Get Real!

William and Robert have invented a game that needs two fair four-sided dice.
Construct accurately, using a ruler and compasses only, the net for a regular tetrahedron with edges of 2 cm.

8 Get Real!

A manufacturer is creating a box for cocktail sticks in the shape of a regular hexagonal prism.

a Sketch a net for the box.

b Using a ruler and compasses only, construct a full-size diagram of the hexagonal base.

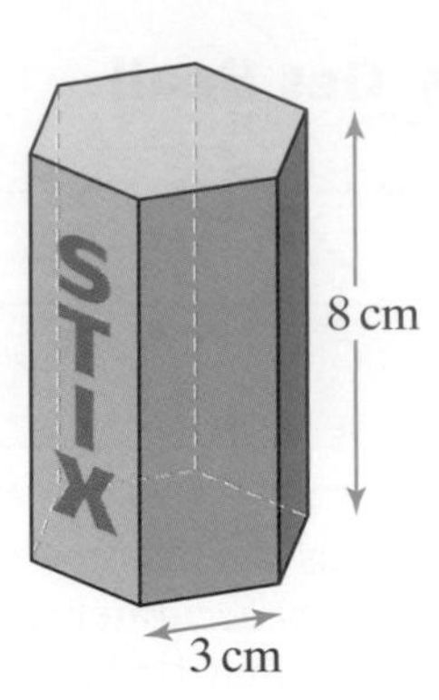

Explore

- Pentominoes are made with five identical squares joined together, edge to edge, in different ways, for example:

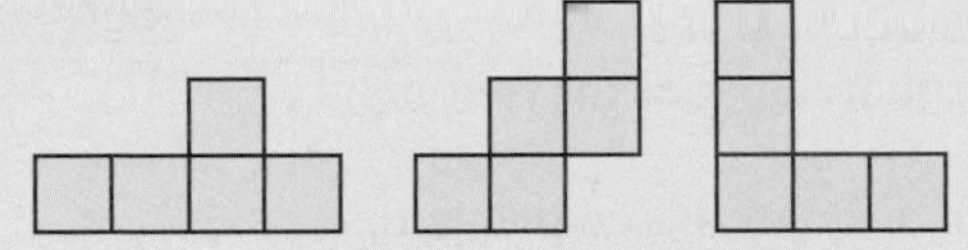

- Can you sketch the 12 different pentominoes? (Reflections and rotations don't count!)
- Which pentominoes could be nets of open boxes?
- Hexominoes are made of six identical squares. How many are the nets of cubes?

Investigate further

Explore

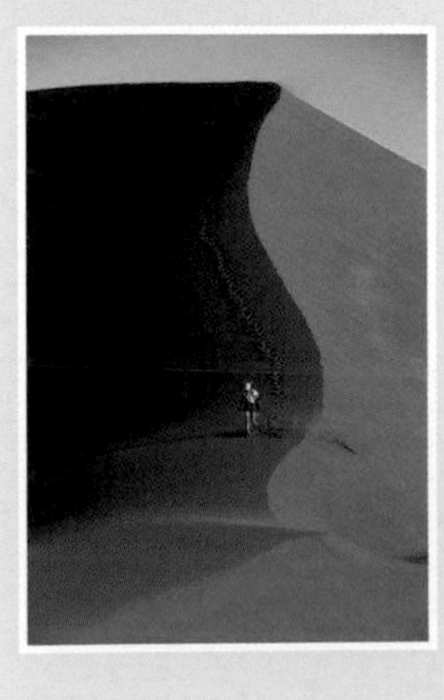

- Platonic solids are 3-D shapes where all the faces are identical regular polygons
- A regular tetrahedron has four equilateral triangles as its faces and a cube has six squares as its faces, so these are both Platonic solids

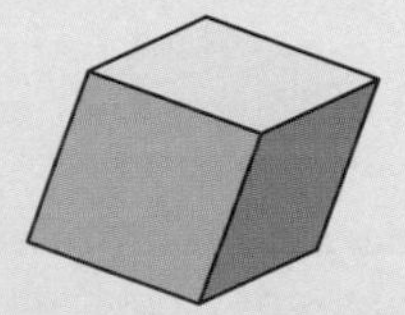

- There are three other Platonic solids – can you find out what they are?
- Draw the nets of the shapes
- Can you construct any of them using only a ruler and compasses?

Investigate further

Construction

ASSESS

The following exercise tests your understanding of this chapter, with the questions appearing in order of increasing difficulty.

1 The diagrams below show some nets.
Name the ones that would form cubes.

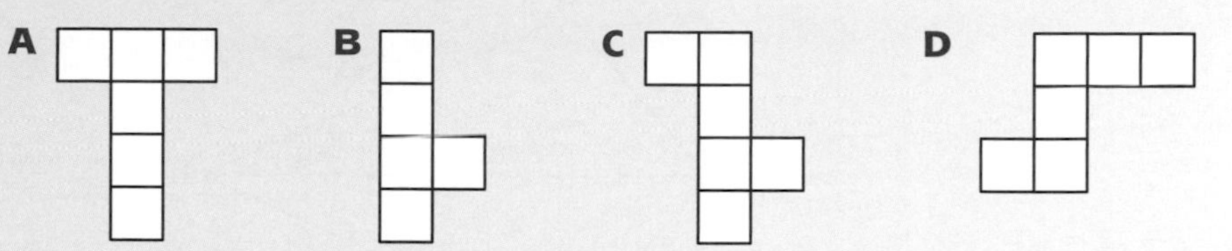

2 Draw a horizontal line 55 mm long. At the right-hand end draw another line of 55 mm at an angle of 108° to the first. Repeat this three more times. What should have happened? Has it?

3 A yacht is taking part in a race. It leaves the start line at P and has to go round two buoys, Q and R, before sailing back to P. Q is 7 miles from P on a bearing of 156°. R is 4.5 miles from Q on a bearing of 038°.

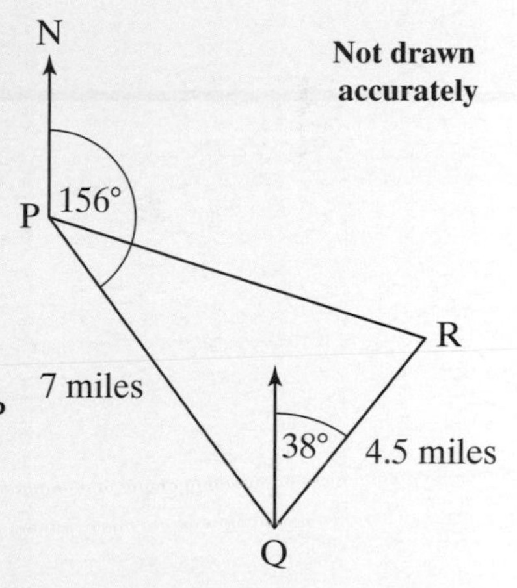

a Draw a scale diagram of the triangle PQR. Use a scale of 1 cm to 1 mile.

b The yacht has just rounded R.

i How far does it have to sail back to P

ii What is the bearing of R from P?

4 X, Y and Z are three observation posts at the vertices of a right-angled triangle, where YZ = 20 miles, angle Z is 30° and angle Y is 90°.

a Use a ruler and compasses only to construct this triangle.

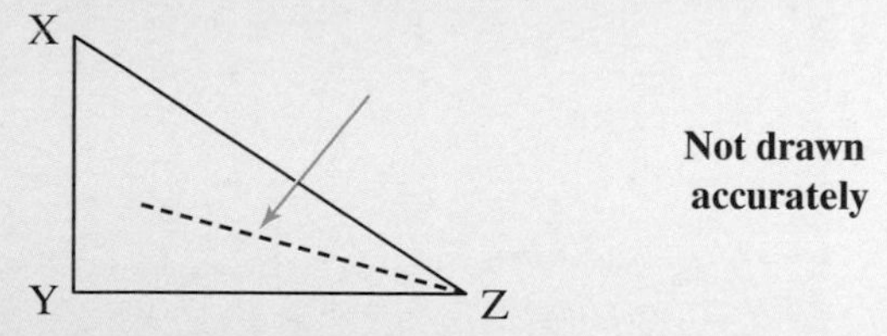

Observers at X and Z simultaneously record a UFO equidistant from them over the midpoint of XZ and moving along its perpendicular bisector. Three seconds later the UFO is lying on the angle bisector of angle Z.

b Using a ruler and compasses only locate the position of the UFO at this time.

c Measure and write down the distance the UFO has travelled.

d Estimate the speed of the UFO in mph.

Try a real past exam question to test your knowledge:

5 The kite PQRS is sketched below.
QR = SR = 6 cm
Angle QRS = 50°
The diagonal PR = 8 cm

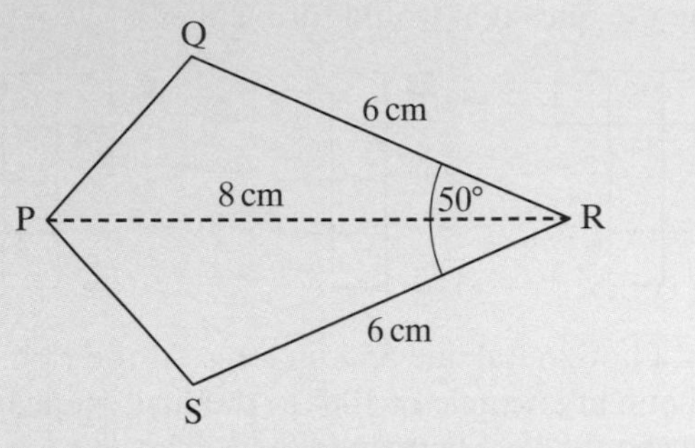

Not drawn accurately

Make an accurate drawing of the kite PQRS.

Spec B, Module 5, Int. Paper 1, Nov 04

17 Graphs of linear functions

OBJECTIVES

Examiners would normally expect students who get an E grade to be able to:

Plot the graphs of straight lines such as $x = 3$ and $y = 4$

Complete a table of values for equations such as $y = 2x + 3$ and draw the graph

Examiners would normally expect students who get a D grade also to be able to:

Solve problems involving graphs, such as finding where the line $y = x + 2$ crosses the line $y = 1$

Examiners would normally expect students who get a C grade also to be able to:

Recognise the equations of straight-line graphs such as $y = -4x + 2$

Find the gradients of straight-line graphs

What you should already know ...

- Plot coordinates in all four quadrants
- Discuss and interpret graphs of real-life situations

VOCABULARY

Gradient – a measure of how steep a line is

$$\text{Gradient} = \frac{\text{change in vertical distance}}{\text{change in horizontal distance}} = \frac{y}{x}$$

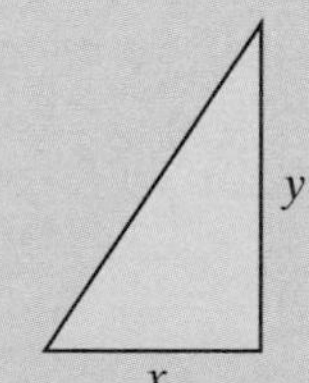

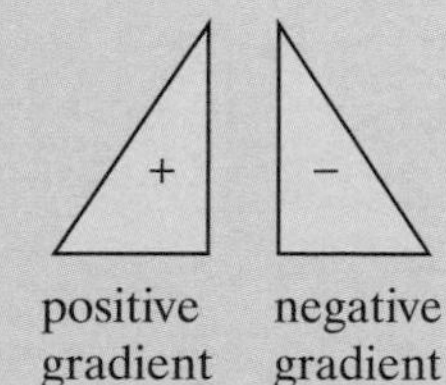

Linear graph – the graph of a linear function of the form $y = mx + c$; if c is zero, the graph is a straight line through the origin (the point (0, 0)) indicating that y is directly proportional to x; if m is zero, the graph is parallel to the x-axis

Intercept – the y-coordinate of the point at which the line crosses the y-axis

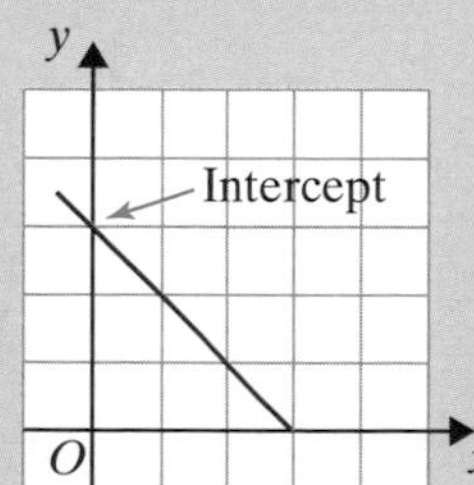

Learn 1 Drawing straight-line graphs

Examples:

a Draw the lines $x = 3$ and $y = -4$ and find where they meet.

b Draw the straight-line graph $y = 2x - 3$ for values of x from -2 to 4.

a $x = 3$ means that the x-coordinate of every point is 3. The y-coordinate can be any number, so some possible points on the line are (3, 0), (3, 5), (3, –2), etc. The line is parallel to the y-axis.

The simplest of all straight-line graphs are those that are parallel to the x-axis or the y-axis

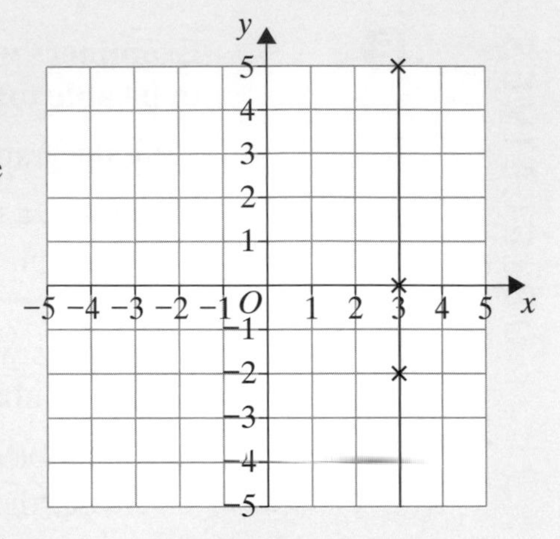

$y = -4$ means that the y-coordinate of every point is –4. The x-coordinate can be any number, so some possible points on the line are (0, –4), (–4, –4), (2, –4), etc. The line is parallel to the x-axis.

The two straight lines meet at the point (3, –4).

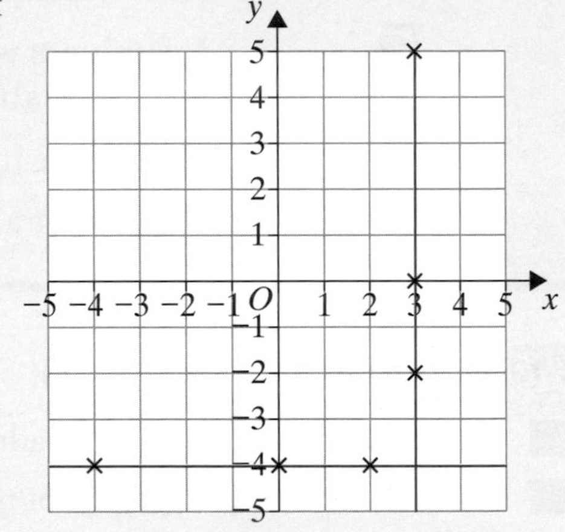

b Use the equation $y = 2x - 3$ to work out some points on the line.

The equation says that you double the x-coordinate and subtract 3 to find the y-coordinate, so choose any value of x from –2 to 4 and work out the corresponding value of y

If x is 0, $y = 2 \times 0 - 3 = 0 - 3 = -3$
So one point on the graph is (0, –3).

It is possible to work out many more pairs of values but three points are enough for a straight line

If x is 4, $y = 2 \times 4 - 3 = 8 - 3 = 5$
So another point on the graph is (4, 5).

If x is –2, $y = 2 \times (-2) - 3 = -4 - 3 = -7$
So a third point on the line is (–2, –7).

For each value of x there is just one value of y that fits the equation

Each pair of (x, y) values is a point on the line

There is an infinite number of points on any straight line

Now mark at least three of the points on squared paper and join them with a straight line.

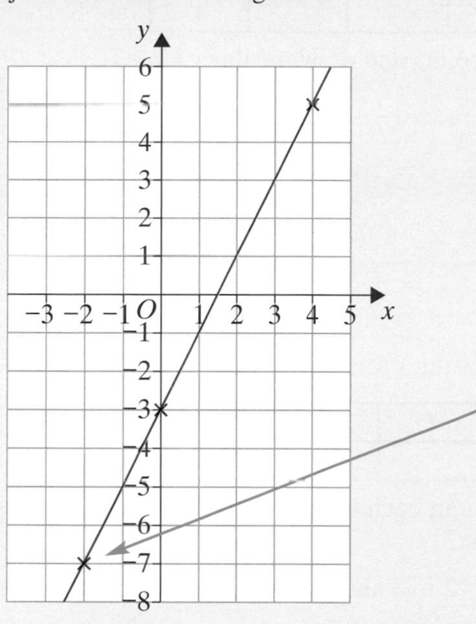

Other coordinates can be found from a table, for example, for the graph $y = 2x - 3$:

x	−2	−1	0	1	2	3	4
y	−7	−5	−3	−1	1	3	5

Apply 1

1 Draw an x-axis and a y-axis, each going from −6 to 6.

a On the axes, draw the lines $x = 2$ and $x = -4$. These lines are parallel to the y-axis

b Draw the lines $y = 5$ and $y = -2$. These lines are parallel to the x-axis

c Where does the line $x = 2$ meet the line $y = -2$?

d Where does the line $x = -4$ meet the line $y = 5$?

e How can you work out the points of intersection without using the graph?

2 Here is a table of values for the line $y = 2x$

x	−3	−2	−1	0	1	2	3
y	−6	−4	−2	0	2	4	6

a Draw an x-axis and a y-axis, the x-axis going from −3 to 3 and the y-axis going from −6 to 6.
Plot the points in the table of values and join them with a straight line.

b Write down the coordinates of the point on the line:

i with an x-coordinate of 0.5

ii with a y-coordinate of 5.

3 **a** Copy and complete this table of x- and y-values for the equation $y = x + 3$

x	−4	−3	−2	−1	0	1
y	−1		1			4

b Draw suitable axes, plot the points and draw the line $y = x + 3$

c Where does the line cut:

i the x-axis

ii the y-axis

iii the line $x = -3$

iv the line $y = 2$?

4 Copy and complete this table of values for the equation $y = 3x - 1$

x	−2	−1	0	1	2	3
y	−7		−1			8

a The x-coordinates go up one unit each time. How many units do the y-coordinates go up each time?

b Draw an x-axis labelled from −2 to 3 and a y-axis labelled from −7 to 8 and plot the points.
Draw a straight line through the points.

c Find the y-coordinate of a point on the line with x-coordinate 1.5

d Find the x-coordinate of a point on the line with a y-coordinate of 4.

5 **a** Make a table of values for $y = 2x + 4$ for values of x from −2 to 3.

b Draw suitable axes, plot the points and draw the line $y = 2x + 4$.

c Write the coordinates of the point where $y = 2x + 4$ crosses:

i the x-axis

ii the y-axis

iii the line $x = -1$

iv the line $y = 3$

6 **a** For the line $y = 3x + 1$, copy and complete these coordinate pairs:
(−3, ...) (0, ...) (3, ...)

b Draw suitable axes on squared paper, mark the three points and join them with a straight line.

c Write down the coordinates of the points where the line $y = 3x + 1$ crosses the x-axis and the y-axis.

7 Draw an x-axis and a y-axis, each going from −10 to 10.

a On these axes, draw the lines:

i $y = 2x + 4$ **ii** $y = \frac{1}{2}x$

b **i** Where does the line $y = \frac{1}{2}x$ cut the axes?

ii How can you find this from the equation?

8 Which of these points lie on the line $y = \frac{1}{3}x$?

(3, 1), (3, 0), (0, 0), (−3, −1), (1, 3), (2, 6), (0.9, 0.3)

Show how you found your answers.

9 **a** Draw the straight line that goes through the points $(-1, -4)$, $(0, 0)$ and $(2, 8)$.

b Which of these is the equation of the line?

$y = x + 6$ $\quad y = x - 3$ $\quad y = 4x$

10 The equation $x + y = 6$ means that the x-coordinates and y-coordinates add up to 6. So one point on this line is $(5, 1)$.

a Find at least two other points on the line.

b Draw an x-axis and a y-axis, each going from -10 to 10. Plot the points and join them with a straight line.

c On the same diagram, draw the line $x + y = 8$

d What is the same and what is different about the two lines?

11 The diagram shows several straight-line graphs.

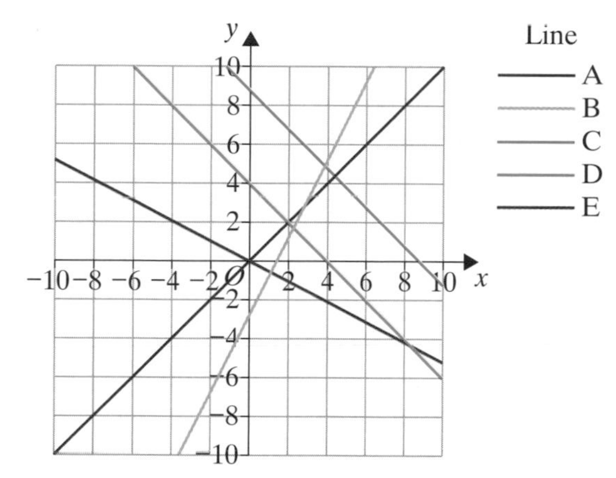

a Write down the letters of two lines that are parallel.

b Write down the letters of two lines that are perpendicular.

c Write down the letters of the lines that go through the origin.

d Write down the letter of the line with the greatest slope.

e Write down the coordinates of three points on Line A.

Explore

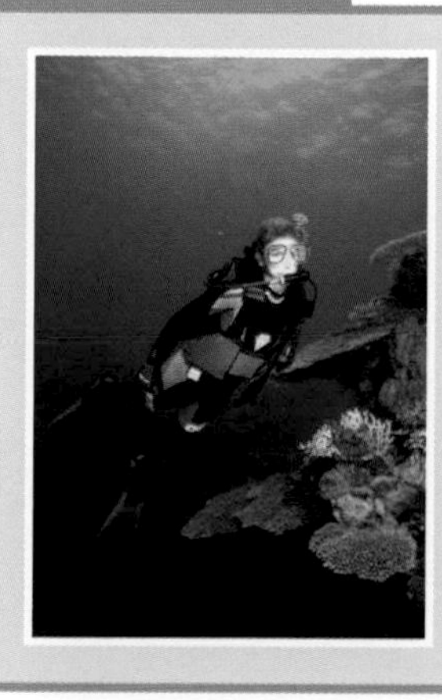

- Draw the lines $x + y = 10$ and $x + y = 5$
- What is the same and what is different about the two lines?
- Draw some more lines of the form $x + y = a$ using different values of the number a (the value of a can be positive, negative or zero – use all three)
- You may want to use a graphical calculator or a spreadsheet program on a computer
- What do you notice about the lines?
- Repeat this investigation for lines of the form $y = x + a$

Investigate further

Learn 2 Recognising the equations of straight-line graphs and finding their gradients

Examples:

a Find the gradient of the line with equation:

i $y = 4x + 2$ **ii** $y = 3x + 6$

b Rearrange the following equations in the form $y = mx + c$

i $x + y = 4$ **ii** $2y + 3x = 12$

a i First think about the line with equation $y = 4x$
Here is the table of values:

The x-values go up in ones and the y-values go up in fours

x	−3	−2	−1	0	1	2
$y = 4x$	−12	−8	−4	0	4	8

Now back to the line with equation $y = 4x + 2$

Here is the table of values:

x	−3	−2	−1	0	1	2
$y = 4x$	−12	−8	−4	0	4	8
$y = 4x + 2$	−10	−6	−2	2	6	10

To find the y-values for $y = 4x + 2$, add 2 to the y values for $y = 4x$

The y-values still go up in fours

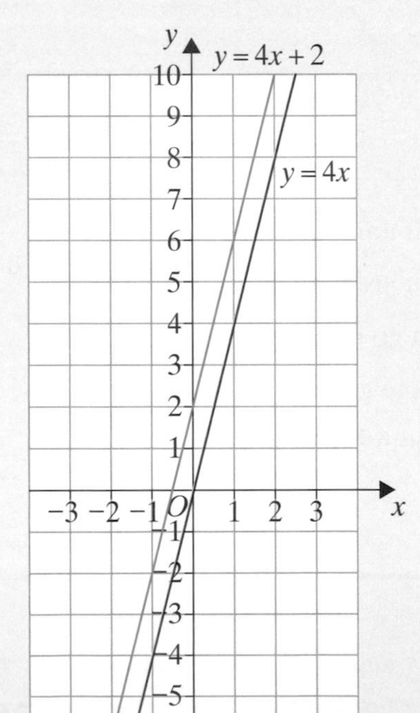

The diagram shows the lines $y = 4x$ and $y = 4x + 2$
The lines are parallel; they both have the same gradient.

Adding 2 to make $y = 4x + 2$ does not change the gradient but moves the line up 2 squares

The gradient of each line is 4 as each line goes up the page 4 units for every unit across the page (Count the squares)

$y = 4x + 2$ ← y-intercept is at +2 so passing through (0, +2)

↑ gradient

ii For the line with equation $y = 3x - 6$

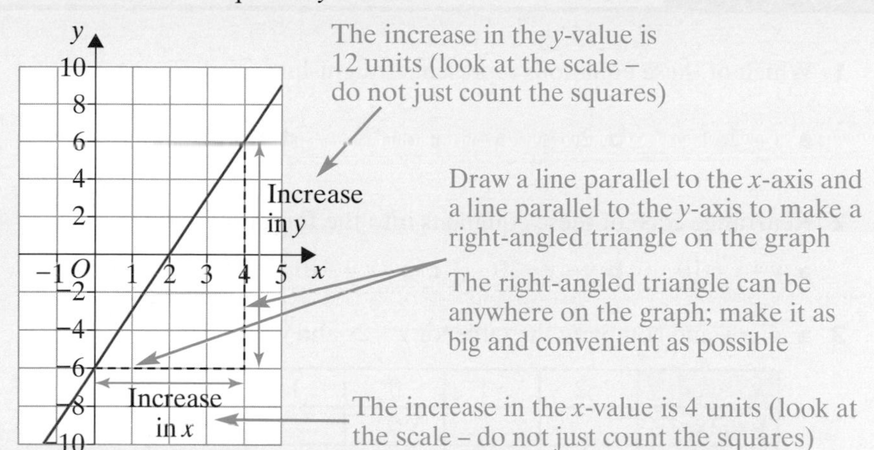

The gradient of the graph is how many units it goes up for every unit across.

So gradient $= \dfrac{\text{increase in } y}{\text{increase in } x} = \dfrac{12}{4} = 3$

The graph goes up 3 units for every unit across.
It crosses the y-axis at -6, as the equation of this graph is $y = 3x - 6$.

$y = 3x - 6$ ← y-intercept is at -6 so passing through $(0, -6)$

↑ gradient

Any graph parallel to $y = 3x - 6$ has gradient 3 and any graph with gradient 3 is parallel to $y = 3x - 6$.

Any equation that can be written in the form $y = mx + c$ is a linear equation and can be drawn as a straight-line graph.

b i

$x + y = 4$	
$y = 4 - x$	Subtract x from both sides
$y = -x + 4$	Rearranging in the form $y = mx + c$

This is a straight line with gradient $= -1$ and y-intercept at 4 so passing throught the point $(0, 4)$.

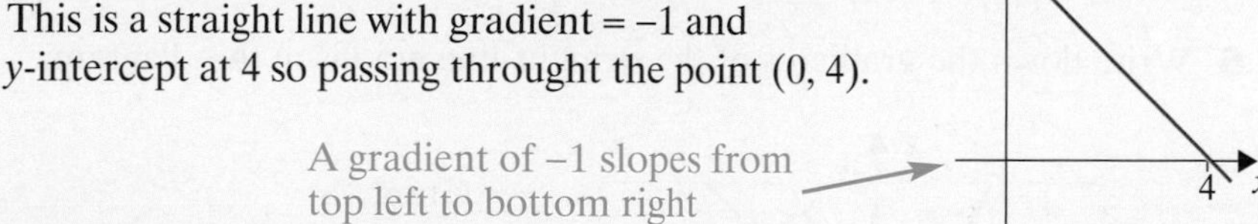

ii

$2y + 3x = 12$	
$2y = 12 - 3x$	Subtract $3x$ from both sides
$y = 6 - \frac{3}{2}x$	Divide each side by 2
$y = -\frac{3}{2}x + 6$	Rearranging in the form $y = mx + c$

This is a straight line with gradient $= -\frac{3}{2}$ and y-intercept at 6 so passing through the point $(0, 6)$.

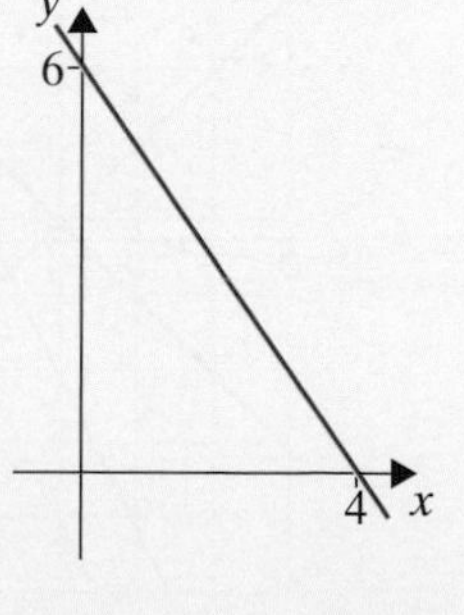

A gradient of $-\frac{3}{2}$ slopes from top left to bottom right

Apply 2

1 Which of these equations represent straight-line graphs?

a $y = 2x + 8$ **b** $2y = x + 8$ **c** $x = 2y$ **d** $y = \frac{x}{2}$ **e** $y = \frac{2}{x}$

2 Rearrange each of these equations into the form $y = mx + c$.

a $y + x = 10$ **b** $y - x = 10$ **c** $y + x = -10$ **d** $y - x = -10$

3 a Copy and complete the table for $y = 2x$ and $y = 2x - 3$.

x	-2	-1	0	1	2
$y = 2x$			0		
$y = 2x - 3$			-3		

b Draw an x-axis and a y-axis, the x-axis going from -2 to 2 and the y-axis going from -8 to 4. On the axes, plot each set of points from the table and join them with a straight line.

c Use your diagram to find the gradient of each line.

4 Repeat question **3** for the lines $y = 5x$ and $y = 5x - 4$. (The y-axis will need to go from -15 to 10.)

5 a Draw these straight lines and work out their gradients.

i $y = 2x$ **ii** $y = \frac{1}{2}x$ **iii** $y = -2x$ **iv** $y = -\frac{1}{2}x$

b Use the straight-line graphs in part **a** to help you to work out the gradients of these lines:

i $y = 2x + 5$ **ii** $y = \frac{1}{2}x + 5$ **iii** $y = -2x + 5$ **iv** $y = -\frac{1}{2}x + 5$

c Work out the gradients of these lines without drawing the graphs.

i $y = \frac{1}{3}x$, $y = -\frac{1}{3}x$, $y = \frac{1}{3}x - \frac{2}{3}$, $y = -\frac{1}{3}x - \frac{2}{3}$

ii $y = -0.2x$, $y = 0.2x - 0.5$, $y = 0.2x$, $y = -0.2x - 0.5$

6 Write down the gradients of the straight-line graphs in this diagram.

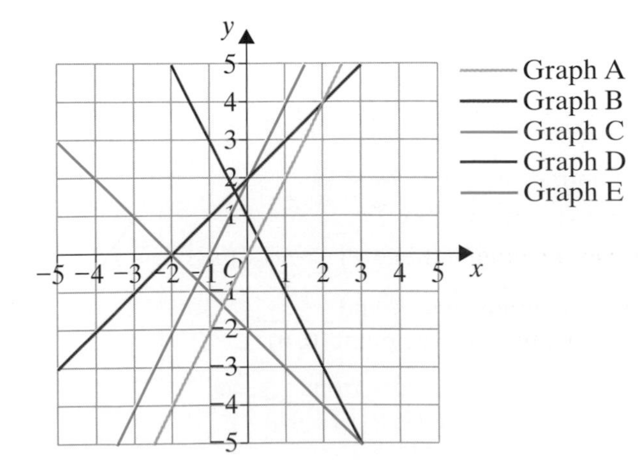

7 a Rearrange each of these equations into the form $y = mx + c$:

i $2y + x = 6$ **v** $3x + 4y = 8$

ii $y - 2x = 6$ **vi** $3x - 4y = 8$

iii $2y - x = -6$ **vii** $-3x + 4y = -8$

iv $y + 2x = -6$ **viii** $3x + 4y + 8 = 0$

b Find the gradients of the lines.

8 The diagram shows the line $y = 3x - 5$
AB is 2 units long.
How long is BC?

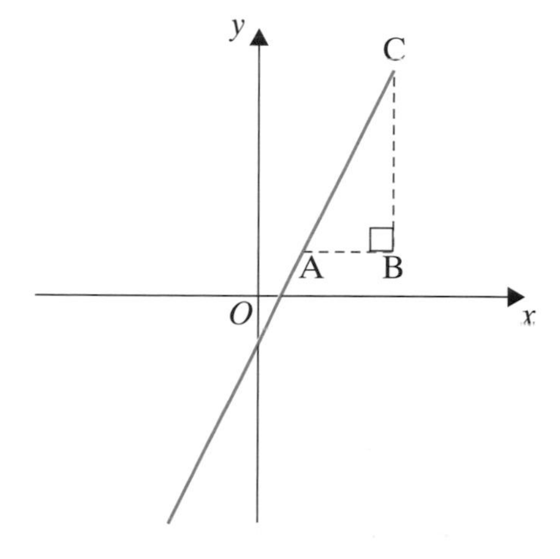

Graphs of linear functions

ASSESS

The following exercise tests your understanding of this chapter, with the questions appearing in order of increasing difficulty.

1 a The x-coordinates of three points on the line $y = 2x - 1$ are given by $x = -3, x = 0$ and $x = 4$. Find the y-coordinates corresponding to these x-values.

b Plot these points on a graph and show that all three points do indeed lie on a straight line.

2 a Draw axes with values of x and y from -10 to 10. On this grid draw the graphs of:

$y = \frac{x}{2}$, $y = 3x - 5$ and $y = 9 - x$

b Write down the coordinates of the points where each pair of lines intersect.

3 a Does the point $(2, -3)$ lie on the line $y = 2x - 7$?

b The point $(1, -4)$ lies on either the line $y = 3 - x$ or the line $y = 2x - 6$
Which line does the point lie on?

4 **a** Write down the gradient of each of the following straight lines:

i $y = 5x + 2$ **iii** $y = 7x - 1$ **v** $4x + 2y = 5$

ii $y = 8 + x$ **iv** $y = 5 - 3x$ **vi** $5x - 2y = 6$

b **i** Nadia says that the line $y = 7x - 1$ crosses the y-axis at $(-1, 0)$.
Is this correct? Give a reason for your answer.

ii Orla says that the line $y = 5 - 3x$ has the same gradient as the line $3x + y = 8$
Is this correct? Give a reason for your answer.

5 Shane sells luxury electrical goods.
He is paid a basic wage each month plus a percentage commission on his sales that month.
If his sales for the month are £20 000 he is paid £1400.
If his sales are £50 000 he is paid £2300.

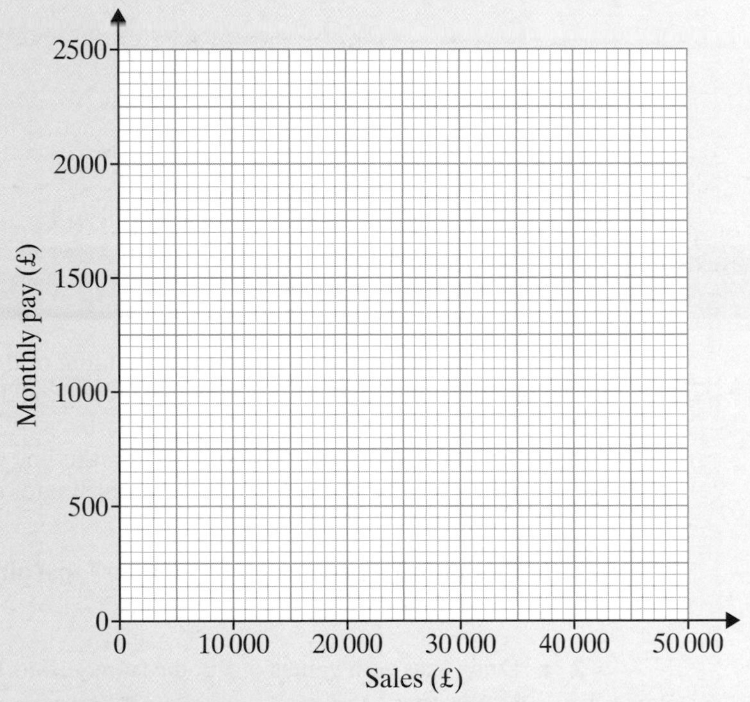

a Use a grid like the one shown above to plot the points (20 000, 1400) and (50 000, 2300).

Draw a straight-line graph through your points and use it to answer the following questions.

b **i** How much is Shane's basic wage?

ii What is the gradient of the graph?

iii What is the percentage rate of his commission?

18 Pythagoras' theorem

OBJECTIVES

Examiners would normally expect students who get a C grade to be able to:

Use Pythagoras' theorem to find the hypotenuse of a right-angled triangle

Use Pythagoras' theorem to find any side of a right-angled triangle

Use Pythagoras' theorem to find the height of an isosceles triangle

Use Pythagoras' theorem in practical problems

What you should already know ...

- Squares of integers up to 15 and the corresponding square roots
- Round numbers with decimals to the nearest integer
- The properties of quadrilaterals and their diagonals
- Calculate areas and volumes

VOCABULARY

Equilateral triangle – a triangle with 3 equal sides and 3 equal angles – each angle is 60°

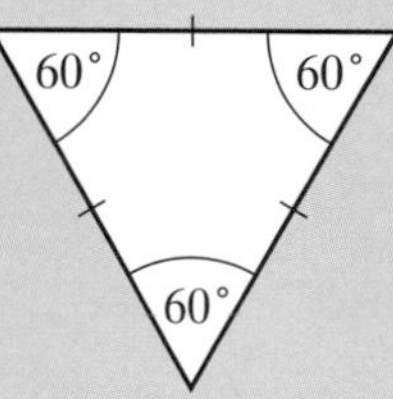

Isosceles triangle – a triangle with 2 equal sides and 2 equal angles; the equal angles are called **base angles**

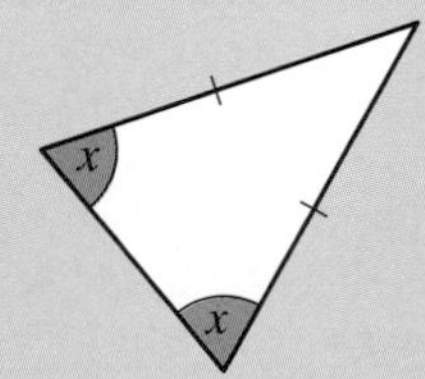

Right-angled triangle – a triangle with one angle of 90°

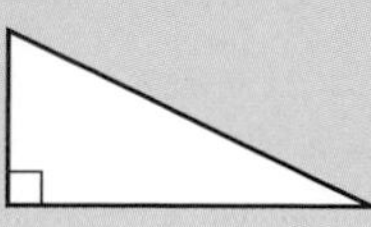

Hypotenuse – the longest side of a right-angled triangle, opposite the right angle

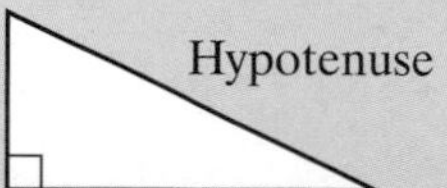

Pythagoras' theorem – in a right-angled triangle, the square of the length of the hypotenuse is equal to the sum of the squares of the lengths of the other two sides

$c^2 = a^2 + b^2$

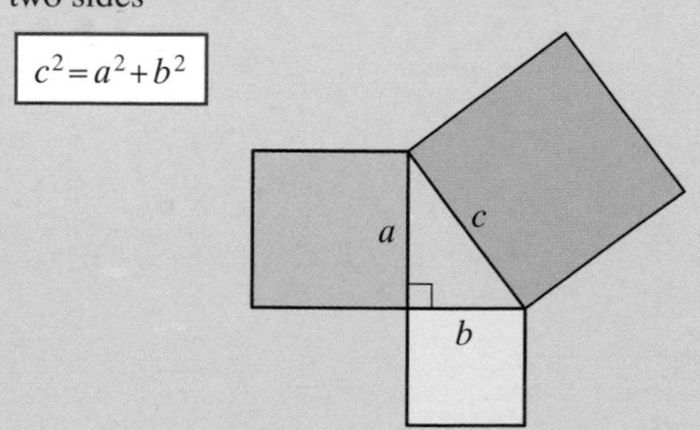

The area of the largest square = the total area of the two smaller squares

Converse of Pythagoras' theorem – in any triangle, if $c^2 = a^2 + b^2$ then the triangle has a right-angled trinagle opposite c; for example, if $c = 17$ cm, $a = 8$ cm, $b = 15$ cm, then $c^2 = a^2 + b^2$, so this is a right-angled triangle

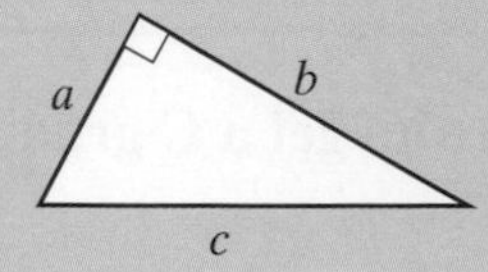

Pythagorean triple – a set of three integers a, b, c that satisfies $c^2 = a^2 + b^2$; for example, 3, 4, 5 ($5^2 = 3^2 + 4^2$), 5, 12, 13 ($13^2 = 5^2 + 12^2$), 6, 8, 10 ($10^2 = 6^2 + 8^2$) and 15, 36, 39 ($39^2 = 15^2 + 36^2$)

Learn 1 Finding the hypotenuse

Example: Find the hypotenuse of this triangle. Give your answer to an appropriate degree of accuracy.

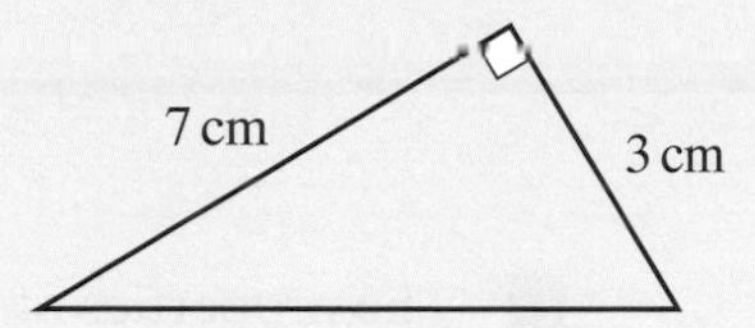

Pythagoras' theorem states that, in a right-angled triangle, the square of the length of the hypotenuse is equal to the sum of the squares of the lengths of the other two sides.

$c^2 = a^2 + b^2$

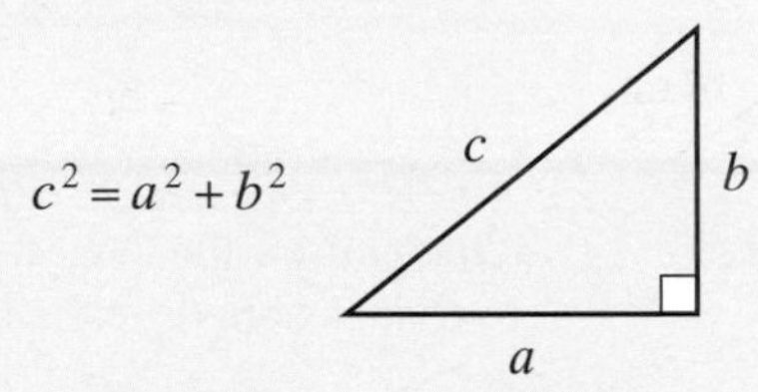

c is the hypotenuse – it is the longest side as it is opposite the right angle

Using Pythagoras' theorem:

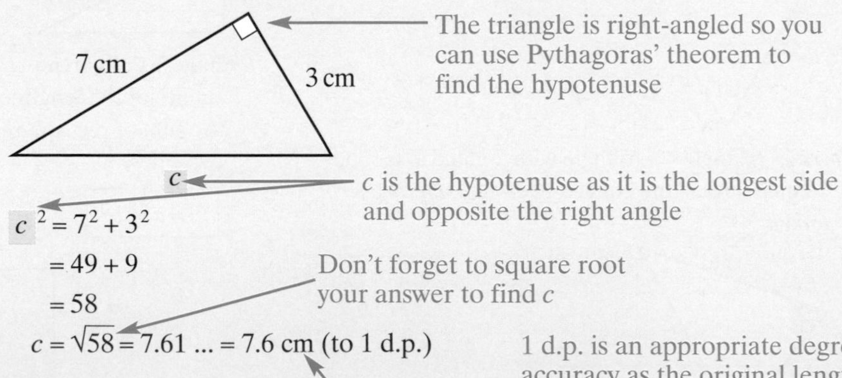

$c^2 = 7^2 + 3^2$

$= 49 + 9$

$= 58$

$c = \sqrt{58} = 7.61 \ldots = 7.6$ cm (to 1 d.p.)

Don't forget to square root your answer to find c

Remember the units

1 d.p. is an appropriate degree of accuracy as the original lengths are given to the nearest whole number

Apply 1 

1 Find the length of the hypotenuse.
Give your answer to an appropriate degree of accuracy.

a
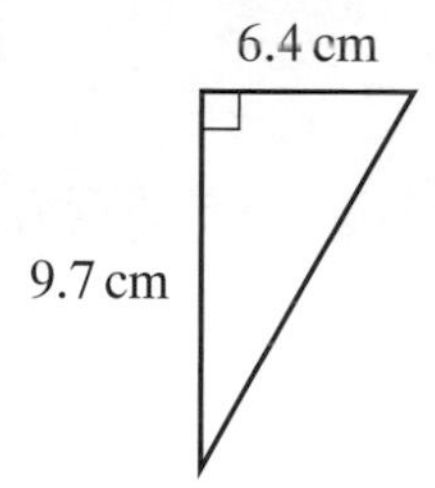

b
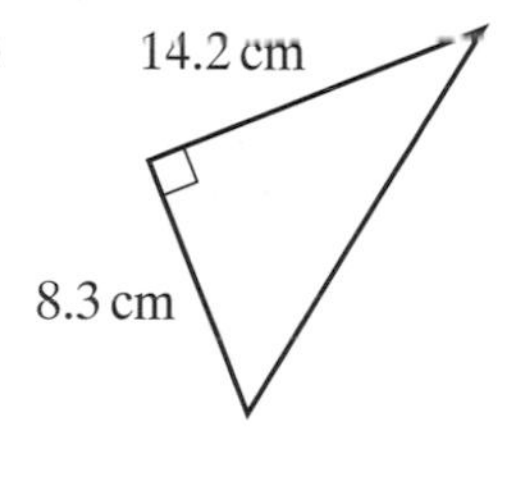

c
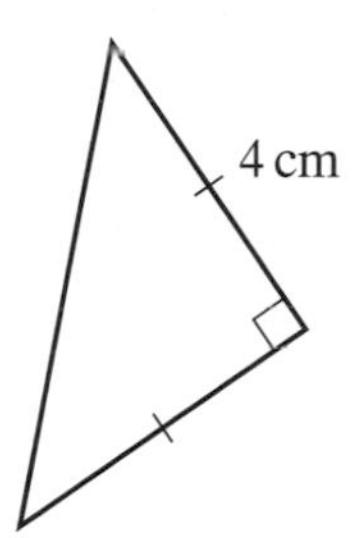

2 Find x.

a
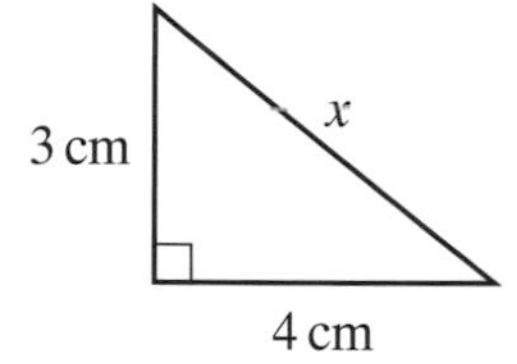

b
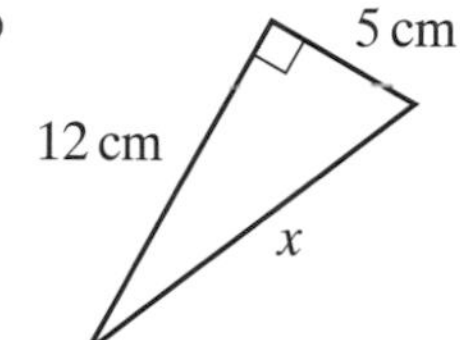

c
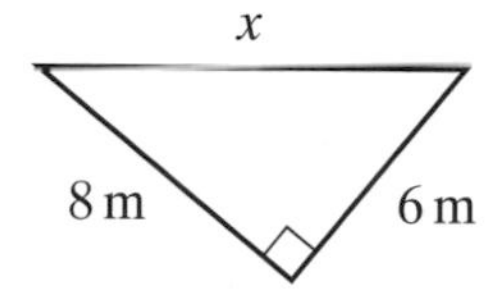

3 Find the length of the missing side.
Give your answer as a square root.

a
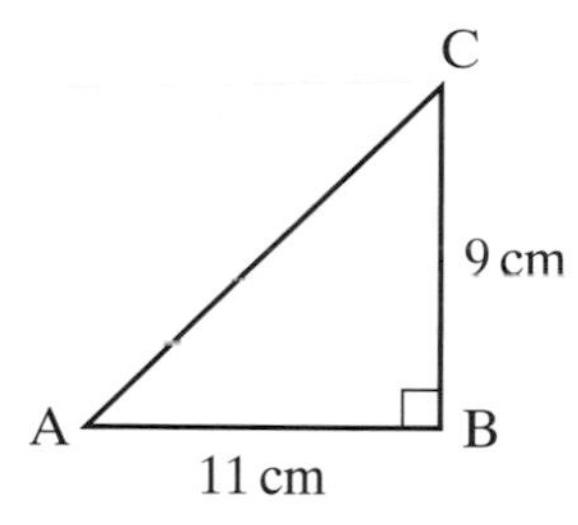

b
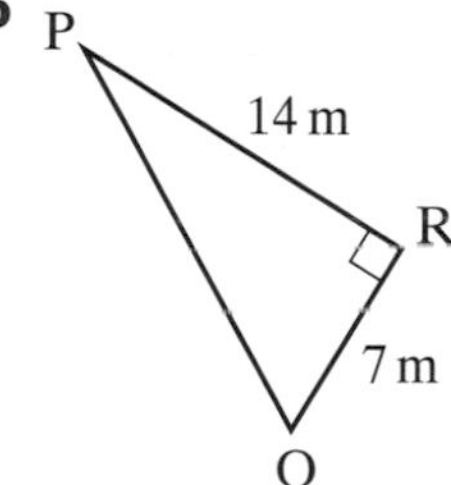

c
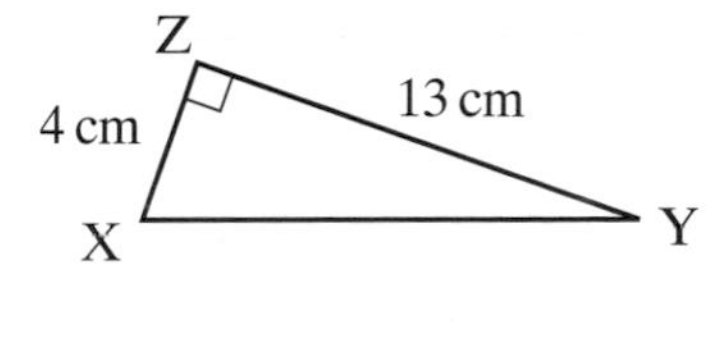

4 Get Real!
A village green has a footpath along its diagonal.
If the village green is a square of side 45 metres, how long is the footpath to the nearest metre?

45 m

5 The diagram shows a rectangle that is 8 cm by 2 cm.
Calculate x to the nearest millimetre.

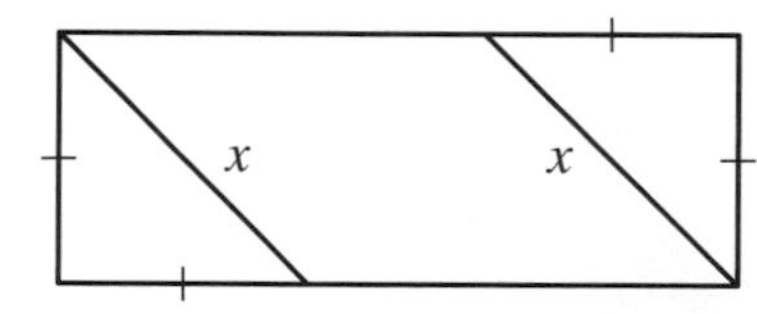

There is more information than you need in this question.
Which bit is unnecessary?

6 Get Real!

Matthew is buying a new television for his grandad, Ken.
Ken wants the screen of his new television to be at least as big as his old screen, which measures 22 inches by 15 inches.
Television screen sizes are given by their diagonal length.
What is the smallest screen size, to the nearest inch, that Matthew should buy for his grandad?

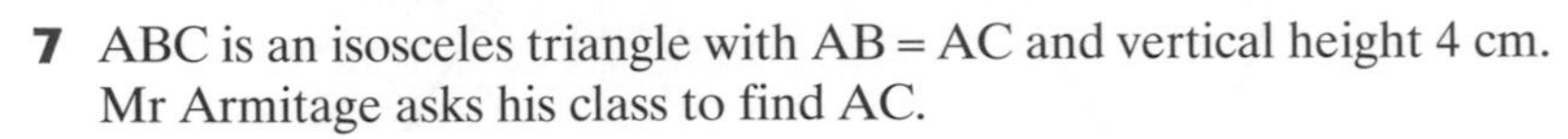

7 ABC is an isosceles triangle with AB = AC and vertical height 4 cm.
Mr Armitage asks his class to find AC.

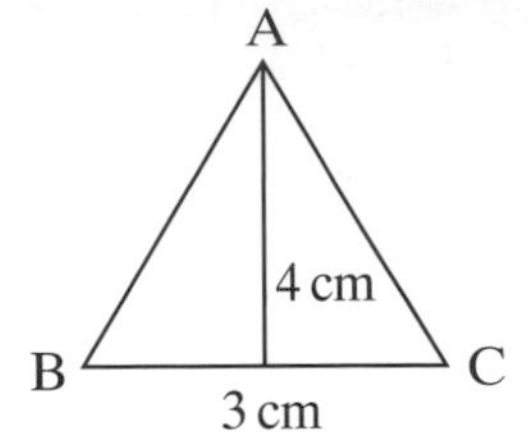

Gareth says that you can't use Pythagoras' theorem because ABC is an isosceles triangle and is not right-angled.

Ruth disagrees. She says that there is a right angle and if you use Pythagoras' theorem you get $AC^2 = 4^2 + 3^2 = 25$, so AC = 5 cm.

Simon says that Ruth has also made a mistake and that $AC = 4^2 + 1.5^2 = 18.25$ cm.

In fact, all three are wrong!

a Explain why there is a right-angled triangle.

b What did Ruth and Simon do wrong?

c Use Pythagoras' theorem correctly to find AC.

8 Get Real!

A ship sets out from a harbour and sails due west for 12.6 km.
The ship then sails due north for 16.8 km to reach an island.

a Draw a sketch of the ship's journey.

b Calculate the shortest distance from the harbour to the island.

9 Get Real!

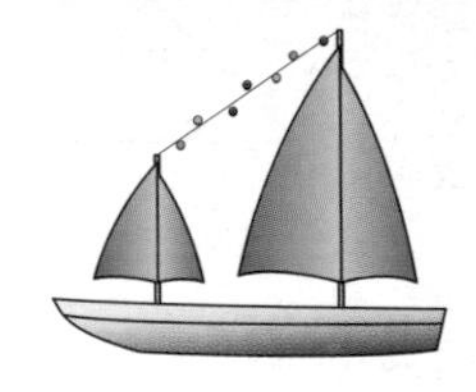

A boat has two masts that are 3.7 m apart.
The first mast is 5.5 m high and the second is 8.5 m high.
Kasim wants to attach a string of fairy lights between the tops of the masts.
What is the shortest string of lights that he will need?

10 At what angle do the diagonals of a rhombus intersect?
Do the diagonals bisect each other?
Draw a sketch of a rhombus with diagonals of 12 cm and 16 cm.
Find the perimeter of the rhombus.

11 Get Real!

A farmer is fencing off a rectangular area of his square field as shown in the diagram.

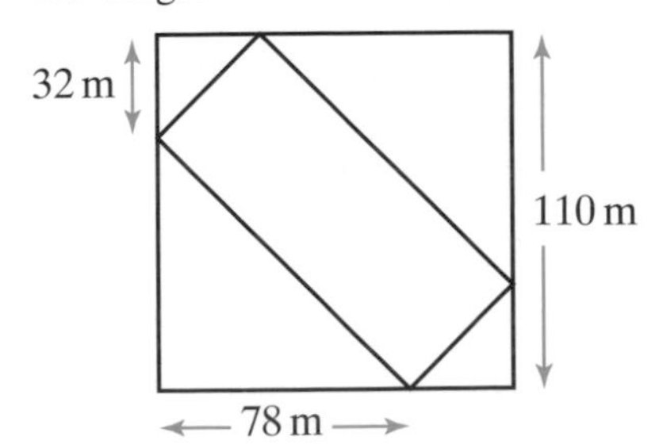

Calculate the total length of fencing he will need.
Give your answer in metres and centimetres.

12 Get Real!

Annette is designing a sun room with a sloping glass roof.
The diagram shows a side elevation of the room with her planned measurements.
Calculate the length of the sloping roof.
Give your answer in metres and centimetres.

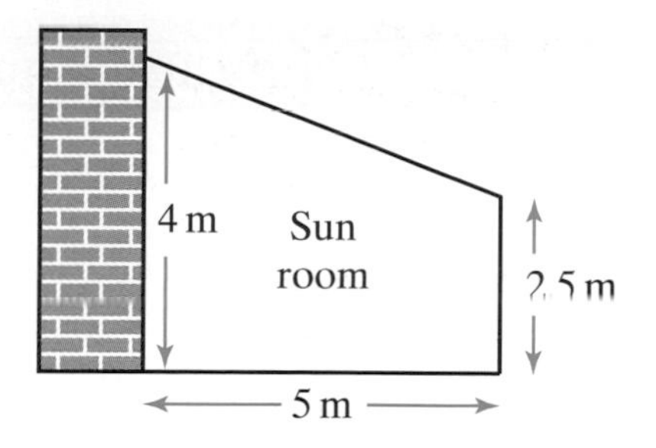

13 Get Real!

The diagram shows the side view of the rain cover on Marianna's toy pram.
Use Pythagoras' theorem in the first triangle to find a^2.
Now use your value of a^2 in the second triangle to find b^2.
Find c^2.
Finally find d, leaving your answer as a square root.

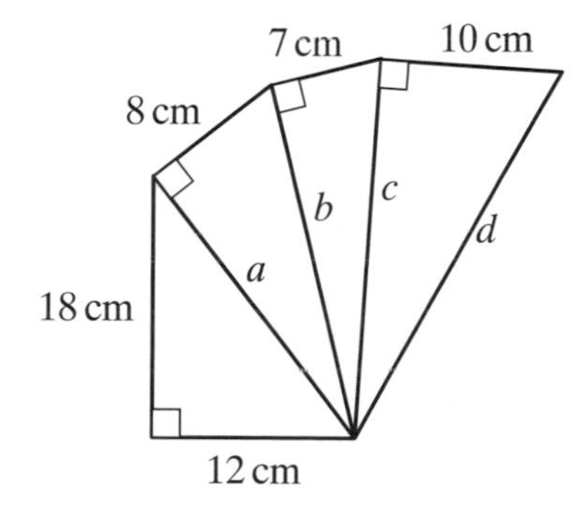

14 Draw accurately a line of length:

a $\sqrt{13}$ cm **b** $\sqrt{17}$ cm.

Explore

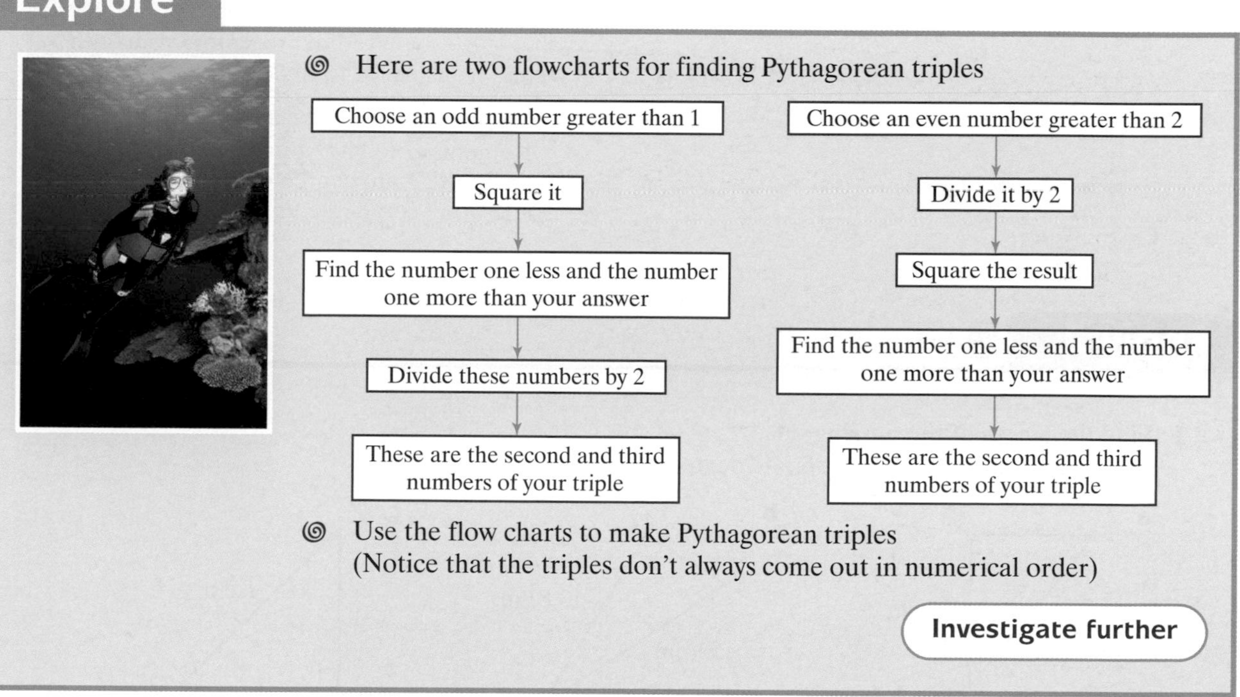

Here are two flowcharts for finding Pythagorean triples

Use the flow charts to make Pythagorean triples
(Notice that the triples don't always come out in numerical order)

Investigate further

Learn 2 Finding any side

Example: Find PQ, giving your answer to an appropriate degree of accuracy.

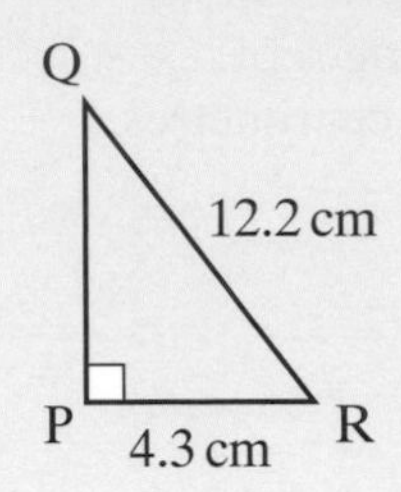

Using Pythagoras' theorem,

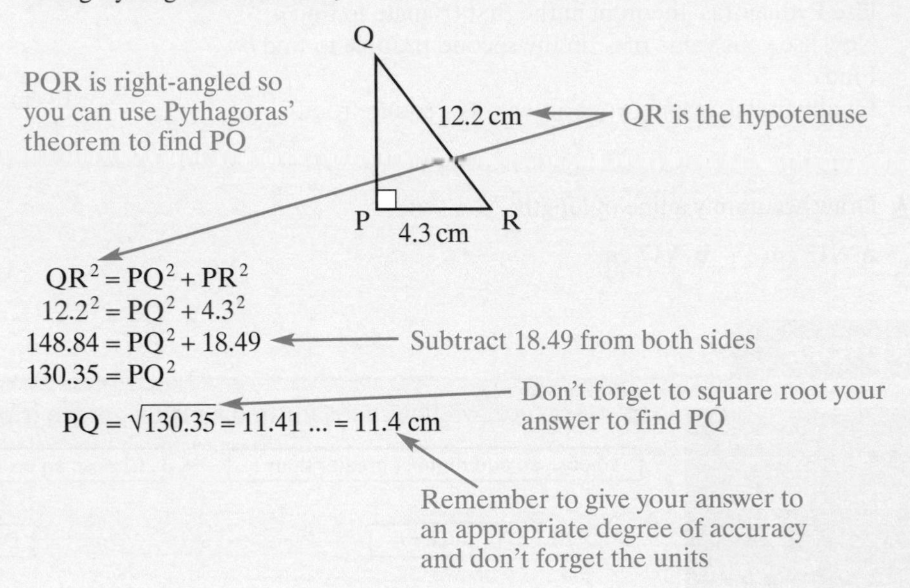

$QR^2 = PQ^2 + PR^2$

$12.2^2 = PQ^2 + 4.3^2$

$148.84 = PQ^2 + 18.49$ ← Subtract 18.49 from both sides

$130.35 = PQ^2$

$PQ = \sqrt{130.35} = 11.41 \ldots = 11.4$ cm ← Don't forget to square root your answer to find PQ

Remember to give your answer to an appropriate degree of accuracy and don't forget the units

Apply 2

1 Find the length of the missing side.
Give your answer to an appropriate degree of accuracy.

a

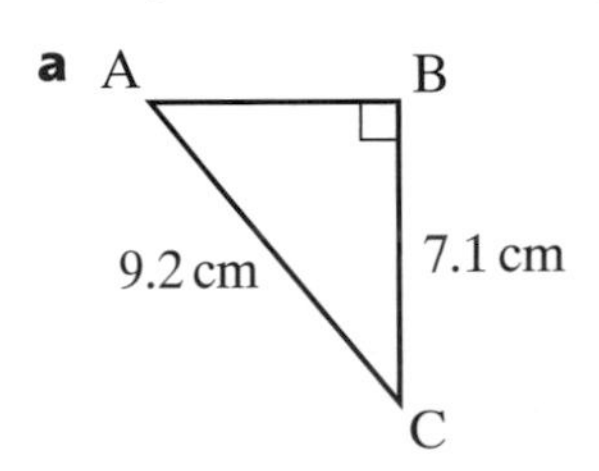

b

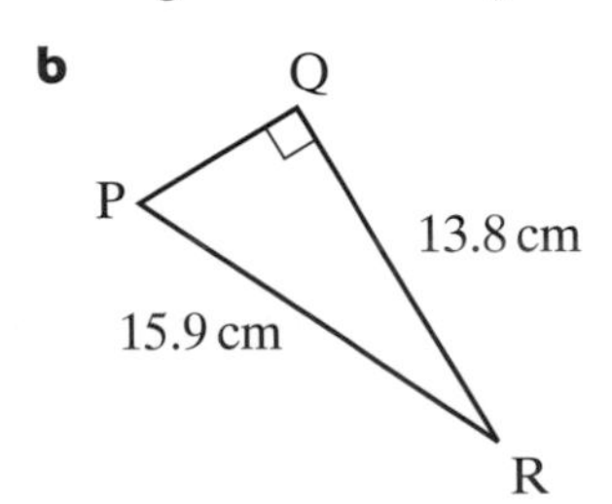

c

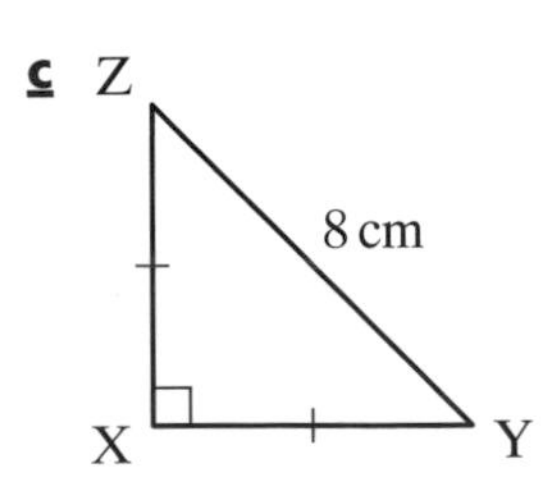

 **2** Find x, leaving your answer as a square root where appropriate.

a

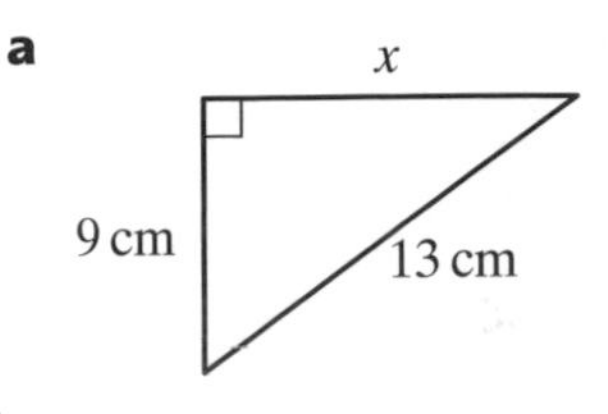

b

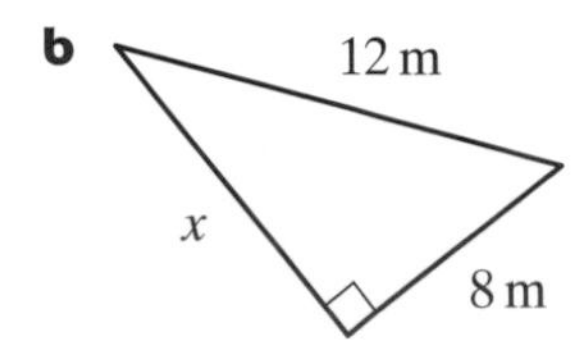

c

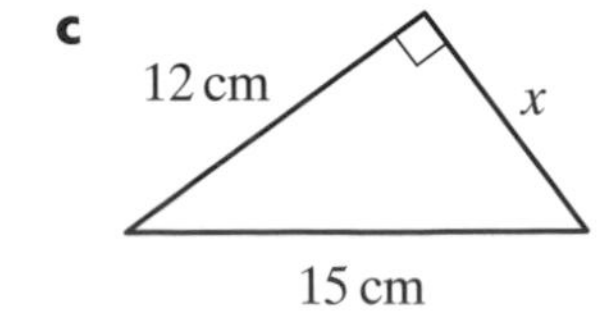

In the remaining questions in this section you might need to find a shorter side or the hypotenuse of a right-angled triangle. Decide carefully which is needed.

3 Find the length of the missing side.
Give your answer to an appropriate degree of accuracy.

a

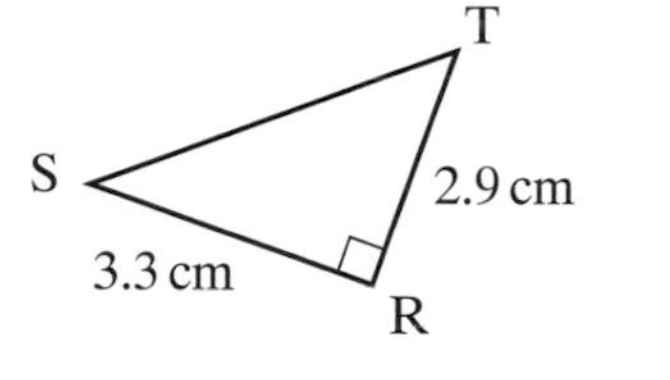

b

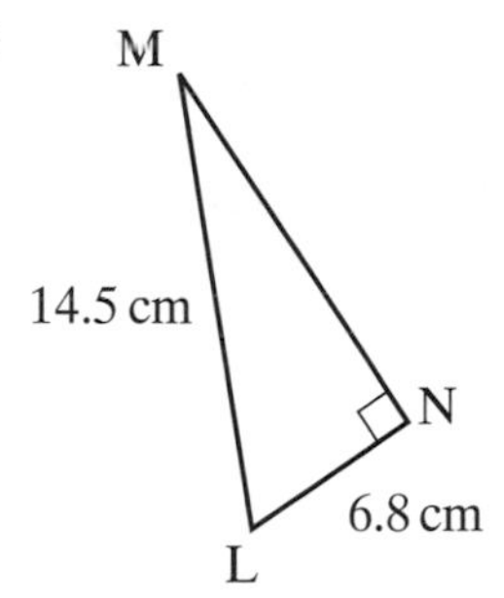

c

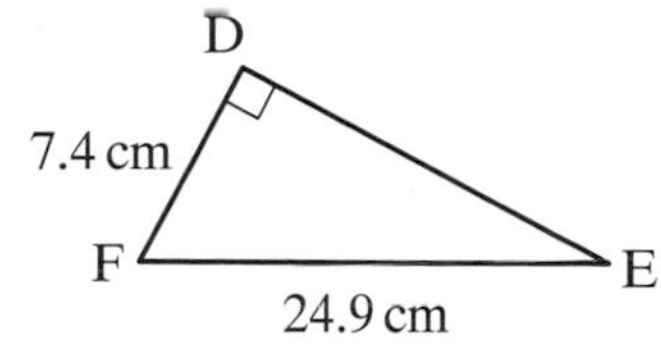

4 
Get Real!

Kwame has got his kite stuck in a tree.
He knows that the string, which just reaches the ground where he is standing, is 10 metres long.
He estimates his distance from the base of the tree as 9 metres.
Approximately how high is the tree?

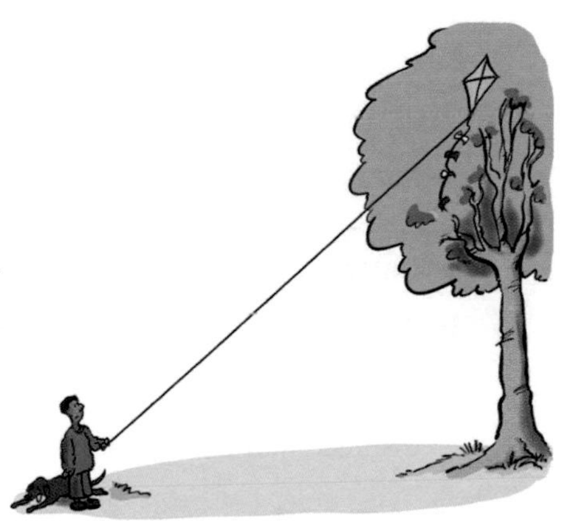

5 Get Real!

ABCD is the lid of a rectangular pencil tin.

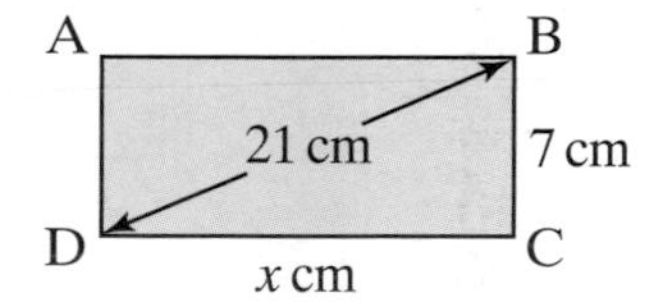

a Find x.

b What is the area of the lid?

6 Albert is trying to find x.

This is his working:

$x^2 = 6^2 + 4^2$
$x^2 = 36 + 16$
$x^2 = 52$
$x = \sqrt{52}$ cm

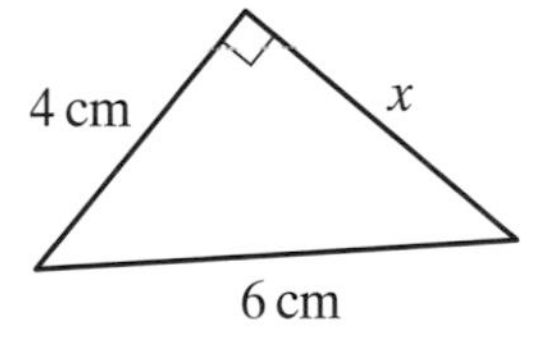

Is Albert correct?
Give a reason for your answer.

7 Get Real!

A sunken wreck, X, is 12 kilometres south and 14 kilometres west of Arton, A.

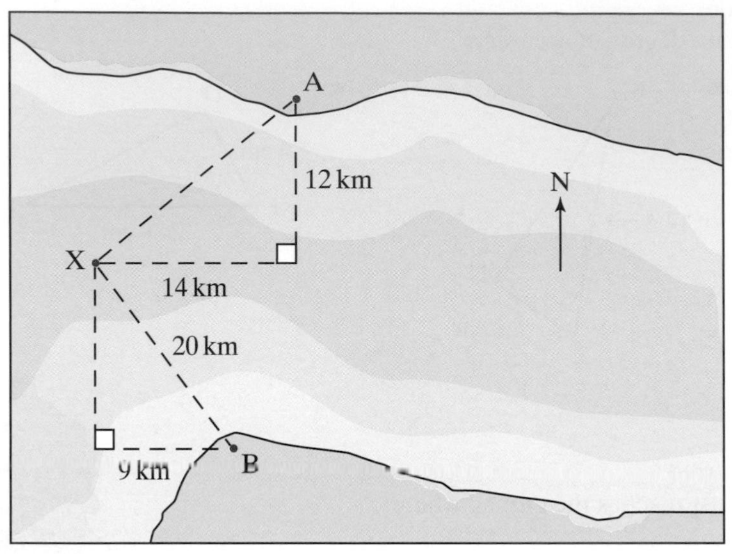

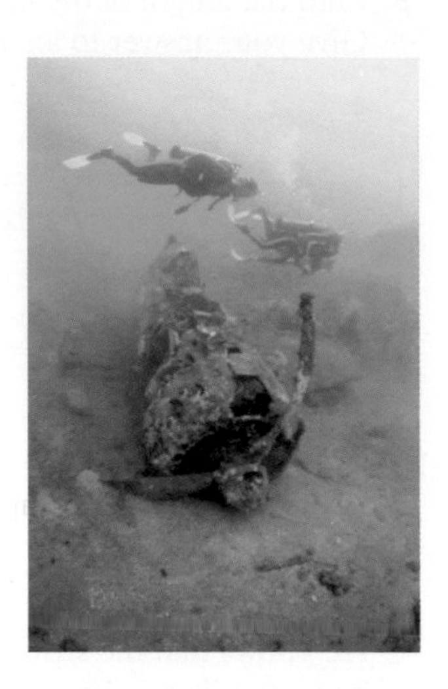

a Calculate the direct distance, AX, of the sunken wreck from Arton.

b A boat sails directly from Briswick, B, to the sunken wreck, a distance of 20 kilometres.
If the sunken wreck is 9 kilometres west of Briswick, calculate the distance that the wreck is north of Briswick.

8 Get Real!

The safety instructions with a 6.1 metre ladder say that the foot of the ladder must be a maximum distance of 1.7 metres and a minimum distance of 1.5 metres from the wall.
To the nearest centimetre, what is the maximum vertical height that the ladder can safely reach?

9 A right-angled triangle has a hypotenuse of length 25 cm.
The lengths of the other two sides are whole numbers of centimetres.
Can you find two different possible triangles?

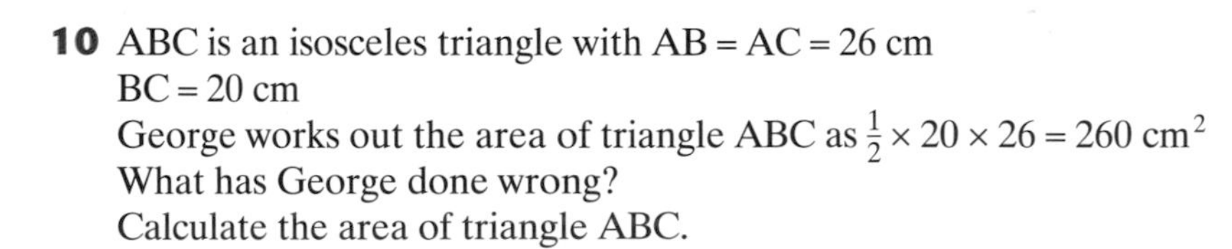

10 ABC is an isosceles triangle with AB = AC = 26 cm
BC = 20 cm
George works out the area of triangle ABC as $\frac{1}{2} \times 20 \times 26 = 260\text{ cm}^2$
What has George done wrong?
Calculate the area of triangle ABC.

11

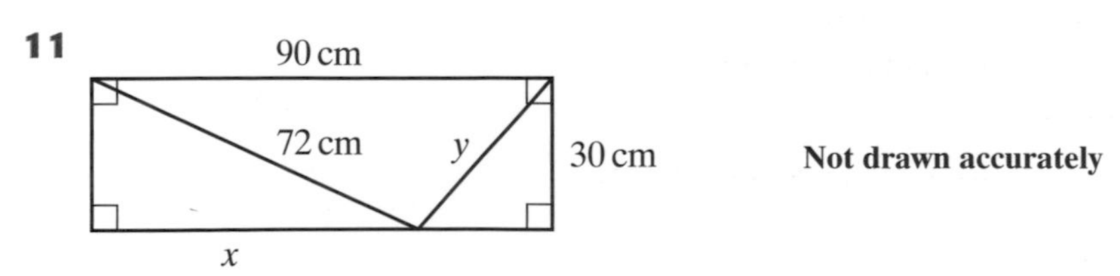

Not drawn accurately

a Calculate x, giving your answer to the nearest millimetre.

b Calculate y, giving your answer to the nearest millimetre.

12 ABCD is an isosceles trapezium, with base DC = 43 cm

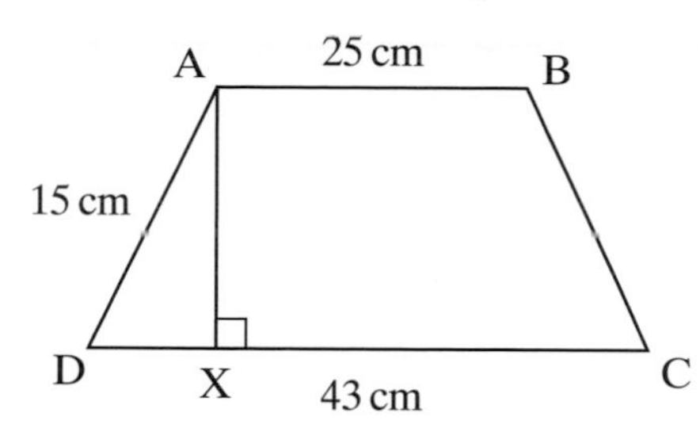

Not drawn accurately

a Calculate AX, the height of the trapezium.

b Calculate the area of ABCD.

13 Get Real!

A transmitter, AB, is supported by two guy ropes, BC and BD.
Angle BAC = 90°
The shorter guy rope, BC, is 10 m.
AC = CD = 4 m

Show that the length of the longer guy rope, BD, is $\sqrt{148}$ m.

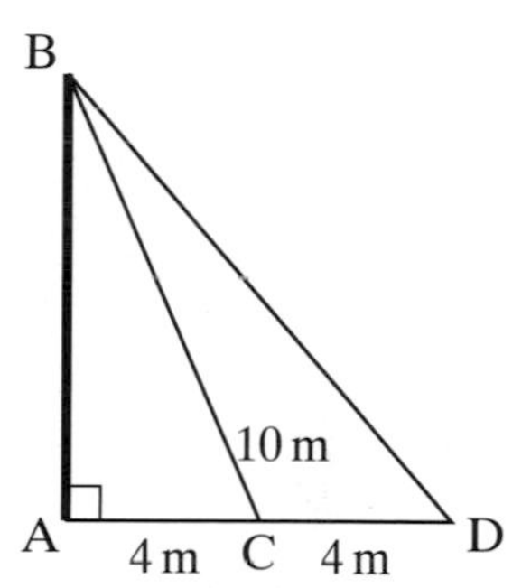

14 Get Real!

Isobel is working out the volume of a chocolate mint that is in the shape of a triangular prism.
She knows that the volume of a prism = area of cross-section × length
She calculates the volume as 3 cm × 3 cm × 0.5 cm = 4.5 cm^3
Is Isobel correct?
Give a reason for your answer.

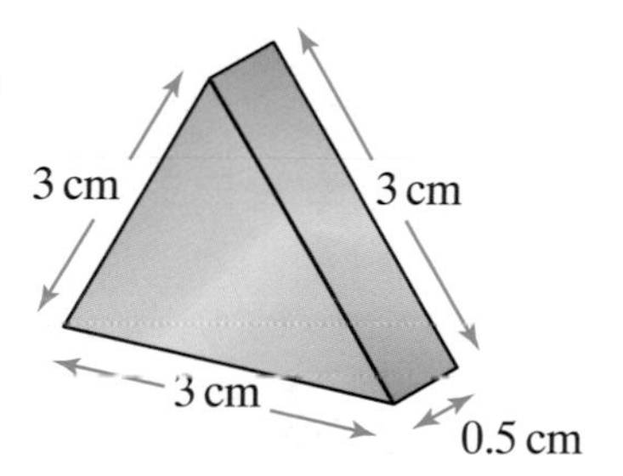

15 Haseeb works out the perimeter of the quadrilateral ABCD.
He uses Pythagoras' theorem in triangle ABC to calculate AC.
Then he uses Pythagoras' theorem in triangle ADC to find DC.
Haseeb says that DC is 2 cm and the perimeter is 24 cm.
Is Haseeb correct?
Give a reason for your answer.

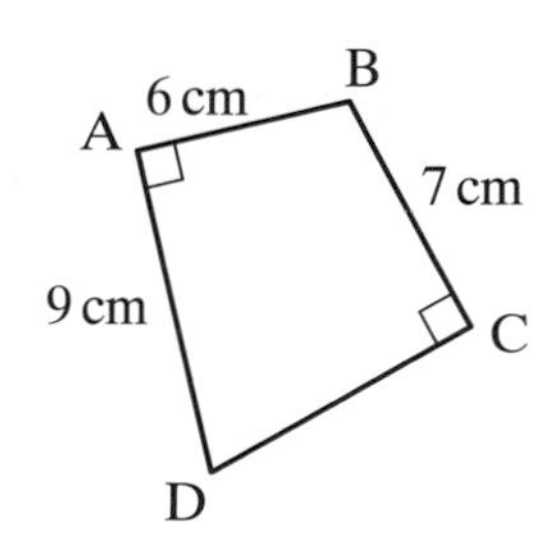

16 Get Real!

Catriona has designed a jade pendant for a necklace.
It is a kite, with the measurements shown in the diagram.
The width of the pendant is 4 cm.

What is the height, x, of the pendant to the nearest millimetre?

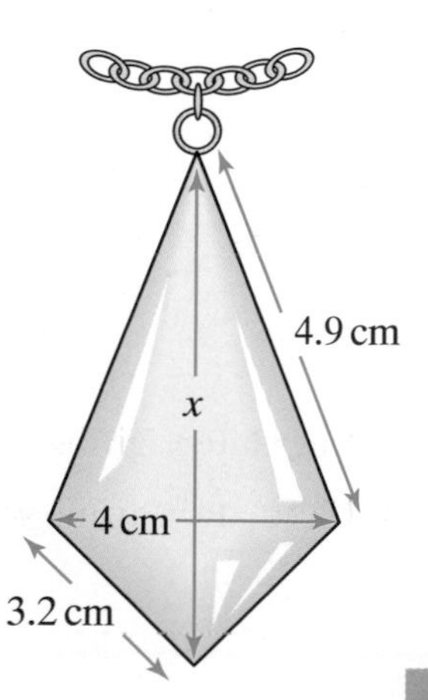

Explore

- A is the point (2, 3) and B is the point (5, 7)

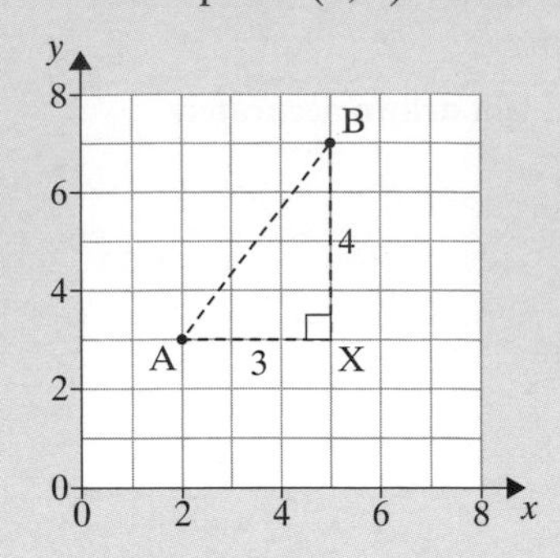

- The distance between the points A and B can be found using Pythagoras' theorem:

  AX = 5 − 2 = 3 units and BX = 7 − 3 = 4 units

  Using Pythagoras' theorem:

$$\begin{aligned} AB^2 &= AX^2 + BX^2 \\ AB^2 &= 3^2 + 4^2 \\ &= 9 + 16 \\ &= 25 \\ AB &= \sqrt{25} = 5 \text{ units} \end{aligned}$$

- Find the distance between the points in these diagrams:

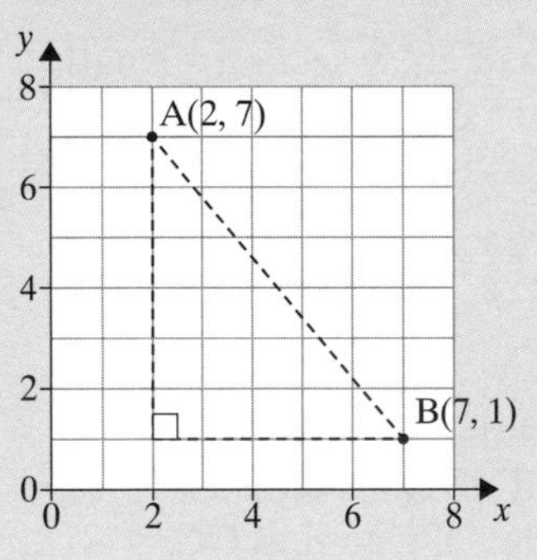

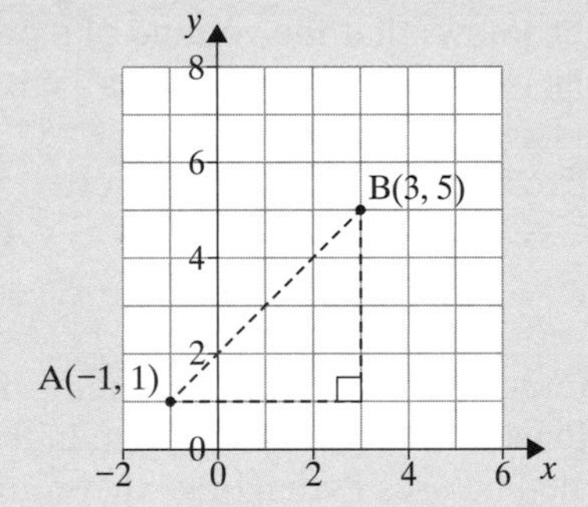

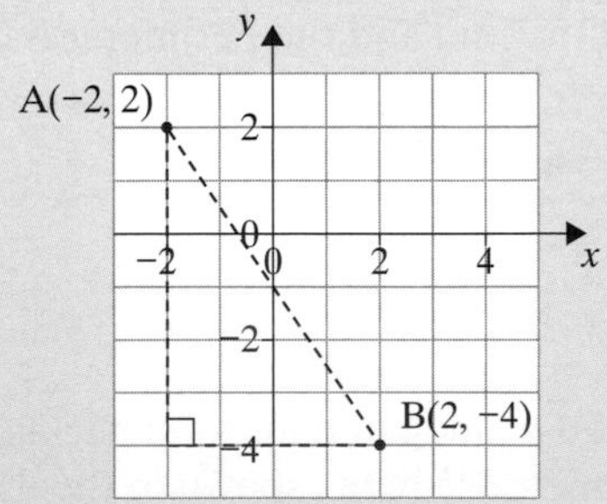

- Sketch a diagram and use Pythagoras' theorem to find the distance between each of the following pairs of points:
  - **i** A(2, 4), B(7, 6)
  - **ii** C(1, 5), D(4, 2)
  - **iii** P(−2, 1), Q(3, 8)
  - **iv** X(−3, 6), Y(1, −2)

- Is there a rule that allows you to work out the distance between $A(x_1, y_1)$ and $B(x_2, y_2)$?

Investigate further

Explore

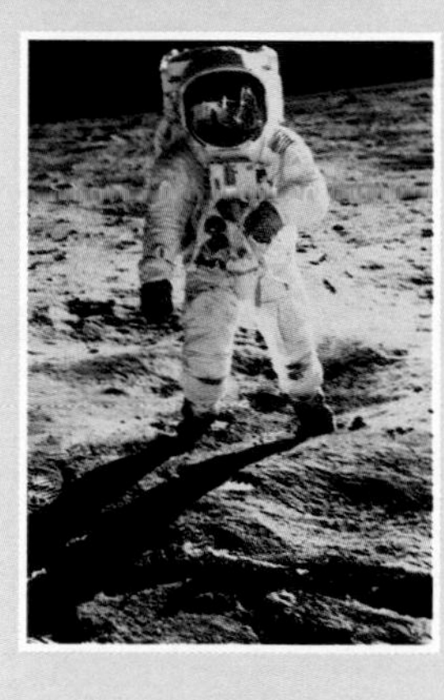

- Construct these triangles accurately using a ruler and a pair of compasses
  - **i** AB = 8 cm, BC = 6 cm, AC = 10 cm
  - **ii** AB = 7.4 cm, BC = 4.6 cm, AC = 8.7 cm
  - **iii** AB = 10.2 cm, BC = 5.1 cm, AC = 8.8 cm
  - **iv** AB = 6.1 cm, AC = BC = 4.3 cm
- What kind of triangles are they? Use a protractor to check
- How could you have shown this would be true without constructing the triangles?
- Using Pythagoras' theorem 'in reverse' is called the **converse** of Pythagoras' theorem
- Work out which of the following lengths would form right-angled triangles
  - **i** 3 cm, 4 cm, 5 cm
  - **ii** 6 cm, 7 cm, 8 cm
  - **iii** 5 cm, 12 cm, 13 cm
  - **iv** 2 cm, $\sqrt{8}$ cm, $\sqrt{12}$ cm
- Take some measurements from objects in the room, for example, the door, the window, a table, a chair, a book, etc.
- Use the lengths of the diagonals and the converse of Pythagoras' theorem to check that they have a vertex that is a right angle

Investigate further

Pythagoras' theorem

ASSESS

The following exercise tests your understanding of this chapter, with the questions appearing in order of increasing difficulty.

 **1** Find the marked lengths in these diagrams.

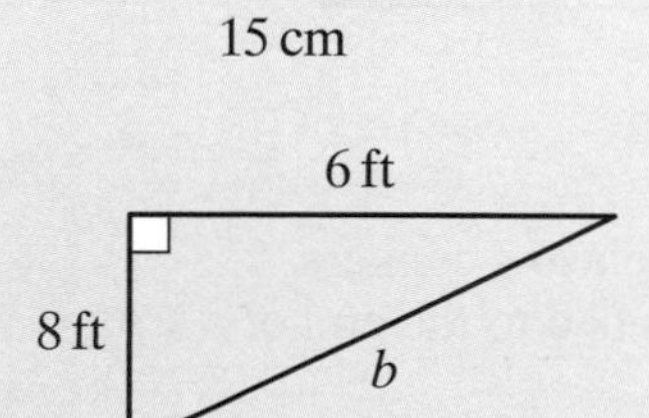

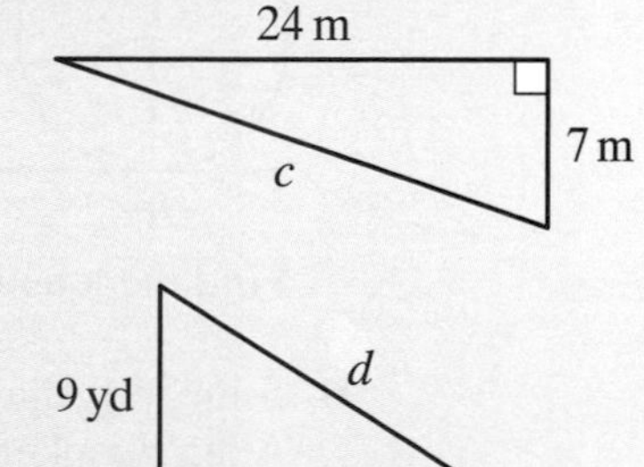

2 Find the marked lengths in these diagrams.
Give your answer to an appropriate degree of accuracy.

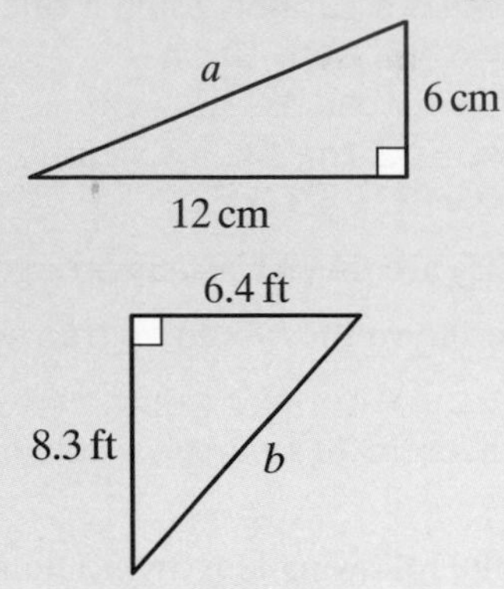

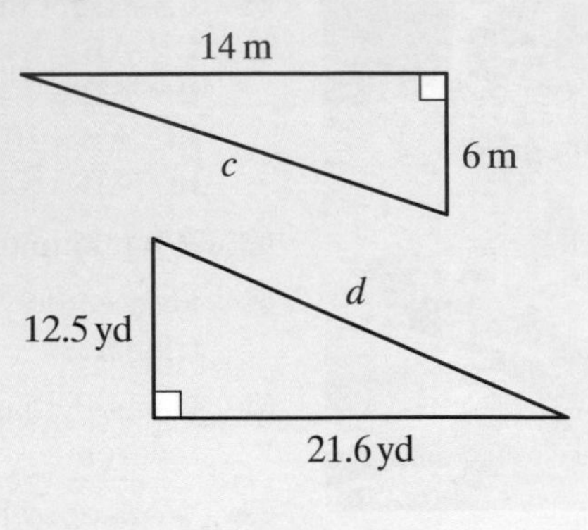

3 Find the marked lengths in these diagrams.
Give your answer to an appropriate degree of accuracy.

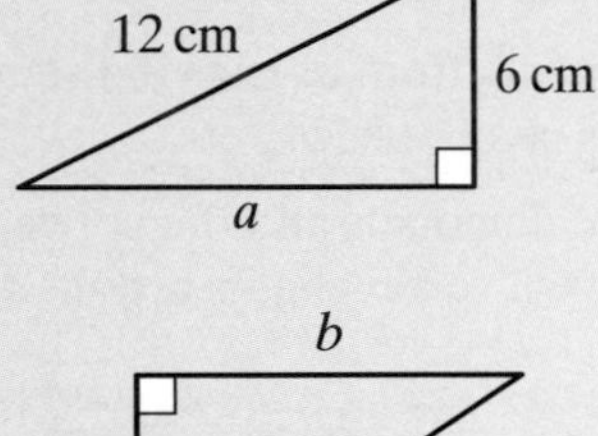

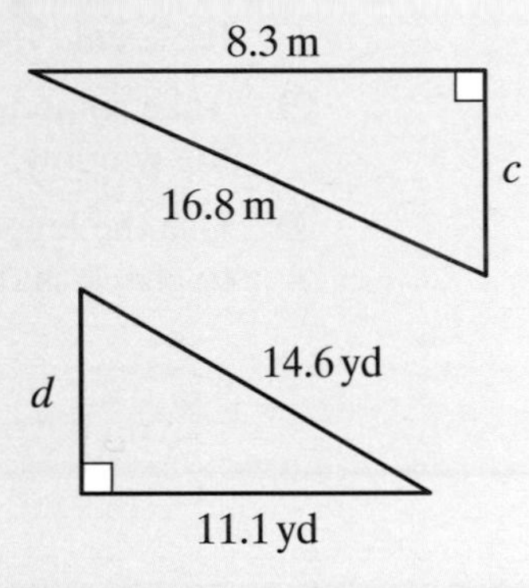

b
23 ft
30 ft

4 Arthur draws a triangle with sides 12, 21 and 24 cm. He says the triangle is right-angled. Is he right? Give a reason for your answer.

5 A tangent, PT, is drawn to a circle of radius 3 m. P is 7.5 m from the centre of the circle.

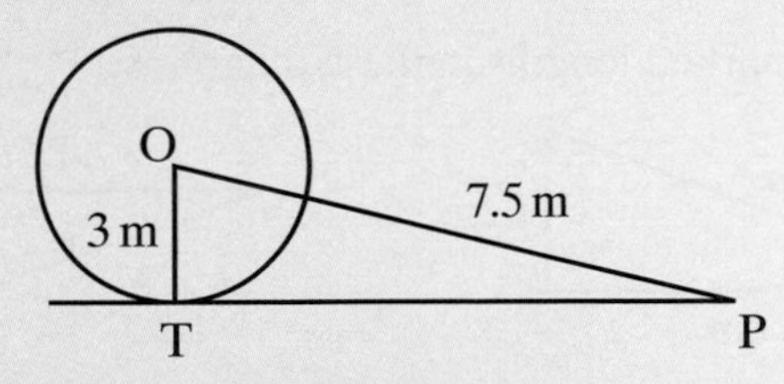

Find the length of TP.

6 John and Allan are playing conkers.
Allan's conker, C, is tied to the end of a string 25 inches long.
He pulls it back from the vertical until it is 11 inches horizontally from its original position.
Calculate the vertical distance, h, that the conker has risen.

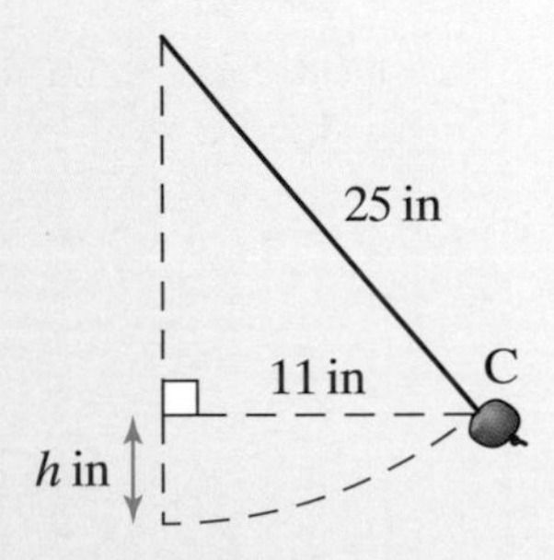

19 Quadratic graphs

OBJECTIVES

D **Examiners would normally expect students who get a D grade to be able to:**

Draw graphs of simple quadratic functions such as $y = 3x^2$ and $y = x^2 + 4$

C **Examiners would normally expect students who get a C grade also to be able to:**

Draw graphs of harder quadratic functions such as $y = x^2 - 2x + 1$

Find the points of intersection of quadratic graphs with lines

Use graphs to find the approximate solutions of quadratic equations

What you should already know ...

- Substitute positive and negative values of x into expressions including squared terms
- Plot graphs from coordinates

VOCABULARY

Variable – a symbol representing a quantity that can take different values such as x, y or z

Quadratic expression – an expression containing terms where the highest power of the variable is 2

Quadratic expressions	Non-quadratic expressions
x^2	x
$x^2 + 2$	$2x$
$3x^2 + 2$	$\frac{1}{x}$
$4 + 4y^2$	$3x^2 + 5x^3$
$(x + 1)(x + 2)$	$x(x + 1)(x + 2)$

Quadratic function – functions like $y = 3x^2$, $y = 9 - x^2$ and $y = 5x^2 + 2x - 4$ are quadratic functions; they include an x^2 term and may also include x terms and constants

The graphs of quadratic functions are always ∪-shaped or ∩-shaped

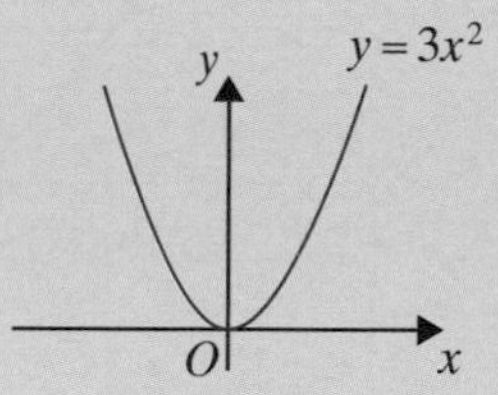

$y = ax^2 + bx + c$ is ∪-shaped when a is positive and ∩-shaped when a is negative

c is the intercept on the y-axis

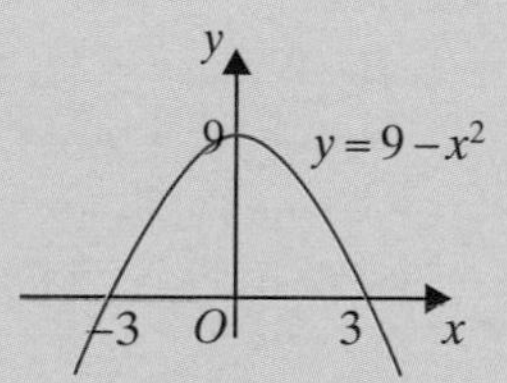

Note that other letters could be used as the variable instead of x (for example, $6t^2 - 3t - 5$ is also a quadratic expression and $h = 30t - 2t^2$ is a quadratic function)

Learn 1 Graphs of simple quadratic functions

Examples:

a Draw the graph of $y = 3x^2$ for values of x from −4 to 4.

b Use the graph to find **i** the value of y when $x = -2.4$
ii the values of x when $y = 35$

a The table below gives the values of y for all integer values of x from −4 to 4.

x	−4	−3	−2	−1	0	1	2	3	4
y	48	27	12	3	0	3	12	27	48

$3x^2$ means $3 \times x^2$ or $3 \times x \times x$ so when $x = -4$, $3x^2 = 3 \times -4 \times -4 = 48$

The y-values are often symmetrical like this

Quadratic functions give curves – you need more points than for linear functions to get the right shape

Check that you can work out the values in the table, with and without your calculator

Plotting these values gives the graph:

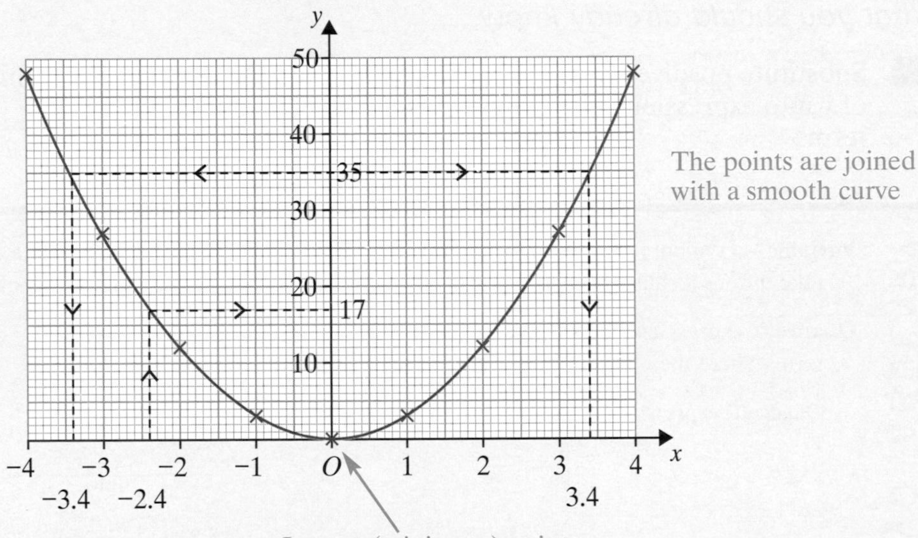

The points are joined with a smooth curve

Lowest (minimum) point

b i From the graph, when $x = -2.4$, $y \approx 17$

Remember $\approx$ means 'is approximately equal to'

ii When $y = 35$, there are two possible values of x:
$x = -3.4$ and $x = 3.4$ (to 1 d.p.)

Use your calculator to work out $3x^2$ with each of these values to check how accurate they are

Apply 1

1 **a** Draw the graph of $y = x^2$ for values of x from –4 to 4.

b Use your graph to estimate the value of:
i 2.5^2 **ii** $(-1.8)^2$

c Use your graph to estimate the square roots of:
i 13 **ii** 5

2 **a** Copy and complete this table for $y = x^2 - 2$

x	–3	–2	–1	0	1	2	3
y	7		–1			2	

b Draw the graph of $y = x^2 - 2$ for values of x from –3 to 3.

c Use your graph to find the value of y when:
i $x = 2.4$ **ii** $x = -1.6$

d Use your graph to find the values of x when:
i $y = 6$ **ii** $y = -0.5$

3 **a** Copy and complete this table for $y = -5x^2$

x	–3	–2	–1	0	1	2	3
y	–45	–20			–5		–45

b Draw the graph of $y = -5x^2$

c Use your graph to find the value of y when:
i $x = 2.4$ **ii** $x = -1.75$

d Use your graph to find the values of x when:
i $y = -40$ **ii** $y = -25$

4 The graph shows the function $y = \frac{x^2}{2}$ for values of x from –4 to 4.

a Use the graph to find the value of y when:
i $x = 3.2$ **ii** $x = -1.6$

b Use the graph to find the values of x when:
i $y = 3.2$ **ii** $y = 6.5$

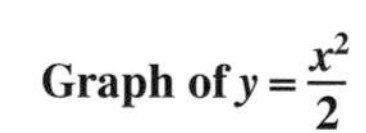

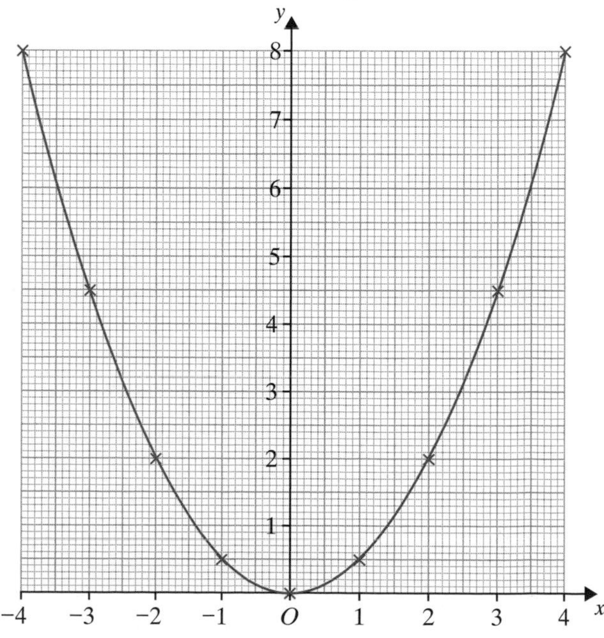

 **5 a** Copy and complete the table below for $y = x^2 + 3$

x	−4	−3	−2	−1	0	1	2	3	4
y	19		7		3			12	

b Draw the graph of $y = x^2 + 3$ for values of x from −4 to 4.

c Give the y-coordinate of the point on the curve with an x-coordinate of:
i 2.5 **ii** −1.5

d Give the x-coordinates of the points on the curve with a y-coordinate of:
i 11 **ii** 16

e Write down the coordinates of the lowest point on the curve.

6 a Copy and complete the table below for $y = 10 - x^2$

x	−4	−3	−2	−1	0	1	2	3	4
x^2	16			1			4	9	
$y = 10 - x^2$	−6			9			6	1	

b Draw the graph of $y = 10 - x^2$ for values of x from −4 to 4.

c Write down the coordinates of the points where the curve crosses the x-axis.

d Write down the coordinates of the highest point on the curve.

7 a Copy and complete the table.

x	−4	−3	−2	−1	0	1	2	3	4
$3x^2$	48	27		3			12		
$y = 3x^2 - 5$	43	22		−2			7		

b Draw the graph of $y = 3x^2 - 5$ for values of x from −4 to 4.

c Use your graph to find the value of y when:
i $x = 1.8$ **ii** $x = -3.4$

d Use your graph to find the values of x when:
i $y = 20$ **ii** $y = 36$

8 a Copy and complete the table below, then use it to draw the graph of $y = (x + 2)(3 - x)$

x	−3	−2	−1	0	0.5	1	2	3	4
$x + 2$	−1		1		2.5				6
$3 - x$	6		4		2.5				−1
$y = (x + 2)(3 - x)$	−6		4		6.25				−6

b Write down the coordinates of the points where the curve crosses the x-axis.

 **9 a** Draw a table and a graph for $y = x(x - 4)$ for values of x from −1 to 5.

b Write down the coordinates of the points where the curve crosses the x-axis.

10 The graph shows the points that Paul has plotted for his graph of $y = x^2$

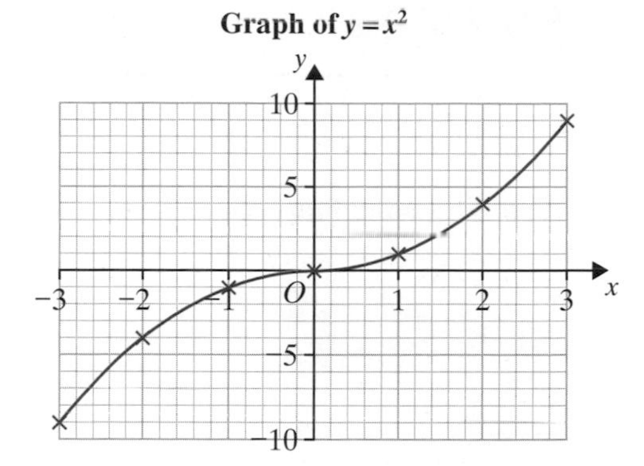

a Is this the shape you would expect?
Give a reason for your answer.

b Complete the table of values for Paul's graph.

x	−3	−2	−1	0	1	2	3
y							

c What mistake has Paul made in calculating the values?

11 The graphs of three quadratic functions are shown in the sketch.

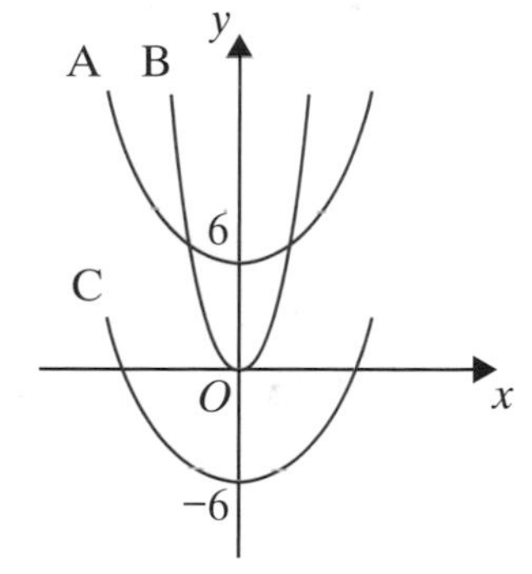

The functions are

$y = 6x^2$ $\quad$ $y = x^2 + 6$ $\quad$ $y = x^2 - 6$

Choose the function that represents each curve.

12 **Get Real!**

The area of the glass in a circular window is given by $A = \pi r^2$, where r is the radius in metres and A is the area in square metres.

a Copy and complete the following table, giving values of A to 2 decimal places.

r (m)	0	0.1	0.2	0.3	0.4	0.5	0.6	0.7	0.8	0.9	1.0
$A = \pi r^2$ (m²)			0.13			0.79			2.01		3.14

b Draw a graph of A against r using 2 cm to represent 0.2 on the r-axis and 0.5 on the A-axis.

c Use your graph to estimate the area of the window when the radius is:
i 0.28 m $\quad$ **ii** 0.75 m

d Use your graph to estimate the radius of the window when the area is:
i 2.8 m² $\quad$ **ii** 1.45 m²

Explore

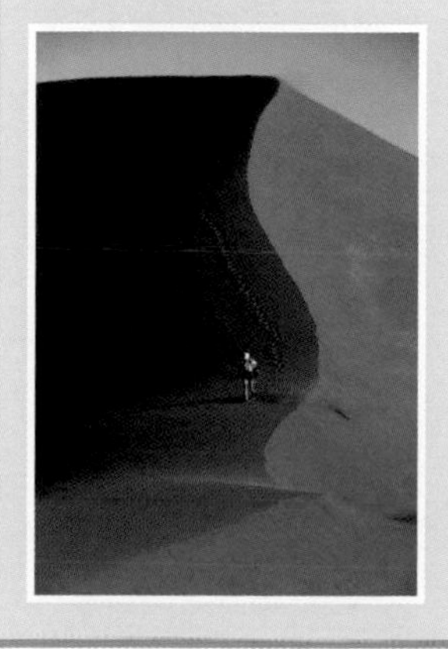

- On the same axes draw the graphs of the functions $y = x^2$ and $y = 2x^2$
  Describe the similarities and differences between the graphs
- On the same axes draw the graphs of the functions $y = 3x^2$ and $y = -3x^2$
  Describe the similarities and differences between the graphs

Investigate further

Explore

On the same axes draw the graphs of the functions $y = x^2 + 1$ and $y = x^2 - 1$
Describe the similarities and differences between the graphs

Investigate further

Learn 2 Graphs of harder quadratic functions

Examples:

a Draw the graph of $y = x^2 - 2x - 1$ for values of x from −2 to 4.

b Use the graph to find the solutions of $x^2 - 2x - 1 = 0$

c i Find the x-coordinates of the points where the curve crosses the line $y = 6$

ii Write down a quadratic equation whose solutions are the answers to part **i**.

a The table below gives values for this function.

x	−2	−1	0	1	2	3	4
y	7	2	−1	−2	−1	2	7

The y-values are often, but not always, symmetrical like this

When $x = -2$, $y = (-2)^2 - 2 \times (-2) - 1$
$= 4 + 4 - 1$
$= 7$

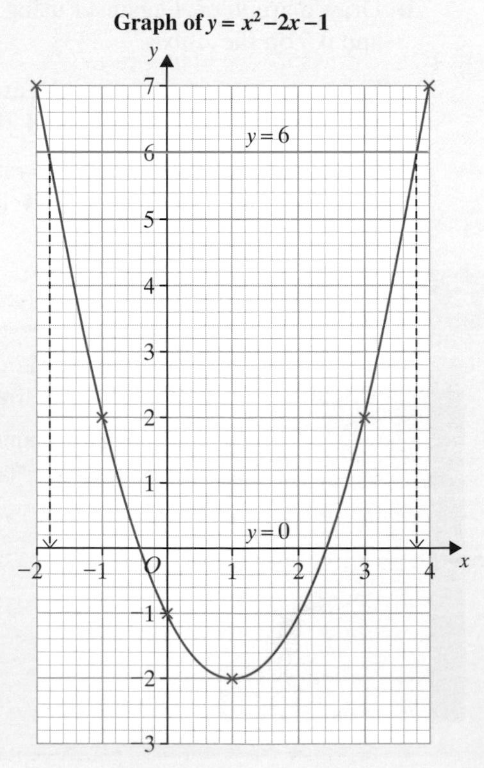

b To find the solutions of $x^2 - 2x - 1 = 0$, look at the points on $y = x^2 - 2x - 1$ where $y = 0$ (that is, where the curve crosses the x-axis).
The solutions are $x = -0.4$ and $x = 2.4$ (to 1 d.p.)

c i The graph shows the line $y = 6$.
It crosses the curve at the points where $x = -1.8$ and $x = 3.8$ (to 1 d.p.)

ii These are the solutions of the equation $x^2 - 2x - 1 = 6$

Apply 2

1 **a** Copy and complete this table for $y = x^2 - 4x$

x	−1	0	1	2	3	4	5
y	5			−4	−3		

b Draw the graph of $y = x^2 - 4x$ for values of x from −1 to 5.

c Use your graph to find the solutions of the equation $x^2 - 4x = 0$

d **i** Draw the line $y = 2$ on your graph.

ii Find the x-coordinates of the points where the line $y = 2$ crosses the curve $y = x^2 - 4x$

iii Write down the quadratic equation whose solutions are the answers to part **ii**.

2 **a** Copy and complete this table for $y = x^2 + 2x + 1$

x	−4	−3	−2	−1	0	1	2
y	9		1	0		4	

b Draw the graph of $y = x^2 + 2x + 1$ for values of x from −4 to 2.

c **i** Write down the x-coordinate of the point where the curve meets the x-axis.

ii Write down the quadratic equation whose solution is the answer to part **i**.

d **i** Draw the line $y = 8$ on your graph.

ii Write down the coordinates of the points where the line $y = 8$ crosses the graph of $y = x^2 + 2x + 1$

3 **a** Copy and complete this table for $y = x^2 - 3x$

x	−1	0	1	2	3	4
y				−2		4

b Also calculate the value of y when $x = 1.5$

c Use your answers to parts **a** and **b** to draw the graph of $y = x^2 - 3x$

d Use your graph to solve the equation $x^2 - 3x = 0$

e **i** Draw the line $y = 3$ on your graph.

ii Find the x-coordinates of the points where the line $y = 3$ crosses the curve $y = x^2 - 3x$

iii Write down the quadratic equation whose solutions are the answers to part **ii**.

4 **a** Copy and complete this table for $y = 5 + x - x^2$

x	−3	−2	−1	0	0.5	1	2	3	4
y	−7				5.25		3		−7

b Draw the graph of $y = 5 + x - x^2$ for values of x from −3 to 4.

c Find the coordinates of the points where the graph crosses:

i the line $y = 0$ **ii** the line $y = 2$ **iii** the line $y = -5$

 **5** **a** Copy and complete this table for $y = 2x^2 - x - 3$

x	−3	−2	−1	0	1	2	3	4
y	18		0		−2		12	25

b Calculate the value of y when $x = 0.25$

c Draw the graph of $y = 2x^2 - x - 3$ for $-3 \leqslant x \leqslant 4$

d **i** Write down the x-coordinates where the curve crosses the x-axis.

ii Write down the quadratic equation whose solutions are the answers to part **i**.

e **i** Write down the x-coordinates where the curve meets the line $y = 10$

ii Write down the quadratic equation whose solutions are the answers to part **i**.

 **6** **Get Real!**

A ball is thrown vertically upwards into the air.
After t seconds its height above the ground, h metres, is given by the function $h = 1 + 14t - 5t^2$

h metres

a Copy and complete the table below giving values of $h = 1 + 14t - 5t^2$

t	0	0.5	1	1.5	2	2.5	3
h		6.75		10.75		4.75	

b Draw the graph of $h = 1 + 14t - 5t^2$ for $0 \leqslant t \leqslant 3$

c Use your graph to find:

i the height of the ball after 0.4 seconds

ii how long the ball takes to reach a height of 9 metres on its way up

iii after how long the ball hits the ground.

 **7** A teacher asks his class to complete a table for $y = 2x^2 - x + 4$

a This is Ella's table but only one of her values for y is correct.

x	−3	−2	−1	0	1	2	3
$2x^2$	36	16	4	0	4	16	36
$-x$	+3	+2	+1	0	−1	−2	−3
$+4$	+4	+4	+4	+4	+4	+4	+4
$y = 2x^2 - x + 4$	43	22	9	4	7	18	37

i Which of Ella's y-values is correct?

ii Explain what Ella has done wrong.

b This is Pete's table.

x	−3	−2	−1	0	1	2	3
$2x^2$	18	8	2	0	2	8	18
$-x$	−3	−2	−1	0	−1	−2	−3
$+4$	+4	+4	+4	+4	+4	+4	+4
$y = 2x^2 - x + 4$	19	10	5	4	5	10	19

i Which of his y-values are correct?

ii Explain any mistakes he has made.

8 **a** Draw a table and a graph for $y = 5 - 2x - 4x^2$ for $-3 \leqslant x \leqslant +3$

b Use your graph to solve these equations:

i $5 - 2x - 4x^2 = 0$ **ii** $5 - 2x - 4x^2 = 4$

c Luke says that the equation $5 - 2x - 4x^2 = 9$ cannot be solved.

i Is he correct?

ii Explain your answer.

9 **a** Draw the graph of $y = 5x^2 + 2x - 4$ for $-3 \leqslant x \leqslant +3$

b **i** Write down the solutions of the equation $5x^2 + 2x - 4 = 0$

ii Explain how you found the solutions and why your method works.

Explore

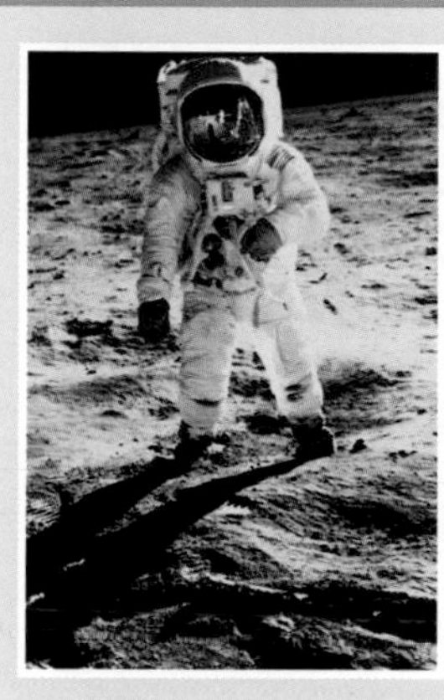

- Draw the graph of the function $y = x^2 + 2x$
- Where does the curve meet the x-axis and y-axis?
- What are the coordinates of the lowest point on the curve?
- Repeat the above for the function $y = x^2 + 4x$

Investigate further

Quadratic graphs

ASSESS

The following exercise tests your understanding of this chapter, with the questions appearing in order of increasing difficulty.

1 **a** Taking values of x from −4 to 4, draw tables of values for the functions $y = x^2$, $y = x^2 + 4$ and $y = x^2 - 3$.

b On the same grid, and using the same axes and scales, draw the graphs of these functions.

c Describe the similarities and differences between the graphs.

2 **a** Taking values of x from −4 to 4, draw tables of values for the functions $y = x^2$, $y = \frac{1}{2}x^2$ and $y = 3x^2$.

b On the same grid, and using the same axes and scales, draw the graphs of these functions.

c Describe the similarities and differences between the graphs.

3 **a** Copy and complete the table of values for the function $y = 2x^2 - 7$

x	−3	−2	−1	0	1	2	3
y	11		−5	−7		1	

b Draw the graph.

c Use your graph to find:

i the coordinates of the lowest point on the curve

ii the value of y when $x = 1.4$

iii the values of x when $y = 6$

iv the solutions of the equation $2x^2 - 7 = 0$

4 The average safe braking distance for vehicles, d yards, is given by the equation $d = \frac{v^2}{50} + \frac{v}{3}$, where v is the speed of the vehicle in mph.

a Copy and complete the table of values for the function $d = \frac{v^2}{50} + \frac{v}{3}$

v (mph)	0	10	20	30	40	50	60	70	80
d (yards)	0	5	15		45	67		121	

b Draw the graph of $d = \frac{v^2}{50} + \frac{v}{3}$, using v as the horizontal axis.

c Use your graph to find the safe braking distance when the vehicle is travelling at:

i 15 mph **ii** 45 mph **iii** 75 mph.

d A driver suddenly sees an obstruction 50 yards ahead. She just stops in time. How fast was she travelling when she first saw it?

Try a real past exam question to test your knowledge:

5 **a** Complete the table of values for $y = 2x^2 - 4x - 1$

x	−2	−1	0	1	2	3
y	15		−1		−1	5

b Draw the graph of $y = 2x^2 - 4x - 1$ for values of x from −2 to 3.

c An approximate solution of the equation $2x^2 - 4x - 1 = 0$ is $x = 2.2$

i Explain how you can find this from the graph.

ii Use your graph to write down another solution of this equation.

Spec A, Higher Paper 1, June 04

20 Loci

OBJECTIVES

D **Examiners would normally expect students who get a D grade to be able to:**

Understand the idea of a locus

C **Examiners would normally expect students who get a C grade also to be able to:**

Construct accurately loci, such as those of points equidistant from two fixed points

Solve loci problems, such as identifying points less than 3 cm from a point P

What you should already know ...

- Measure a line accurately (within 2 millimetres)
- Measure and draw an angle accurately (within 2 degrees)
- Construct the perpendicular bisector of a line
- Construct the bisector of an angle
- Construct and interpret a scale drawing

VOCABULARY

Locus – the path followed by a moving point; for example, the locus of a point, A, keeping at a fixed distance of 5 cm from a point B, is a circle, centre B, radius 5 cm, or, in 3 dimensions, a sphere, centre B, radius 5 cm

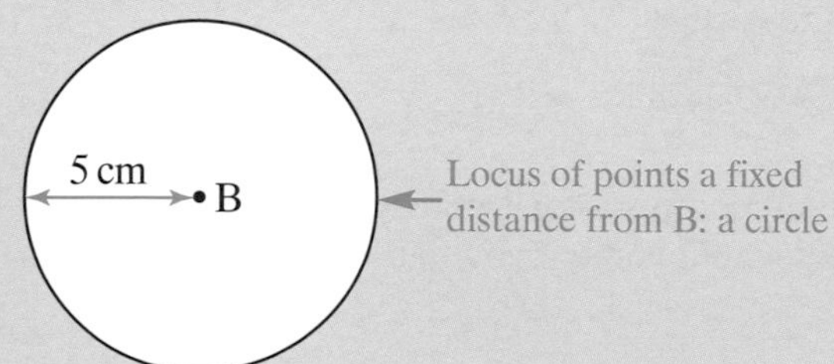

Loci – the plural of locus

Perpendicular – at right angles to; two lines at right angles to each other are perpendicular lines

Bisect – to divide into two equal parts

Equidistant – the same distance; if A is equidistant from B and C, then AB and AC are the same length

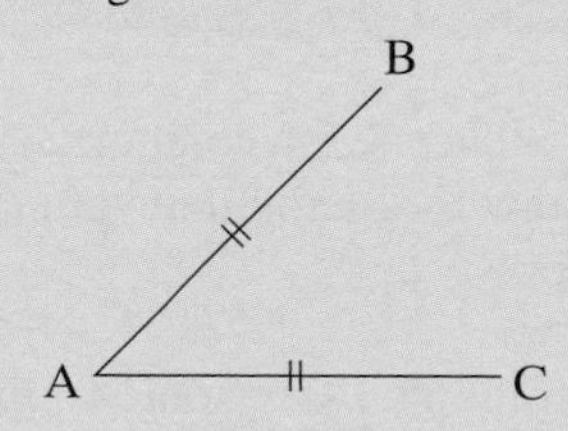

Learn 1 Describing a locus

Example: Find the locus of a child on a swing.

The child moves in an arc, which has a radius equal to the length of the swing and whose centre is the top of the swing.

Apply 1

1 Explore these loci practically.

a Find the path followed by the corner of a square as the square is rotated about an opposite corner.
Draw the locus.

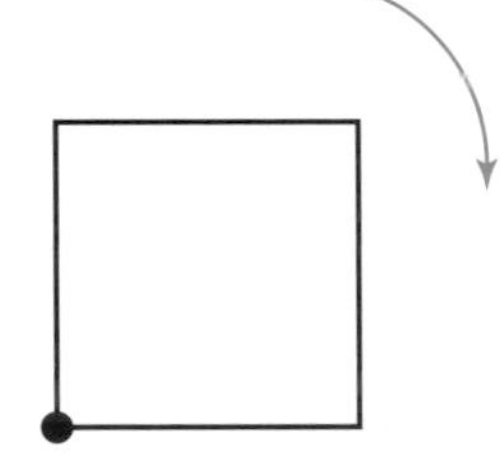

b Find the path of a corner of a square of card as it is rotated against a ruler.
Draw the locus.

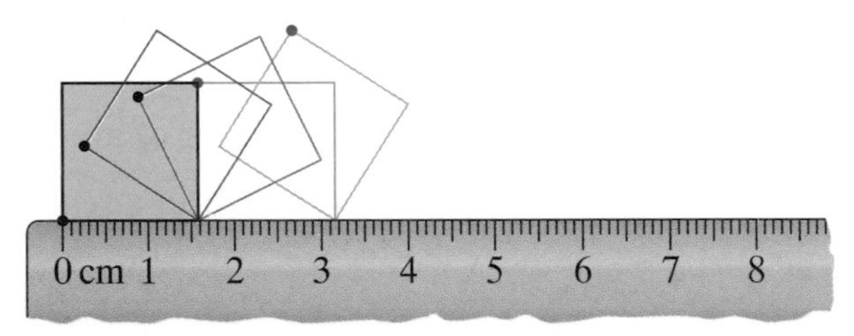

c Find the locus of a mark on the edge of a coin as it rolls along a ruler.

2 Mark a point, X. Place some counters so that their centres are all exactly 5 cm from X. Draw the shape that your counters would make if you had an unlimited supply.

3 Draw a line 10 cm long. Place some counters so that their centres are all exactly 5 cm from the line.
Draw the shape that your counters would make if you had an unlimited supply.

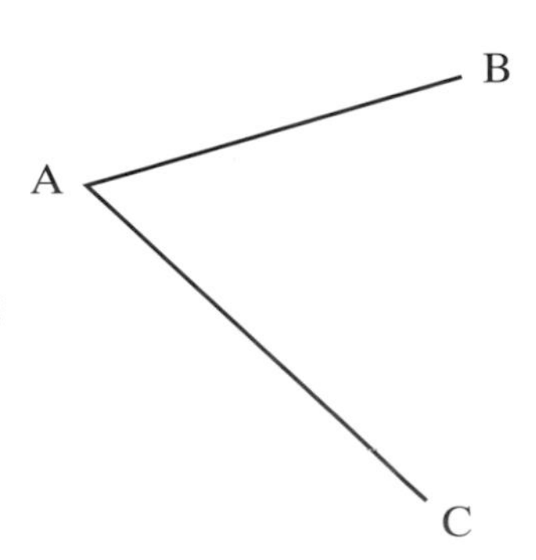

4 Draw two lines, AB and AC, that meet at A. Place some counters so that their centres are all the same distance from AB as they are from AC.
Draw the shape that your counters would make if you had an unlimited supply.

5 Draw a line AB and a point C. Place some counters so that their centres are all the same distance from AB as they are from C.
Draw the shape that your counters would make if you had an unlimited supply.

A

C ×

B

6 A square has sides of length 6 cm.
Sam says that if an ant moves round the outside of the square, always staying 3 cm from the square, it walks in a square with sides of length 12 cm.
Is Sam correct?
Give a reason for your answer.

7 **Get Real!**
You are going to find the path of a man, halfway up a ladder, as the foot of the ladder slips until the ladder is resting on the ground.

a Cut a strip of card, 10 cm long, to represent the ladder.

b Make a small hole exactly in the middle to represent where the man is.

c Draw two lines at right angles, each at least 10 cm long, to represent the ground and the wall.

d Put the ladder almost vertical, leaning against the wall, and mark the position of the man through the hole in the card.

e Move the ladder as if it has slipped slightly, and again mark where the man is.

f Repeat until the ladder is horizontal.

g Join up the marks you have made to complete the locus.

h Draw the locus.

i What would happen if the man was three quarters of the way up the ladder?
Make a new hole in your card 'ladder' and repeat the experiment.
Draw the locus.

Explore

- You need two discs or coins of the same size
- Hold one still, and roll the other one all the way around it
- How many times does the disc rotate as it goes round the stationary disc once?
- What if one disc has a diameter twice the size of the other?

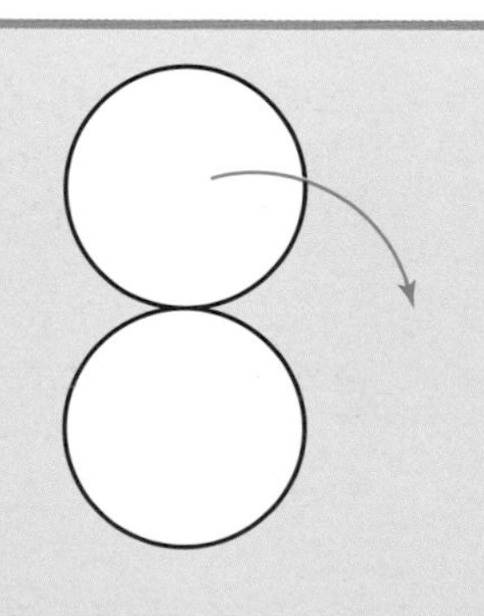

Investigate further

Learn 2 Constructing loci

Examples:

Draw accurately:

a the locus of points 1 cm from a point A

b the locus of points 1 cm from a line AB

c the locus of points the same distance from two points A and B

d the locus of points the same distance from two lines, AB and AC.

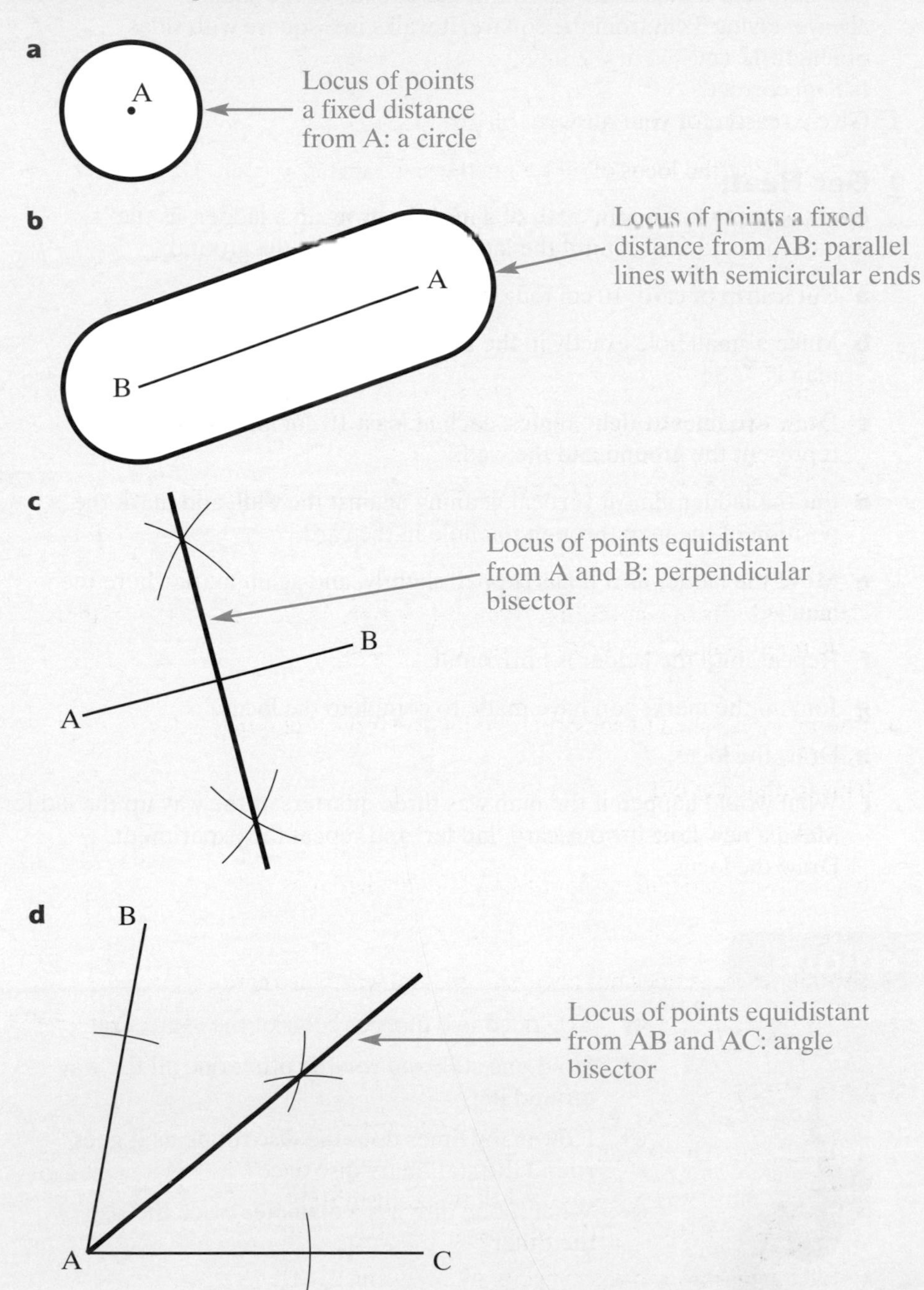

Apply 2

1 Get Real!

A goat lives in a field 12 m long and 8 m wide. In the centre of the field is a post. The goat is tethered to the post with a rope that is 4 m long. Using a scale of 1 cm to represent 1 m:

a make a scale drawing of the field

b mark the position of the post

c shade in the area of the field that the goat can reach.

2 a Construct a square ABCD with all sides 8 cm long.

b Construct the locus of all points the same distance from AB and AD.

c Construct the locus of all points 3 cm from A.

d Shade in all the points that are closer to AB than AD and less than 3 cm from A.

3 Get Real!

Tommy builds a toy train track. He draws a rectangle measuring 8 m by 6 m. He builds the track outside the rectangle, so that the track is always exactly 1 m from the rectangle.
Using a scale of 1 cm to represent 1 m, make a drawing of the rectangle and the track.

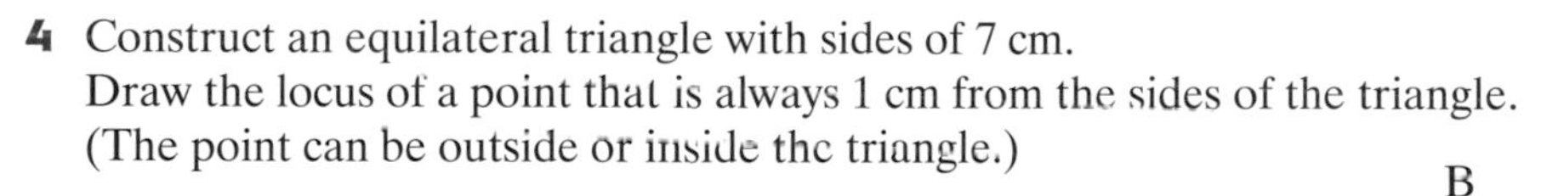

4 Construct an equilateral triangle with sides of 7 cm.
Draw the locus of a point that is always 1 cm from the sides of the triangle.
(The point can be outside or inside the triangle.)

5 Philippa was asked to construct the locus of points closer to AB than AC.
This is what she did.

a She measured the distance BC.

b She marked the midpoint of BC, and labelled it M.

c She drew in the line AM as her answer.

Explain why this does not always work, and write instructions telling Philippa the correct method.

B
M
C
A

6 Get Real!

A large island has two radio transmitters, M and N, marked × on the diagram.

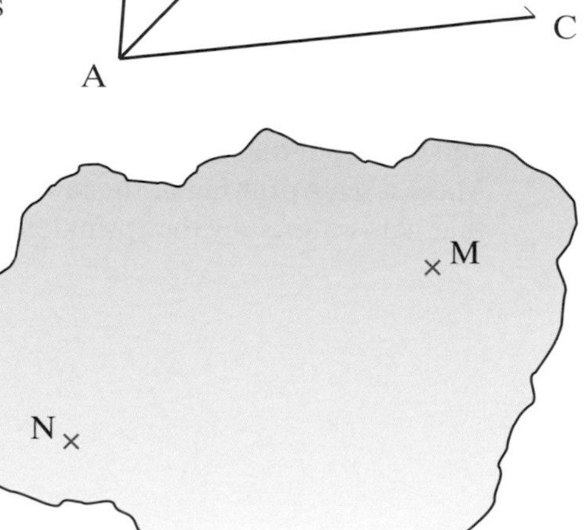

a Draw your own island, and mark two transmitters as in the diagram, so that M and N are 6 cm apart.

b Each transmitter has a range of 5 km. Using a scale of 1 cm to represent 1 km, mark all the points that are in range of transmitter M.

c Shade in the area that is in the range of both transmitters.

7 **a** Construct an isosceles triangle ABC, with AB = AC = 10 cm and BC = 7 cm

b Construct the bisector of angle ABC.

c Construct the perpendicular bisector of AB.

d Shade in all the points inside the triangle that are closer to AB than BC, and closer to B than A.

8 Toby was asked to draw the locus of points 2 cm from the rectangle ABCD.
Toby drew this:

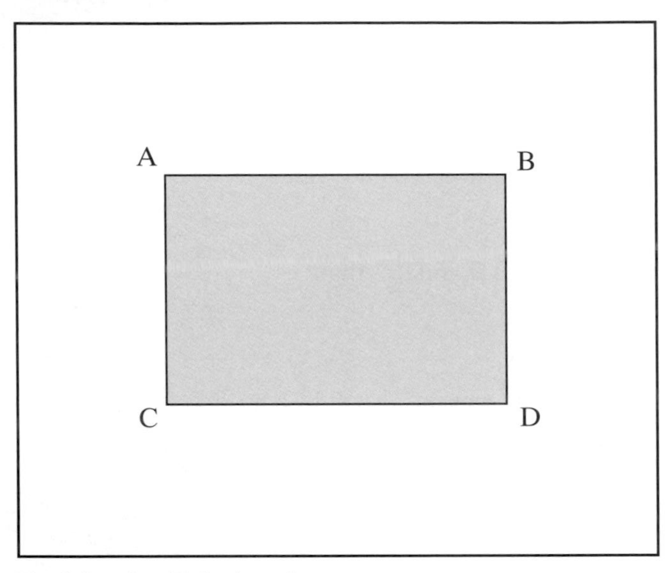

Explain what Toby has done wrong.

9 Mark two points A and B, which are 8 cm apart.

a Construct the locus of all points that are exactly 6 cm from B.

b Construct the locus of all points equidistant from A and B.

c Shade in all the points that are less than 6 cm from B, and closer to A than B.

10 **Get Real!**

A rectangular lawn is 12 m long and 10 m wide.
A gardener has three sprinklers, which spray water in a circle with a radius of 3 m.
She puts a sprinkler at A, which is 3 m from the top edge and 3 m from the left edge; she puts another at B, which is 3 m from the top edge and 3 m from the right edge; and the third at C, which is 3 m from the bottom edge and 6 m from the left edge.
Make a scale drawing of the lawn, and shade in the part that gets watered by the sprinklers.

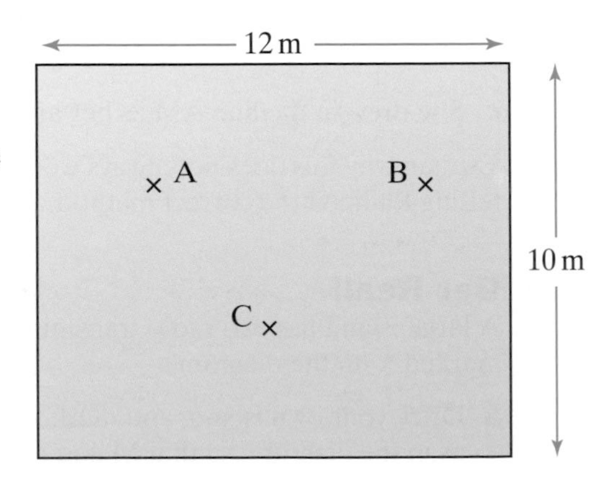

11 Get Real!

A field measures 24 m long and 16 m wide. In one corner, there is a shed, 8 m long and 6 m wide.

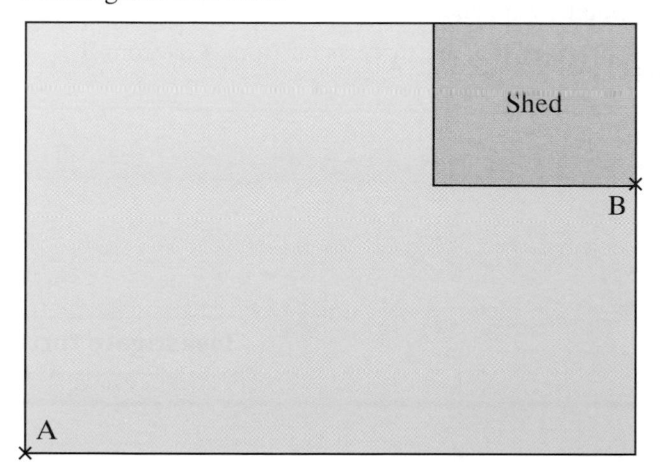

a Using a scale of 1 cm to 2 m, make a scale drawing of the field and the shed.

b A goat is tethered to corner A with a rope, 14 m long.
Using the same scale, shade the area that the goat can reach.

c After a week, the goat is tethered to point B, where the shed meets the edge of the field.
Shade the area the goat can reach now.

12 Get Real!

Three men discover an island. Each wants to claim it as his own. Each man plants a flag in part of the island.

a Draw your own sketch of the island.

b Label three points, A, B and C.

The men agree to divide the island between them, so that they keep the part that is closer to them than to the others.

c Construct the locus of points the same distance from A as from B.

d Construct the locus of points equidistant from A and C.

e Construct the locus of points equidistant from B and C.

f Shade in all points closer to A than B or C.

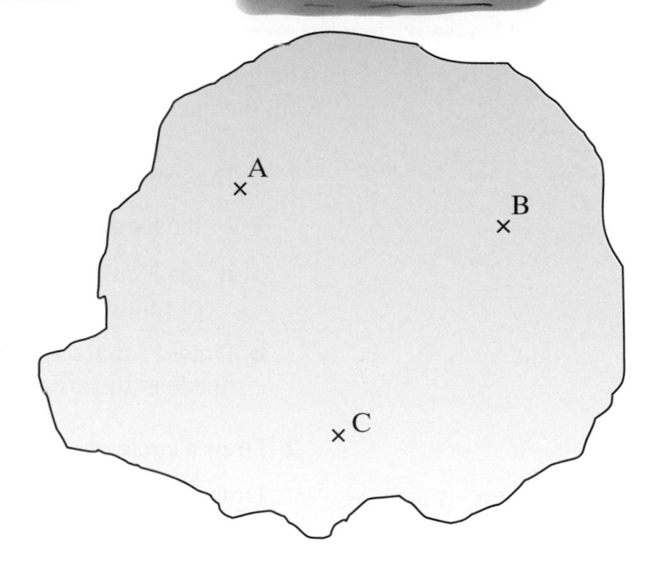

Explore

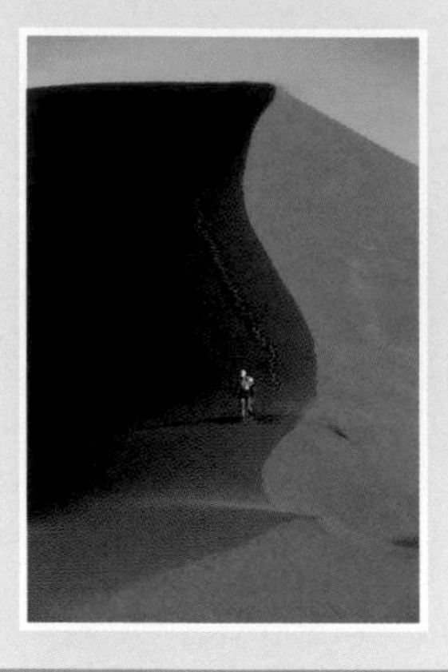

- A and B are two points, 9 cm apart
- Find the locus of all points that are twice as far from A as from B

Investigate further

Explore

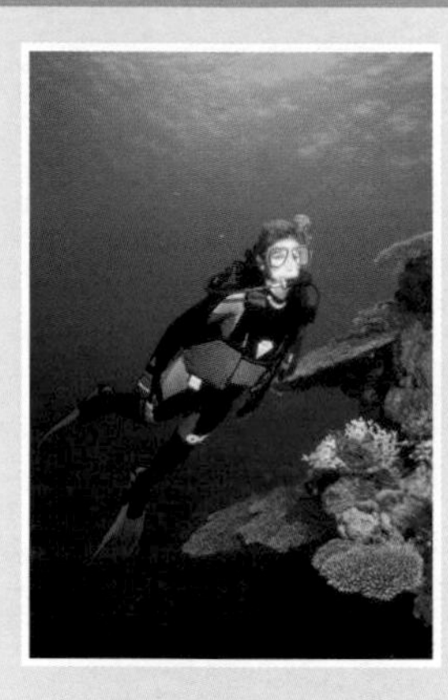

- A is a point 8 cm from a line BC
- Find the locus of all points that are twice as far from BC as they are from A

Investigate further

Loci

ASSESS

The following exercise tests your understanding of this chapter, with the questions appearing in order of increasing difficulty.

1 **a** Describe the following loci:

i the locus of all points that are 4 cm from a fixed point, P

ii the locus of all points that are 3 cm from the circumference of a circle of radius 4 cm

b Draw a square ABCD.
Shade in the area where points are closer to A than they are to B, C or D.

2 Draw a circle of radius 5 cm.

Draw a number of chords of this circle, each 8 cm long, and mark their midpoints.

Describe the locus of the midpoints of all such chords.
Give a reason for your answer.

3 Draw any large triangle.
Construct the perpendicular bisector of each side.
The three bisectors should all meet at one point, label it C.
This point is called the 'circumcentre' of the triangle.
With compasses, using C as the centre, draw a circle that passes through all the vertices of the triangle.
Explain why this happens.

4 Draw any large triangle.
Construct the bisector of each angle.
The three bisectors should all meet at one point; label it I.
This point is called the 'incentre' of the triangle.
With compasses, using I as the centre, draw a circle that fits exactly inside the triangle – i.e. each side of the triangle is a tangent to this circle.
Explain why this happens.

5 Long John Hypotenuse has buried his treasure on Irregular Island.
His enemy, Isosceles Scarface, is looking for it.

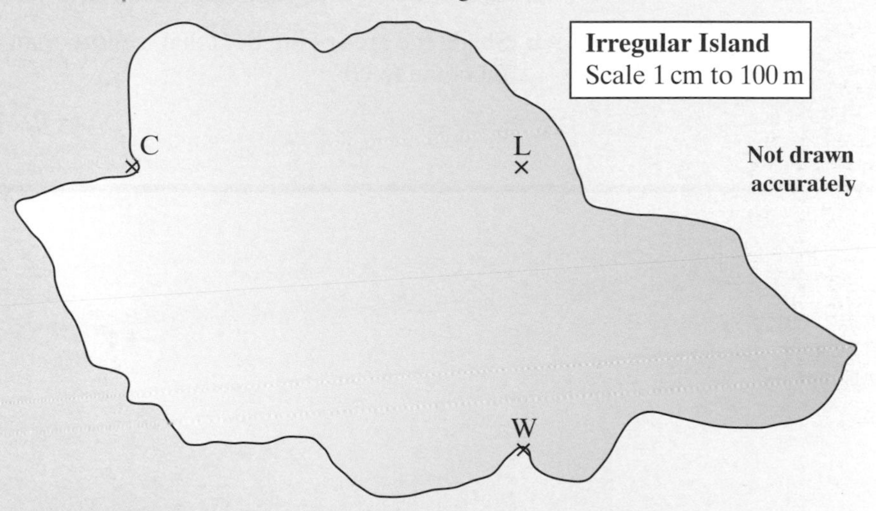

The maps shows a look-out point, L, which is 650 metres due east of a cave, C.
An old wreck, W, is 500 metres due south of the look-out point.
Scarface knows that the treasure is less than 300 m from the wreck and within 250 m of the ridge that runs across the island from the cave to the look-out point.

a Using the same scale as the map, make an accurate drawing of the positions of C, L and W.

b Use loci to shade in the area where Scarface should dig.

Try a real past exam question to test your knowledge:

6 The quadrilateral DEFG is a scale drawing of a field.
The line GH bisects angle DGF.

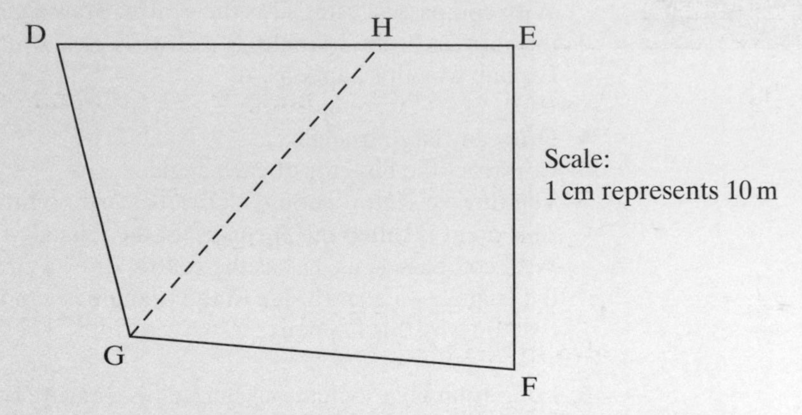

a Construct the locus of points in the field that are 40 m from E.

b Shade the area of the field that is more than 40 m from E **and** nearer to DG than to GF.

Spec B, Module 5, Int Paper 2, Nov 03

21 3-D solids

OBJECTIVES

G **Examiners would normally expect students who get a G grade to be able to:**

Recognise and name three-dimensional (3-D) solids

Sketch three-dimensional (3-D) solids

E **Examiners would normally expect students who get an E grade also to be able to:**

Draw a cuboid on an isometric grid and mark its dimensions

D **Examiners would normally expect students who get a D grade also to be able to:**

Draw plans and elevations of three-dimensional (3-D) solids

What you should already know ...

- Find the surface area of simple shapes
- Calculate the volume and surface area of cuboids, prisms and shapes made from cubes and cuboids
- Draw the nets of 3-D solids

VOCABULARY

Vertex (pl. vertices) – the point where two or more edges meet

Face – one of the flat surfaces of a solid

Edge – a line segment that joins two vertices of a solid

Triangular prism – a 3-D shape with 6 vertices, 9 edges, 2 triangular faces and 3 rectangular faces

Tetrahedron – a solid with four triangular faces

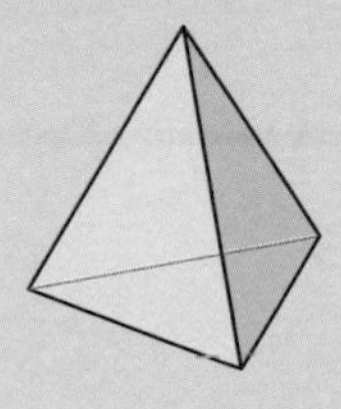

Pyramid – a solid with a polygon as the base and one other vertex; all the vertices of the base are joined to this vertex forming triangular faces. Pyramids are named according to their base, for example,

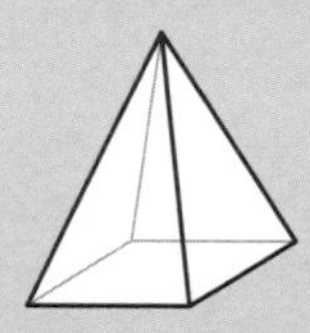

Square pyramid

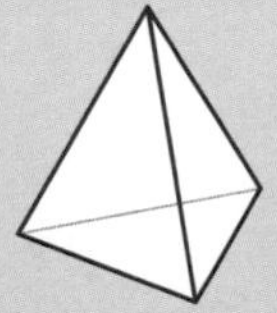

Triangular pyramid

Cone – a pyramid with a circular base and a curved surface rising to a vertex

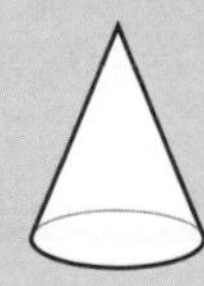

Hemisphere – a half sphere

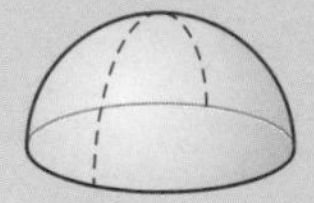

Isometric diagram – a 2-D drawing of a 3-D solid drawn on an isometric grid or isometric dotty paper; the prism shown in the diagrams below is 4 cm long, and its cross-section is an isosceles right-angled triangle with sides of length 2 cm

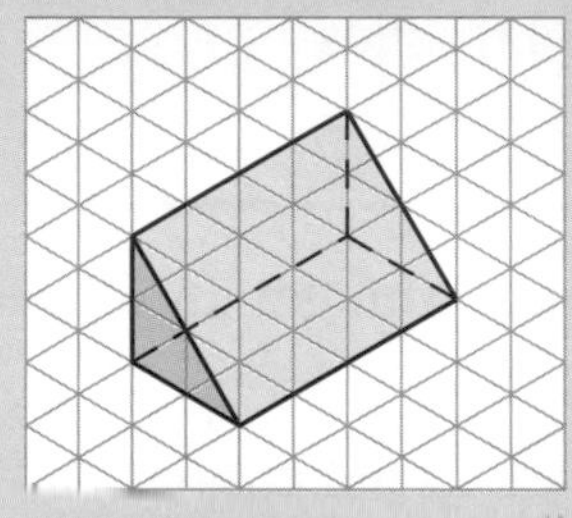

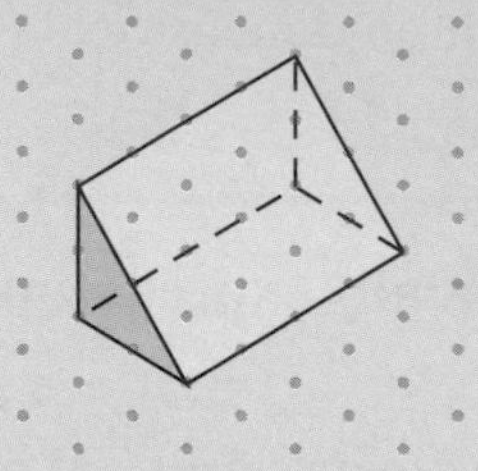

Plan – a diagram of a 3-D solid showing the view from above; these diagrams show a square-based pyramid and its plan

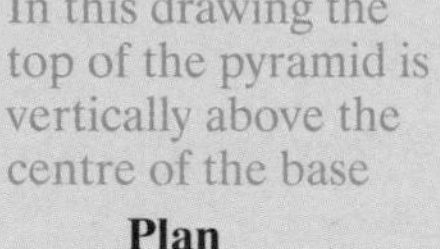

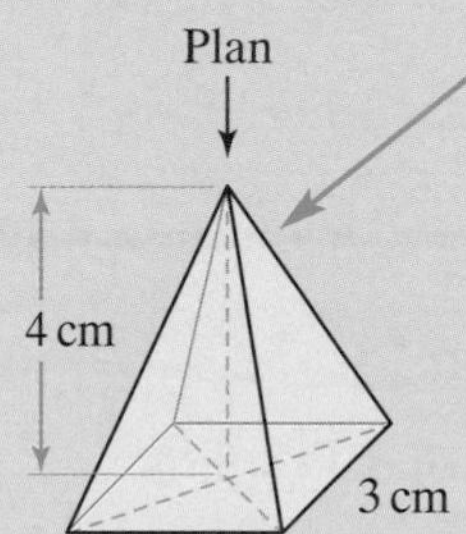

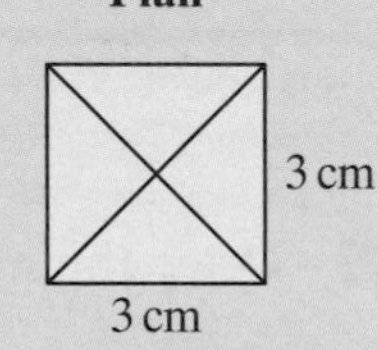

Front elevation – a diagram of a 3-D solid showing the view from the front, for example,

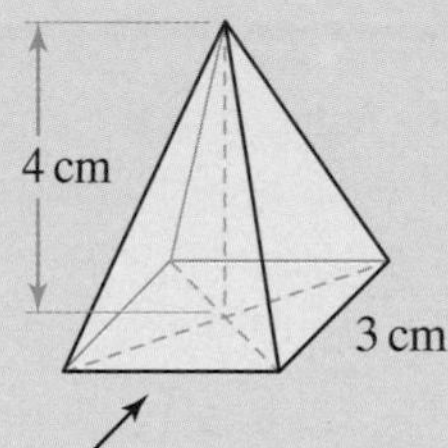

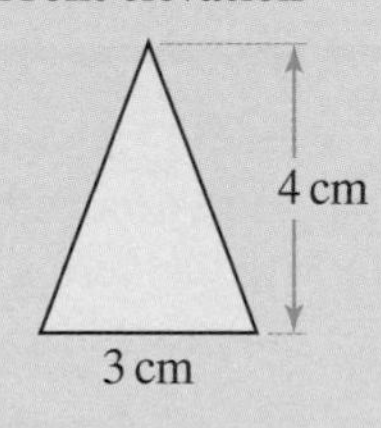

In some cases, as in this prism, the elevation from the front is the same as its cross-section

They are congruent trapezia

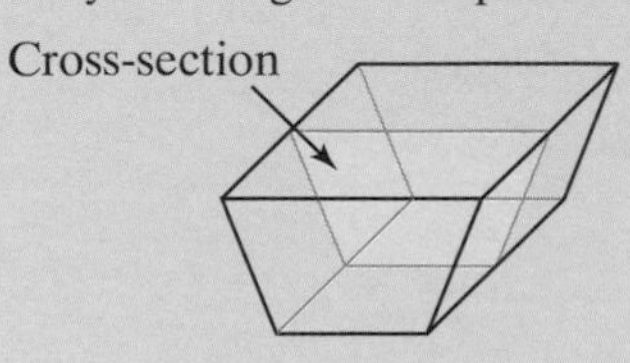

Side elevation – a diagram of a 3-D solid showing the view from the side; sometimes the elevation from the side of the shape is the same as the front elevation, for example,

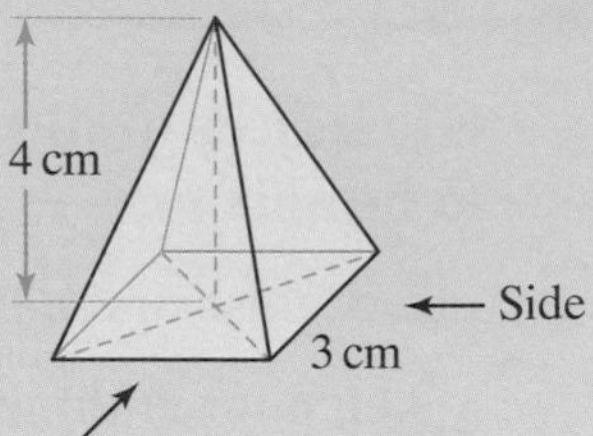

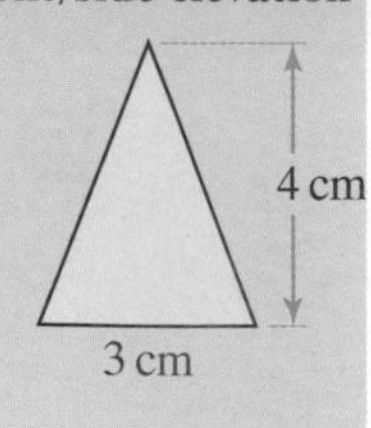

Usually, however, they will be different, for example,

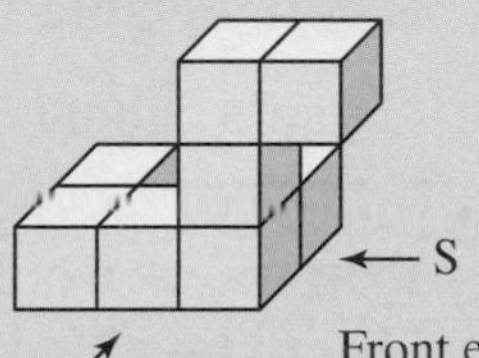

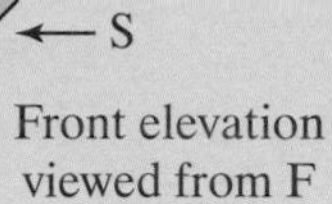

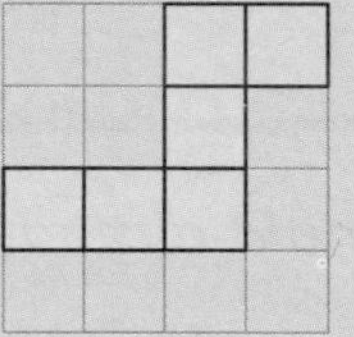

Side elevation viewed from S

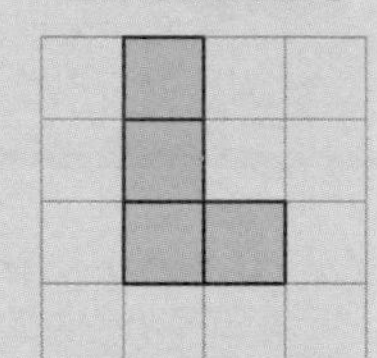

Unseen edges are shown as dotted lines, for example, the side elevation of this cylindrical container has three unseen edges. (The front elevation is the same)

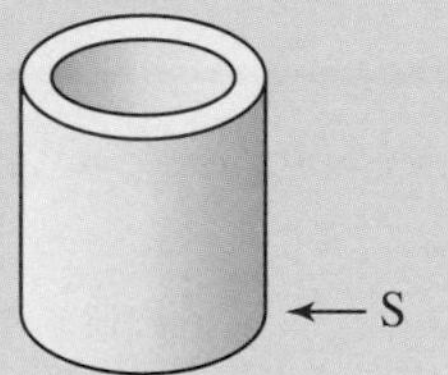

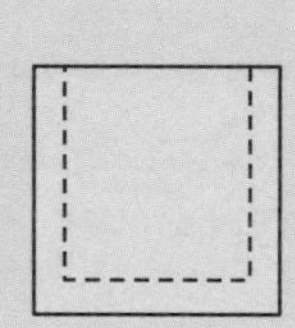

Learn 1 Cubes and cuboids

Example: Use isometric paper to draw a cuboid with dimensions 2 cm by 2 cm by 4 cm.

A good way to start is to draw a rough sketch:

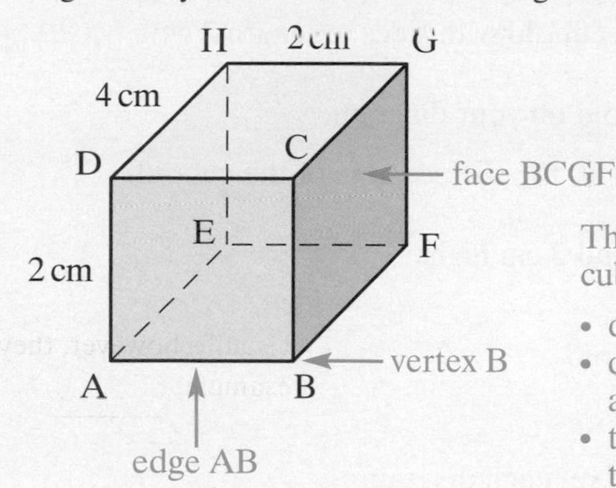

All cubes and cuboids have 6 faces, 8 vertices and 12 edges

The easiest way to sketch a cube or cuboid is to:

- draw the front face first (ABCD)
- draw the identical back face to the right and up (EFGH)
- then draw edges that are perpendicular to the front (BF, CG, DH, AE)

Using isometric paper where the sides of the triangles are 1 cm long, draw the cuboid:

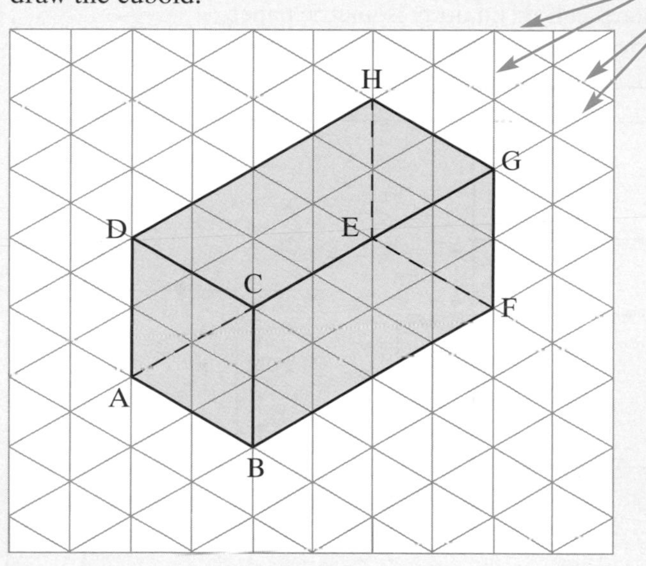

The edges are drawn along the gridlines to show the dimensions.

In both diagrams the hidden edges are shown by dotted lines.

Apply 1

1 A cuboid is shown here.

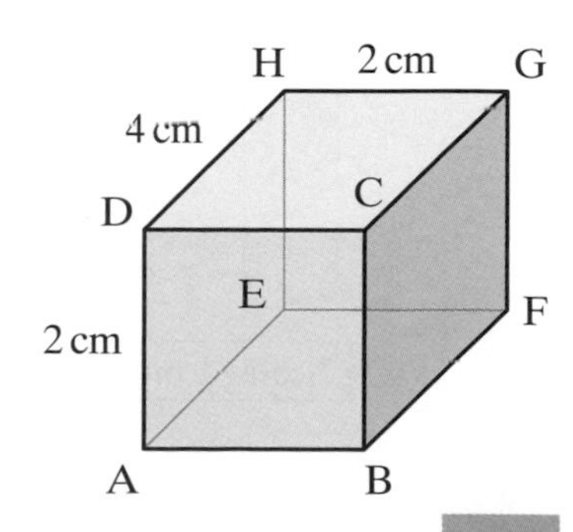

a What shape is:

i face ABCD **ii** face BCGF?

b i Write down the size of the angles of triangle ABD.

ii Name a triangle that is congruent to triangle ABD.

iii Name a pair of congruent right-angled triangles that are not isosceles.

2 **a** Sketch cubes with sides of length:

i 2 cm **ii** 3 cm **iii** 6 cm.

Show the dimensions of each cube on your sketch.

b Draw a table to give the volume and surface area of each cube.

3 **a** Draw an isometric diagram of a cuboid with sides of length 2 cm, 3 cm and 4 cm.
Give the dimensions of the cuboid on your diagram.

b Calculate: **i** the volume **ii** the surface area of the cuboid.

4 A cuboid is 5 cm long, 4 cm wide and 2 cm high.
Draw:

a an isometric diagram of the cuboid

b a net of the cuboid.

Show the dimensions of the cuboid on each diagram.

5 The nets of two cuboids are drawn below.
Draw a diagram of each cuboid on dotty isometric paper.
Label each diagram to show the dimensions of the cuboid.

a

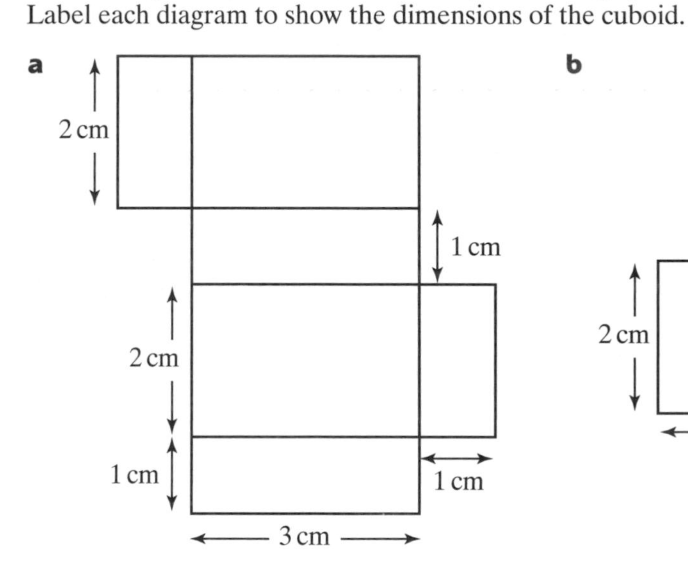

b

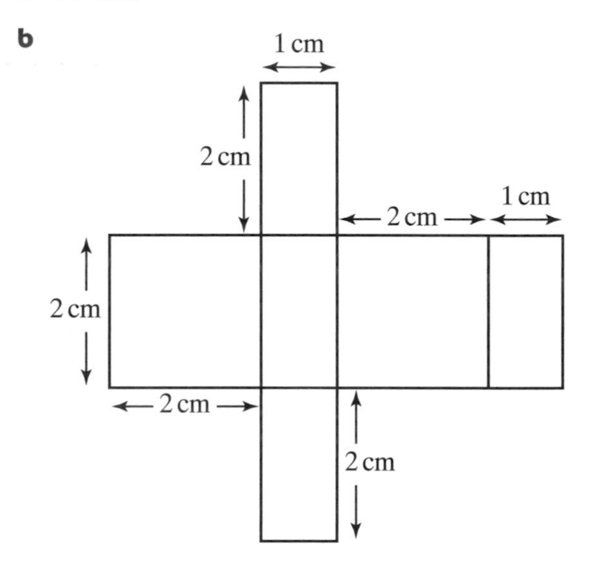

6 **Get Real!**
The isometric diagrams below are scale drawings of two rooms in a school.
Sketch each room and give the room's dimensions on your sketch.

a

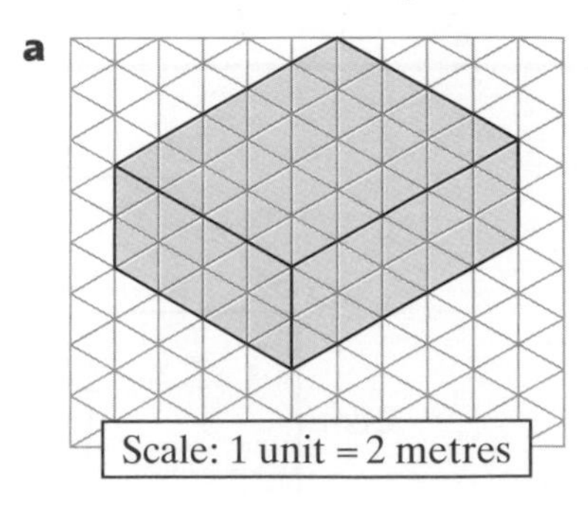

b

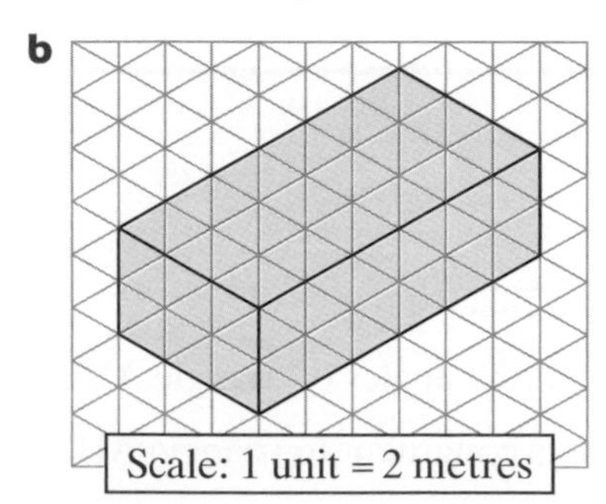

7 A cuboid is 25 cm long, 15 cm wide and 20 cm high.

a Draw an isometric diagram of this cuboid using 1 unit on the grid to represent 5 cm.

b Sketch a net of the cuboid.

8 A teacher asks students to draw a sketch of a cube with sides of length 2 cm, including unseen edges as dotted lines. Two of the attempts are shown below. What is wrong with each diagram?

a

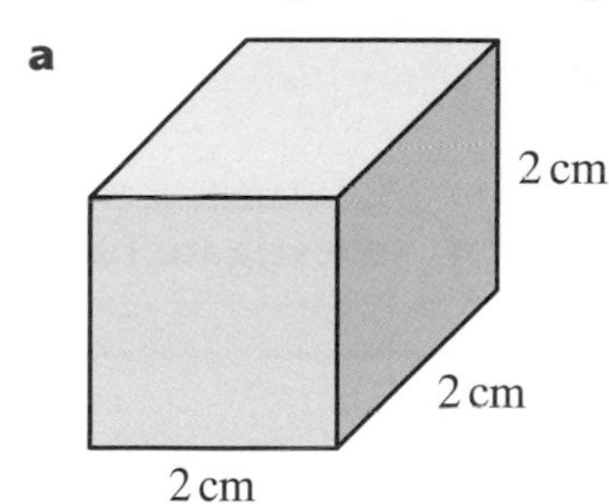

b

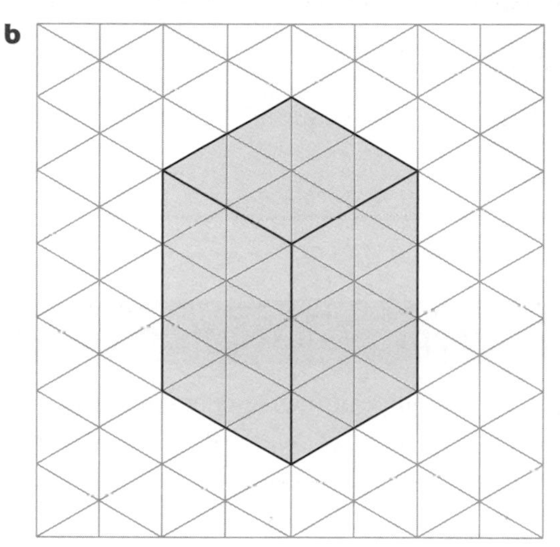

9 On an isometric grid, draw two **different** cuboids that have a volume of 36 cm^3.

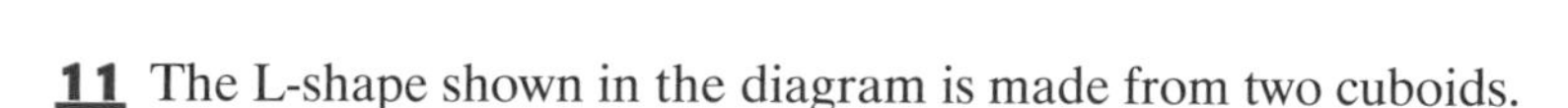

10 Get Real!

A factory needs a water tank that will hold 64 m^3 of water.

Draw an isometric diagram to show the possible dimensions of a tank if it is in the shape of:

a a cuboid **b** a cube.

11 The L-shape shown in the diagram is made from two cuboids.

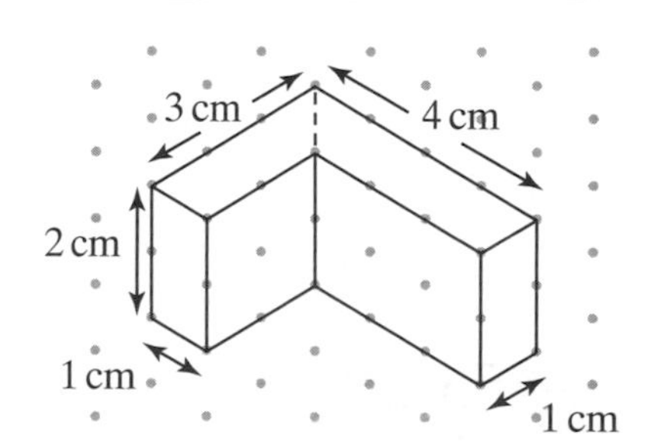

a Sketch the shape.

b Calculate:

i its volume

ii its total surface area.

c Write down the number of:

i faces

ii edges

iii vertices.

Explore

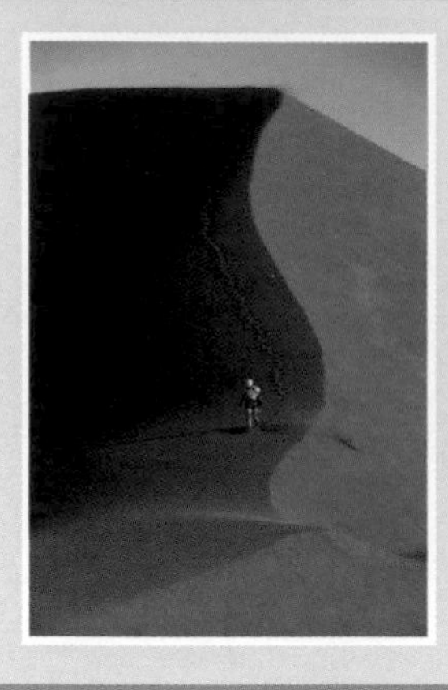

- A soft drinks company wants to design a carton in the shape of a cube or cuboid that will hold 1 litre (1000 cm^3)
- Draw isometric diagrams to show the dimensions of three possible cartons
- Calculate the surface area of each of the cartons you have drawn

Investigate further

Learn 2 2-D representations of 3-D solids

Example: Eight centimetre-cubes are glued together to make this 3-D solid.

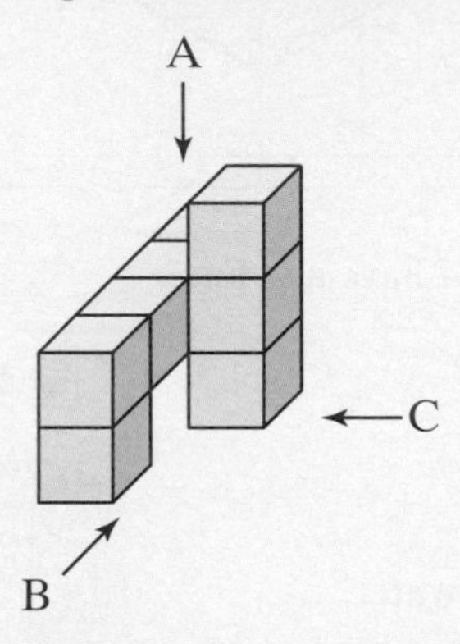

Show its plan, front and side elevations on a grid.

Plan viewed from A | Front elevation viewed from B | Side elevation viewed from C

Apply 2

1 Draw a sketch of:

a a cone

b a prism with a semicircular base

c a tetrahedron.

2 Sketch a prism whose cross-section is:

a a right-angled triangle

b a parallelogram

c a regular hexagon.

3 Each of the 3-D solids below is made from six centimetre-cubes. On a centimetre grid, draw the plan of each solid and its elevations from the directions marked F and S.

a

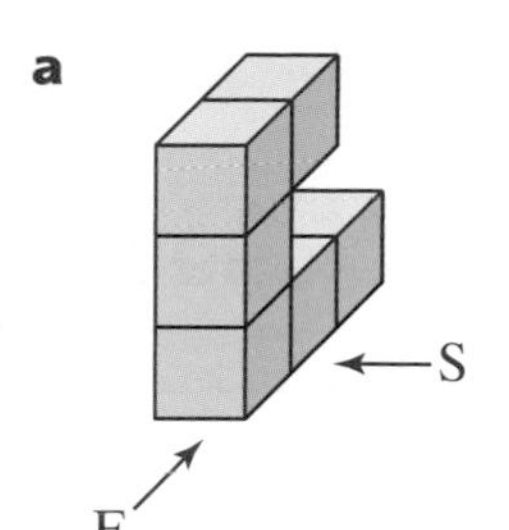

b

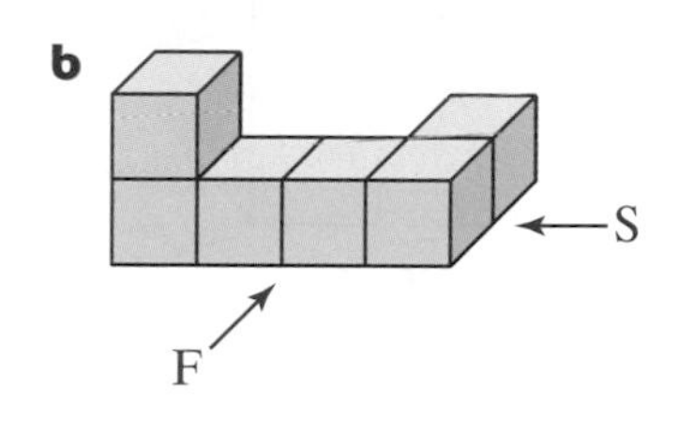

c

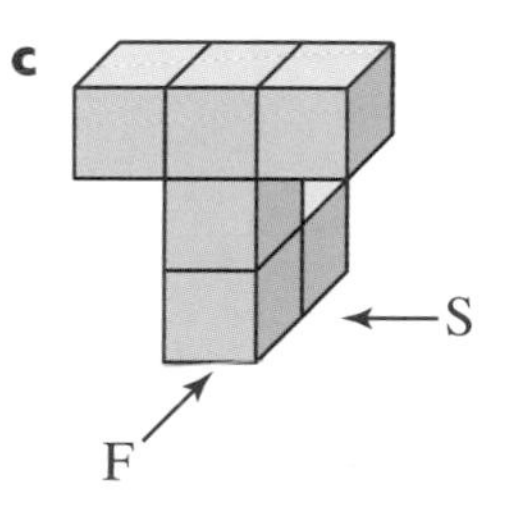

4 Nine centimetre-cubes are arranged into a three by three square block, and another three cubes are placed in front as shown in the diagram.

On a grid draw:

a the plan of this 3-D solid

b the front elevation as viewed from A

c the side elevation as viewed from B.

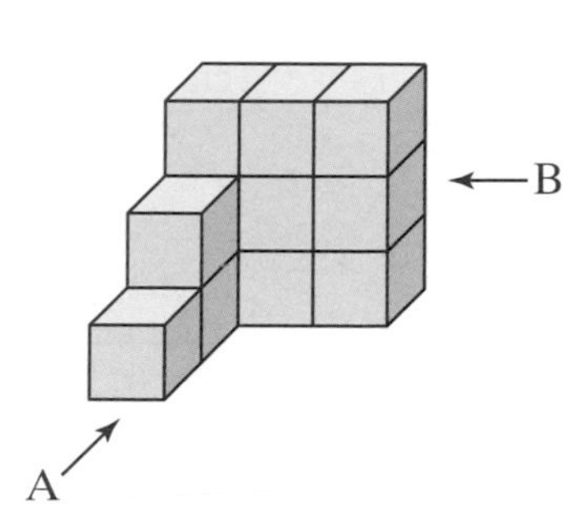

5 Get Real!
The diagram shows the dimensions of a small can of baked beans. Draw an accurate plan and elevation. Leave out unseen edges.

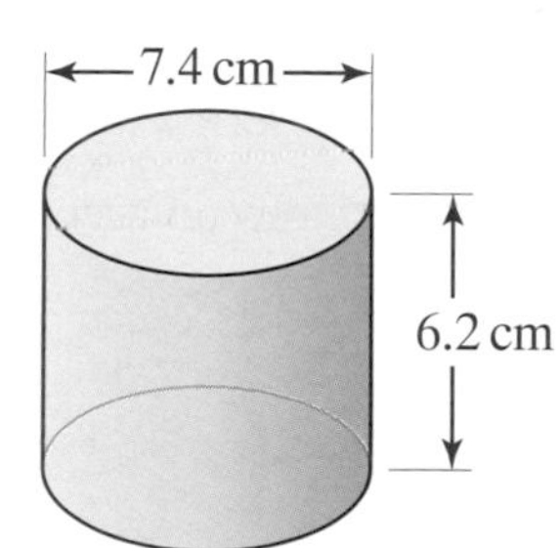

6 For each prism below, sketch the plan, front and side elevations from the directions shown by the arrows. Remember to show unseen edges as dotted lines.

a Triangular prism

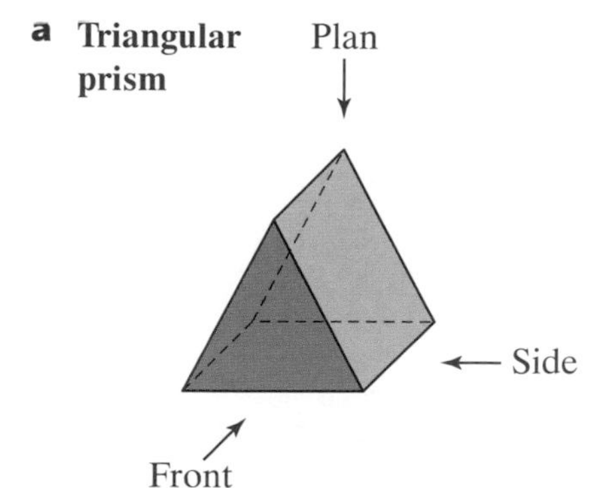

b Pentagonal prism

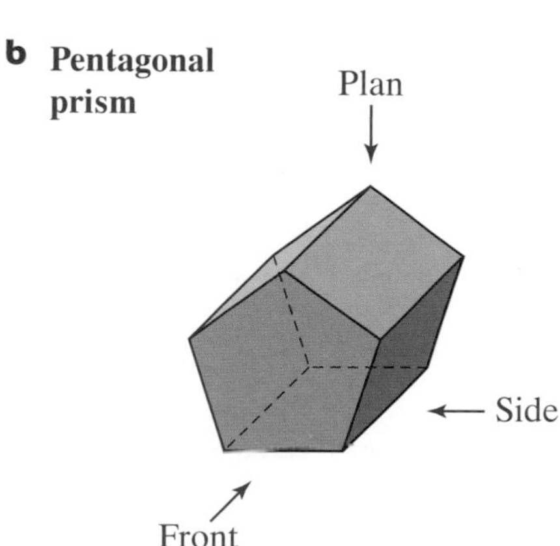

7 Each solid below is a cuboid with part removed.
The dimensions are given in centimetres.
For each object, draw full-size plan, and front and side elevations on a centimetre grid.

a

2
1
3
S
3
F

b

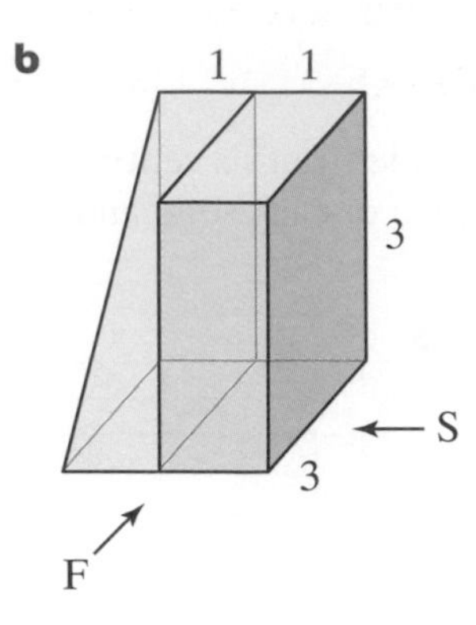

8 Sara has drawn a plan and a side elevation (from S) of the 3-D solid shown below.

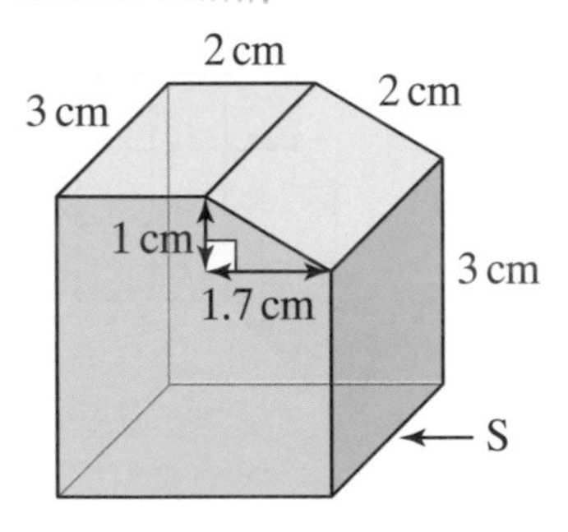

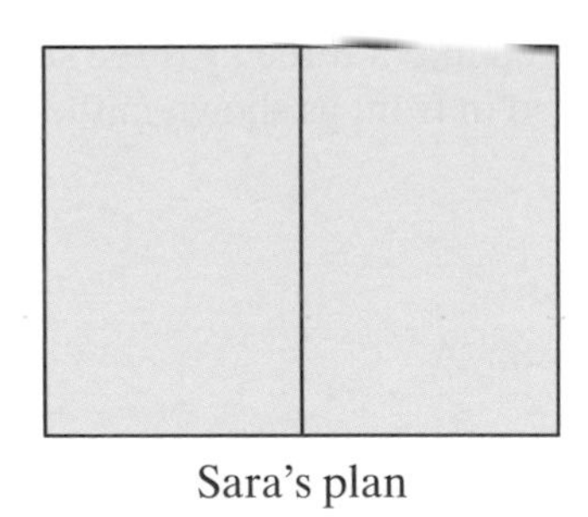

Sara's plan

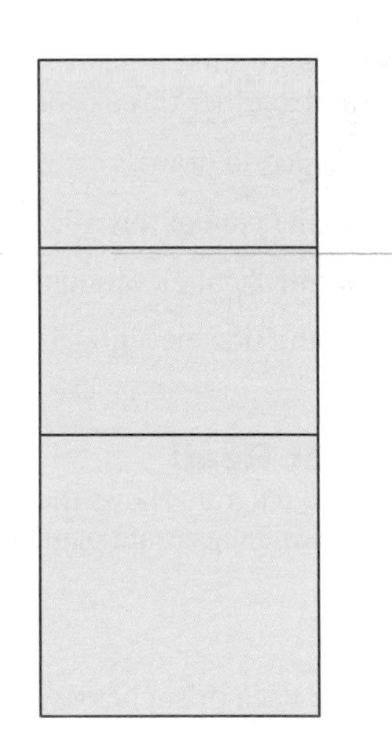

Sara's side elevation

a Describe what is wrong with Sara's diagrams.

b Draw an accurate plan and side elevation.

9 Plans and elevations of three objects are shown below. Sketch each object.
You do not need to show hidden edges.

a Plan

Front Side

b

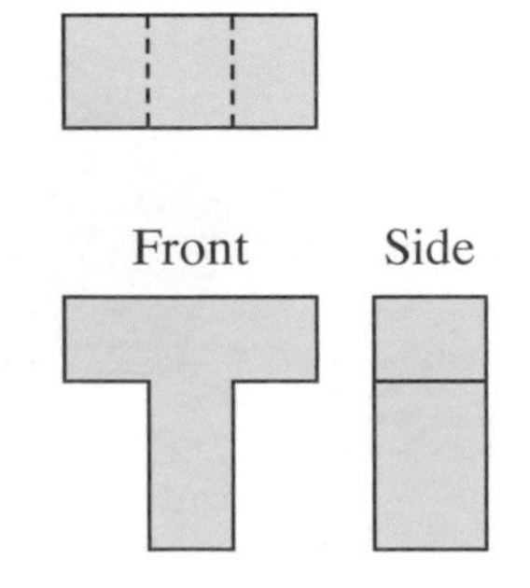

c

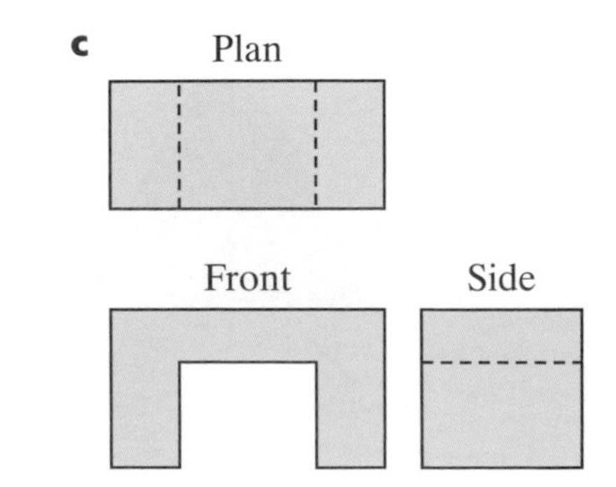

10 Get Real!

The diagram shows a stool. Its top is in the shape of a regular hexagon. It has a cylindrical leg under each corner of the top.

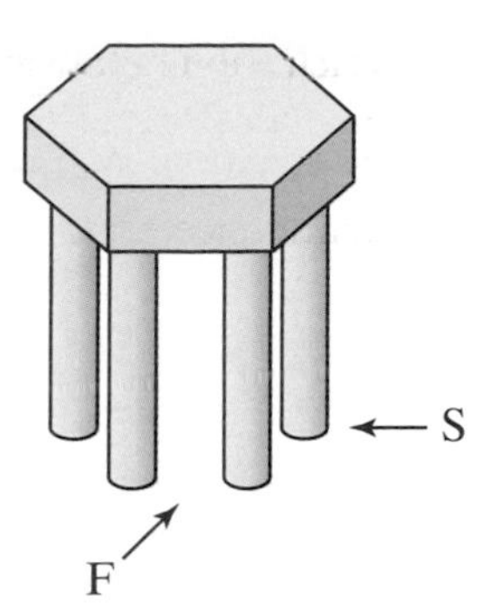

a Sketch a plan of the stool. Show unseen edges with dotted lines.

b Sketch an elevation of the stool when it is viewed from the front, F.

c Sketch an elevation of the stool when it is viewed from the side, S.

Explore

Copy and complete this table

Solid	Number of faces	Number of vertices	Number of edges
Cube			
Cuboid			
Pyramid			
Triangular prism			
Pentagonal prism			

Add some other 3-D shapes

Investigate further

3-D solids

ASSESS

The following exercise tests your understanding of this chapter, with the questions appearing in order of increasing difficulty.

1 Sketch cuboids with the following dimensions:

a 2 cm, 4 cm and 6 cm **b** 3 cm, 4 cm and 5 cm

Find the volume and surface area of each cuboid.

2 Sketch a triangular prism where the triangle is an equilateral triangle of side 4 cm and the length of the prism is 6 cm.

3 On dotty isometric paper, draw cuboids with the following dimensions:

a 3 cm, 5 cm and 7 cm **b** 1 cm, 4 cm and 8 cm

Find the volume and surface area of each cuboid.

4 On dotty isometric paper, draw a triangular prism where the triangle is an equilateral triangle of side 5 cm and the length of the prism is 8 cm.

5 Draw the plan and elevations of the shapes shown below.

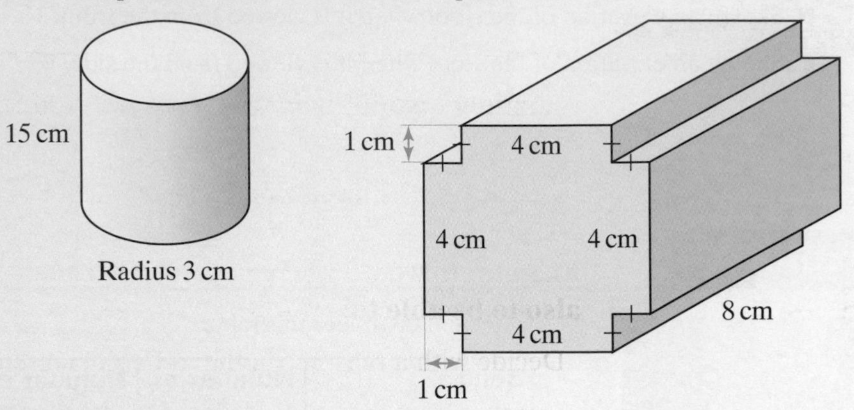

Try a real past exam question to test your knowledge:

6 The diagram shows a solid shape made from 8 cubes.

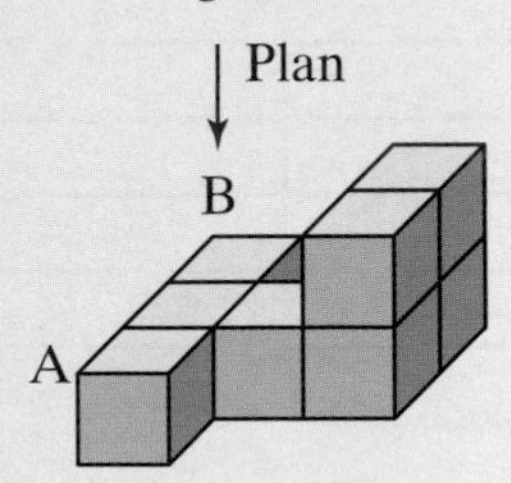

Copy and complete the plan view of the shape on the grid below.

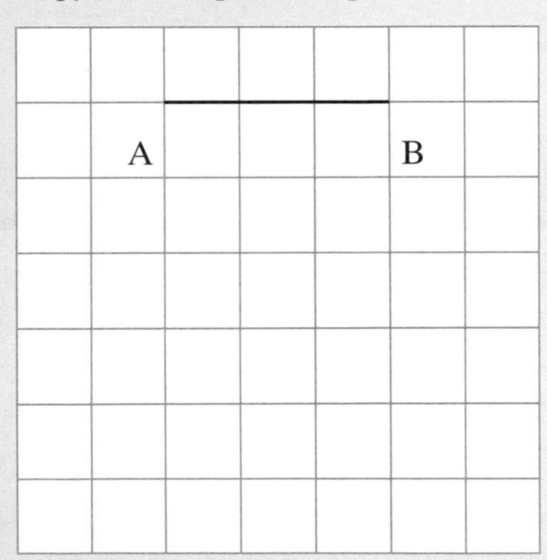

Spec A, Int Paper 1, Nov 03

22 Algebraic proofs

OBJECTIVES

E **Examiners would normally expect students who get an E grade to be able to:**

Decide with a reason whether a simple statement is true or false

D **Examiners would normally expect students who get a D grade also to be able to:**

Decide with a reason whether a harder statement is true or false

Identify a counter example

C **Examiners would normally expect students who get a C grade also to be able to:**

Understand the difference between a demonstration and a proof

Show step-by-step deductions in providing a basic algebraic explanation

What you should already know …

- Odd and even numbers
- Prime numbers
- Factors and multiples
- Square and cube numbers

VOCABULARY

Sum – the result of adding together two (or more) numbers, variables, terms or expressions

Product – the result of multiplying together two (or more) numbers, variables, terms or expressions

Consecutive numbers – whole numbers that are next to each other on the number line in sequence

Counter example – an example that disproves a statement. For example, the statement 'All prime numbers are odd', is not true because 2 is a prime number and 2 is not odd – this is a counter example

Proof – a series of logical mathematical steps that confirms the truth of a mathematical statement

Prove – to provide a proof

Theorem – a mathematical statement that can be demonstrated to be true using proof

Learn 1 Algebraic proofs

Examples:

a If R is an odd number and S is an even number, state whether each of the following is odd or even and give a reason for your answer.

i $R + S$
ii $R \times S$

i $R + S$ is odd because an odd number + an even number is always odd

Try out different values of R and S to show the result

In general: $O + O = E$
$O + E = O$
$E + O = O$
$E + E = E$

ii $R \times S$ is even because an odd number × an even number is always even

Try out different values of R and S to show the result

In general: $O \times O = O$
$O \times E = E$
$E \times O = E$
$E \times E = E$

b Trevor says that when you square a number, the answer is always greater than or equal to the original number. Give an example to show that Trevor is wrong.

Try positive then negative then decimal (fraction) numbers

Try 1 $1^2 = 1$ so true when the number = 1
Try 2 $2^2 = 4$ so true when the number = 2
Try 3 $3^2 = 9$ so true when the number = 3
Try –1 $(-1)^2 = 1$ so true when the number = –1
Try –2 $(-2)^2 = 4$ so true when the number = –2
Try 0.5 $0.5^2 = 0.25$ so **not** true when the number = 0.5

Sometimes it will be necessary to try out a few examples to find the solution

Trevor is wrong as $0.5^2 = 0.25$ which is smaller than the original number.

Giving a numerical example that disproves a statement is often called a counter example

Numbers such as 0 and 1 as well as prime numbers, negative numbers and fractional numbers are all useful examples to try out when trying to find a counter example in questions like this

Apply 1

1 If n is a positive integer:

a explain why $2n$ must be an even number

b explain why $2n + 1$ must be an odd number.

2 r and s are both even numbers.
r is greater than s.
Is $r - s$ always odd, always even or could it be either?
Give a reason for your answer.

3 Is the value of $3n - 1$ always an even number?
Give a reason for your answer.

4 Is the value of $5n + 1$ always an even number?
Give a reason for your answer.

5 If p is an odd number explain why $p^2 + 1$ is an even number.

6 If p is an odd number and q is an even number explain why $(p - q)(p + q)$ is an odd number.

7 P is a prime number.
Q is an odd number.

State whether each of the following is:

i always even

ii always odd

iii could be either even or odd.

a PQ

b P(Q + 1)

8 k is an odd number.
Gail says that $\frac{1}{2}k + \frac{1}{2}$ is always even.
Give a counter example to show that Gail is wrong.

9 Christine says that all numbers have an even number of factors.
Give a counter example to show that Christine is wrong.

10 The product of two consecutive numbers is divisible by 12.
Write down two numbers that satisfy this statement.

11 The product of two consecutive numbers is divisible by 20.
Write down two numbers that satisfy this statement.

12 Duncan says that the square of a number is always bigger than the number itself.
Is Duncan correct?
Give a reason for your answer.

13 Prove that the sum of two consecutive numbers is always an odd number.

14 Prove that the difference between any two even numbers is always an even number.

15 Prove that the sum of three consecutive numbers is equal to three times the middle number.

16 Prove that the sum of two consecutive odd numbers is always a multiple of 4.

17 Prove that the sum of three consecutive odd numbers is always a multiple of three.

18 Paul says that an odd number squared is always an odd number.
Is Paul correct?
Give a reason for your answer.

19 Mick says that when you square a number the answer is always greater than zero.
Is Mick correct?
Give a reason for your answer.

20 Part of a number grid is shown below:

1	2	3	4	5	6	7	8
9	10	11	12	13	14	15	16
17	18	19	20	21	22	23	24
25	26	27	28	29	30	31	32
33	34	35	36	37	38	39	40
41	42	43	44	45	46	47	48
49	50	51	52	53	54	55	56
57	58	59	60	61	62	63	64

The shaded shape is called T_{12} because it has 12 in the middle of the top row.
The sum of the numbers in T_{12} is 56.

a This is T_n

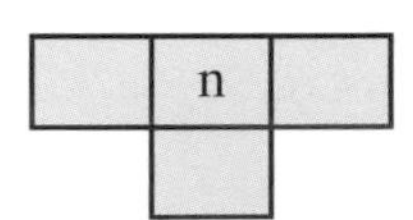

Copy and fill in the empty boxes on T_n

b Find the sum of all the numbers in T_n in terms of n.
Give your answer in its simplest form.

c Explain why the sum of all the numbers in T_n is always divisible by 4.

21 Part of a number grid is shown below:

1	2	3	4	5	6	7	8
9	10	11	12	13	14	15	16
17	18	19	20	21	22	23	24
25	26	27	28	29	30	31	32
33	34	35	36	37	38	39	40
41	42	43	44	45	46	47	48
49	50	51	52	53	54	55	56
57	58	59	60	61	62	63	64

The shaded shape is called B_{11} because it has 11 in the top left-hand corner.

a This is B_n

n	

Copy and fill in the empty boxes on B_n

b Amara notices that $11 \times 20 = 220$
and that $19 \times 12 = 228$

Show, using algebra, that the difference of the products of the diagonals is always 8.

Explore

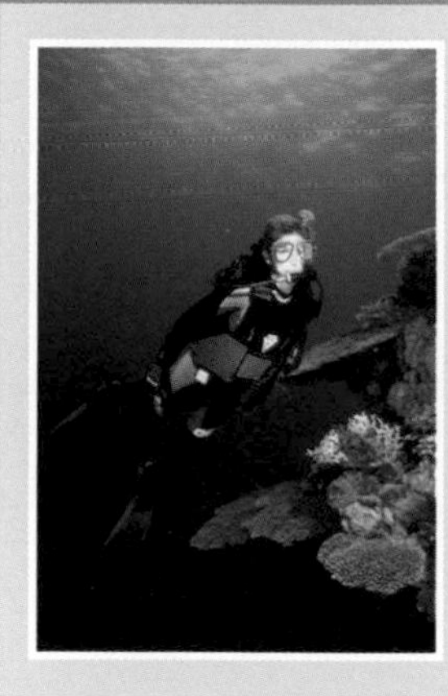

What numbers can be written as the sum of two consecutive numbers, three consecutive numbers, four consecutive numbers ... ?

Investigate further

Explore

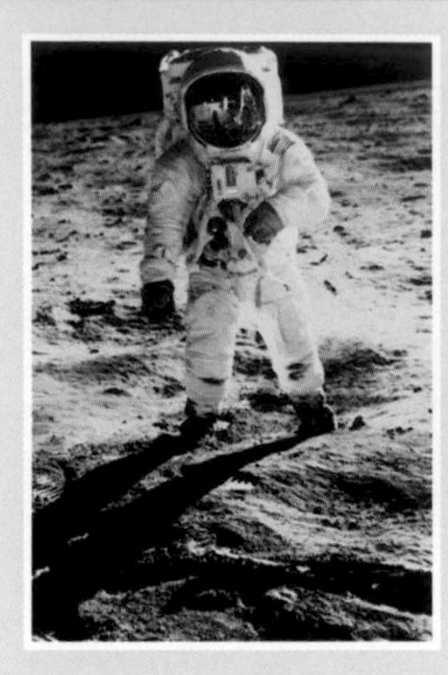

- A quick way to add the numbers from 1 to 10 is shown below:

Write out the numbers	$1 + 2 + 3 + 4 + ... + 7 + 8 + 9 + 10$
Reverse the numbers	$10 + 9 + 8 + 7 + ... + 4 + 3 + 2 + 1$
Adding the two series	$11 + 11 + 11 + 11 + ... + 11 + 11 + 11 + 11$

Each pair of numbers adds up to 11 and there are 10 pairs of numbers
So the sum of all the numbers written down is 10×11 (that is, 10 lots of the total of 11)

However this is twice the total so

$$\begin{aligned}1 + 2 + 3 + 4 + ... + 7 + 8 + 9 + 10 &= \tfrac{1}{2} \times 10 \times 11\\ &= \tfrac{1}{2} \times 110\\ &= 55\end{aligned}$$

- Can you add the numbers from 1 to 20 using this method?
- Can you add the numbers from 1 to 100 using this method?

Investigate further

Algebraic proofs

ASSESS

The following exercise tests your understanding of this chapter, with the questions appearing in order of increasing difficulty.

1 a and b are both odd numbers.
Is ab always odd, always even or could it be either?
Give a reason for your answer.

2 Ross says that the product of two prime numbers is always odd.
Give a counter example to show that Ross is wrong.

3 P is a prime number.
Q is an even number.

State whether each of the following is:

i always even

ii always odd

iii could be either even or odd.

a PQ

b P(Q − 1)

4 Prove that the sum of four consecutive numbers is always an even number.

5 Prove that the sum of three consecutive even numbers is always a multiple of six.

6 Zoë says that when you double a prime number and add one, the answer is another prime.
Is Zoë correct?
Give a reason for your answer.

7 Prove that the mean of three consecutive numbers is equal to the middle number.

8 Part of a number grid is shown below:

1	2	3	4	5	6	7
8	9	10	11	12	13	14
15	16	17	18	19	20	21
22	23	24	25	26	27	28
29	30	31	32	33	34	35
36	37	38	39	40	41	42
43	44	45	46	47	48	49

The shaded shape is called L_{16} because it has 16 in the corner.
The sum of the numbers in L_{16} is 60.

a This is L_n

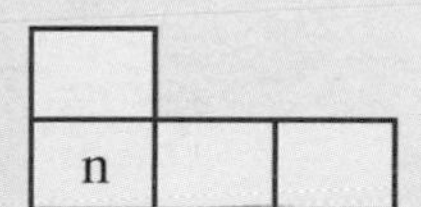

Copy and fill in the empty boxes on L_n

b Find the sum of all the numbers in L_n in terms of n.
Give your answer in its simplest form.

c Explain why the sum of all the numbers in L_n is always divisible by 4.

Try some real past exam questions to test your knowledge:

9 p is an even number.
q is an odd number.

a Is pq

i an odd number **ii** an even number **iii** could be either?

b Is p
i a prime number **ii** not a prime number **iii** could be either?

c Is $q \div p$
i an integer **ii** not an integer **iii** could be either?

Spec A, Foundation specimen paper, 2006

10 a Find the value of $3x + 5y$ when $x = -2$ and $y = 4$.

b Find the value of $3a^2 + 5$ when $a = 4$.

c k is an even number.
Jo says that $\frac{1}{2}k + 1$ is always even.
Give an example to show that Jo is wrong.

d The letters a and b represent prime numbers.
Give an example to show that $a + b$ is **not** always an even number.

Spec A, Foundation Paper 2, June 2004

Glossary

Acute angle – an angle between 0° and 90°

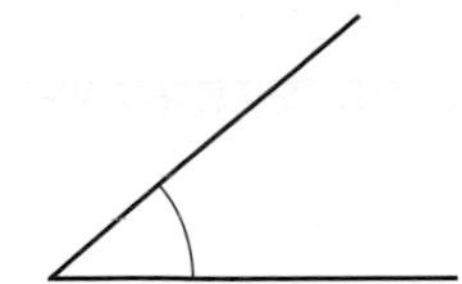

Algebraic expression – a collection of terms separated by + and – signs such as $x + 2y$ or $a^2 + 2ab + b^2$

Alternate angles – the angles marked a, which appear on opposite sides of the transversal

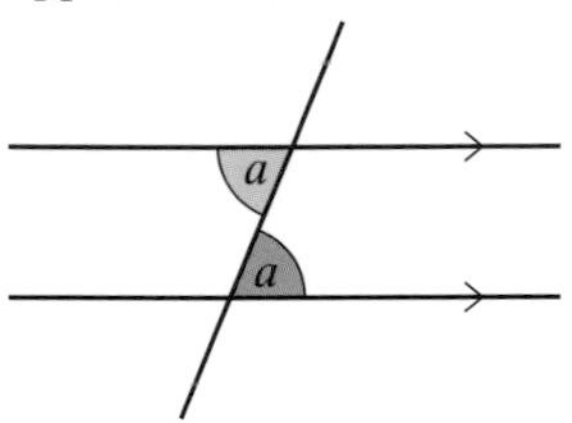

Angle bisector – a line that divides an angle into two equal parts

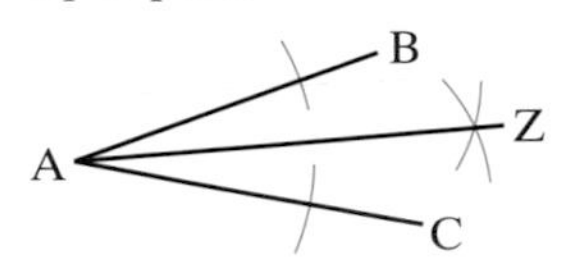

AZ is the angle bisector of angle BAC

Arc (of a circle) – part of the circumference of a circle; a minor arc is less than half the circumference and a major arc is greater than half the circumference

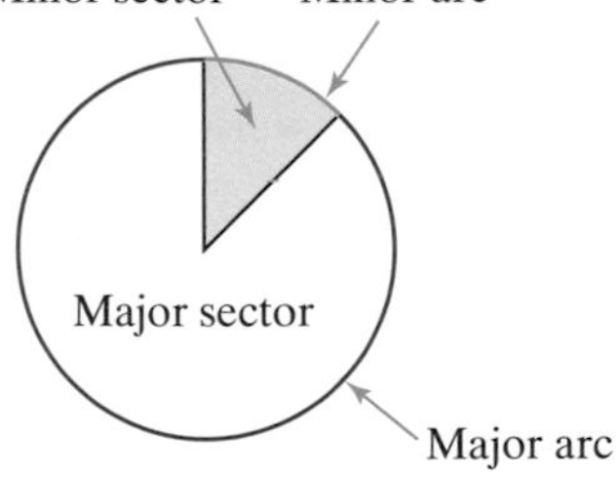

Axis (pl. **axes**) – the lines used to locate a point in the coordinates system; in two dimensions, the x-axis is horizontal, and the y-axis is vertical. This system of Cartesian coordinates was devised by the French mathematician and philosopher René Descartes

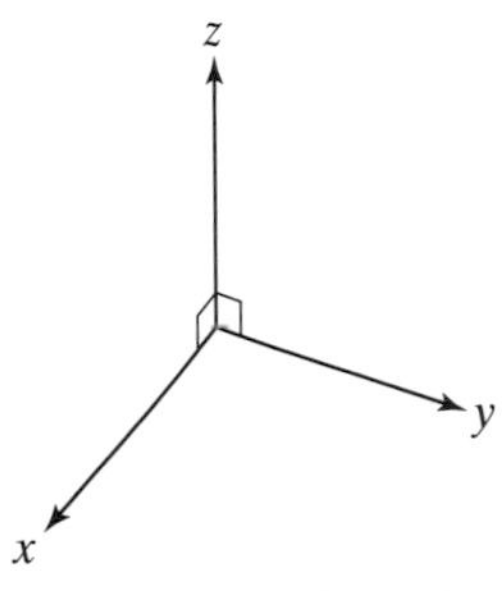

In three dimensions, the x- and y-axes are horizontal and at right angles to each other and the z-axis is vertical

Axis of symmetry – the mirror line in a reflection

Bearing – an angle measured clockwise from North; all bearings should be written as three figure numbers, for example, 125° or 045°

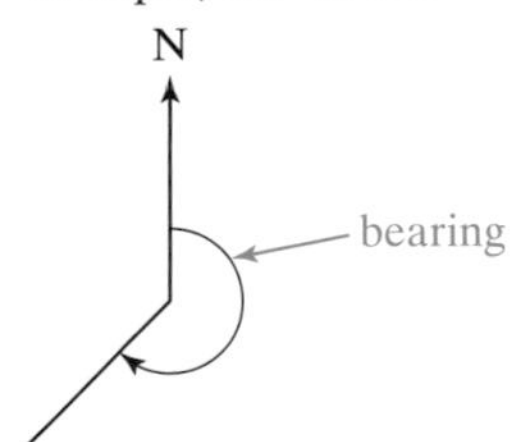

Bisect – to divide into two equal parts

Brackets – these show that the terms inside should be treated alike, for example,

$2(3x + 5) = 2 \times 3x + 2 \times 5 = 6x + 10$

Capacity – the amount of liquid a hollow container can hold, commonly measured in litres (1 litre = 1000 cm^3)

Centre of rotation – the fixed point around which the object is rotated

Chord – a straight line joining two points on the circumference of a circle

Circle – a shape formed by a set of points that are always the same distance from a fixed point (the centre of the circle)

Circumference – the perimeter of a circle

Coefficient – the number (with its sign) in front of the letter representing the unknown, for example:

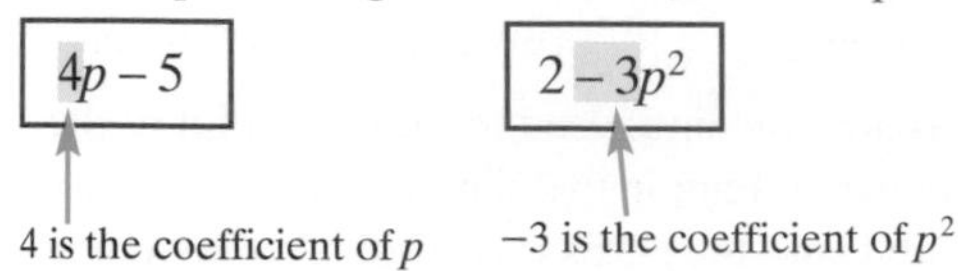

Collect like terms – to group together terms of the same variable, for example, $2x + 4x + 3y = 6x + 3y$

Compound measure – a measure formed from two or more other measures, for example,
speed $(= \frac{\text{distance}}{\text{time}})$, density $(= \frac{\text{mass}}{\text{volume}})$,
population density $(= \frac{\text{population}}{\text{area}})$

Concave polygon – a polygon with at least one interior reflex angle

Cone – a pyramid with a circular base and a curved surface rising to a vertex

Congruent – exactly the same size and shape; one of the shapes might be rotated or flipped over

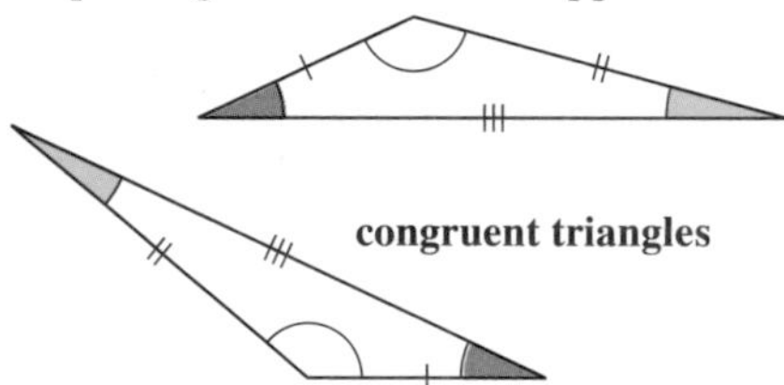

Consecutive numbers – whole numbers that are next to each other on the number line in sequence

Converse of Pythagoras' theorem – in any triangle, if $c^2 = a^2 + b^2$ then the triangle has a right angle opposite c; for example, if c = 17 cm, a = 8 cm, b = 15 cm, then $c^2 = a^2 + b^2$, so this is a right-angled triangle

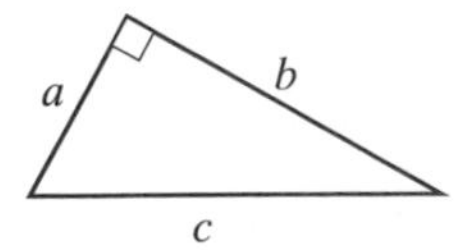

Conversion factor – the number by which you multiply or divide to change measurements from one unit to another. The approximate conversion factors that you should know are:

Length	Mass	Capacity
1 foot ≈ 30 cm 5 miles ≈ 8 km	1 kg ≈ 2.2 pounds	1 gallon ≈ 4.5 litres 1 litre ≈ 1.75 pints

The table below gives conversion factors for metric units of length, mass and capacity

Metric system

Length	Mass	Capacity
1 cm = 10 mm	1 g = 1000 mg	1 ℓ = 100 cℓ or 1000 mℓ
1 m = 100 cm or 1000 mm	1 kg = 1000 g	
1 km = 1000 m	1 t = 1000 kg	

Conversion graph – a graph used to convert one unit into another unit, for example, pounds to kilograms, litres to pints

Convex polygon – a polygon with no interior reflex angles

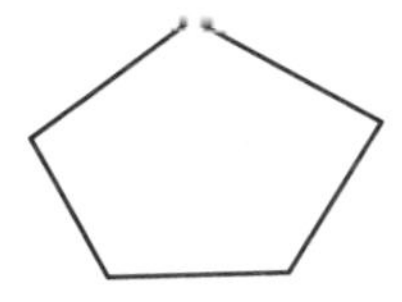

Coordinates – a system used to identify a point; an x-coordinate and a y-coordinate give the horizontal and vertical positions

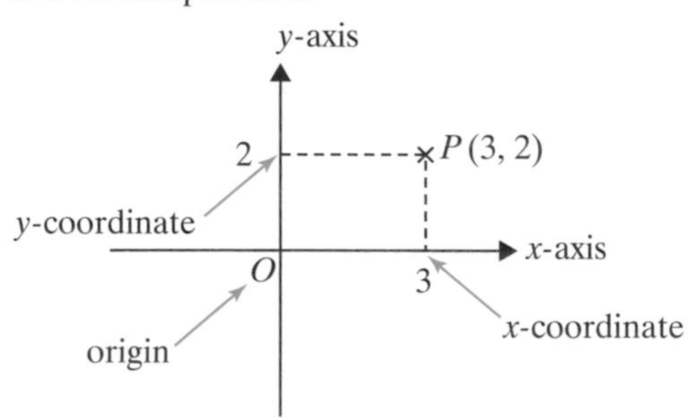

Corresponding angles – the angles marked c, which appear on the same side of the transversal

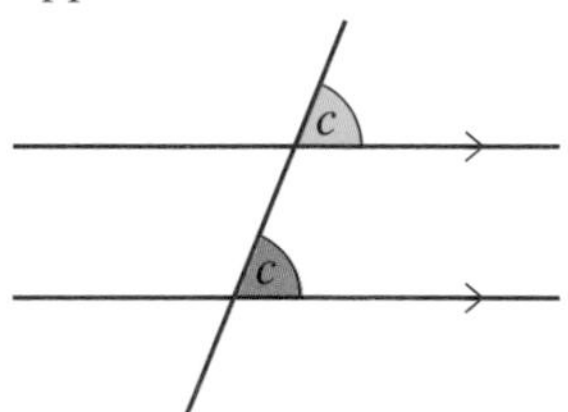

Counter example – an example that disproves a statement. For example, the statement 'All prime numbers are odd', is not true because 2 is a prime number and 2 is not odd – this is a counter example

Cross-section – a cut at right angles to a face and usually at right angles to the length of a prism

Cube – a solid with six identical square faces

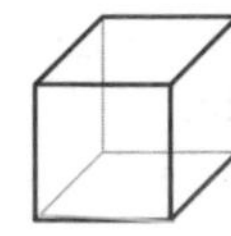

Cube number – a cube number is the outcome when a whole number is multiplied by itself then multiplied by itself again; cube numbers are 1, 8, 27, 64, 125, ...

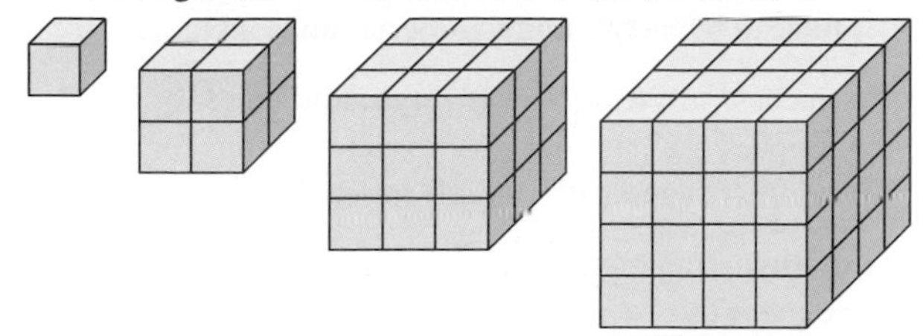

The rule for the nth term of the cube numbers is n^3

Cube root – the cube root of a number such as 125 is a number whose outcome is 125 when multiplied by itself then multiplied by itself again

Cubic centimetre (cm^3) and **cubic metre (m^3)** – commonly used units of measurement for volume; $1 \text{ cm}^3 = 1000 \text{ mm}^3$, $1 \text{ m}^3 = 1\,000\,000 \text{ cm}^3$

Cuboid – a solid with six rectangular faces (two or four of the faces can be squares)

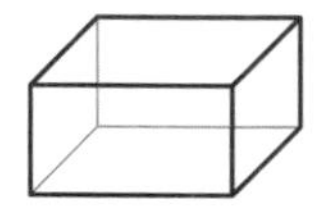

Cylinder – a prism with a circle as a cross-sectional face

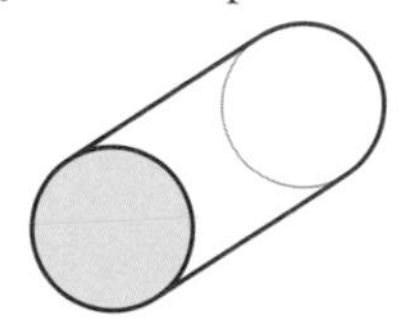

Decagon – a polygon with ten sides

Decimal places – the digits to the right of a decimal point in a number, for example, in the number 23.657, the number 6 is the first decimal place (worth $\frac{6}{10}$), the number 5 is the second decimal place (worth $\frac{5}{100}$), and 7 is the third decimal place (worth $\frac{7}{1000}$); the number 23.657 has 3 decimal places

Density – to calculate density, divide the mass of the object by the volume of the object. It is usually given in grams per cubic centimetre (g/cm^3) or kilograms per cubic metre (kg/m^3)

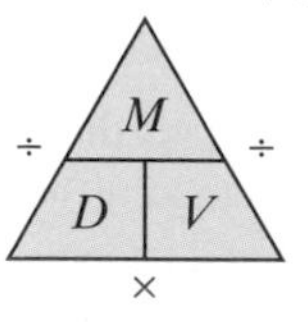

In the triangle, cover the item you want, then the rest tells you what to do

$$\text{Density} = \frac{\text{mass}}{\text{volume}}$$

$\frac{\text{kilograms}}{\text{cubic metres}}$ gives kg/m^3

Mass = density × volume

$$\text{Volume} = \frac{\text{mass}}{\text{density}}$$

Diameter – a chord passing through the centre of a circle; the diameter is twice the length of the radius

Dimension – the measurement between two points on the edge of a shape

Edge – a line segment that joins two vertices of a solid

Enlargement – an enlargement changes the size of an object (unless the scale factor is 1); but not its shape; it is defined by giving the centre of enlargement and the scale factor; the object and the image are similar

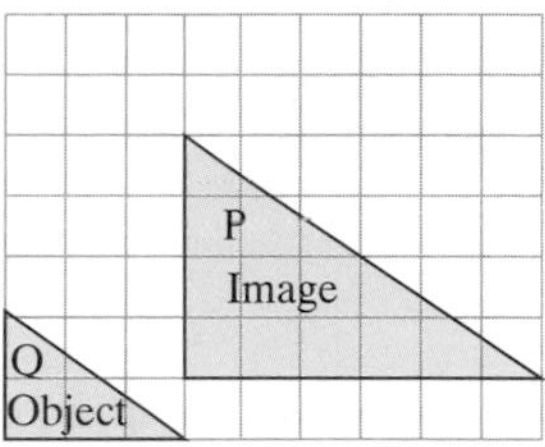

Triangle P is an enlargement of triangle Q
All the lines have doubled in size
The scale factor of the enlargement is 2

Equation – a statement showing that two expressions are equal, for example, $2y - 7 = 15$

Equidistant – the same distance; if A is equidistant from B and C, then AB and AC are the same length

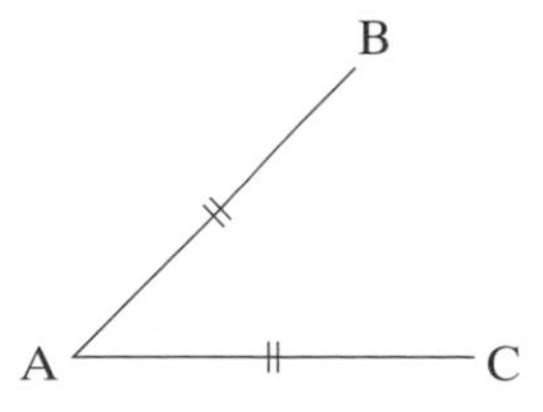

Equilateral triangle – a triangle with 3 equal sides and 3 equal angles – each angle is 60°

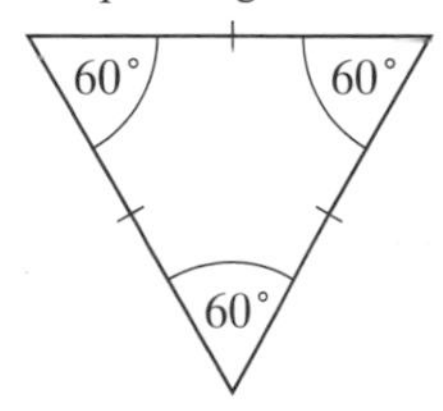

Expand – to remove brackets to create an equivalent expression (expanding is the opposite of factorising)

Expression – a mathematical statement written in symbols, for example, $3x + 1$ or $x^2 + 2x$

Exterior angle – the angle between one side of a polygon and the extension of the adjacent side

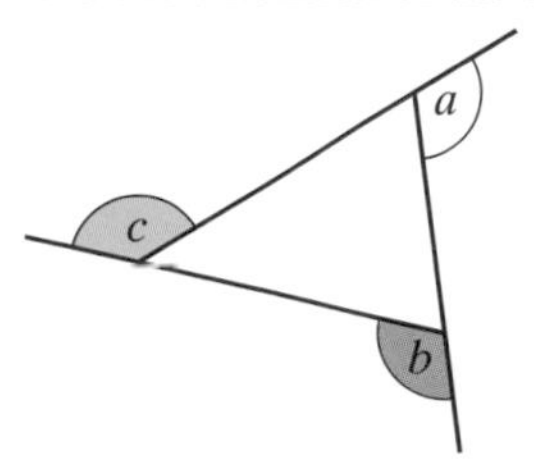

a, b and c are exterior angles

Face – one of the flat surfaces of a solid

Factorise – to include brackets by taking common factors (factorising is the opposite of expanding)

Formula – an equation showing the relationship between two or more variables, for example, $E = mc^2$

Front elevation – a diagram of a 3-D solid showing the view from the front, for example,

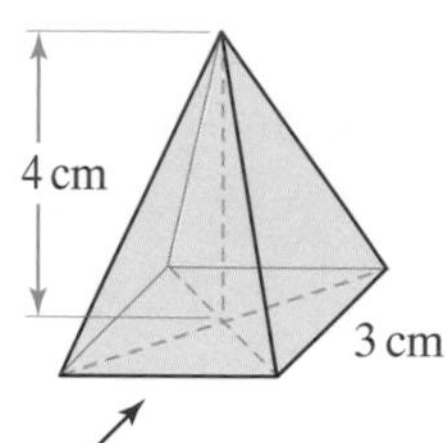

Front elevation

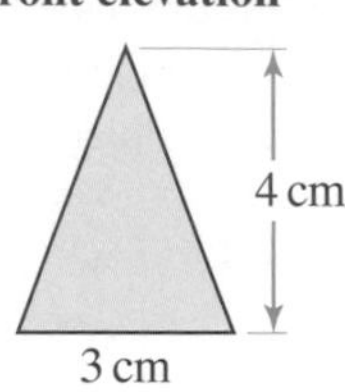

In some cases, as in this prism, the elevation from the front is the same as its cross-section

They are congruent trapezia.

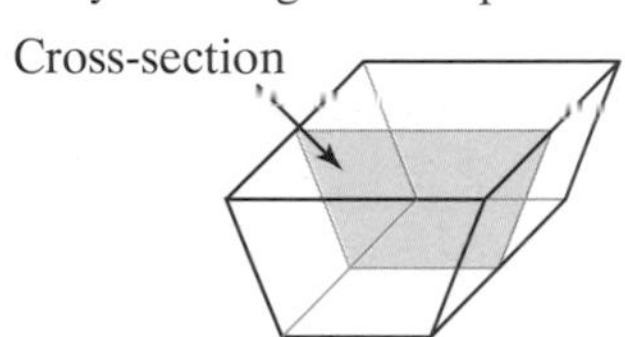

Gradient – a measure of how steep a line is

$$\text{Gradient} = \frac{\text{change in vertical distance}}{\text{change in horizontal distance}}$$
$$= \frac{y}{x}$$

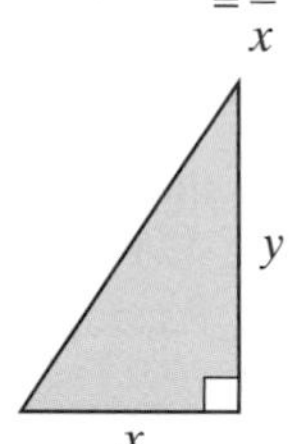

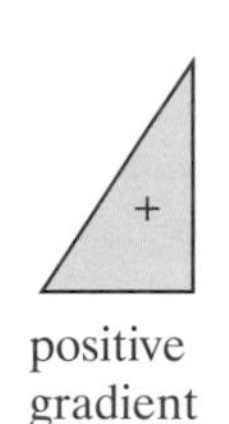

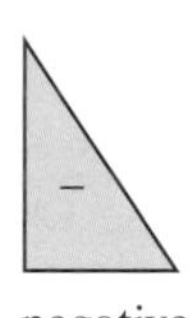

Hemisphere – a half sphere

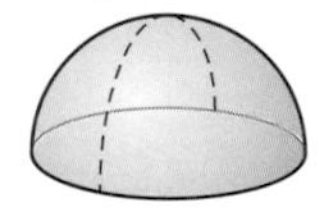

Heptagon – a polygon with seven sides

Hexagon – a polygon with six sides

Horizontal – from left to right; parallel to the horizon

Horizontal

Hypotenuse – the longest side of a right-angled triangle, opposite the right angle

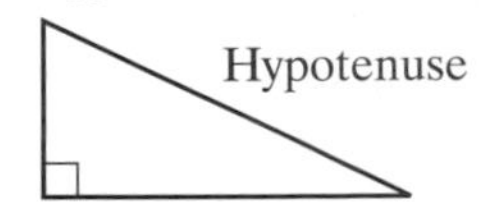

Identity – two expressions linked by the ≡ sign are true for all values of the variable, for example, $3x + 3 \equiv 3(x + 1)$

Image – the shape after it undergoes a transformation, for example, reflection, rotation, translation or enlargement

Imperial units – these are units of measurement historically used in the United Kingdom and other English-speaking countries; they are now largely replaced by metric units. Imperial units include:

- inches (in), feet (ft), yards (yd) and miles for lengths
- ounces (oz), pounds (lb), stones and tons for mass
- pints (pt) and gallons (gal) for capacity

Inequality – statements such as $a \neq b$, $a \leqslant b$ or $a > b$ are inequalities

Inequality signs – < means less than, ⩽ means less than or equal to, > means greater than, ⩾ means greater than or equal to

Integer – any positive or negative whole number or zero, for example, −2, −1, 0, 1, 2 ...

Intercept – the y-coordinate of the point at which the line crosses the y-axis

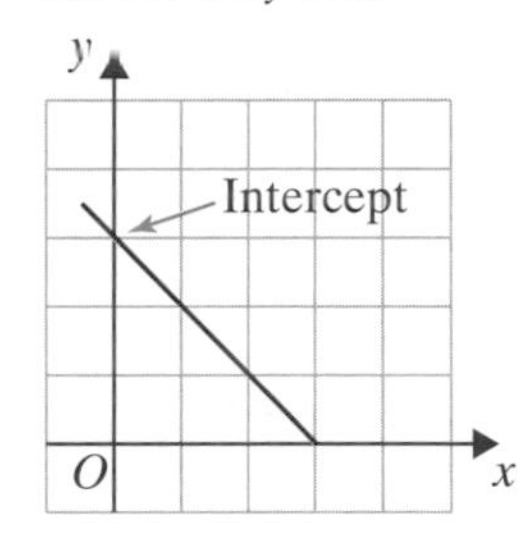

Interior angle – an angle inside a polygon

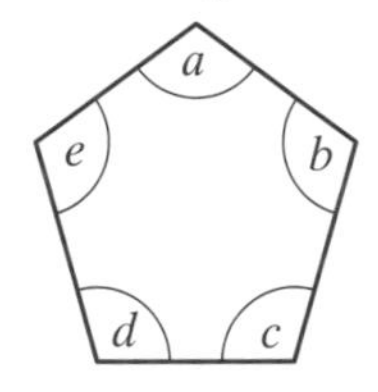

a, b, c, d and *e* are interior angles

Inverse operation – the operation undoes or reverses a previous operation, for example subtract is the inverse of add:

15 + 8 = 23 Add 8
23 − 8 = 15 Subtract 8 to return to the starting number 15

Irregular polygon – a polygon whose sides and angles are not all equal (they do not all have to be different)

Isometric diagram – a 2-D drawing of a 3-D solid drawn on an isometric grid or isometric dotty paper; the prism shown in the diagrams below is 4 cm long, and its cross-section is an isosceles right-angled triangle with sides of length 2 cm

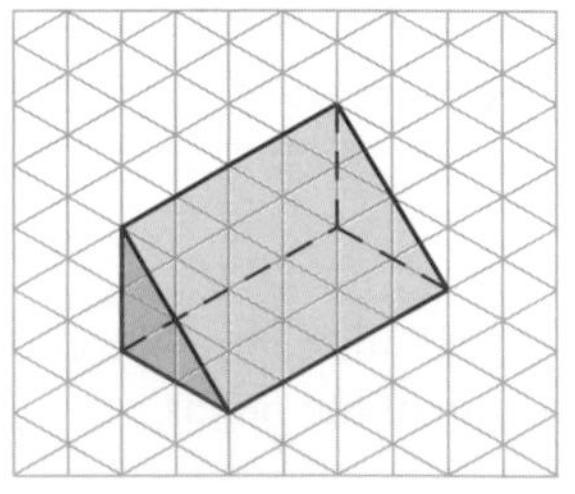

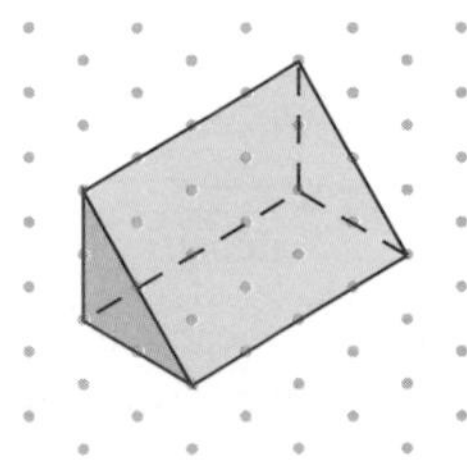

Isosceles trapezium – a quadrilateral with one pair of parallel sides. Non-parallel sides are equal

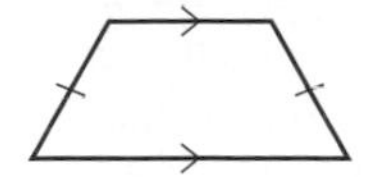

Isosceles triangle – a triangle with 2 equal sides and 2 equal angles; the equal angles are called **base angles**

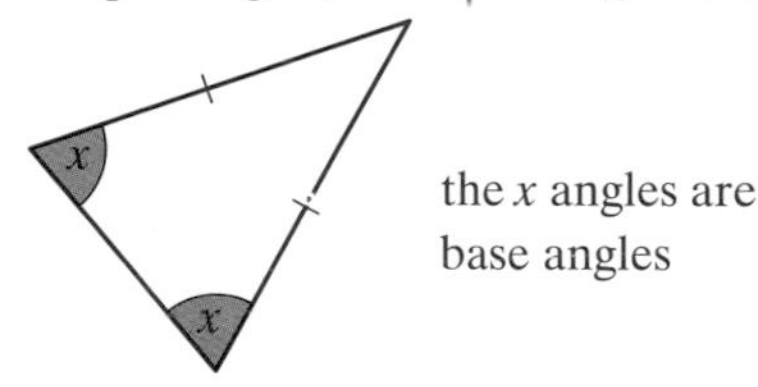

Kite – a quadrilateral with two pairs of equal adjacent sides

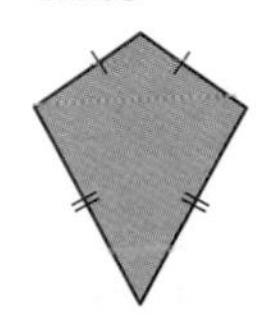

Linear equation – an equation where the highest power of the variable is 1; for example, $3x + 2 = 7$ is a linear equation but $3x^2 + 2 = 7$ is not

Linear expression – a combination of terms where the highest power of the variable is 1

Linear expressions	**Non-linear expressions**
x	x^2
$x + 2$	$\frac{1}{x}$
$3x + 2$	$3x^2 + 2$
$3x + 4y$	$(x + 1)(x + 2)$
$2a + 3b + 4c + \ldots$	x^3

Linear graph – the graph of a linear function of the form $y = mx + c$; if c is zero, the graph is a straight line through the origin (the point $(0, 0)$) indicating that y is directly proportional to x; if m is zero, the graph is parallel to the x-axis

Line of symmetry – a shape has reflection symmetry about a line through its centre if reflecting it in that line gives an identical-looking shape

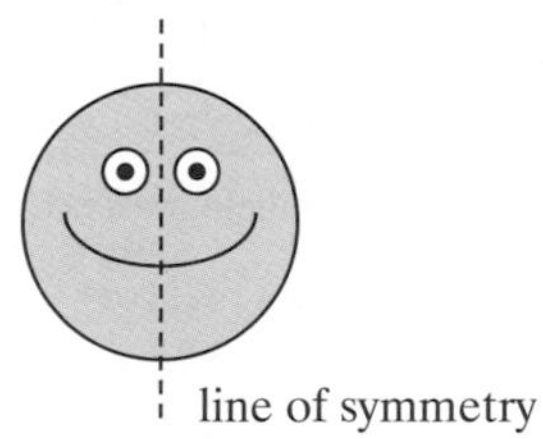

Line segment – the part of a line joining two points

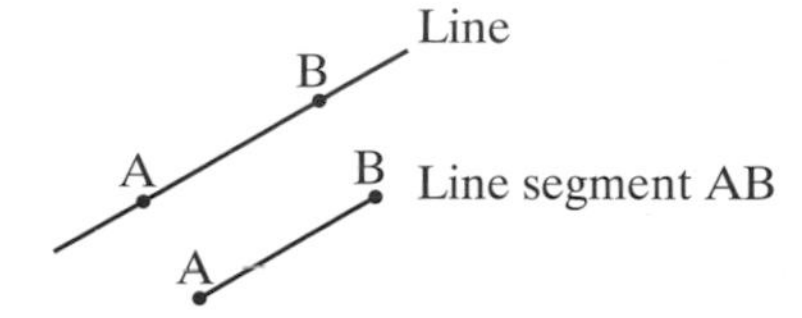

Loci – the plural of locus

Locus – the path followed by a moving point; for example, the locus of a point, A, keeping at a fixed distance of 5 cm from a point B, is a circle, centre B, radius 5 cm, or, in 3 dimensions, a sphere, centre B, radius 5 cm

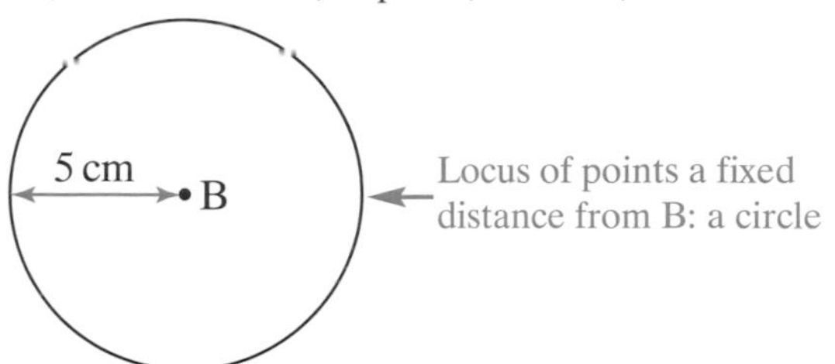

Lower bound – this is the minimum possible value of a measurement, for example, if a length is measured as 37 cm correct to the nearest centimetre, the lower bound of the length is 36.5 cm

Mapping – a transformation or enlargement is often referred to as a mapping with points on the object mapped onto points on the image

Metric units – these are related by multiples of 10 and include:

- metres (m), millimetres (mm), centimetres (cm) and kilometres (km) for lengths
- grams (g), milligrams (mg), kilograms (kg) and tonnes (t) for mass
- litres (ℓ), millilitres (mℓ) and centilitres (cℓ) for capacity

Midpoint – the middle point of a line

Net – a two-dimensional shape made of polygons that can be folded to make a three-dimensional solid, for example,

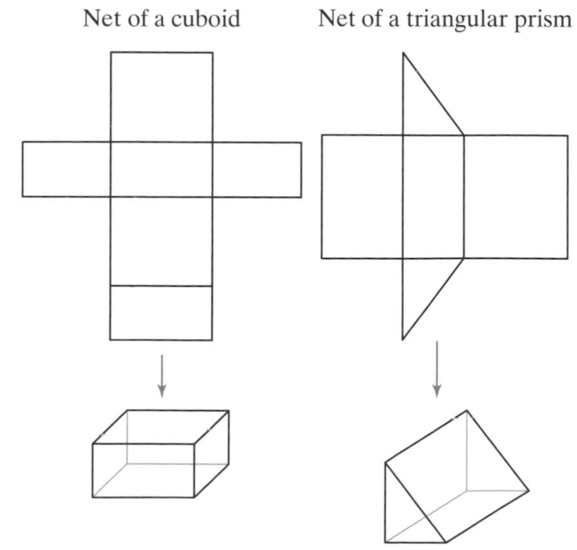

Nonagon – a polygon with nine sides

***n*th term** – this phrase is often used to describe a 'general' term in a sequence; if you are given the *n*th term, you can use this to find the terms of a sequence

Number line – a line where numbers are represented by points upon it; simple inequalities can be shown on a number line

Object – the shape before it undergoes a transformation, for example, translation or enlargement

Obtuse angle – an angle greater than 90° but less than 180°

Octagon – a polygon with eight sides

Operation – a rule for combining two numbers or variables, such as add, subtract, multiply or divide

Order of rotation symmetry – the number of ways a shape would fit on top of itself as it is rotated through 360°

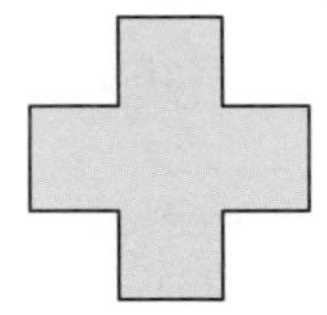

rotation symmetry order 4

(Shapes that are not symmetrical have rotation symmetry of order 1 because a rotation of 360° always produces an identical-looking shape)

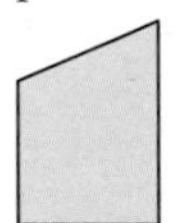

rotation symmetry order 1 (i.e. not symmetrical)

Origin – the point (0, 0) on a coordinate grid

Parallel lines – two lines that never meet and are always the same distance apart

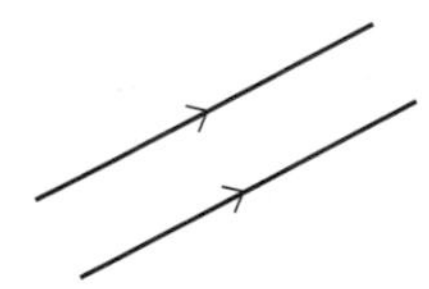

Parallelogram – a quadrilateral with opposite sides equal and parallel

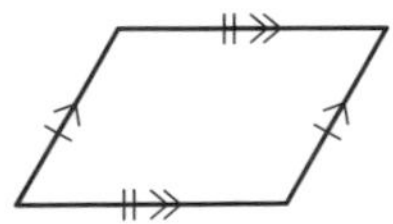

Pentagon – a polygon with five sides

Perimeter – the distance around an enclosed shape

Perpendicular – at right angles to; two lines at right angles to each other are perpendicular lines

Perpendicular bisector – a line at right angles to a given line that also divides the given line into two equal parts

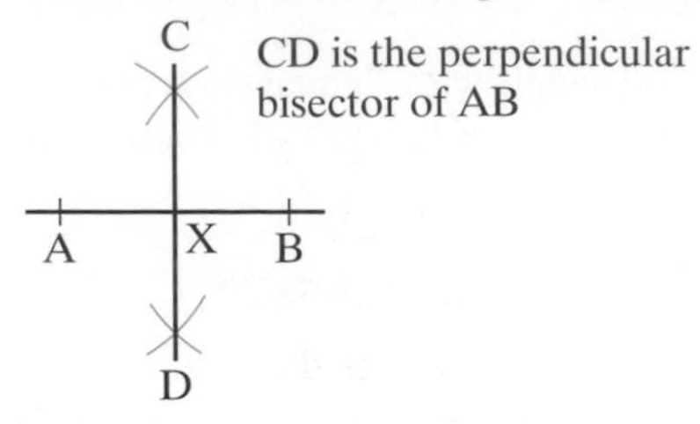

Perpendicular lines – two lines at right angles to each other

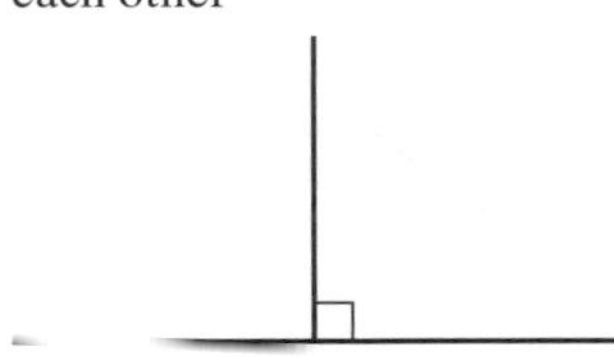

Plan – a diagram of a 3-D solid showing the view from above; these diagrams show a square-based pyramid and its plan

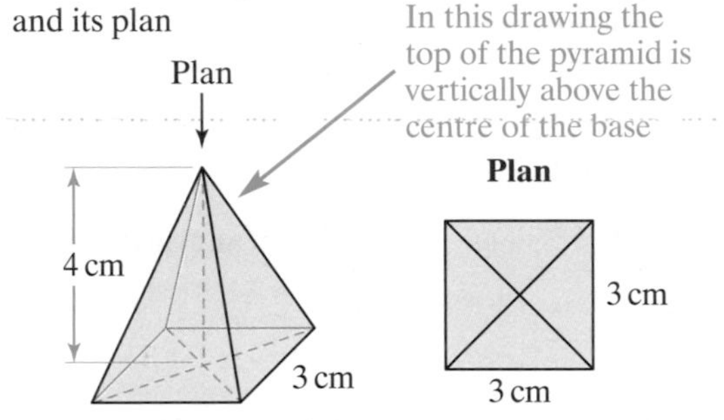

Polygon – a closed two-dimensional shape made from straight lines

Prism – a three-dimensional solid with two cross-sectional faces that are identical polygons, parallel to each other; all other faces are either parallelograms or rectangles

Prisms are named according to the cross-sectional face; for example,

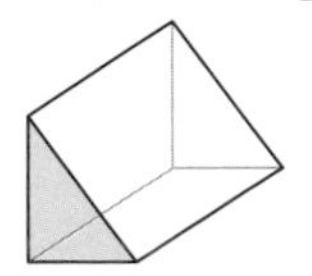

Triangular prism

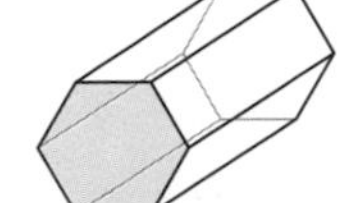

Hexagonal prism

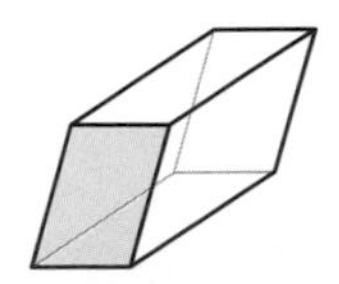

Parallelogram prism

Product – the result of multiplying together two (or more) numbers, variables, terms or expressions

Proof – a series of logical mathematical steps that confirms the truth of a mathematical statement

Prove – to provide a proof

Pyramid – a solid with a polygon as the base and one other vertex; all the vertices of the base are joined to this vertex forming triangular faces. Pyramids are named according to their base, for example,

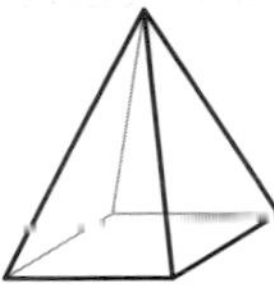

Square pyramid

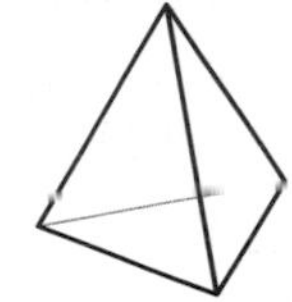

Triangular pyramid

Pythagoras' theorem – in a right-angled triangle, the square of the length of the hypotenuse is equal to the sum of the squares of the lengths of the other two sides

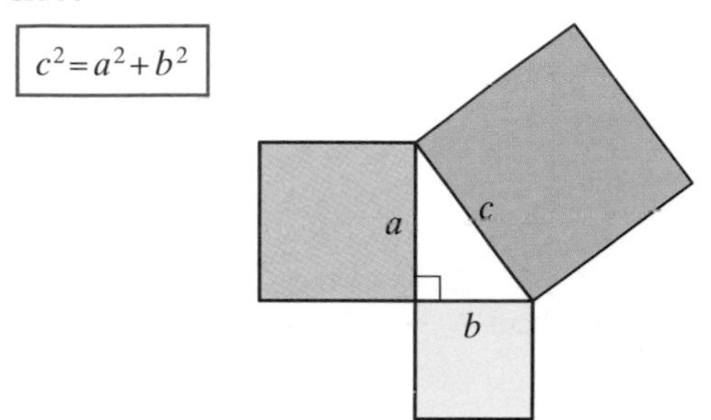

The area of the largest square = the total area of the two smaller squares

Pythagorean triple – a set of three integers a, b, c that satisfies $c^2 = a^2 + b^2$; for example, 3, 4, 5 ($5^2 = 3^2 + 4^2$), 5, 12, 13 ($13^2 = 5^2 + 12^2$), 6, 8, 10 ($10^2 = 6^2 + 8^2$) and 15, 36, 39 ($39^2 = 15^2 + 36^2$)

Quadrant – one of the four regions formed by the x- and y-axes in the Cartesian coordinate system

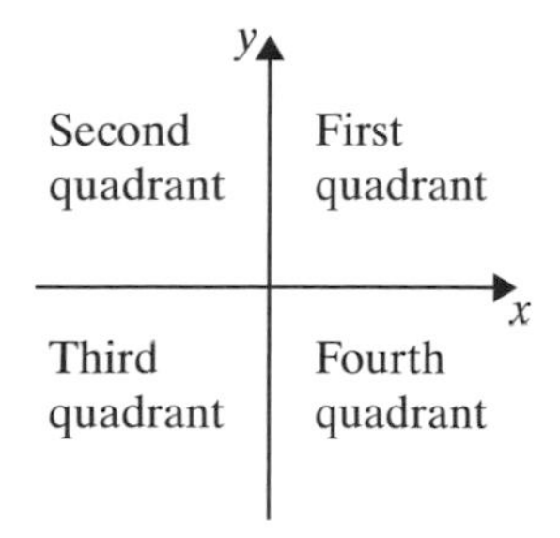

Quadrant (of a circle) – one quarter of a circle

Quadratic expression – an expression containing terms where the highest power of the variable is 2

Quadratic expressions	**Non-quadratic expressions**
x^2	x
$x^2 + 2$	$2x$
$3x^2 + 2$	$\frac{1}{x}$
$4 + 4y^2$	$3x^2 + 5x^3$
$(x + 1)(x + 2)$	$x(x + 1)(x + 2)$

Quadratic function – functions like $y = 3x^2$, $y = 9 - x^2$ and $y = 5x^2 + 2x - 4$ are quadratic functions; they include an x^2 term and may also include x terms and constants

The graphs of quadratic functions are always ∪-shaped or ∩-shaped

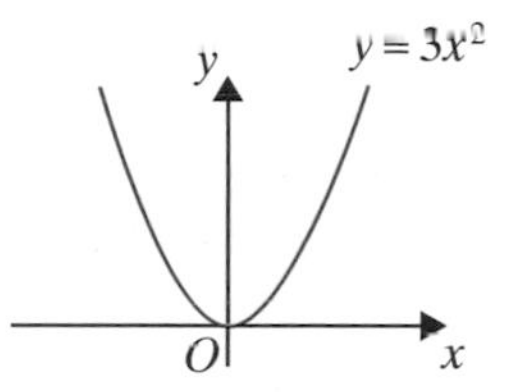

$y = ax^2 + bx + c$ is ∪-shaped when a is positive and ∩-shaped when a is negative

c is the intercept on the y-axis

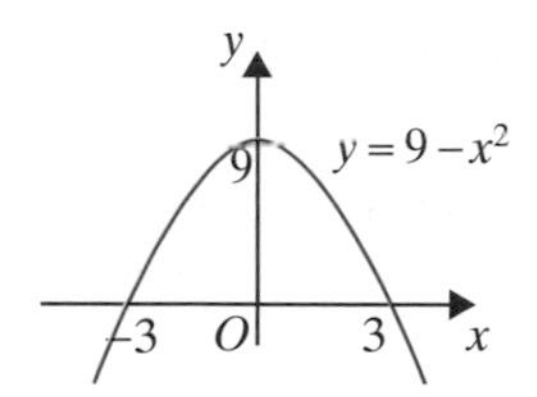

Note that other letters could be used as the variable instead of x (for example, $6t^2 - 3t - 5$ is also a quadratic expression and $h = 30t - 2t^2$ is a quadratic function)

Quadrilateral – a polygon with four sides

Radius – the distance from the centre of a circle to any point on the circumference

Rectangle – a quadrilateral with four right angles, and opposite sides equal in length

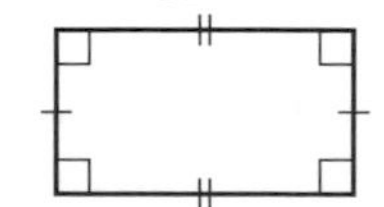

Reflection – a transformation involving a mirror line (or axis of symmetry), in which the line from the shape to its image is perpendicular to the mirror line. To describe a reflection fully, you must describe the position or give the equation of its mirror line, for example, the triangle A is reflected in the mirror line $y = 1$ to give the image B

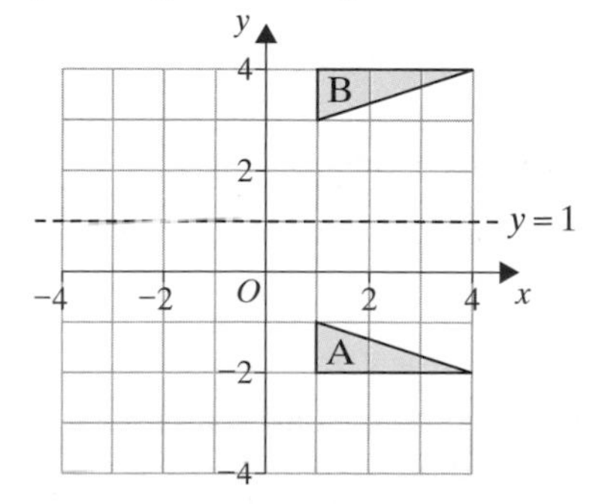

Reflex angle – an angle greater than 180° but less than 360°

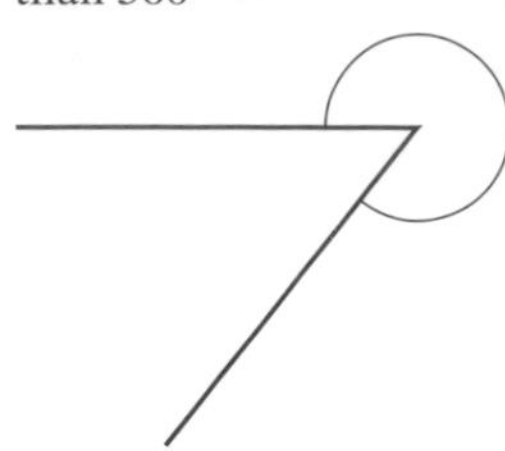

Regular polygon – a polygon with all sides and all angles equal

Regular tetrahedron – a triangular pyramid with equilateral triangles as its faces

Revolution – one revolution is the same as a full turn or 360°

Rhombus – a quadrilateral with four equal sides and opposite sides parallel

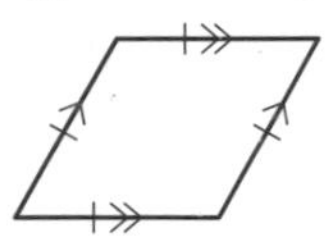

Right angle – an angle of 90°

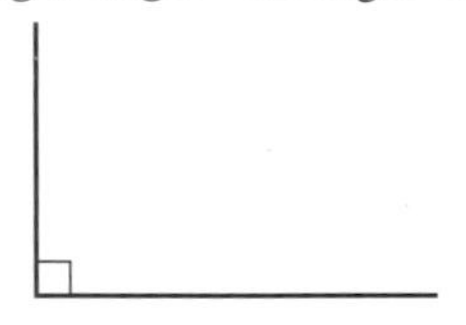

Right-angled triangle – a triangle with one angle of 90°

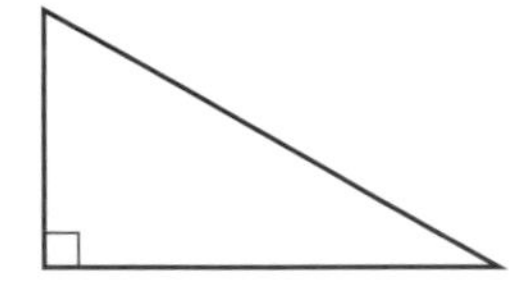

Rotation – a transformation in which the shape is turned about a fixed point called the centre of rotation. To describe a rotation fully, you must give the centre, angle and direction (a *positive angle* is *anticlockwise* and a *negative angle* is *clockwise*), for example, the triangle A is rotated about the origin through 90° anticlockwise to give the image C

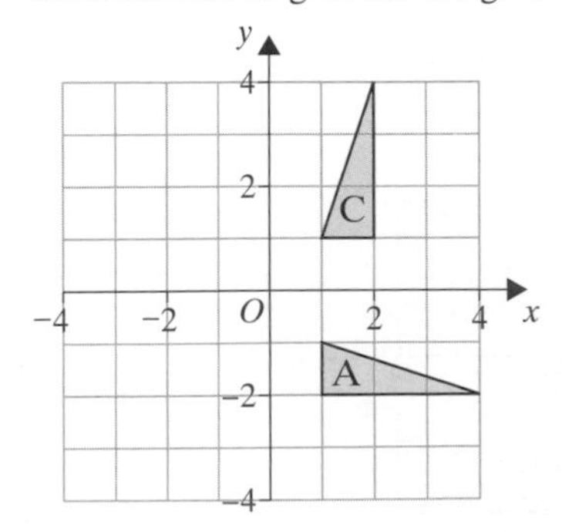

Scale factor – the ratio of corresponding sides usually expressed numerically so that:

$$\text{Scale factor} = \frac{\text{length of line on the enlargement}}{\text{length of line on the original}}$$

Semicircle – one half of a circle

Sequence – a list of numbers or diagrams that are connected in some way

In this sequence of diagrams, the number of squares is increased by one each time:

The dots are included to show that the sequence continues

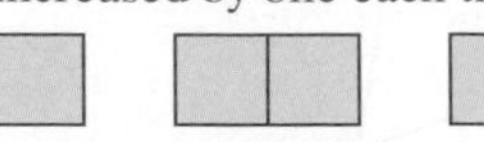

Shape – an enclosed space

Side elevation – a diagram of a 3-D solid showing the view from the side; sometimes the elevation from the side of the shape is the same as the front elevation, for example,

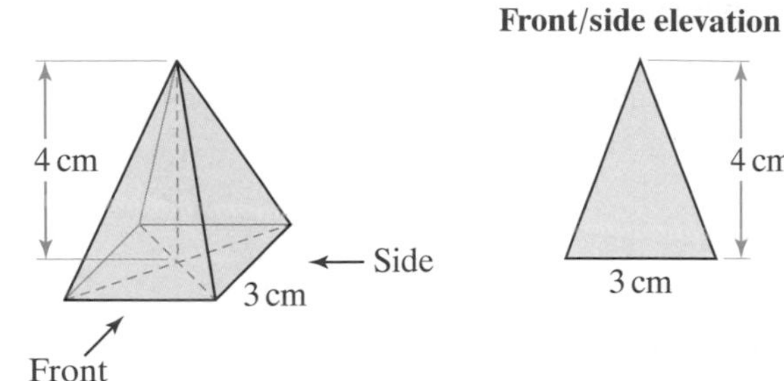

Usually, however, they will be different, for example,

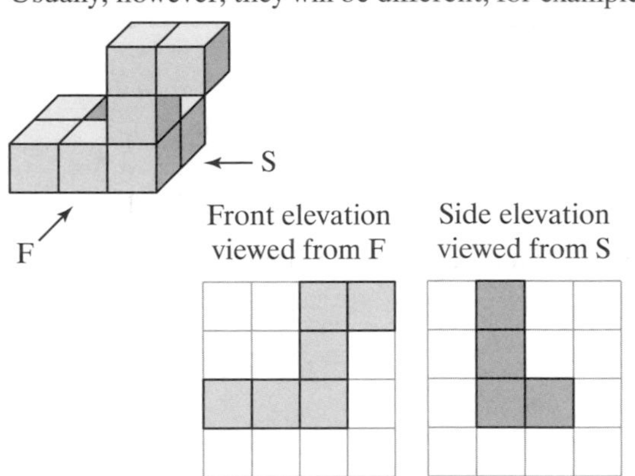

Unseen edges are shown as dotted lines, for example, the side elevation of this cylindrical container has three unseen edges. (The front elevation is the same)

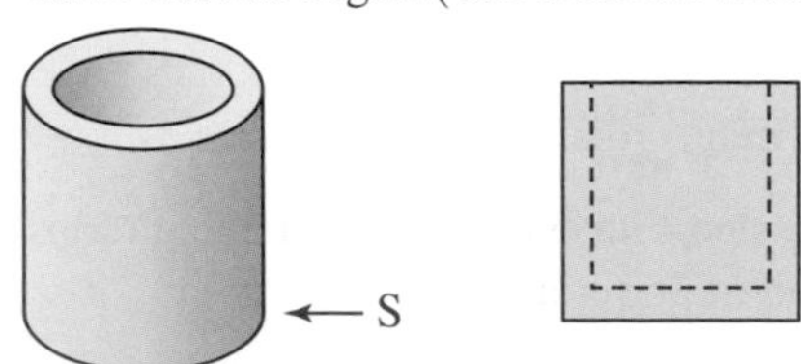

Similar – shapes are similar if their corresponding angles are equal *and* their corresponding sides are in the same ratio, so that one shape may be an enlargement of the other

Simplify – to make simpler by collecting like terms

Solid – a three-dimensional shape

Solution – the value of the unknown quantity, for example, if the equation is $3y = 6$, the solution is $y = 2$

Speed – the rate of change of distance with respect to time. To calculate the average speed, divide the total distance moved by the total time taken. It is usually given in metres per second (m/s) or kilometres per hour (km/h) or miles per hour (mph)

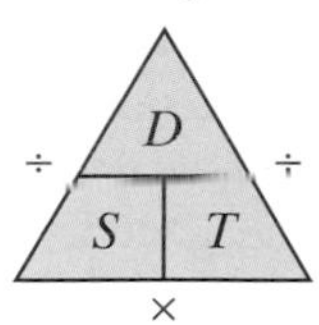

In the triangle, cover the item you want, then the rest tells you what to do

$\text{Speed} = \dfrac{\text{distance}}{\text{time}}$ $\dfrac{\text{metres}}{\text{seconds}}$ gives m/s

$\text{Distance} = \text{speed} \times \text{time}$

$\text{Time} = \dfrac{\text{distance}}{\text{speed}}$

Square – a quadrilateral with four equal sides and four right angles

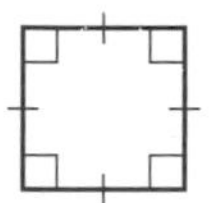

Square number – a square number is the outcome when a number is multiplied by itself; square numbers are 1, 4, 9, 16, 25, ...

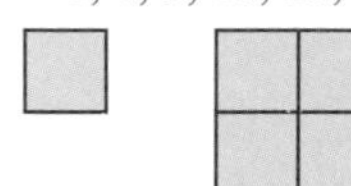
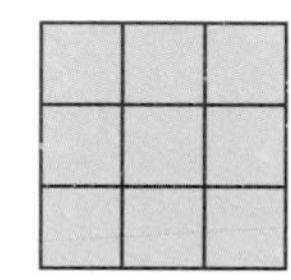
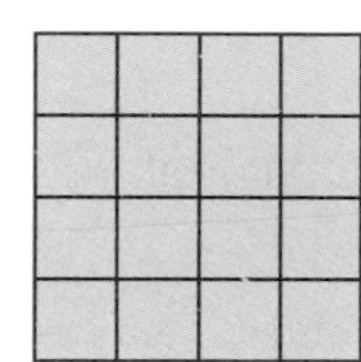

The rule for the nth term of the square numbers is n^2

Square root – the square root of a number such as 16 is a number whose outcome is 16 when multiplied by itself

Subject of a formula – in the formula $P = 2(l + w)$, P is the subject of the formula

Sum – the result of adding two (or more) numbers, variables, terms or expressions

Symmetrical – a shape that has symmetry

Symmetry (reflection) – a shape has (reflection) symmetry if a reflection through a line passing through its centre produces an identical-looking shape. The shape is said to be symmetrical

Term – a number, variable or the product of a number and a variable(s) such as 3, x or $3x$

Tessellation – a pattern where one or more shapes are fitted together repeatedly leaving no gaps

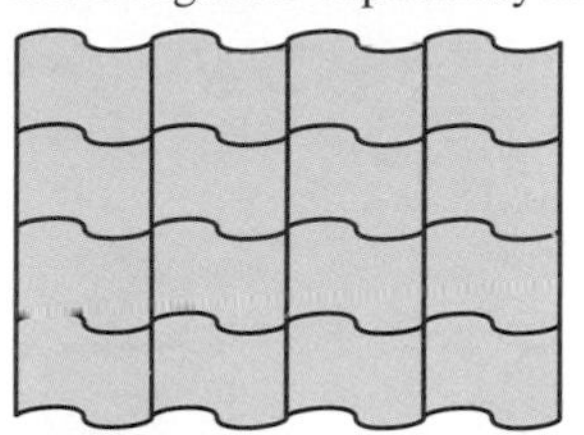

Tetrahedron – a solid with four triangular faces

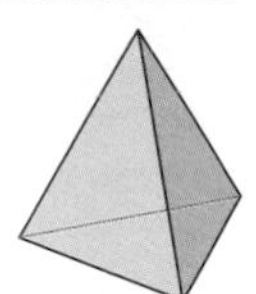

Theorem – a mathematical statement that can be demonstrated to be true using proof

Transformation – reflections, rotations, translations and enlargements are examples of transformations as they transform one shape onto another

Translation – a transformation where every point moves the same distance in the same direction so that the object and the image are congruent

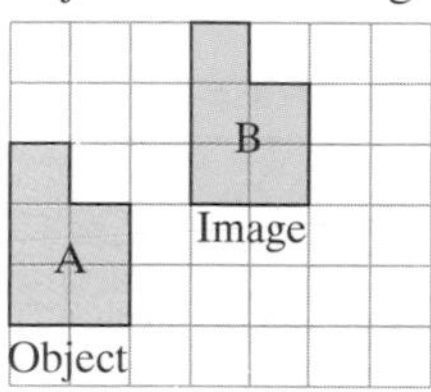

Shape A has been mapped onto Shape B by a translation of 3 units to the right and 2 units up

The vector for this would be $\begin{pmatrix} 3 \\ 2 \end{pmatrix}$

A translation is defined by the distance and the direction (vector)

Transversal – a line drawn across parallel lines

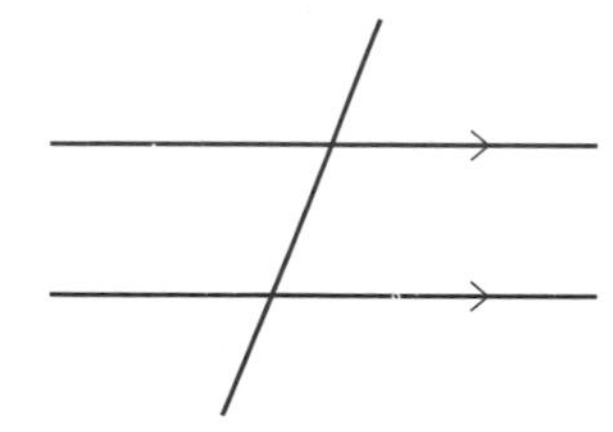

Trapezium (pl. **trapezia**) – a quadrilateral with one pair of parallel sides

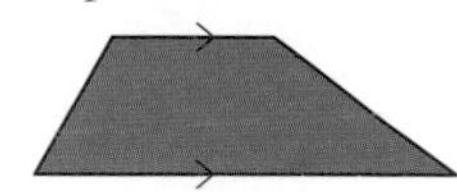

Trial and improvement – a method for solving algebraic equations by making an informed guess then refining this to get closer and closer to the solution

Triangle – a polygon with three sides

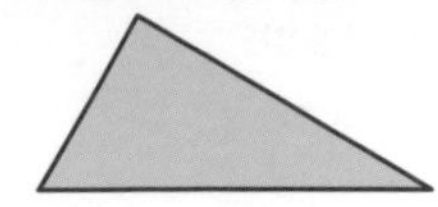

Triangle numbers – 1, 3, 6, 10, 15, ...

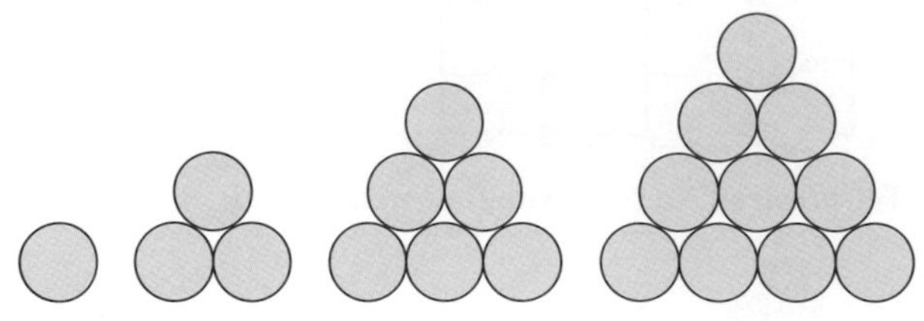

The rule for the nth term of the triangle numbers is $\frac{1}{2}n(n+1)$

Triangular prism – a 3-D shape with 6 vertices, 9 edges, 2 triangular faces and 3 rectangular faces

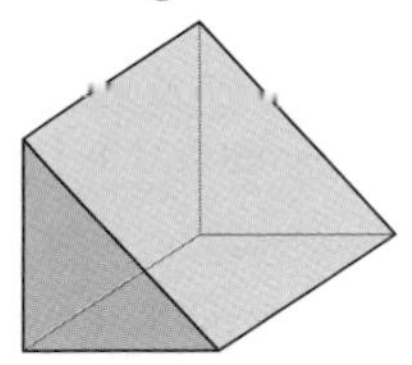

Unit – a standard used in measuring, for example, a metre is a unit of length

Unknown – the letter in an equation such as x or y

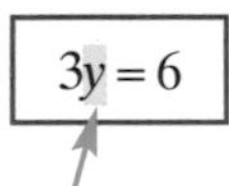

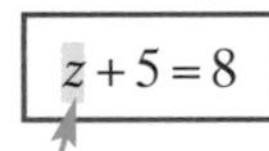

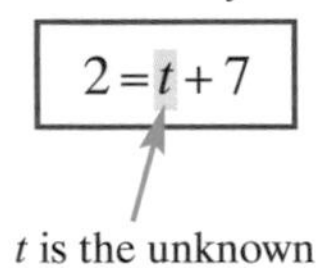

y is the unknown z is the unknown t is the unknown

Upper bound – this is the maximum possible value of a measurement, for example, if a length is measured as 37 cm correct to the nearest centimetre, the upper bound of the length is 37.5 cm

Variable – a symbol representing a quantity that can take different values such as x, y or z

Vector – a quantity with direction and magnitude (size)

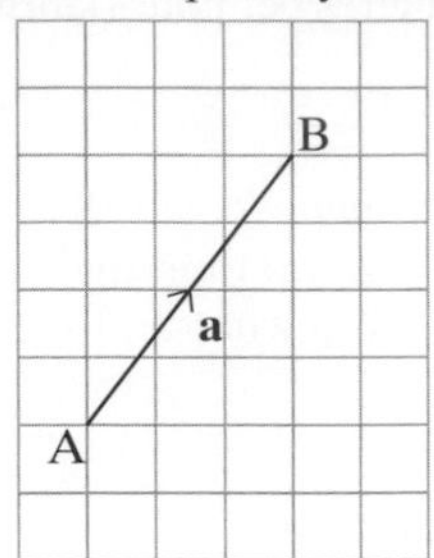

In this diagram, the arrow represents the direction and the length of the line represents the magnitude

In print, this vector can be written as **AB** or **a**

In handwriting, this vector is usually written as $\overrightarrow{\mathbf{AB}}$ or $\underset{\sim}{a}$

The vector can also be described as a column vector $\begin{pmatrix} 3 \\ 4 \end{pmatrix}$

where $\begin{pmatrix} x \\ y \end{pmatrix}$ ← x is the horizontal displacement, ← y is the vertical displacement

Vertex (pl. **vertices**) – the point where two or more edges meet

Vertical – directly up and down; perpendicular to the horizontal

Vertical

Volume – a measure of how much space fills a solid, commonly measured in cubic centimetres (cm^3) or cubic metres (m^3)